Advances in Computational Coupling and Contact Mechanics

Computational and Experimental Methods in Structures

ISSN: 2044-9283

Series Editor: Ferri M. H. Aliabadi *(Imperial College London, UK)*

This series will include books on state-of-the-art developments in computational and experimental methods in structures, and as such it will comprise several volumes covering the latest developments. Each volume will consist of single-authored work or several chapters written by the leading researchers in the field. The aim will be to provide the fundamental concepts of experimental and computational methods as well as their relevance to real world problems.

The scope of the series covers the entire spectrum of structures in engineering. As such it will cover both classical topics in mechanics, as well as emerging scientific and engineering disciplines, such as: smart structures, nanoscience and nanotechnology; NEMS and MEMS; micro- and nano-device modelling; functional and smart material systems.

Published:

Vol. 11 *Advances in Computational Coupling and Contact Mechanics*
edited by Luis Rodríguez-Tembleque (Universidad de Sevilla, Spain) and M. H. Ferri Aliabadi (Imperial College London, UK)

Vol. 10 *Vibration-Based Techniques for Damage Detection and Localization in Engineering Structures*
edited by Ali S. Nobari (Imperial College London, UK & Amirkabir University of Technology, Iran) and M. H. Ferri Aliabadi (Imperial College London, UK)

Vol. 9 *Buckling and Postbuckling Structures II: Experimental, Analytical and Numerical Studies*
edited by B. G. Falzon (Queen's University Belfast, UK) and M. H. Ferri Aliabadi (Imperial College London, UK)

Vol. 8 *Structural Health Monitoring for Advanced Composite Structures*
edited by M. H. Aliabadi and Z. Sharif Khodaei (Imperial College London, UK)

Vol. 7 *Active Control of Aircraft Cabin Noise*
by I. Dimino (CIRA, The Italian Aerospace Research Centre, Italy) and M. H. Aliabadi (Imperial College London, UK)

More information on this series can also be found at http://www.worldscientific.com/series/cems

(Continued at end of book)

Computational and Experimental Methods in Structures – Vol. 11

Advances in Computational Coupling and Contact Mechanics

Editors

Luis Rodríguez-Tembleque
Universidad de Sevilla, Spain

M H Ferri Aliabadi
Imperial College London, UK

World Scientific

NEW JERSEY • LONDON • SINGAPORE • BEIJING • SHANGHAI • HONG KONG • TAIPEI • CHENNAI • TOKYO

Published by

World Scientific Publishing Europe Ltd.

57 Shelton Street, Covent Garden, London WC2H 9HE

Head office: 5 Toh Tuck Link, Singapore 596224

USA office: 27 Warren Street, Suite 401-402, Hackensack, NJ 07601

Library of Congress Cataloging-in-Publication Data
Names: Rodríguez-Tembleque, Luis, editor. | Aliabadi, M. H., editor.
Title: Advances in computational coupling and contact mechanics / edited by
Luis Rodríguez-Tembleque (Universidad de Sevilla, Spain), M.H. Ferri
Aliabadi (Imperial College London, UK).
Description: [Hackensack?] New Jersey : World Scientific, 2018. | Series:
Computational and experimental methods in structures ; volume 11 |
Includes bibliographical references.
Identifiers: LCCN 2017053360 | ISBN 9781786344779 (hc : alk. paper)
Subjects: LCSH: Contact mechanics--Mathematics. | Joints
(Engineering)--Mathematical models.
Classification: LCC TA353 .A37 2018 | DDC 620.1/05--dc23
LC record available at https://lccn.loc.gov/2017053360

British Library Cataloguing-in-Publication Data
A catalogue record for this book is available from the British Library.

Desk Editors: Chandrima Maitra/Jennifer Brough/Shi Ying Koe

Typeset by Stallion Press
Email: enquiries@stallionpress.com

About the Editors

L. Rodríguez-Tembleque is an Associate Professor in the Department Continuum Mechanics and Theory of Structures at the Escuela Técnica Superior de Ingeniería (ETSI) at the Universidad de Sevilla, Spain. His research interests are focused on Computational Mechanics (based on the Finite Element Method and the Boundary Element Method), with emphasis on computational contact mechanics (including friction, rolling and wear simulation), substructure coupling techniques using Lagrange multipliers and integrity problems in advanced multifield materials (i.e. piezoelectric and magnetoelectroelastic materials) or fiber reinforced materials (i.e. carbon nanotube reinforced composites). He has co-edited books on wear and contact mechanics and he has published several chapters and papers in scientific journals and conference proceedings.

M. H. Ferri Aliabadi is a Professor of Aerospace Structures and Head of Aerostructures, Department of Aeronautics, Imperial College, London (UK) and has served as a Professor of Computational Mechanics and Director of Research, Department of Engineering, Queen Mary, University of London (UK); Reader and Head of Damage Tolerance Division, Wessex Institute of Technology, Southampton; Director of the Aerospace Engineering Programme, Queen Mary, London; and is currently Editor of the *International Journal for Structural Integrity and Durability*, Editor in Chief of *Computational and Experimental Methods in Structures*, and Editor of *Electronic Journal for Boundary Elements*. He has published numerous books, chapters and papers in technical journals and conference proceedings and obtained several grants from different institutions.

Preface

This volume presents recent advances in the field of computational coupling and contact mechanics with particular emphasis on the numerical methodologies necessary to solve advanced engineering applications. As such, the book covers formulations based on the finite element method (FEM) and their variants (e.g. isogeometric analysis or standard and generalized high-order FEM: hp-FEM and GFEM, respectively), the boundary element method (BEM), the material point method (MPM) or the recently proposed finite block method (FBM). A number of topics are addressed, such as computational contact modelling in large deformation, high-order mortar-based methods, treatment of nonmatching interfaces in partitioned fluid–structure interaction (FSI) problems, thermoelastic frictional contact or multifield magneto-electro-elastic contact modelling, indentation of functionally graded media (FGM), cohesive-frictional contact for microstructural analysis or recent approaches of MPM with applications on impact or fragmentation and comminution processes. Contributions from leading experts in these fields and active researchers give a detailed overview of different modern numerical schemes that can be considered by main numerical methodologies to simulate interaction problems in continuum mechanics. All chapters are self-contained and can be read independently.

This book is written for Ph.D. students in computational mechanics but serves also as an introduction into recent numerical interaction techniques for researchers and practicing engineers. The first chapter provides a detailed introduction to 3D frictional mortar contact formulation for nonmatching interfaces coupling techniques, explained in the context of finite strains with inelastic material behaviour. Chapter 2 presents a higher-order mortar-based finite element approximation for 2D standard and generalized high-order FEM, i.e. hp-FEM and GFEM, respectively. The numerical treatment of large deformation contact problems by isogeometric analysis is presented in Chapter 3. Chapter 4 presents the dynamic non-matching contact interaction of incompressible fluids with flexible structures. More particular methodologies in the context of advanced engineering applications are presented in Chapters 5–10. Chapter 5 proposes an Eulerian-based thermo-flexible multi-body approach in order to simulate ring testing of disc brake systems. Chapter 6 presents the recently proposed FBM to study the frictional indentation response of functionally graded materials. Chapter 7 introduces the MPM and gives a state-of-the-art review of the coupled finite element material point (CFEMP) method for large deformation problems such as contact and impact problems. New approaches and applications of MPM to model complex behaviours such as fragmentation and comminution are presented in Chapter 8. A cohesive-frictional grain-boundary element technique for the analysis of degradation and failure in polycrystalline materials is reviewed in Chapter 9. Finally, an interface BEM contact formulation for multifield piezoelectric and magneto-electro-elastic materials is presented in Chapter 10.

The editors are grateful to all the contributors and all the staff at World Scientific Publishing who provided all the encouragement, help and support during the process of producing this book.

Luis Rodríguez-Tembleque
M. H. Ferri Aliabadi

List of Contributors

T. Doca
Department of Mechanical Engineering, Faculty of Technology
University of Brasilia, Campus Darcy Ribeiro, 70910-900
Brasilia, DF, Brazil
doca@unb.br

F. M. A. Pires
Department of Mechanical Engineering, Faculty of Engineering
University of Porto, Rua Dr. Roberto Frias, 4200-465 Porto, Portugal
fpires@fe.up.pt

A. P. C. Dias
Department of Integrated Systems, Faculty of Mechanical Engineering
University of Campinas, 13083-860, Campinas, SP, Brazil
apcdias@fem.unicamp.br

S. P. B. Proenca
Structural Engineering Department, São Carlos School of Engineering
University of São Paulo, Av Trabalhador Sãocarlense, 400, CEP
13566-590, São Carlos, SP, Brazil
persival@sc.usp.br

M. L. Bittencourt
Department of Integrated Systems, Faculty of Mechanical Engineering
University of Campinas, 13083-860, Campinas, SP, Brazil
mlb@fem.unicamp.br

J. Kopačka
Institute of Thermomechanics of the CAS v. v. i.
Dolejškova 1402/5, 182 00 Praha 8, Czech Republic
kopacka@it.cas.cz

D. Gabriel
Institute of Thermomechanics of the CAS v. v. i.
Dolejškova 1402/5, 182 00 Praha 8, Czech Republic
gabriel@it.cas.cz

R. Kolman
Institute of Thermomechanics of the CAS v. v. i.
Dolejškova 1402/5, 182 00 Praha 8, Czech Republic
kolman@it.cas.cz

J. Plešek
Institute of Thermomechanics of the CAS v. v. i.
Dolejškova 1402/5, 182 00 Praha 8, Czech Republic
plesek@it.cas.cz

J. A. González
Escuela Técnica Superior de Ingeniería, Universidad de Sevilla
Camino de los Descubrimientos s/n, E-41092, Sevilla, Spain
japerez@us.es

K. C. Park
Department of Aerospace Engineering Sciences and
Center for Aerospace Structures
University of Colorado Boulder, CO 80309, USA
kcpark@colorado.edu

N. Strömberg
Örebro University, Department of Mechanical Engineering
SE–71 82 Örebro, Sweden
niclas.stromberg@oru.se

J. Li
School of Mathematics and Statistics
Changsha University of Technology, 410004, China
lijingnew@126.com

J. Jin
School of Communication and Transportation Engineering
Changsha University of Technology, 410004, China

J. J. Yang
School of Communication and Transportation Engineering
Changsha University of Technology, 410004, China
yangjianjun01@126.com

T. Huang
School of Communication and Transportation Engineering
Changsha University of Technology, 410004, China
huangtuomao@163.com

P. H. Wen
School of Engineering and Materials Science
Queen Mary University of London, London, E1 4NS, UK
p.h.wen@qmul.ac.uk

Y. Lian
Northwestern University, Mechanical Department
2145 Sheridan Road, IL 60208, USA
lianyp@northwestern.edu

X. Zhang
Tsinghua University, School of Aerospace
Beijing 100084, China
xzhang@tsinghua.edu.cn

M. A. Homel
Atmospheric, Earth and Energy Division, Lawrence Livermore National Laboratory (LLNL), Livermore, CA 94550, USA
homel1@llnl.gov

E. B. Herbold
Atmospheric, Earth and Energy Division, Lawrence Livermore National Laboratory (LLNL), Livermore, CA 94550, USA
herbold1@llnl.gov

I. Benedetti
Dipartimento di Ingegneria Civile, Ambientale
Aerospaziale, dei Materiali DICAM
Università degli Studi di Palermo
Viale delle Scienze, 90128, Palermo, Italy
ivano.benedetti@unipa.it

L. Rodríguez-Tembleque
Escuela Técnica Superior de Ingeniería, Universidad de Sevilla
Camino de los Descubrimientos s/n Sevilla E-41092, Spain
luisroteso@us.es

F. C. Buroni
Escuela Técnica Superior de Ingeniería, Universidad de Sevilla
Camino de los Descubrimientos s/n Sevilla E-41092, Spain
fburoni@us.es

A. Sáez
Escuela Técnica Superior de Ingeniería, Universidad de Sevilla
Camino de los Descubrimientos s/n Sevilla E-41092, Spain
andres@us.es

M. H. Ferri Aliabadi
Department of Aeronautics, Faculty of Engineering
Imperial College of London, South Kensington Campus
London SW7 2AZ, UK
m.h.aliabadi@imperial.ac.uk

Contents

Chapter 1

Frictional Mortar Formulation for Large Inelastic Deformation Problems

T. Doca* and F. M. A. Pires

A frictional mortar contact formulation for the simulation of solids undergoing finite strains with inelastic material behaviour is presented. The framework includes the modelling of finite strain inelastic deformation, the numerical treatment of frictional contact conditions and specific finite element technology. Several well-established and recent models are employed for each of these building blocks to capture the distinct physical aspects of the deformation behaviour. The approach is based on a mortar formulation, and the enforcement of contact constraints is realized with dual Lagrange multipliers. The introduction of nonlinear complementarity functions into the frictional contact conditions combined with the global equilibrium leads to a system of nonlinear equations, which is solved in terms of the semi-smooth Newton method. The resulting method can be interpreted as a primal–dual active set strategy (PDASS) which deals with contact nonlinearities, material and geometrical nonlinearities in one iterative scheme. The consistent linearization of all building blocks of the framework yields a robust and highly efficient approach for the analysis of metal forming problems. The effect of finite inelastic strains on the solution behaviour of the PDASS method is examined in detail based on the complementarity parameters. A comprehensive set of numerical examples is presented to demonstrate the accuracy and efficiency of the approach against the traditional node-to-segment (NTS) penalty contact formulation.

1. Introduction

In a wide range of problems involving deformable bodies undergoing finite strains and experiencing inelastic material behaviour, frictional contact is

*Corresponding author.

inevitably present. These types of problems are very common in engineering practice and are of the utmost importance in forming operations, machine design, structural engineering and collision of vehicles. Therefore, the numerical strategy developed for the analysis of this variety of applications requires appropriate consideration of both theoretical and algorithmic issues. In particular, the modelling of finite strain inelastic deformation, the numerical treatment of frictional contact conditions and element technology capable of dealing with plastic incompressibility deserve careful attention. Due to the strong nonlinearities involved, the solution of these problems using implicit methods is still a challenging task.

This chapter presents an overview of contact formulations for problems experiencing large inelastic strains, and it is structured as follows. In Section 2, a state-of-the-art review is presented while in Section 3 the description of a two-body frictional contact problem in the context of finite deformations is undertaken. In Section 4, the boundary value problem (BVP) is converted into a weak formulation. The numerical integration of the rate constitutive equations that leads to the incremental BVP is presented in Section 5. The spatial discretization of the contact virtual work and the nonlinear contact constraints based on dual Lagrange multipliers is described in Section 6 together with the integration of the mortar matrices. In Section 7, the F-bar formulation employed is briefly reviewed. Then, in Section 8 the solution procedures employed, based on the so-called primal–dual active set strategy (PDASS), with and without complementary functions, are presented. Various examples in Section 9 examine the effects of the complementarity parameters on the solution behaviour and demonstrate the robustness and accuracy over existing approaches. Finally, conclusion remarks are given in Section 10.

2. State-of-the-art review

The modelling of nonlinear solid mechanics problems with the finite element method has reached a rather mature level and a considerable number of algorithms have already been developed, at both the kinematic and constitutive levels, for the analysis of large inelastic deformations.

The formulation of finite inelastic constitutive models and the associated numerical integration algorithms have gained widespread acceptance and are currently adopted for large strain analysis. Notable contributions in this area are presented in Refs. [1–5]. By performing the multiplicative split of the deformation gradient into elastic and plastic contributions and employing logarithmic stretches as strain measures, a simple model for large inelastic deformations at finite strains is obtained. Procedures for integration of the constitutive equations typically rely on the general operator split methodology. The constitutive behaviour at small strains is usually described by a conventional internal variable-based structure, and the set of ordinary differential equations in time is integrated with a number of integration algorithms.[6–8]

The field of computational contact mechanics has achieved significant progress over the last decade and several algorithms have been proposed for the numerical treatment of frictional contact conditions.[9,10] The contact formulations were initially based on the node-to-segment (NTS) approach, which has been extended and generalized by numerous authors.[11–14] Although quite popular and widely used, there are well-known limitations in the robustness of NTS formulations. It has been shown in Ref. [15] that the so-called single-pass algorithms do not satisfy the contact patch test and there is a degradation of spatial convergence rates, which has been demonstrated in the kinematically linear context in Ref. [16]. Furthermore, non-physical jumps in the contact forces can occur when finite sliding situations occur due to the non-smoothness of the discretized contact surfaces. Several strategies have been proposed to overcome these problems[17–19] leading to more elaborated algorithms. Alternative methods for the discretization of the contact surface are the so-called segment-to-segment (STS) approaches, where the mortar method has become very popular. This method was originally introduced in the context of domain decomposition techniques[20] and is particularly suited for the exchange of information between non-conforming meshes. It introduces the continuity condition at the interface in an integral form, rather than as nodal constrains. Of particular importance is the fact that the mortar finite element method preserves optimal convergence rates from the finite element method as long as suitable mortar spaces are chosen.

All the aforementioned contact formulations lead to the mathematical structure of variational inequalities.[21,22] The numerical treatment of these nonlinear complementarity constraints, i.e. the enforcement of the contact constraints, is mainly accomplished by three methods[9,10]: the Penalty method, the Lagrange method and the Augmented Lagrangian method. However, when these regularization schemes are employed with the mortar finite element method, unwanted side effects occur such as: ill conditioning, additional equations in the global system or a significant increase in the computational time for solution.

In order to circumvent these shortcomings, Wohlmuth (2001)[23] has proposed the use of dual spaces for the Lagrange multipliers allowing the local elimination of the contact constraints. As a consequence, the Lagrangian multipliers can be conveniently condensed and no additional equations are needed for the solution of the global system of equations.[24,25] The remaining problem is positive definite and the unknowns are the displacements only. The reformulation of the frictional contact constraints into so-called complementarity functions[26–28] allows rewriting the inequality constraints as equalities. The combination of this set of equalities with the equilibrium equations leads to a system of nonlinear equations that can be solved with the semi-smooth Newton method.[22,27,29] The resulting method can be interpreted as a PDASS,[29,30] which has been introduced for small deformation frictional contact in Ref. [31]. This strategy was latter applied to nonlinear material problems at small strains in Refs. [32, 33], to the solution of dynamic contact problems in Ref. [34], extended to finite deformation frictionless 2D contact in Ref. [35], to 2D frictional contact in Ref. [36] and 3D finite deformation contact in Refs. [37, 38].

In recent years, the modelling of finite strain inelastic material behaviour coupled with finite frictional sliding has also been addressed.[39,40] In this case, the degree of nonlinearity increases significantly and can greatly affect the convergence behaviour of the overall strategy. In addition, such contact formulations require element technology capable of dealing with plastic incompressibility while avoiding volumetric locking.[41,42] The structure of a frictional mortar formulation for large inelastic deformation problems is presented in the following sections.

3. Problem description

The kinematical description of the contact problem with finite deformations and finite sliding is described in Fig. 1. The motion, $\mathbf{x}$, is described by a mapping φ between the reference configuration $\mathbf{X}$ (at time 0) and current configuration $\mathbf{x}$ (at time t),

$$\mathbf{x} = \varphi(\mathbf{X}, t). \tag{1}$$

Therefore, the vector of nodal displacements, $\mathbf{u}$, is obtained from the reference configuration as follows:

$$\mathbf{u}(\mathbf{X}, t) = \mathbf{x}(\mathbf{X}, t) - \mathbf{X}. \tag{2}$$

Solid bodies in the reference configuration are denoted by Ω_0^s and Ω_0^m, $\{\Omega_0^s \cup \Omega_0^m = \Omega_0 : \Omega \subset \mathbb{R}^{\{n_d\}}\}$, the dimension of the solids is represented by

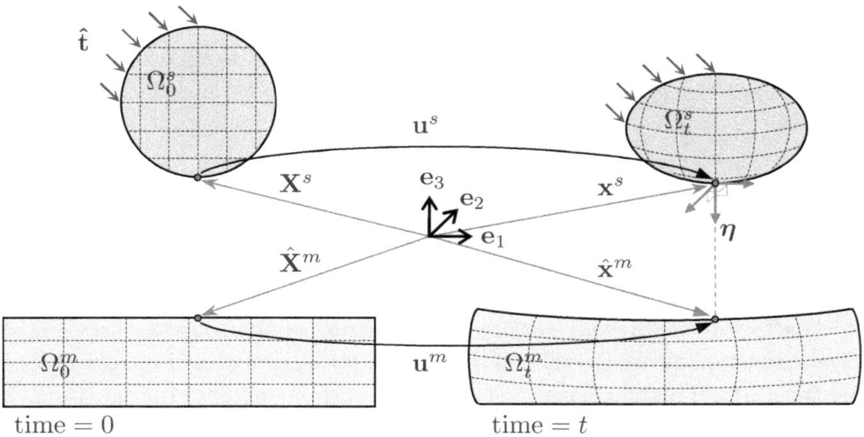

Fig. 1. Illustration of a finite deformation contact problem.

$n_d = 2, 3$. The superscripts s and m represent the common nomenclature employed in contact mechanics of a *Slave* and a *Master* surface/body. The boundaries of subset Ω_0 are divided in a contact zone, $\Gamma_c = \{\Gamma_c^s \cup \Gamma_c^m\}$, a Neumann boundary, Γ_N and a Dirichlet boundary Γ_D, $\{\Gamma_c \cup \Gamma_N \cup \Gamma_D = \Gamma \; : \; \Gamma \subset \mathbb{R}\}$.

The spatial counterparts of the three boundaries are denoted by γ_c, γ_N and γ_D, respectively. Furthermore, it is assumed that the bodies are only subjected to body forces, $\mathbf{b}$, and applied boundary tractions, $\hat{\mathbf{t}}$. Therefore, the BVP, in terms of the displacement vector, $\mathbf{u}$ and the Cauchy stress tensor, $\boldsymbol{\sigma}$, is defined as follows:

$$\begin{aligned} \operatorname{div}\left(\boldsymbol{\sigma}^i(\mathbf{u})\right) + \mathbf{b}^i &= 0 \;\text{ in } \Omega_t^i, \\ \boldsymbol{\sigma}^i(\mathbf{u})\boldsymbol{\eta}^i &= \hat{\mathbf{t}}^i \;\text{ on } \gamma_N^i, \\ \mathbf{u}^i &= \hat{\mathbf{u}}^i \text{ on } \gamma_D^i, \quad i = s, m, \end{aligned} \tag{3}$$

where the prescribed displacement on the Dirichlet boundary is represented by $\hat{\mathbf{u}}^i$ and the current outward unit normal vector on γ_N is represented by $\boldsymbol{\eta}^i$. Index i emphasizes that the BVP condition must be attained for all bodies involved. The deformation gradient, $\mathbf{F}$, with respect to the reference configuration, Ω_0, is given by

$$\mathbf{F} = \nabla\varphi(\mathbf{X}, t) = \frac{\partial\varphi(\mathbf{X}, t)}{\partial\mathbf{X}} = \frac{\partial\mathbf{x}}{\partial\mathbf{X}}. \tag{4}$$

The determinant of the deformation gradient has to be greater than zero, $J = \det(\mathbf{F})$, in order for the motion of the bodies to be meaningful.

3.1. *Contact constraints*

For enforcing the contact constraints at normal and tangential directions, the surface tractions defined on the slave surface, $\mathbf{t}_c^s$, are decomposed as follows:

$$p_\eta = \mathbf{t}_c^s \cdot \boldsymbol{\eta}, \quad \mathbf{t}_\tau = [\mathbf{I} - \boldsymbol{\eta} \otimes \boldsymbol{\eta}]\, \mathbf{t}_c^s, \tag{5}$$

where $\boldsymbol{\eta}$ is now defined on the slave surface, γ_c^s.

In the normal direction, the non-penetration condition is fulfilled by evaluating the relative distance (*gap*) between, $\mathbf{x}^s$, on the slave surface and, $\hat{\mathbf{x}}^m$, over the master surface in the current configuration — see Fig. 1. The caret represents that $\hat{\mathbf{x}}^m$ is the closest projection of $\mathbf{x}^s$ onto the master surface, γ_c^m. This gap vector is obtained, in the current configuration, as follows:

$$\mathbf{g}(\mathbf{X}^s, t) = [\mathbf{x}^s(\mathbf{X}^s, t) - \hat{\mathbf{x}}^m(\hat{\mathbf{X}}^m, t)]. \tag{6}$$

Then, the computation of the normal distance between $\mathbf{x}^s$ and $\hat{\mathbf{x}}^m$ yields the definition of the scalar-valued gap function

$$g(\mathbf{X}^s, t) = -\boldsymbol{\eta}\left(\mathbf{x}^s(\mathbf{X}^s, t)\right) \cdot \mathbf{g}(\mathbf{X}^s, t). \tag{7}$$

Together with the definition of a non-positive normal contact traction, p_η, the gap function creates a set of inequalities known as the Karush–Kuhn–Tucker (KKT) conditions

$$g(\mathbf{X}^s, t) \geq 0, \;\; p_\eta \leq 0, \;\; p_\eta g(\mathbf{X}, t) := 0. \tag{8}$$

Therefore, the KKT conditions are responsible for enforcing a non-penetration relation on the normal direction of γ_c.

In the tangential direction, the frictional conditions can be given by Coulomb's law:

$$\Phi := \|\mathbf{t}_\tau\| - \mu |p_\eta| \leq 0, \tag{9}$$

$$\boldsymbol{\vartheta}_\tau\left(\mathbf{X}^s, t\right) + \beta \mathbf{t}_\tau = 0, \tag{10}$$

with the relative tangential velocity, $\boldsymbol{\vartheta}_\tau(\mathbf{X}^s, t)$, defined as the rate of change:

$$\dot{(*)} = \frac{d(*)}{dt} \approx \frac{(*)_t - (*)_{t-\Delta t}}{\Delta t}, \tag{11}$$

of the gap vector in the tangential direction, Eq. (6),

$$\boldsymbol{\vartheta}_\tau(\mathbf{X}^s, t) = [\mathbf{I} - \boldsymbol{\eta} \otimes \boldsymbol{\eta}]\, \dot{\mathbf{g}}(\mathbf{X}^s, t). \tag{12}$$

Tangential stresses below the Coulomb threshold, $\Phi < 0$, Eq. (9), imply a tangential velocity, $\|\boldsymbol{\vartheta}_\tau\|$, equal to zero, commonly called as stick/static state. When tangential stresses are at the Coulomb limit, they are opposed by the relative tangential velocity. In this case, we have a slip/kinetic state. The association of these two definitions generates the following state equation:

$$\Phi\beta = 0. \tag{13}$$

3.2. *Constitutive modelling for finite inelastic deformations*

The modelling of nonlinear solid material behaviour has undergone substantial development over the last decades and, as result, a wide range of material models, incorporating elastic, viscoelastic and viscoplastic material behaviour, are currently available. A class of isotropic hyperelastic based finite strain inelastic constitutive models, formulated in the spatial configuration, is considered here. A more detailed discussion of the approach, which is well established and widely accepted for the description of finitely deforming solids, can be found in Refs. [6, 8].

The main hypothesis underlying this class of models is the multiplicative decomposition of the deformation gradient, $\mathbf{F}$, into an elastic, $\mathbf{F}^e$, and a plastic, $\mathbf{F}^p$, contribution

$$\mathbf{F} = \mathbf{F}^e\mathbf{F}^p, \tag{14}$$

which was first introduced by Lee (1969)[43] and admits the existence of a local unstressed intermediate configuration defined by the plastic deformation gradient, $\mathbf{F}^p$. The polar decomposition of the elastic and plastic deformation gradients leads to

$$\mathbf{F}^e = \mathbf{R}^e\mathbf{U}^e = \mathbf{V}^e\mathbf{R}^e, \quad \mathbf{F}^p = \mathbf{R}^p\mathbf{U}^p = \mathbf{V}^p\mathbf{R}^p, \tag{15}$$

where $\mathbf{U}^e$ and $\mathbf{U}^p$ are the elastic and the plastic right stretch tensors, $\mathbf{V}^e$ and $\mathbf{V}^p$ are the elastic and plastic left stretch tensors and $\mathbf{R}^e$ and $\mathbf{R}^p$ the elastic and plastic rotation tensors.

The velocity gradient, $\mathbf{L} = \dot{\mathbf{F}}\mathbf{F}^{-1}$, can be decomposed additively as

$$\mathbf{L} = \mathbf{L}^e + \mathbf{F}^e\mathbf{L}^p[\mathbf{F}^e]^{-1}, \tag{16}$$

where the elastic, $\mathbf{L}^e$, and plastic, $\mathbf{L}^p$, velocity gradients are defined by

$$\mathbf{L}^e = \dot{\mathbf{F}}^e[\mathbf{F}^e]^{-1}, \quad \mathbf{L}^p = \dot{\mathbf{F}}^p[\mathbf{F}^p]^{-1}. \tag{17}$$

The plastic stretch, D^p, and the plastic spin, W^p, tensors can also be defined as

$$\mathsf{D}^p = \mathrm{sym}(\mathbf{L}^p), \;\; \mathsf{W}^p = \mathrm{skew}(\mathbf{L}^p). \tag{18}$$

The rotation of the plastic stretch, D^p, to the deformed configuration leads us to

$$\mathsf{d}^p = \mathbf{R}^e \mathsf{D}^p [\mathbf{R}^e]^T = \mathbf{R}^e \mathrm{sym}(\dot{\mathbf{F}}^p [\mathbf{F}^p]^{-1})[\mathbf{R}^e]^T. \tag{19}$$

Following the formalism of thermodynamics with internal variables, an isotropic hyperelastic constitutive equation can be obtained:

$$\mathbf{T} = \bar{\rho}\frac{\partial \psi^e}{\partial \boldsymbol{\varepsilon}^e} = \mathsf{D}^e : \boldsymbol{\varepsilon}^e = [2G\mathsf{I} + \varsigma\,(\mathbf{I} \otimes \mathbf{I})] : \boldsymbol{\varepsilon}^e, \tag{20}$$

where $\mathbf{T} = J\boldsymbol{\sigma}$ is the Kirchhoff stress, $\bar{\rho}$ is the reference mass density, D^e denotes the fourth-order isotropic constant elastic tensor with $\mathbf{I}$ and I given in component form as $\mathbf{I}_{ij} = \delta_{ij}$ and $\mathsf{I}_{ijkl} = \frac{1}{2}[\delta_{ik}\delta_{jl} + \delta_{il}\delta_{jk}]$. Equation (20) is derived from the so-called Henky strain energy function, $\psi^e(\boldsymbol{\varepsilon}^e)$, which is generally accepted for a wide range of applications, given by,

$$\psi^e(\varsigma_1^e, \varsigma_2^e, \varsigma_3^e) = G\left[\ln(\varsigma_1^e)^2 + \ln(\varsigma_2^e)^2 + \ln(\varsigma_3^e)^2 + \frac{1}{2}\varsigma \ln(J^e)^2\right], \tag{21}$$

where G and ς_i^e represent the bulk modulus and the principal stretches, respectively, and $J^e = \varsigma_1^e\varsigma_2^e\varsigma_3^e$ is the Jacobian. The Eulerian logarithmic elastic strain tensor is employed as strain measure, which can be defined by,

$$\varepsilon^e = \ln(\mathbf{V}^e) = \frac{1}{2}\ln(\mathbf{B}^e) = \frac{1}{2}\ln\left(\mathbf{F}^e[\mathbf{F}^e]^T\right), \tag{22}$$

where $\ln(\cdot)$ denotes the tensorial logarithm of $(\cdot)$ and $\mathbf{B}^e = [\mathbf{V}^e]^2$ is the elastic Cauchy–Green strain tensor.

The evolution law for the plastic deformation gradient, $\mathbf{F}^p$, adopted here, is defined in terms of the plastic multiplier, $\dot{\gamma}^p$, and a generic plastic flow potential, $\psi(\mathbf{T}, A)$, expressed as a function of the Kirchhoff stress and the thermodynamic force set, A

$$\mathsf{d}^p = \dot{\gamma}^p \frac{\partial \psi(\mathbf{T}, A)}{\partial \mathbf{T}}. \tag{23}$$

Combining the definition of the rotated plastic stretch, Eq. (19), with the constitutive law, Eq. (23), and the assumption of zero plastic spin, $\mathsf{W}^p = 0$, the following evolution for the plastic deformation gradient can be obtained

$$\dot{\mathbf{F}}^p[\mathbf{F}^p]^{-1} = \dot{\gamma}^p[\mathbf{R}^e]^T \frac{\partial \psi}{\partial \mathbf{T}}\mathbf{R}^e. \tag{24}$$

The evolution equation for the set of internal variables is given by

$$\dot{\alpha} = \dot{\gamma}^p \mathbf{H}(\mathbf{T}, A), \tag{25}$$

where $\mathbf{H}(\mathbf{T}, A)$ represents a constitutive function and $\dot{\gamma}^p$ is the plastic multiplier. In order to define the onset of plastic flow, a general yield function, $\Phi^p(\mathbf{T}, A)$, is introduced. Together with the plastic multiplier, the yield function must comply with the standard complementarity relations (loading/unloading criterion)

$$\Phi^p \leq 0, \; \dot{\gamma}^p \geq 0, \quad \dot{\gamma}^p \Phi^p = 0. \tag{26}$$

The finite strain hyperelastic-based model described here permits a convenient extension of the general isotropic infinitesimal elasto-plastic models to the finite strain range.

4. Weak form

Having defined the contact constraints and the constitutive model employed, the next step consists in deriving the weak formulation of the BVP with contact conditions, Eq. (3). Therefore, the vector valued solution space, $\mathcal{U}^i$, containing the admissible displacement field $\mathbf{u} \subset \{\mathbf{u}^s, \mathbf{u}^m\}$ is defined on the Sobolev space of functions, $H^1(\Omega^i)$, as

$$\mathcal{U}^i := \{\mathbf{u}^i_j \in [H^1(\Omega^i)] : \mathbf{u}^i_j|_{\Gamma^i_D} = \hat{\mathbf{u}}^i_j, \; j = 1, \ldots, n_d\}, \tag{27}$$

while the weighting space, $\mathcal{V}^i$, for the virtual displacements, $\delta\mathbf{u}$, is given by

$$\mathcal{V}^i := \{\delta\mathbf{u}^i_j \in [H^1(\Omega^i)] : \delta\mathbf{u}^i_j|_{\Gamma^i_D} = 0, \; j = 1, \ldots, n_d\}. \tag{28}$$

Using the principle of virtual work in the current configuration, the problem is reduced to finding $\mathbf{u}^i \in \mathcal{U}^i$, such that

$$\delta\Pi(\mathbf{u}, \delta\mathbf{u}) = \delta\Pi_{\text{int,ext}}(\mathbf{u}, \delta\mathbf{u}) + \delta\Pi_c(\mathbf{u}, \delta\mathbf{u}) = 0, \quad \forall \, \{\delta\mathbf{u}^i \in \mathcal{V}^i\}, \tag{29}$$

where $\delta\Pi_{\text{int,ext}}(\mathbf{u}, \delta\mathbf{u})$ is the standard virtual work from the internal and external forces while, $\delta\Pi_c(\mathbf{u}, \delta\mathbf{u})$, represents the contact virtual work. The former can be computed for each body as

$$\delta\Pi^i_{\text{int,ext}}(\mathbf{u}^i, \delta\mathbf{u}^i) = \int_{\varphi(\Omega^i_0)} \boldsymbol{\sigma}^i \cdot \text{grad}\,(\delta\mathbf{u}^i) \; d\Omega - \int_{\varphi(\Omega^i_0)} \mathbf{b}^i \cdot \delta\mathbf{u}^i \; d\Omega$$
$$- \int_{\varphi(\Gamma^i_N)} \hat{\mathbf{t}}^i \cdot \delta\mathbf{u}^i \; d\gamma_N . \tag{30}$$

Using the balance of linear momentum at the contact interface, $\mathbf{t}_c^s d\gamma_c^s = \mathbf{t}_c^m d\gamma_c^m$, it is possible to obtain the contact virtual work as the integral — over the *slave side* — of the work done by the contact traction $\mathbf{t}_c^s$,

$$\delta\Pi_c\left(\mathbf{u}, \delta\mathbf{u}\right) = -\int_{\gamma_c^s} \mathbf{t}_c^s \cdot [\delta\mathbf{u}^s - \delta\mathbf{u}^m] d\gamma_c^s. \tag{31}$$

In order to formulate the problem using the Lagrange multiplier method, the negative contact tractions are identified with the Lagrange multiplier vector, such that,

$$\boldsymbol{\lambda} = -\mathbf{t}_c^s \quad \text{therefore } \lambda_\eta = -p_\eta, \quad \boldsymbol{\lambda}_\tau = \mathbf{t}_\tau. \tag{32}$$

Introducing the Lagrangian multipliers into Eq. (31), the final expression for contact virtual work is obtained,

$$\delta\Pi_c\left(\mathbf{u}, \delta\mathbf{u}, \boldsymbol{\lambda}\right) = \int_{\gamma_c^s} \boldsymbol{\lambda} \cdot (\delta\mathbf{u}^s - \delta\mathbf{u}^m) d\gamma_c^s. \tag{33}$$

A weak form of the non-penetration condition, Eq. $(8)_1$, can then be derived as

$$\delta g = \int_{\gamma_c^s} \delta\lambda_\eta \left[g(\mathbf{X}^s, t)\right] \, d\gamma_c^s \geq 0, \quad \forall \{\delta\lambda_\eta \in \mathscr{M}\}, \tag{34}$$

where $\mathscr{M}$ represents the chosen space for the Lagrangian multipliers. In a similar way, the contact conditions in the tangential direction, Eq. (10), are weakly enforced

$$\int_{\gamma_c^s} \delta\boldsymbol{\lambda}_\tau \left[(\boldsymbol{\vartheta}_\tau(\mathbf{X}^s, t) - \beta\boldsymbol{\lambda}_\tau)\right] d\gamma_c^s = 0, \quad \forall \{\delta\boldsymbol{\lambda}_\tau \in \mathscr{M}\}. \tag{35}$$

The remaining contact conditions in both normal and tangential directions are enforced pointwise. Therefore, the KKT conditions, after inserting the Lagrangian multipliers, lead to the following set of normal conditions:

$$\delta g(\mathbf{X}^s, t) \geq 0, \quad \lambda_\eta \geq 0, \quad \lambda_\eta \, g(\mathbf{X}^s, t) = 0. \tag{36}$$

In the tangential direction, the remaining constraints are enforced as follows:

$$\Phi := \|\boldsymbol{\lambda}_\tau\| - \mu|\lambda_\eta| \leq 0, \quad \beta \geq 0, \;\; \Phi\beta = 0. \tag{37}$$

5. Incremental form

The solution of the constitutive model, described in Section 3.2, defined by the corresponding rate constitutive equations and a set of initial conditions is not usually known for complex deformation paths. Therefore, the use of a numerical algorithm for integration of the rate constitutive equations is essential.

Algorithms based on the operator split methodology are particularly suitable for numerical integration of the evolution problem and have been widely used in computational plasticity.[6,8] Here, such an operator split method is used in the numerical integration of the elasto-plastic constitutive equations. An essential point in the derivation of the algorithm is the exponential approximation employed in the discretization of the plastic flow rule, in the plastic corrector stage, which was firstly employed in the computational literature by.[1] It leads to the following incremental evolution from the pseudo-time t_n to t_{n+1}:

$$\begin{aligned}\mathbf{F}^p_{n+1} &= \exp\left(\Delta\gamma^p \left[\mathbf{R}^e_{n+1}\right]^T \left.\frac{\partial\Phi}{\partial\mathbf{T}}\right|_{n+1} \mathbf{R}^e_{n+1}\right)\mathbf{F}^p_n \\ &= \left[\mathbf{R}^e_{n+1}\right]^T \exp\left(\Delta\gamma^p \left.\frac{\partial\Phi}{\partial\mathbf{T}}\right|_{n+1}\right)\mathbf{R}^e_{n+1}\mathbf{F}^p_n. \end{aligned} \tag{38}$$

In addition, a one-step backward Euler scheme is used to integrate the evolution equation for the internal variable,

$$\bar{\varepsilon}^p_{n+1} = \bar{\varepsilon}^p_\eta - \Delta\gamma^p \left.\frac{\partial\Phi^p}{\partial\mathbf{T}}\right|_{n+1}. \tag{39}$$

It can be shown that the approximation of Eq. (38) results in the following simpler update formula in terms of the logarithm eulerian strain tensor:

$$\boldsymbol{\varepsilon}^e_{n+1} = \boldsymbol{\varepsilon}^{e\ \mathrm{trial}}_{n+1} - \Delta\gamma^p \left.\frac{\partial\Phi^p}{\partial\mathbf{T}}\right|_{n+1}, \tag{40}$$

which is valid whenever the elastic right Cauchy–Green tensor $\mathbf{C}^{e\ \mathrm{trial}}_{n+1} = [\mathbf{F}^{e\ \mathrm{trial}}_{n+1}]^T\mathbf{F}^{e\ \mathrm{trial}}_{n+1}$ and $\frac{\partial\Phi^p}{\partial\mathbf{T}}|_{n+1}$ commute. Using Eq. (40) and imposing the elastic law at t_{n+1}, the following expression for the stress tensor consistent with the exponential approximation, Eq. (38), is obtained,

$$\mathbf{T}_{n+1} = \mathsf{D}^e : \left[\boldsymbol{\varepsilon}^{e\ \mathrm{trial}}_{n+1} - \Delta\gamma^p \left.\frac{\partial\Phi}{\partial\mathbf{T}}\right|_{n+1}\right]. \tag{41}$$

Consequently, the Cauchy stress is given by

$$\boldsymbol{\sigma}_{n+1} = \mathbf{R}^e_{n+1}\mathbf{T}_{n+1}[\mathbf{R}^e_{n+1}]/\det(\mathbf{U}^e_{n+1}). \tag{42}$$

Therefore, due to the use of logarithmic strains to describe elasticity together with the exponential approximation, the stress updating procedure can be written in the same format as the classical return mapping schemes of infinitesimal elasto-plasticity, see Refs. [7, 8, 44]. Here, the numerical integration of the small strain von Mises constitutive equations was undertaken with the backward Euler scheme.

Assuming that the values of $\mathbf{F}^p$ and $\bar{\varepsilon}^p$ are known at a certain time t_n, the deformation φ_{n+1} at a subsequent time t_{n+1} will determine uniquely the value of $\boldsymbol{\sigma}$ at time t_{n+1}, through the integration algorithm. This defines the incremental constitutive relation

$$\boldsymbol{\sigma}_{n+1} = \hat{\boldsymbol{\sigma}}\left(\mathbf{F}^p_n, \bar{\varepsilon}^p_n, \varphi_{n+1}\right). \tag{43}$$

Including Eq. (43) in the weak form of the equilibrium, the incremental BVP can be stated as follows: given $\mathbf{F}^p$ and $\bar{\varepsilon}^p$ at time t_n and given the body force and surface traction fields at time t_{n+1}, find the kinematically admissible configuration $\varphi(\Omega_0)|_{n+1}$ such that

$$\sum_i^{s,m}\left\{\int\limits_{\varphi(\Omega_0^i)|_{n+1}} \hat{\boldsymbol{\sigma}}^i \cdot \operatorname{grad}\left(\delta\mathbf{u}^i\right)\, d\Omega - \int\limits_{\varphi(\Omega_0^i)|_{n+1}} \mathbf{b}^i_{n+1}\cdot\delta\mathbf{u}^i\, d\Omega - \int\limits_{\varphi\left(\Gamma_N^i\right)|_{n+1}} \hat{\mathbf{t}}^i_{n+1}\cdot\delta\mathbf{u}^i\, d\gamma_N\right\}$$
$$- \int\limits_{\varphi(\Gamma_c^s)|_{n+1}} \boldsymbol{\lambda}_{n+1}[\delta\mathbf{u}^s - \delta\mathbf{u}^m]\, d\gamma_c^s = 0,\ \forall\ \{\delta\mathbf{u}^i \in \mathcal{V}^i\}. \tag{44}$$

6. Spatial discretization

In order to solve the incremental boundary value described in Section 5, a finite element discretization is introduced. Therefore, the functional sets $\mathcal{U}^i$, $\mathcal{V}^i$ and $\mathscr{M}$ are replaced by their discrete counterparts, $\{\mathcal{U}^i\}^h$, $\{\mathcal{V}^i\}^h$ and $\{\mathscr{M}\}^h$.

Remark 1. In what follows, the notation related with time/pseudo-time will be omitted for the sake of simplicity since all parameters and variables are defined at t_{n+1}.

Discretization of the bodies is performed using 4-noded elements with bilinear shape functions for 2D problems and 8-noded elements with trilinear shape functions for the 3D problems. This leads to a piecewise linear interpolation along the surface. Consequently, the geometrical coordinates of the slave and master surface elements are approximated by interpolation of the their nodal coordinates, $\mathbf{x}^i$, as follows,

$$\mathbf{x}^s \approx \{\mathbf{x}^s\}^h|_{\{\Gamma_c^s\}^h} = \sum_{k=1}^{n_s} N_k^s \mathbf{x}_k^s, \tag{45}$$

$$\mathbf{x}^m \approx \{\mathbf{x}^m\}^h|_{\{\Gamma_c^m\}^h} = \sum_{l=1}^{n_m} N_l^m \mathbf{x}_l^m \tag{46}$$

and the displacement field interpolation is obtained from the nodal displacements, $\mathbf{d}^i$, in a similar way,

$$\mathbf{u}^s \approx \{\mathbf{u}^s\}^h|_{\{\Gamma_c^s\}^h} = \sum_{k=1}^{n_s} N_k^s \mathbf{d}_k^s, \tag{47}$$

$$\mathbf{u}^m \approx \{\mathbf{u}^m\}^h|_{\{\Gamma_c^m\}^h} = \sum_{l=1}^{n_m} N_l^m \mathbf{d}_l^m. \tag{48}$$

Here, n_s and n_m denote the number of nodes on the slave and master boundary side, respectively. The slave and master element shape functions are represented by N^i.

The set of Lagrangian multipliers $\boldsymbol{\lambda}$ is approximated by a 1D discretized space, $\mathscr{M}^h \subset \mathscr{M}$, on the slave surface, such that,

$$\boldsymbol{\lambda}^h = \sum_{j=1}^{n_s} \boldsymbol{\phi}_j \mathbf{z}_j, \tag{49}$$

where, $\boldsymbol{\phi}$ is the so-called dual shape functions,[23] $\mathbf{z}_j$ is the Lagrangian multiplier on the slave node $j = \{1, \ldots, n_s\}$ and n_s represents the number of nodes on the slave side.

6.1. *Discretization of the contact virtual work*

Replacing the displacement, $\mathbf{u}$, the incremental displacement, $\delta\mathbf{u}$, and the vector of Lagrangian multipliers, $\boldsymbol{\lambda}$, in Eq. (33) by their respective interpolations yields the discretized form of the contact virtual work,

$$\delta\Pi_c \approx \delta\Pi_c^h = \int_{\{\gamma_c^s\}^h} \boldsymbol{\lambda}^h \cdot [\{\delta\mathbf{u}^s\}^h - \{\delta\mathbf{u}^m\}^h] d\gamma_c^s$$

$$= \sum_{j=1}^{n_s} \left\{ \sum_{k=1}^{n_s} [\delta \mathbf{d}_k^s]^T \int_{\{\gamma_c^s\}^h} \boldsymbol{\phi}_j N_k^s \, d\gamma_c^s \right.$$

$$\left. - \sum_{l=1}^{n_m} [\delta \mathbf{d}_l^m]^T \int_{\{\gamma_c^s\}^h} \boldsymbol{\phi}_j N_l^m \, d\gamma_c^s \right\} \mathbf{z}_j. \tag{50}$$

Moreover, by introducing the mortar matrices, $\mathbf{D} \in \mathbb{R}^{\{(n_d \ n_s)\times(n_d \ n_s)\}}$ and $\mathbf{M} \in \mathbb{R}^{\{(n_d \ n_s)\times(n_d \ n_m)\}}$,

$$\mathbf{D}_{[j,k]} = \int_{\{\gamma_c^s\}^h} \boldsymbol{\phi}_j N_k^s \, d\gamma_c^s \, \mathbf{I}_{n_d}, \tag{51}$$

$$\mathbf{M}_{[j,l]} = \int_{\{\gamma_c^s\}^h} \boldsymbol{\phi}_j N_l^m \, d\gamma_c^s \, \mathbf{I}_{n_d}, \tag{52}$$

it is possible to rewrite Eq. (50), in its algebraic form,

$$\delta \Pi_c^h = \{[\delta \mathbf{d}^s]^T \mathbf{D} - [\delta \mathbf{d}^m]^T \mathbf{M}^T\} \mathbf{z}, \tag{53}$$

where $\mathbf{I}_{n_d}$ is the identity matrix in $\mathbb{R}^{\{n_d \times n_d\}}$. The nodal test function values and the discrete nodal values are conveniently expressed in matrix notation by the global vectors $\delta \mathbf{d}^s$, $\delta \mathbf{d}^m$ and $\mathbf{z}$.

6.2. *Discretization of the contact constraints*

To complete the description of the contact problem, a discretized form of the contact constraints is needed. In the normal direction, the introduction of the interpolation of displacements and Lagrangian multipliers into Eq. (34) yields the following discretized form of the weak non-penetration condition,[35]

$$\delta g \approx \delta g^h = - \int_{\{\gamma_c^s\}^h} \{\delta \lambda_\eta\}^h \boldsymbol{\eta} [\{\mathbf{x}^s\}^h - \{\hat{\mathbf{x}}^m\}^h] d\gamma_c^s$$

$$= - \sum_{j=1}^{n_s} \left\{ \sum_{k=1}^{n_s} \{\delta z_\eta\}_j \int_{\{\gamma_c^s\}^h} \boldsymbol{\phi}_j N_k^s \boldsymbol{\eta}_j^T \, d\gamma_c^s \, \mathbf{x}_k^s \right\}$$

$$+ \sum_{j=1}^{n_s} \left\{ \sum_{l=1}^{n_m} \{\delta z_\eta\}_j \int_{\{\gamma_c^s\}^h} \boldsymbol{\phi}_j N_l^m \boldsymbol{\eta}_j^T \, d\gamma_c^s \, \mathbf{x}_l^m \right\} \geq 0. \tag{54}$$

Then, introducing a vector of discretized weighted gaps, $\tilde{\mathbf{g}} \in \mathbb{R}^{\{n_d \times n_s\}}$, and a vector of nodal values of the normal contact stresses, $\delta \mathbf{z} \in \mathbb{R}^{\{n_d \times n_s\}}$, yields the algebraic form of Eq. (54),

$$\delta \tilde{g}^h = [\delta \mathbf{z}]^T \tilde{\mathbf{g}}, \tag{55}$$

where a single entry, $\tilde{g}_j$, of $\tilde{\mathbf{g}}$ is given by,

$$\tilde{g}_j = -\,[\boldsymbol{\eta}_j]^T \left\{ \mathbf{D}_{[j,j]}[\mathbf{x}_j^s] - \sum_{l=1}^{n_m} \mathbf{M}_{[j,l]}[\mathbf{x}_l^m] \right\}. \tag{56}$$

The discretization of the tangential contact constraints described here has been presented in Refs. [36, 39, 40] and is only briefly addressed. By employing an appropriate measure for the tangential velocity and a backward Euler scheme, it is possible to obtain the condensed discretized form of the weak tangential condition, for each slave node j, as follows:

$$\{\tilde{\mathbf{u}}_\tau\}_j - \{\tilde{\beta}_\tau\}_j \{\mathbf{z}_\tau\}_j = 0. \tag{57}$$

6.3. *Integration of mortar matrices*

The evaluation of the mortar matrices $\mathbf{D}$ and $\mathbf{M}$, previously introduced in expressions (51) and (52), is addressed in this section. While the numerical integration of matrix $\mathbf{D}$ is relatively straightforward due to the biorthogonality, the computation of the matrix $\mathbf{M}$ is far more complex. In this case, the integrand of expression (52) has got terms defined on two different surfaces and, therefore, the mortar integral must be carefully subdivided into contact segments for the 2D case (Fig. 2),

$$\mathbf{D}_{[j,k]} = \sum_{gp=1}^{n_{gp}} w_{gp} \phi_j(\xi^s(\zeta_{gp})) N_k^s(\xi^s(\zeta_{gp})) J(\xi^s(\zeta_{gp})) \frac{\xi_b^s - \xi_a^s}{2}, \tag{58}$$

$$\mathbf{M}_{[j,l]} = \sum_{gp=1}^{n_{gp}} w_{gp} \phi_j(\xi^s(\zeta_{gp})) N_k^m(\xi^m(\zeta_{gp})) J(\xi^s(\zeta_{gp})) \frac{\xi_b^s - \xi_a^s}{2}, \tag{59}$$

or triangular cells for the 3D case (Fig. 3),

$$\mathbf{D}_{[j,k]} = \sum_{gp=1}^{n_{gp}} w_{gp} \phi_j(\boldsymbol{\xi}^s(\boldsymbol{\zeta}_{gp})) N_k^s(\boldsymbol{\xi}^s(\boldsymbol{\zeta}_{gp})) J_{\text{cell}}, \tag{60}$$

$$\mathbf{M}_{[j,l]} = \sum_{gp=1}^{n_{gp}} w_{gp} \phi_j(\boldsymbol{\xi}^s(\boldsymbol{\zeta}_{gp})) N_k^m(\boldsymbol{\xi}^m(\boldsymbol{\zeta}_{gp})) J_{\text{cell}}, \tag{61}$$

where n_{gp} is the number of Gaussian integration points, w_{gp} is the respective quadrature weight, $J(\cdot)$ is the slave element Jacobian determinant and the

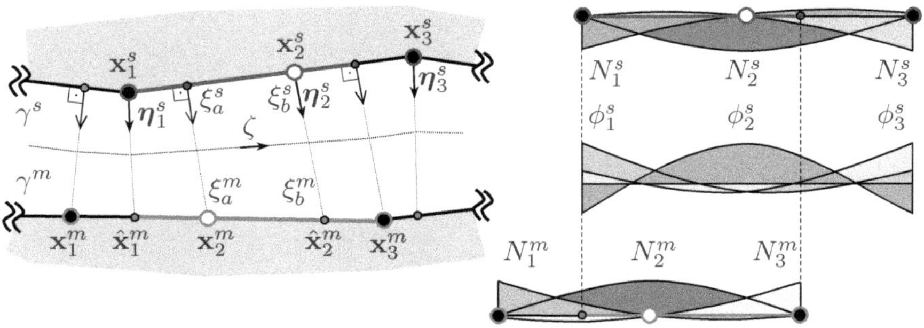

Fig. 2. Continuous normal field mortar segmentation method.

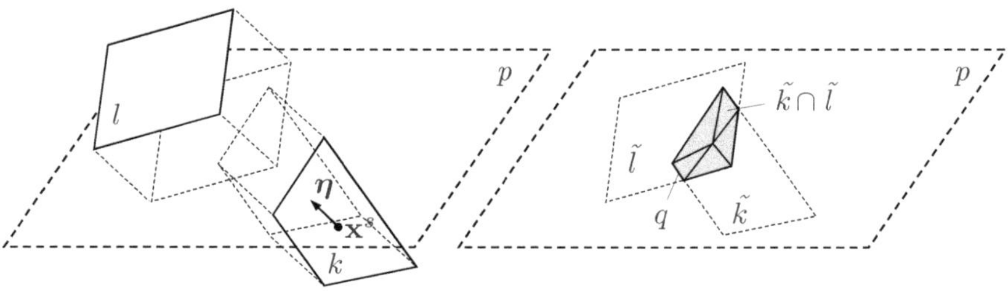

Fig. 3. Clipping procedure for a mortar numerical integration.

scalar J_{cell} denotes the Jacobian determinant of the triangular integration cell under analysis, see Fig. 3. For additional information regarding, the integration of mortar matrices, the reader is referred to Refs. [37, 45–47].

6.4. *Discretized problem*

In order to make the notation clearer, the finite element nodes of the problem can be grouped into three subsets: the set $\mathcal{M}$ of potential contact nodes on the mortar side, the set $\mathcal{S}$ of potential contact nodes on the slave side and the set $\mathcal{N}$ of all remaining nodes. Therefore, the vector of global nodal displacements can be represented by

$$\mathbf{d} = [\mathbf{d}_{\mathcal{N}},\ \mathbf{d}_{\mathcal{M}},\ \mathbf{d}_{\mathcal{S}}]^T. \tag{62}$$

An equivalent notation is employed for the geometry and virtual displacements. Consequently, the discretized contact virtual work, Eq. (53), can be expressed in matrix form

$$\delta\Pi_c^h = [\delta\mathbf{d}]^T[\mathbf{D} - \mathbf{M}^T]\mathbf{z}, \tag{63}$$

from where the contact forces can be retrieved

$$\mathbf{f}_c(\mathbf{d}, \mathbf{z}) = [\mathbf{0},\ -\mathbf{M}^T,\ \mathbf{D}]\mathbf{z}. \tag{64}$$

The weak form of equilibrium, Eq. (44), can then be expressed by the following nonlinear algebraic system:

$$\mathbf{f}_{\text{int}}(\mathbf{d}) - \mathbf{f}_{\text{ext}} + \mathbf{f}_c(\mathbf{d}, \mathbf{z}) = 0, \tag{65}$$

where $\mathbf{f}_{\text{int}}$ and $\mathbf{f}_{\text{ext}}$ are, respectively, internal and external global force vectors resulting from assemblage of the element vectors for each body

$$\mathbf{f}_{\text{int}}^{i(e)} = \int_{\varphi(\Omega_0^{i(e)})} [\mathsf{B}]^T \mathsf{S}_{n+1}\ d\Omega^i, \tag{66}$$

$$\mathbf{f}_{\text{ext}}^{i(e)} = \int_{\varphi(\Omega_0^{i(e)})} [\mathsf{N}]^T \mathbf{b}_{n+1}\ d\Omega^i + \int_{\varphi(\Omega_0^{i(e)})} [\mathsf{N}]^T \hat{\mathbf{t}}_{n+1}\ d\gamma_N^i, \tag{67}$$

with B and N being the standard discrete symmetric gradient operators and the interpolation matrix of the element (e), respectively. The vector S_{n+1} contains the Cauchy stress components delivered by the algorithmic constitutive function.

It is important to remark that Eq. (65) is highly nonlinear. From one side, the internal force vector is a nonlinear function of the displacements due to the finite inelastic deformations involved. On the other hand, the contact force vector also depends nonlinearly on the displacements since the mortar matrices are computed over the deformed configuration.

The overall discretized problem is composed by the discretized equilibrium equations, Eq. (65), the discretized KKT conditions

$$\{\tilde{g}\}_j \geq 0,\ \{z_\eta\}_j \geq 0,\ \{z_\eta\}_j\{\tilde{g}\}_j = 0 \tag{68}$$

and the discretized relation for the tangential contact stresses

$$\begin{aligned} \Phi_j &= \|\{\mathbf{z}_\tau\}_j\| - \mu|\{z_\eta\}_j| \leq 0, \\ \{\tilde{\mathbf{u}}_\tau\}_j &- \{\tilde{\beta}_\tau\}_j\{\mathbf{z}_\tau\}_j = 0, \\ \{\tilde{\beta}_\tau\}_j &\geq 0,\ \Phi_j\{\tilde{\beta}_\tau\}_j = 0. \end{aligned} \tag{69}$$

The contact constraints Eqs. (68) and (69) are expressed by a set of inequality conditions which need to be solved with an appropriate solution strategy.

7. F-bar formulation

In order to improve the performance of the 4-node quadrilateral displacement-based finite element near the incompressibility limit, the F-bar formulation proposed in Ref. [41] will be employed here. In this formulation, the standard deformation gradient, $\mathbf{F}$, is replaced with an assumed modified (F-bar) deformation gradient. This deformation gradient is defined such that the (near-) incompressibility constraint is relaxed, overcoming the spurious locking which frequently occurs as a consequence of the inability of low-order interpolation polynomials to adequately represent general volume preserving fields. Considering the analysis of large deformation elasto-plastic frictional contact problems addressed in this work, these effects are even more critical. In the F-bar formulation, the deformation gradient $\mathbf{F}$, at the Gauss point of interest and the deformation gradient at the centroid of the finite element, $\mathbf{F}_0$, are split into a volumetric and deviatoric components:

$$\mathbf{F} = \mathbf{F}_{\text{iso}}\mathbf{F}_v; \quad \mathbf{F}_0 = \{\mathbf{F}_0\}_{\text{iso}}\{\mathbf{F}_0\}_v. \tag{70}$$

An F-bar element is obtained from the standard finite element by replacing the volumetric component of the deformation gradient, $\mathbf{F}_v$ by the volumetric component of the deformation gradient at the centroid of the element, $\{\mathbf{F}_0\}_v$. Therefore, the F-bar deformation gradient, at a specific Gauss point, is the product of the isochoric component of $\mathbf{F}$ with the volumetric component of $\mathbf{F}_0$:

$$\bar{\mathbf{F}} = \mathbf{F}_{\text{iso}}\{\mathbf{F}_0\}_v = \left[\frac{\det(\mathbf{F}_0)}{\det(\mathbf{F})}\right]^{1/3}\mathbf{F}. \tag{71}$$

Particularly important in the present context is the fact that this methodology preserves the displacement-based structure of the finite element equations as well as the strains-driven format of standard algorithms for integration of the inelastic constitutive equations.

8. Solution procedure

The contact problem, which is described by the discretized system of equations (65) and constraints (68) and (69), is essentially a minimization problem with inequality constraints. This problem can be solved with a PDASS proposed by Hintermüller *et al.*[29] and first applied to contact problems by Hüeber and Wohlmuth (2005).[48] It consist in finding the set of *active* contacting nodes, $\mathcal{A}$, and the set of inactive nodes, $\mathcal{I}$, of the slave set $\mathcal{S} = \{\mathcal{A} \cup \mathcal{I}\}$ while minimizing the contact constraints. The search for the sets $\mathcal{A}$ and $\mathcal{I}$ is performed by solving the non-penetration function for every slave node, j,

$$\tilde{g}_j = 0, \quad \forall \{j \in \mathcal{A}\}. \tag{72}$$

This search can be extended to include the subset of active contact nodes sticking, $\mathcal{Q}$, and the subset of active contact nodes sliding, $\mathcal{L}$, by solving the frictional condition, Eq. $(69)_1$, for every slave node, j.

The determination of the sets $\mathcal{I}$ and $\mathcal{A}$, together with the subsets $\mathcal{Q}$ and $\mathcal{L}$, allows the problem to be stated with equality conditions:

$$\begin{aligned}
\mathbf{r} = \mathbf{f}_{\text{int}}(\mathbf{d}) - \mathbf{f}_{\text{ext}} + \mathbf{f}_c(\mathbf{d}, \mathbf{z}) = \mathbf{0}, \\
\tilde{g}_j = 0,\ \forall\{j \in \mathcal{A}\}, \\
\{z_\eta\}_j = 0,\ \forall\{j \in \mathcal{I}\}, \\
\{\tilde{\beta}_\tau\}_j = 0,\ \forall\{j \in \mathcal{Q}\} \\
\Phi_j = \|\{\mathbf{z}_\tau\}_j\| - \mu|\{z_\eta\}_j| = 0,\ \forall\{j \in \mathcal{S}\} \\
\{\tilde{\mathbf{u}}_\tau\}_j - \{\tilde{\beta}_\tau\}_j\{\mathbf{z}_\tau\}_j = 0,\ \forall\{j \in \mathcal{L}\}.
\end{aligned} \tag{73}$$

To solve the frictional problem with a semi-smooth Newton method, two complementarity functions for the frictional constraints must be introduced. The complementarity functions are non-continuous relations used to verify the frictional state of the contacting nodes. They implicitly incorporate the distinction between inactive nodes, active sticking nodes and active sliding nodes. This concept was first introduced to a dual mortar formulation by Hüeber *et al.* (2008),[31] for the Tresca frictional case. Later, complementarity functions for Coulomb's friction were proposed by Gitterle *et al.*[36] The complementarity function for the normal contact conditions, which is described in detail in Ref. [35], is defined for each slave node $j \in \mathcal{S}$ as

$$\{C_\eta\}_j(\mathbf{d}, \mathbf{z}_j) = \{z_\eta\}_j - \max(0, \{z_\eta\}_j - c_\eta \tilde{g}_j), \tag{74}$$

For the tangential contact conditions, the complementarity functions given in Ref. [36] are defined for each slave node $j \in \mathcal{S}$ as

$$\begin{aligned}
\{\mathbf{C}_\tau\}_j(\mathbf{d}, \mathbf{z}_j) = \{z_\tau\}_j \max(\mu[\{z_\eta\}_j - c_\eta \tilde{g}_j], |\{\mathbf{z}_\tau\}_j + c_\tau\{\tilde{\mathbf{u}}_\tau\}_j|) \\
- \mu[\{\mathbf{z}_\tau\}_j + c_\tau \tilde{\mathbf{u}}_j] \max(0, [\{z_\eta\}_j - c_\tau \tilde{g}_j]).
\end{aligned} \tag{75}$$

The parameters c_η and c_τ are positive constants that do not affect the solution results but can influence the convergence behaviour. In the numerical examples, the impact of these parameters in the solution of the class of problems addressed in this work will be studied in detail.

The combination of the equilibrium equation (65) with the complementarity functions for the contact conditions in the normal, expression (74), and tangential, expression (75), directions leads to the following nonlinear system of equations:

$$\begin{aligned} \mathbf{r} = \mathbf{f}_{\text{int}}(\mathbf{d}) - \mathbf{f}_{\text{ext}} + \mathbf{f}_c(\mathbf{d}, \mathbf{z}) = \mathbf{0}, \\ \{C_\eta\}_j(\mathbf{d}, \mathbf{z}_j) = 0, \ \forall\{j \in \mathcal{S}\}, \\ \{\mathbf{C}_\tau\}_j(\mathbf{d}, \mathbf{z}_j) = 0, \ \forall\{j \in \mathcal{S}\}. \end{aligned} \tag{76}$$

Due to the max-operator and the Euclidian norm employed in the two complementarity functions, they are continuous but non-smooth and not uniquely differentiable. Therefore, this justifies the application of the semi-smooth Newton method.[27,29] The generalized derivative for the max-function,[35] is defined as

$$f(x) = \max(a, x) \rightarrow \Delta f(x) = \begin{cases} 0, & \text{if } x \leq a, \\ 1, & \text{if } x > a. \end{cases} \tag{77}$$

Therefore, the solution of the system of Eq. (76) treats material and geometrical nonlinearities as well as nonlinearities emerging from contact and friction.

8.1. *Primal–dual active set algorithms*

The solution of the contact problem, with the so-called PDASS, can be obtained by solving either the system of Eqs. (73) or (76).

In the first case, two nested iterative schemes are necessary. An outer loop for solving the active set and an inner iteration for the nonlinear finite element problem having the active set fixed, see Algorithm 1. This strategy has been employed for the solution of dynamic problems,[34] and nonlinear material behaviour with multigrid methods in Ref. [32], among others. In the second case, due to the possibility of linearizing the system of Eq. (76), an algorithm with a single iteration loop for the solution of all sources of nonlinearity is obtained, see Algorithm 2. This strategy has been employed by several authors and proven to be very efficient.[31,35,36] Here, both PDASS will be applied to the simulation of frictional contact problems undergoing finite strains and inelastic material behaviour.

A comparison of both approaches will be conducted, in the numerical examples, in order to assess their performance in the solution of large inelastic deformations and metal forming problems.

Algorithm 1 Nested iterative strategy within a load step.

1: Set $\mathbb{I} = 0$ and $\mathbb{k} = 0$. Initialize $\mathbf{d}^0, \mathbf{z}^0$, $\mathcal{A}^0$ and $\mathcal{I}^0$, such that $\mathcal{S} = \mathcal{A}^0 \cup \mathcal{I}^0$, $\mathcal{A}^0 \cap \mathcal{I}^0 = \emptyset$ and $\mathcal{A}^0 = \mathcal{Q}^0 \cup \mathcal{L}^0$.

2: Loop over the active set $\mathbb{k}$,

3: Loop over $\mathbb{I}$,

4: Solve the linearization of (73) for $(\Delta \mathbf{d}^{\mathbb{I}+1}, \mathbf{z}^{\mathbb{I}+1})$:

$$\Delta \mathbf{r}\left(\mathbf{d}^{\mathbb{I}}, \mathbf{z}^{\mathbb{I}}\right) = \mathbf{r}^{\mathbb{I}},$$
$$\Delta \tilde{g}_j^{\mathbb{I}} = -\tilde{g}_j^{\mathbb{I}}, \ \forall \{j \in \mathcal{A}^{\mathbb{k}}\},$$
$$\{z_\eta\}_j^{\mathbb{I}+1} = 0, \ \forall \{j \in \mathcal{I}^{\mathbb{k}}\},$$
$$\left\{\tilde{\beta}_\tau\right\}_j^{\mathbb{I}} = 0, \ \forall \{j \in \mathcal{Q}^{\mathbb{k}}\},$$
$$\{\tilde{u}_\tau\}_j^{\mathbb{I}} - \left\{\tilde{\beta}_\tau\right\}_j^{\mathbb{I}} \{z_\tau\}_j^{\mathbb{I}+1} = 0, \ \forall \{j \in \mathcal{L}^{\mathbb{k}}\}.$$

5: Update $\mathbf{d}^{\mathbb{I}+1} = \mathbf{d}^{\mathbb{I}} + \Delta \mathbf{d}^{\mathbb{I}+1}$

6: Update the sets $\mathcal{A}^{\mathbb{k}+1}$, $\mathcal{I}^{\mathbb{k}+1}$, $\mathcal{Q}^{\mathbb{k}+1}$ and $\mathcal{L}^{\mathbb{k}+1}$ with

$$\mathcal{A}^{\mathbb{k}+1} := \left\{ j \in \mathcal{S} \;\middle|\; \{z_\eta\}_j^{\mathbb{I}+1} \geq \left[\Delta \tilde{g}_j^{\mathbb{I}+1} - \mathbf{d}^{\mathbb{I}+1} \cdot \boldsymbol{\eta}\right] \right\}$$
$$\mathcal{I}^{\mathbb{k}+1} := \mathcal{S} / \mathcal{A}^{\mathbb{k}+1},$$
$$\mathcal{Q}^{\mathbb{k}+1} := \left\{ j \in \mathcal{A}^{\mathbb{k}+1} \;\middle|\; \mu \left[\{z_\eta\}_j^{\mathbb{I}+1}\right] \leq \{z_\tau\}_j^{\mathbb{k}+1} \right\},$$
$$\mathcal{L}^{\mathbb{k}+1} := \mathcal{A}^{\mathbb{k}+1} / \mathcal{Q}^{\mathbb{k}+1}.$$

7: Active set check

IF $\mathcal{I}^{\mathbb{k}+1} = \mathcal{I}^{\mathbb{k}}$, $\mathcal{Q}^{\mathbb{k}+1} = \mathcal{Q}^{\mathbb{k}}$ and $\mathcal{L}^{\mathbb{k}+1} = \mathcal{L}^{\mathbb{k}}$

THEN GOTO 8

ELSE Set $\mathbb{I} = 0$, $\mathbb{k} = \mathbb{k} + 1$ and *GOTO* 2.

8: Convergence check

IF $\|\mathbf{r}_{\text{tot}}\| \leq \mathcal{E}_r$, *GOTO* Next increment

ELSE set $\mathbb{I} := \mathbb{I} + 1$ and *GOTO* 3.

8.2. *Matrix representation*

In order to solve the nonlinear system of Eq. (76) with a semi-smooth Newton algorithm, it is necessary to perform the full linearization of $\mathbf{r}$, $\mathbf{C}_\eta$

Algorithm 2 Single iterative strategy within a load step.

1: Set $\Bbbk = 0$ and initial conditions $\mathbf{d}^0, \mathbf{z}^0$. Initialize $\mathcal{A}^0$ and $\mathcal{I}^0$, such that $\mathcal{S} = \mathcal{A}^0 \cup \mathcal{I}^0$, $\mathcal{A}^0 \cap \mathcal{I}^0 = \emptyset$ and $\mathcal{A} = \mathcal{Q}^0 \cup \mathcal{L}^0$.

2: Loop over $\Bbbk$.

3: Solve the linearization of (76) for $(\Delta\mathbf{d}^{\Bbbk+1}, \mathbf{z}^{\Bbbk+1})$:

$$\begin{aligned}
&\Delta\mathbf{r}\left(\mathbf{d}^{\Bbbk}, \mathbf{z}^{\Bbbk}\right) = -\mathbf{r}^{\Bbbk},\\
&\mathbf{z}_j^{\Bbbk+1} = 0,\ \forall\{j \in \mathcal{I}^{\Bbbk}\},\\
&\Delta\tilde{g}_j^{\Bbbk} = -\tilde{g}_j^{\Bbbk},\ \forall\{j \in \mathcal{A}^{\Bbbk}\},\\
&\Delta\left\{C_\tau\right\}_j^{\Bbbk} = -\left\{C_\tau\right\}_j^{\Bbbk},\ \forall\left\{j \in \mathcal{L}^{\Bbbk}\right\} \vee \{j \in \mathcal{Q}^{\Bbbk}\},
\end{aligned}$$

4: Update $\mathbf{d}^{\Bbbk+1} = \mathbf{d}^{\Bbbk} + \Delta\mathbf{d}^{\Bbbk+1}$.

5: Set $\mathcal{A}^{\Bbbk+1}$, $\mathcal{I}^{\Bbbk+1}$, $\mathcal{Q}^{\Bbbk+1}$ and $\mathcal{L}^{\Bbbk+1}$ to

$$\begin{aligned}
\mathcal{A}^{\Bbbk+1} &:= \left\{j \in \mathcal{S} \;\middle|\; \{z_\eta\}_j^{\Bbbk+1} > c_\eta \tilde{g}_j^{\Bbbk+1}\right\},\\
\mathcal{I}^{\Bbbk+1} &:= \mathcal{S}/\mathcal{A}^{\Bbbk+1},\\
\mathcal{Q}^{\Bbbk+1} &:= \left\{j \in \mathcal{A}^{\Bbbk+1} \;\middle|\; \mu\left[\{z_\eta\}_j^{\Bbbk+1} - c_\eta\tilde{g}_j^{\Bbbk+1}\right]\right.\\
&\qquad \left. > \left|\{z_\tau\}_j^{\Bbbk+1} + c_\tau\left\{\tilde{u}_\tau\right\}_j^{\Bbbk+1}\right|\right\},\\
\mathcal{L}^{\Bbbk+1} &:= \left\{j \in \mathcal{A}^{\Bbbk+1} \;\middle|\; \mu\left[\{z_\eta\}_j^{\Bbbk+1} - c_\eta\tilde{g}_j^{\Bbbk+1}\right]\right.\\
&\qquad \left. \leq \left|\{z_\tau\}_j^{\Bbbk+1} + c_\tau\left\{\tilde{u}_\tau\right\}_j^{\Bbbk+1}\right|\right\}.
\end{aligned}$$

6: Convergence check

IF $\mathcal{I}^{\Bbbk+1} = \mathcal{I}^{\Bbbk}$, $\mathcal{Q}^{\Bbbk+1} = \mathcal{Q}^{\Bbbk}$, $\mathcal{L}^{\Bbbk+1} = \mathcal{L}^{\Bbbk}$ and $\left\|\mathbf{r}_{\text{tot}}^{\Bbbk}\right\| \leq \mathcal{E}_r$,

THEN GOTO Next increment

ELSE set $\Bbbk := \Bbbk + 1$ and *GOTO* 2.

and $\mathbf{C}_\tau$ and then convert the linear system into an algebraic form. In the following, the matrix representation of these directional derivatives is provided for completeness. More information regarding the linearization of each term can be found in Refs. [31, 35, 36]. The linearization of $\mathbf{r}$ includes the derivation of the internal force vector $\mathbf{f}_{\text{int}}(\mathbf{d})$ and the contact force vector $\mathbf{f}_c(\mathbf{d}, \mathbf{z})$ since we assume that the external loads are independent of the displacement $\mathbf{d}$. Therefore, the derivation of the internal force vector $\mathbf{f}_{\text{int}}(\mathbf{d})$ leads to the corresponding tangent stiffness matrix $\mathbf{K} \in \mathbb{R}^{\{(n_d\ n_{\text{nod}})\times(n_d\ n_{\text{nod}})\}}$ which, in the case of the F-bar formulation employed in this work, is obtained by the assemblage of the element stiffness matrices[41]

$$\mathbf{K}^e = \int_{\varphi(\Omega_0^e)} \mathsf{G}^T[\mathsf{a}][\mathsf{G}]d\Omega + \int_{\varphi(\Omega_0^e)} \mathsf{G}^T[\mathsf{q}][\mathsf{G}_0 - \mathsf{G}]d\Omega, \tag{78}$$

where n_{nod} is the number of nodes of the dicretized solids, G is the standard discrete spatial gradient operator, G_0, is the gradient operator at the element centroid, a denotes the matrix form of the fourth-order spatial elasticity tensor evaluated at $\mathbf{F} = \bar{\mathbf{F}}$,

$$[\mathsf{a}]_{ijkl} = \frac{1}{\det(\mathbf{F})} F_{jp} F_{tq} A_{ipkq}, \tag{79}$$

and q is the matrix form of the fourth-order tensor defined by:

$$\mathsf{q} = \frac{1}{3}\mathsf{a} : [\mathbf{I} \otimes \mathbf{I}] - \frac{2}{3}[\sigma \otimes \mathbf{I}], \tag{80}$$

also computed at $\mathbf{F} = \bar{\mathbf{F}}$. In expression (79), A_{ipkq} denotes the components of the first elasticity tensor.

The assemblage of the vector of unknowns (primal–dual pair), the right-hand side vector (residual forces, non-penetration functions and complementarity functions) and the stiffness matrix leads to the following iterative global system of equations:

$$\mathbf{K}^{\Bbbk}[\Delta\mathbf{d}^{\Bbbk} + \mathbf{z}^{\Bbbk+1}] = -[\mathbf{r} + \tilde{\mathbf{g}} + \mathbf{C}_{\mathcal{T}}]^{\Bbbk}, \tag{81}$$

where $\Delta\mathbf{d}$ represents the global displacement incremental vector and $\mathbf{z}$ is the vector of nodal Lagrangian multipliers. Then, subdividing the surface domain into the five different subsets $\{\mathcal{N}, \mathcal{M}, \mathcal{I}, \mathcal{Q}, \mathcal{L}\}$, the following equation system is obtained:

$$\begin{bmatrix}
\mathbf{K}_{\mathcal{N}\mathcal{N}} & \mathbf{K}_{\mathcal{N}\mathcal{M}} & \mathbf{K}_{\mathcal{N}\mathcal{I}} & \mathbf{K}_{\mathcal{N}\mathcal{Q}} & \mathbf{K}_{\mathcal{N}\mathcal{L}} & \mathbf{0} & \mathbf{0} & \mathbf{0} \\
\mathbf{K}_{\mathcal{M}\mathcal{N}} & \tilde{\mathbf{K}}_{\mathcal{M}\mathcal{M}} & \tilde{\mathbf{K}}_{\mathcal{M}\mathcal{I}} & \tilde{\mathbf{K}}_{\mathcal{M}\mathcal{Q}} & \tilde{\mathbf{K}}_{\mathcal{M}\mathcal{L}} & -\mathbf{M}_{\mathcal{I}}^T & -\mathbf{M}_{\mathcal{Q}}^T & -\mathbf{M}_{\mathcal{L}}^T \\
\mathbf{K}_{\mathcal{I}\mathcal{N}} & \tilde{\mathbf{K}}_{\mathcal{I}\mathcal{M}} & \tilde{\mathbf{K}}_{\mathcal{I}\mathcal{I}} & \tilde{\mathbf{K}}_{\mathcal{I}\mathcal{Q}} & \tilde{\mathbf{K}}_{\mathcal{I}\mathcal{L}} & \mathbf{D}_{\mathcal{I}} & \mathbf{0} & \mathbf{0} \\
\mathbf{K}_{\mathcal{Q}\mathcal{N}} & \tilde{\mathbf{K}}_{\mathcal{Q}\mathcal{M}} & \tilde{\mathbf{K}}_{\mathcal{Q}\mathcal{I}} & \tilde{\mathbf{K}}_{\mathcal{Q}\mathcal{Q}} & \tilde{\mathbf{K}}_{\mathcal{Q}\mathcal{L}} & \mathbf{0} & \mathbf{D}_{\mathcal{Q}} & \mathbf{0} \\
\mathbf{K}_{\mathcal{L}\mathcal{N}} & \tilde{\mathbf{K}}_{\mathcal{L}\mathcal{M}} & \tilde{\mathbf{K}}_{\mathcal{L}\mathcal{I}} & \tilde{\mathbf{K}}_{\mathcal{L}\mathcal{Q}} & \tilde{\mathbf{K}}_{\mathcal{L}\mathcal{L}} & \mathbf{0} & \mathbf{0} & \mathbf{D}_{\mathcal{L}} \\
\mathbf{0} & \mathbf{0} & \mathbf{0} & \mathbf{0} & \mathbf{0} & \mathbf{I}_{\mathcal{I}} & \mathbf{0} & \mathbf{0} \\
\mathbf{0} & \tilde{\mathbf{M}}_{\mathcal{A}} & \tilde{\mathbf{D}}_{\mathcal{A}\mathcal{I}} & \tilde{\mathbf{D}}_{\mathcal{A}\mathcal{Q}} & \tilde{\mathbf{D}}_{\mathcal{A}\mathcal{L}} & \mathbf{0} & \mathbf{0} & \mathbf{0} \\
\mathbf{0} & \mathbf{H}_{\mathcal{Q}} & \mathbf{F}_{\mathcal{Q}\mathcal{I}} & \mathbf{F}_{\mathcal{Q}\mathcal{Q}} & \mathbf{F}_{\mathcal{Q}\mathcal{L}} & \mathbf{0} & \mathbf{Q}_{\mathcal{Q}} & \mathbf{0} \\
\mathbf{0} & \mathbf{J}_{\mathcal{L}} & \mathbf{G}_{\mathcal{L}\mathcal{I}} & \mathbf{G}_{\mathcal{L}\mathcal{Q}} & \mathbf{G}_{\mathcal{L}\mathcal{L}} & \mathbf{0} & \mathbf{0} & \mathbf{L}_{\mathcal{L}}
\end{bmatrix} \cdot \begin{bmatrix} \Delta\mathbf{d}_{\mathcal{N}} \\ \Delta\mathbf{d}_{\mathcal{M}} \\ \Delta\mathbf{d}_{\mathcal{I}} \\ \Delta\mathbf{d}_{\mathcal{Q}} \\ \Delta\mathbf{d}_{\mathcal{L}} \\ \mathbf{z}_{\mathcal{I}} \\ \mathbf{z}_{\mathcal{Q}} \\ \mathbf{z}_{\mathcal{L}} \end{bmatrix}$$

$$= \left[\mathbf{r}_{\mathcal{N}},\ \mathbf{r}_{\mathcal{M}},\ \mathbf{r}_{\mathcal{I}},\ \mathbf{r}_{\mathcal{Q}},\ \mathbf{r}_{\mathcal{L}},\ \mathbf{0},\ \tilde{\mathbf{g}}_{\mathcal{A}},\ \{\mathbf{C}_{\mathcal{T}}\}_{\mathcal{Q}},\ \{\mathbf{C}_{\mathcal{T}}\}_{\mathcal{L}}\right]^T. \tag{82}$$

The matrices $\mathbf{K}$ and $\tilde{\mathbf{C}}$ contain the linearization of the internal force vector and the contact force vector, respectively. The blocks $\tilde{\mathbf{K}}_{ij}$ are the sum of the aforementioned matrices, for different combinations of subsets

$\{\imath, \jmath \in \mathcal{M} \cup \mathcal{S}\}$,

$$\tilde{\mathbf{K}}_{\imath\jmath} = \mathbf{K}_{\imath\jmath} + \tilde{\mathbf{C}}_{\imath\jmath}. \tag{83}$$

The matrices **F**, **H**, **Q**, **G**, **J** and **L** contain the directional derivatives obtained from the linearization of the Eq. (75). The first five rows of the system (82) represent the force equilibrium, Eq. (65). The sixth row enforces the contact constraints over the inactive nodes, while the seventh row is concerned with the non-penetration condition on the active nodes. Matrices $\tilde{\mathbf{D}}$ and $\tilde{\mathbf{M}}$ are obtained from the linearization of Eq. (51) and Eq. (52), respectively. Eight and ninth rows enforce the tangential conditions for both sticking, $\mathcal{Q}$, and sliding, $\mathcal{L}$, contact nodes.

The global system depicted in Eq. (82) displays an increase of bandwidth (Lagrangian multipliers) from the initial non-constrained form. However, the key asset of introducing a dual basis for the Lagrangian multipliers is that they can be locally removed from Eq. (82), such that,

$$\mathbf{z} = -\mathbf{D}^{-1}[\tilde{\mathbf{K}}_{\mathcal{SN}}\Delta\mathbf{d}_{\mathcal{N}} + \tilde{\mathbf{K}}_{\mathcal{SM}}\Delta\mathbf{d}_{\mathcal{M}} + \tilde{\mathbf{K}}_{\mathcal{SS}}\Delta\mathbf{d}_{\mathcal{S}} - \mathbf{r}_{\mathcal{S}}]. \tag{84}$$

Hence, the equations related to the enforcement of contact constraints on the inactive set can be eliminated from the system and a pure displacement problem is obtained,

$$\begin{bmatrix} \mathbf{K}_{\mathcal{NN}} & \mathbf{K}_{\mathcal{NM}} & \mathbf{K}_{\mathcal{NI}} & \mathbf{K}_{\mathcal{NQ}} & \mathbf{K}_{\mathcal{NL}} \\ \hat{\mathbf{K}}_{\mathcal{AN}} & \hat{\mathbf{K}}_{\mathcal{AM}} & \hat{\mathbf{K}}_{\mathcal{AI}} & \hat{\mathbf{K}}_{\mathcal{AQ}} & \hat{\mathbf{K}}_{\mathcal{AL}} \\ \mathbf{K}_{\mathcal{IN}} & \tilde{\mathbf{K}}_{\mathcal{IM}} & \tilde{\mathbf{K}}_{\mathcal{II}} & \tilde{\mathbf{K}}_{\mathcal{IQ}} & \tilde{\mathbf{K}}_{\mathcal{IL}} \\ \mathbf{0} & \tilde{\mathbf{M}}_{\mathcal{A}} & \tilde{\mathbf{D}}_{\mathcal{AI}} & \tilde{\mathbf{D}}_{\mathcal{AQ}} & \tilde{\mathbf{D}}_{\mathcal{AL}} \\ \check{\mathbf{K}}_{\mathcal{QN}} & \check{\mathbf{K}}_{\mathcal{QM}} - \mathbf{H}_{\mathcal{Q}} & \check{\mathbf{K}}_{\mathcal{QI}} - \mathbf{F}_{\mathcal{QI}} & \check{\mathbf{K}}_{\mathcal{QQ}} - \mathbf{F}_{\mathcal{QQ}} & \check{\mathbf{K}}_{\mathcal{QL}} - \mathbf{F}_{\mathcal{QL}} \\ \check{\mathbf{K}}_{\mathcal{LN}} & \check{\mathbf{K}}_{\mathcal{LM}} - \mathbf{J}_{\mathcal{L}} & \check{\mathbf{K}}_{\mathcal{LI}} - \mathbf{G}_{\mathcal{LI}} & \check{\mathbf{K}}_{\mathcal{LQ}} - \mathbf{G}_{\mathcal{LQ}} & \check{\mathbf{K}}_{\mathcal{LL}} - \mathbf{G}_{\mathcal{LL}} \end{bmatrix},$$

$$\cdot \begin{bmatrix} \Delta\mathbf{d}_{\mathcal{N}} \\ \Delta\mathbf{d}_{\mathcal{M}} \\ \Delta\mathbf{d}_{\mathcal{I}} \\ \Delta\mathbf{d}_{\mathcal{Q}} \\ \Delta\mathbf{d}_{\mathcal{L}} \end{bmatrix} = - \begin{bmatrix} \mathbf{r}_{\mathcal{N}} \\ \mathbf{r}_{\mathcal{M}} + \left[\mathbf{D}_{\mathcal{A}}^{-1}\mathbf{M}_{\mathcal{A}}\right]^T \mathbf{r}_{\mathcal{A}} \\ \mathbf{r}_{\mathcal{I}} \\ \tilde{\mathbf{g}}_{\mathcal{A}} \\ \mathbf{Q}_{\mathcal{Q}}\mathbf{D}_{\mathcal{Q}}^{-1}\mathbf{r}_{\mathcal{Q}} - \{\mathbf{C}_{\mathcal{T}}\}_{\mathcal{Q}} \\ \mathbf{L}_{\mathcal{L}}\mathbf{D}_{\mathcal{L}}^{-1}\mathbf{r}_{\mathcal{L}} - \{\mathbf{C}_{\mathcal{T}}\}_{\mathcal{L}} \end{bmatrix}, \tag{85}$$

where the two matricial functions, $\hat{\mathbf{K}}_{\mathcal{A}\jmath}$ and $\check{\mathbf{K}}_{\imath\jmath}$, are introduced to facilitate the notation of this condensed system. They are constructed from the primary matrices in Eq. (82) and described by suitable combinations of the subsets $\imath$ and $\jmath$, as follows:

$$\hat{\mathbf{K}}_{\mathcal{A}\jmath} = \begin{cases} \mathbf{K}_{\mathcal{M}\jmath} + \mathbf{D}_{\mathcal{A}}^{-1}\mathbf{M}_{\mathcal{A}}^{T}\mathbf{K}_{\mathcal{A}\jmath} & \text{for } \{\jmath = \mathcal{N}\}, \\ \tilde{\mathbf{K}}_{\mathcal{M}\jmath} + \mathbf{D}_{\mathcal{A}}^{-1}\mathbf{M}_{\mathcal{A}}^{T}\tilde{\mathbf{K}}_{\mathcal{A}\jmath} & \text{for } \{\jmath \neq \mathcal{N}\}. \end{cases} \tag{86}$$

$$\check{\mathbf{K}}_{\imath\jmath} = \begin{cases} \mathbf{Q}_{\imath}\mathbf{D}_{\imath}^{-1}\mathbf{K}_{\imath\jmath} & \text{for } \{\imath = \mathcal{Q}\} \wedge \{\jmath = \mathcal{N}\}, \\ \mathbf{Q}_{\imath}\mathbf{D}_{\imath}^{-1}\tilde{\mathbf{K}}_{\imath\jmath} & \text{for} \{\imath = \mathcal{Q}\} \wedge \{\jmath \neq \mathcal{N}\}, \\ \mathbf{L}_{\imath}\mathbf{D}_{\imath}^{-1}\mathbf{K}_{\imath\jmath} & \text{for } \{\imath = \mathcal{L}\} \wedge \{\jmath = \mathcal{N}\}, \\ \mathbf{L}_{\imath}\mathbf{D}_{\imath}^{-1}\tilde{\mathbf{K}}_{\imath\jmath} & \text{for} \{\imath = \mathcal{L}\} \wedge \{\jmath \neq \mathcal{N}\}. \end{cases} \tag{87}$$

9. Numerical examples

In this section, five numerical examples are presented to demonstrate the accuracy and efficiency of the dual mortar approach against the traditional NTS penalty contact formulation, widely used in commercial finite element codes. Furthermore, the performance of the solution methods described in Section 8.1 will also be addressed for frictional contact problems with deformable bodies undergoing finite strains with inelastic material behaviour.

In the first two problems, the contact forces are predominantly normal to the surfaces involved, and therefore friction plays a minor role. The last three examples focus on the computation of the tangential forces in the presence of finite strains. The examples are presented in such a way that the degree of nonlinearity (i.e. deformable bodies, friction, inelastic material behaviour, etc.) increases problem by problem.

Four materials were employed during the simulations: Pure aluminium, an aluminium alloy, a tungsten alloy and mild steel. Their mechanical properties and coefficients a and b for the Ludwik–Hollomon equation ($\sigma = \sigma_y + a\epsilon^b$) are listed in Table 1. In these examples we employed the 4-node quadrilateral F-bar finite element, described in Section 7, for the discretization of the geometry and the 2-node linear mortar element for the

Table 1. Material properties.

	Mechanical properties					
Materials	E (GPa)	G (GPa)	ν	σ_y (MPa)	a (MPa)	b
Pure aluminium	68.96	26	0.32	31	1574	0.220
Aluminium alloy	71.15	26	0.30	370	550	0.223
Mild steel	210	76	0.33	830	600	0.210
Tungsten alloy	400	85	0.33	880	700	0.205

discretization of the contact surfaces. The only exception was example 9.2 where 8-node hexahedral elements are used together with 4-node quadrilateral mortar elements. All the simulations presented in this section were performed with an implicit quasi-static finite element framework. Unless otherwise mentioned, the values chosen for the complementarity parameters were $c_\eta = c_\tau = 1$ and for the Penalty multipliers were $\epsilon_\eta = \epsilon_\tau = 10^4$.

9.1. *Pressurized spheres*

The first example is the 2D simplification of the pressurized spheres contact problem presented in Ref. [49]. In this problem, the contact surfaces are tied. Therefore, as long the contact forces are transmitted properly, the two spheres are expected to behave as a single structure.

The simulation is performed assuming y-axial symmetry, see Fig. 4. A distributed pressure of 690MPa is applied to the inner sphere within 50 equal pseudo-time increments of $\Delta P = 13.8$ MPa. At the final step, the whole structure is expected to be under plastic deformation. The material chosen for both spheres was the aluminium alloy, and no friction is considered in this example.

The results for both methods (i.e. penalty method and dual mortar single step strategy) have a good agreement during most part of the simulation. Figure 5 shows the results obtained for the equivalent stress distribution at the last step of the simulation with the single-step dual mortar method.

There was no appreciable difference in the equivalent stress contour between both methods. Nevertheless, the same conclusion does not hold for the enforcement of contact constraints. In fact, there was no relative displacement between the contact surfaces while the deformation remained

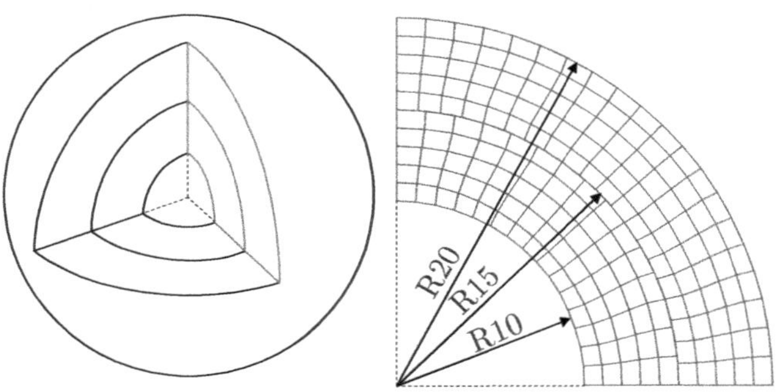

Fig. 4. Pressurized spheres: geometry and finite element mesh (dimensions in mm).

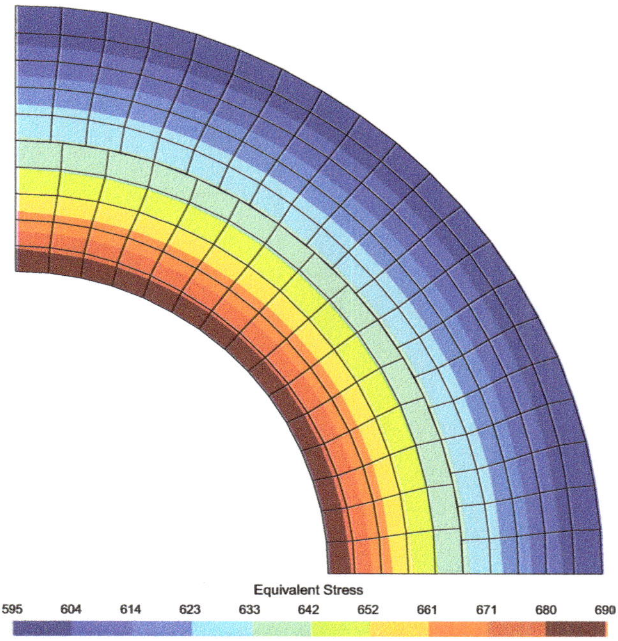

Fig. 5. Pressurized spheres: equivalent stress (MPa) distribution for the dual mortar method.

within the elastic domain, which occurs for loads below 310 MPa. However, the value initially chosen for the penalty multiplier is no longer large enough to enforce the contact constraints as soon as plastic strains are reached. This led to the small penetration shown in Fig. 6.

Increasing the normal penalty multiplier, ϵ_η to 10^7 improves the results obtained by the method, which become very similar to the ones obtained by the dual mortar method, see Fig. 6. Nevertheless, due to this increase of the normal penalty multiplier, the first increments presented convergence difficulties since the stiffness matrix becomes slightly ill-conditioned.

In this example, it became clear that the dual mortar method was able to enforce the contact constraints more accurately and efficiently over the entire loading path. In fact, the total number of iterations required to solve the problem with the Penalty method was equal to 251 and for the dual mortar scheme equal to 189. An increase of 25% on the total number of iterations required throughout the whole simulation was observed in the majority of the numerical examples presented in this work.

(a) (b)

Fig. 6. Pressurized spheres: enforcement of contact constraints. Dual mortar method (a) and Penalty method (b).

9.2. *Cylinder-to-flat*

This numerical example is a classical benchmark in contact mechanics since has an analytical solution within the elastic domain. This example has been extensively discussed in Ref. [50] and it's only briefly addressed here. A cylinder (20 mm radius and 20 mm length) is pressed against a flat disc (60 mm diameter and 20 mm thickness). The cylinders are made with three different materials brass, aluminium alloy and steel (see Table 1). The flat disc is made of hard iron and coated with a 2-mm layer of tungsten carbide alloy in order to increase the Young modulus of the surface and minimize its deformation. The properties of these materials are listed in Table 1.

The initial contact area is very small (non-conforming contact line) and a curved contact surface is present. The simulation is conducted with a 3D model and the results are compared with experimental data. The geometry of the problem and finite element mesh, employed to discretize the geometry, are depicted in Fig. 7. A total of 1728 hexahedral linear F-bar finite elements for the discretization of the cylindrical specimen, while 1080 were employed for the flat surface. This choice of discretization is to emphasize the fact that the mortar formulation employed can provide good results even for relatively coarse meshes.

The axial load of 80 kN, which is applied to the top surface of the half-cylinder, is divided into 89 pseudo-time steps. The bottom of the flat surface is fully restrained. Frictionless contact is assumed, and the complementarity parameters used are $c_\eta = 10$, $c_\tau = 0$. The analytical solution for this

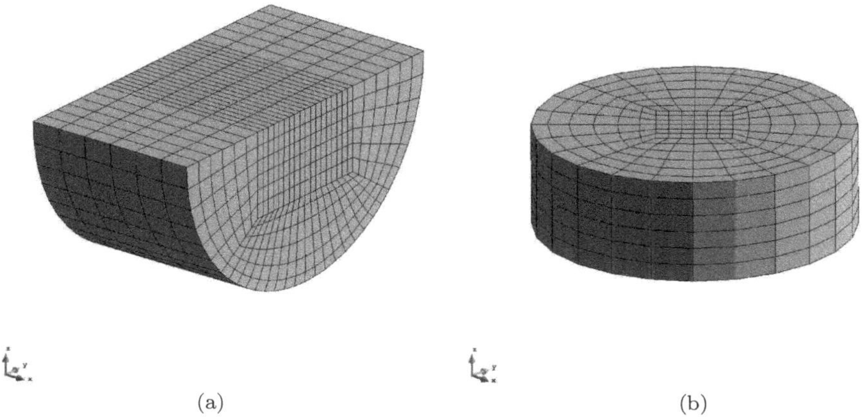

Fig. 7. Cylinder-to-flat: finite element discretization. Cylinder (a) and flat disc (b).

problem was derived from the Hertzian contact formulae[51,52] for two cylinders, which defines the maximum contact pressure, P_0, the contact half-width, a, and the contact pressure along x-coordinate P as

$$P_0 = \sqrt{\frac{FE^*}{\pi R^*}}, \tag{88}$$

$$a = \sqrt{\frac{8FR^*}{\pi BE^*}}, \tag{89}$$

$$P(x) = P_0\sqrt{1 - \left(\frac{x}{a}\right)^2}, \tag{90}$$

where B is the length of the cylinder, and the combined elasticity modulus, E^*, is obtained from the material parameters of the cylinder (E_c) and the block (E_b) as follows:

$$E_{eq} = \frac{2E_pE_b}{E_c(1-\nu_b^2) + E_b(1-\nu_c^2)}. \tag{91}$$

The combined radius, R^*, is evaluated from the radius of the cylinder, R_1, and block, R_2, in a similar way. Nevertheless, since $R_2 \to \infty$, the combined radius is reduced to the cylinder's radius

$$R^* = \lim_{R_2\to\infty} \frac{R_1R_2}{R_1+R_2} = \lim_{R_2\to\infty} \frac{R_1}{R_1/R_2+1} = R_1. \tag{92}$$

Table 2. Cylinder-to-flat: cylinder's behaviour at the elastic domain.

	a(mm)			P(MPa)		
F(kN)	Ana.	Exp.	Num.	Ana.	Exp.	Num.
			Brass			
0.1	0.035	0.038	0.036	90.67	91.11	90.16
0.2	0.049	0.055	0.052	128.23	130.90	131.20
0.3	0.060	0.062	0.059	157.05	157.15	156.80
			Aluminium			
0.1	0.042	0.043	0.041	75.45	75.69	76.00
0.2	0.059	0.061	0.066	106.70	107.12	107.77
0.3	0.073	0.074	0.078	130.68	133.33	131.58
			Steel			
0.1	0.027	0.028	0.031	117.31	118.05	119.22
0.2	0.038	0.041	0.046	165.91	166.14	166.27
0.3	0.047	0.047	0.048	203.20	204.01	204.97

The comparisons between analytical solution, numerical results and experimental data are presented for two stages: elastic strain and plastic strain stage.

9.2.1. *Elastic domain*

During the elastic stage of deformation, the analytical solution of the problem can be used as a validation tool. Therefore, the results obtained with the analytical, the experimental and the numerical methods are compared. To ensure that only elastic deformation is in place, only the first 3 load steps ($F = 0.1$ kN, 0.2 kN and 0.3 kN) are considered at this point. The results for the contact zone's half-width, a, and pressure, P, at $x = 0$ mm are listed in Table 2. The results obtained with all three methods show a close agreement, for all the materials employed in the tests, with a maximum variation, for both the half-width and the maximum pressure, remaining below 4.6%. These results clearly confirm the ability of the formulation to capture elastic deformation and endorse the applicability of the model shown in Fig. 7.

9.2.2. *Plastic domain*

Since the materials selected for the analysis can be regarded as ductile metals, during the tests, the specimens are mainly undergoing plastic deformation. Therefore, a detailed analysis of the results at this specific regime of deformation is of paramount importance since the Hertzian solution will be no longer applicable. The brass specimen is the first to reach its yield stress at an axial force $F = 0.5$ kN (5th time step), followed by the aluminium specimen, $F = 2$ kN (11th time step) and the steel specimen, $F = 3$ kN (12th time step). The maximum values for the effective plastic strain, obtained from the numerical simulation, are: 0.14 for the brass specimen, 0.17 for the aluminium specimen and 0.09 for the steel specimen.

The evolution of the contact half-width is shown in Fig. 8. It shows a comparison of the results obtained with the analytical equations (88)–(89), the numerical simulations and the experimental measurements during the plastic phase of the tests. It can be observed that the analytical results can no longer provide a realistic evolution of the half-width, a, for all materials. The analytical equations are based on the principle that the bodies are still respecting Hooke's law. Therefore, it predicts deformations much smaller than the real results. The numerical results show a similar evolution when compared to the experimental measurements. The growth of the contact zone occurs very rapidly for the brass and steel specimen, until a compression limit is achieved. At this point, the structural stiffness of the specimen imposes a slower evolution of the contact half-width (i.e. deformation Y-direction) while also promoting a small growth of the length of the contact area (i.e. X-direction). At the end of the test, the aluminium specimen had just reached its compression limit while the remaining specimens show this behaviour after the 39th load step ($F = 30$ kN).

A close relation between the Young modulus of the specimens' material and the evolution of the contact area is observed. The aluminium specimen displayed the highest final value of half-width, $a = 7.78$ mm, followed by the brass specimen, $a = 5.76$ mm. As expected, due to its higher yield stress and Young modulus, the steel specimen presented a much smaller contact area and deformation. The final value for its contact half-width reached only $a = 2.27$ mm.

The results for the evolution of the maximum contact pressure, P_0, are displayed in Fig. 9. Once more, the analytical results diverge from the experimental and the numerical ones. Therefore, they were omitted here. On the other hand, the numerical results are consistent with the experimental data. The contact pressure for the brass and the steel specimen remained fairly constant after the 19th load step ($F = 10$ kN). However, the aluminium specimen shows a small variation during this phase, which is consistent with the contact area growth mentioned earlier. The average difference of

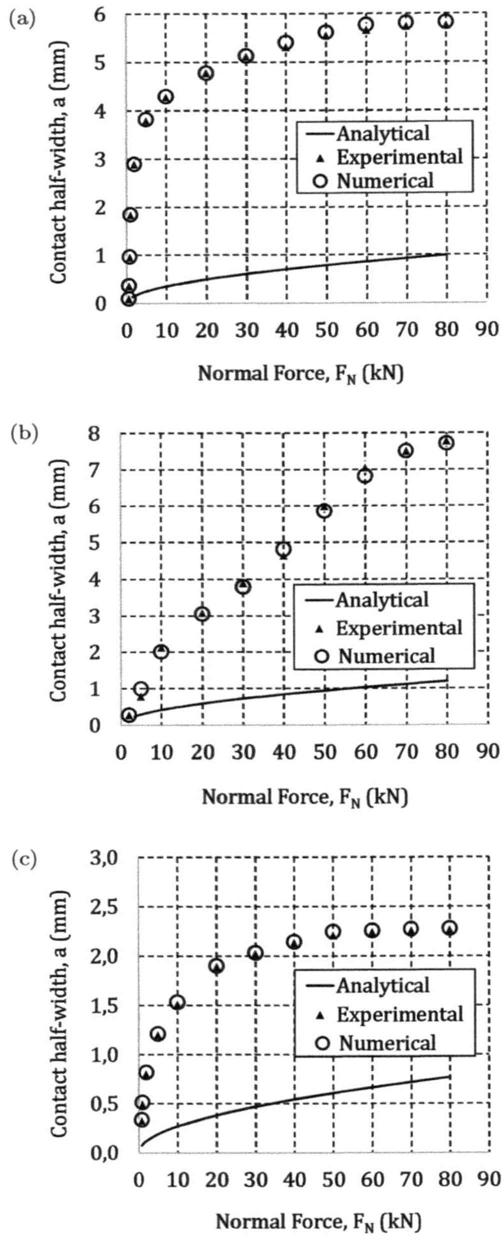

Fig. 8. Cylinder-to-flat: contact half-width evolution. Brass (a), aluminium (b) and steel (c).

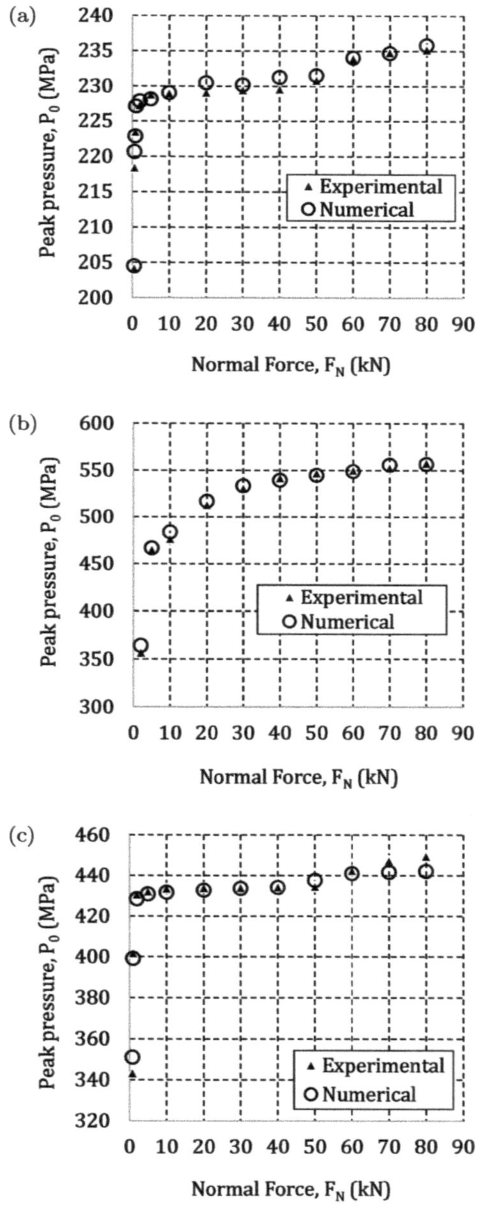

Fig. 9. Cylinder-to-flat: maximum pressure evolution. Brass (a), aluminium (b) and steel (c).

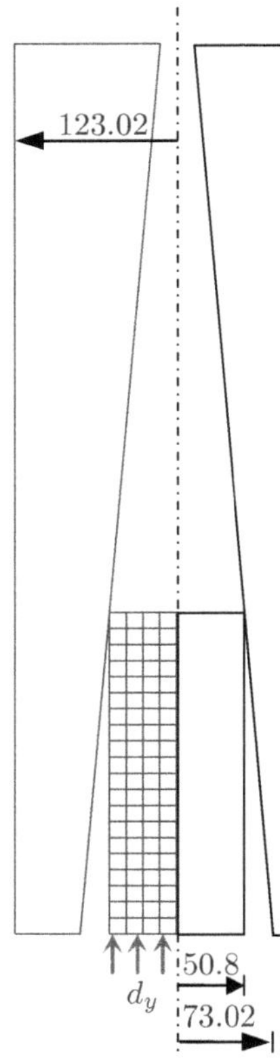

Fig. 10. Extrusion: geometry and finite element discretization (dimensions in mm).

the experimental/numerical results range from 3.7% to 5.1% depending of the material employed.

9.3. *Extrusion*

In this example, the frictional elasto-plastic stress analysis of an aluminium cylindrical billet[53] is presented. The billet is pushed across a total distance of 177.8 mm through a rigid conical die, which has a wall angle of 5°, see Fig. 10. Due to the presence of high frictional contact forces and the development of finite plastic strains in the billet, this example will allow the assessment of the performance of both methods under finite frictional sliding coupled with finite strain inelastic material behaviour. In particular, the displacement field of the billet, the evolution of the extrusion and tangential forces, and the distribution of the plastic strain, will be analysed for both

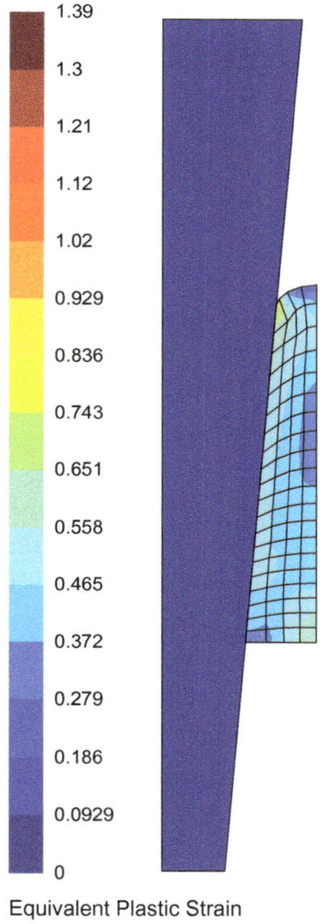

Fig. 11. Extrusion: effective plastic strain contour at the the final load step.

methods. Material properties of the pure aluminium billet are presented in Table 1.

Sliding forces are the principal cause of deformation in this example, and therefore shear effects over the contact surface are expected. The die is regarded as a rigid body and the coefficient of friction adopted was $\mu = 0.1$. The values for the penalty multipliers selected for this problem were $\epsilon_\eta = 10^8$ and $\epsilon_\tau = 10^5$. In the simulation, the displacement load applied was divided into 200 increments. The equivalent plastic strain distribution at the final step obtained by the dual mortar method is shown in Fig. 11.

As expected, the higher plastic deformation level is found at the upper left side of the billet with a maximal value of 1.39. There was no appreciable difference in the equivalent plastic strain contour obtained by the

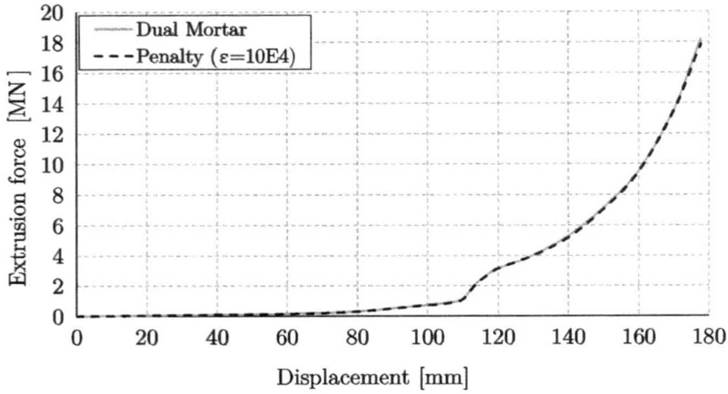

Fig. 12. Extrusion: evolution of the extrusion force.

two methods. The final contact surface length is equal to 257.33 mm. A graphical representation of the extrusion force, measured from the reaction at the billet bottom, is shown in Fig. 12. The difference between the two methods is negligible. The convergence rate of both methods is listed in Table 3 for the increments 1, 100, 150 and 200.

Nevertheless, the evolution of the frictional force obtained with the Penalty method, see Fig. 13, shows the typical non-physical oscillation due to the non-smoothness of finite element discretization along the contact surface. This drawback is mitigated by the continuous normal field provided by the mortar segmentation, where no oscillations are observed. One advantage of performing the simulation with a rigid body is that, since the non-mortar surface is composed by a single segment of the die, the segmentation does not change, even when plastic strains are in place. Therefore, the active set is kept the same during the whole simulation, which provides optimal convergence rates for the dual mortar method.

9.4. *Frictional beam*

The analysis of an elasto-plastic frictional beam contact problem[46] is presented next. The objective of this analysis is to investigate the influence of the parameters c_η, c_τ, employed in the definition of the non-complementarity conditions of the dual mortar method, together with the parameters ϵ_η and ϵ_τ, employed in the enforcement of contact constraints of the NTS–Penalty method, when finite frictional sliding and finite strain elasto-plastic material behaviour are present. Due to the different sources of nonlinearity involved (geometrical, material and contact), this is also a good problem to test the robustness of the approach presented.

Table 3. Extrusion: convergence evolution.

t_n	$\mathbb{k}$	Penalty $\epsilon_\eta = \epsilon_\tau = 10^4$	Dual mortar $c_\eta = c_\tau = 10^0$
		Relative residual norm (%)	
001	1	0.612567E + 02	0.110718E + 01
	2	0.179956E + 02	0.806609E + 00
	3	0.845863E + 00	0.173512E − 02
	4	0.233792E − 02	0.990222E − 07
	5	0.950268E − 06	0.192286E − 11
	6	0.326588E − 11	—
100	1	0.856247E + 03	0.280271E + 01
	2	0.625483E + 01	0.158554E + 01
	3	0.215140E − 02	0.436638E − 01
	4	0.269871E − 04	0.274675E − 04
	5	0.112124E − 08	0.246126E − 10
200	1	0.425783E + 02	0.862577E + 01
	2	0.914758E + 00	0.150099E + 00
	3	0.667821E − 02	0.546990E − 01
	4	0.299631E − 04	0.734911E − 04
	5	0.141457E − 06	0.149939E − 09
	6	0.784265E − 08	—
$\sum(\mathbb{k})$[a]		1215	981

Note: [a]Total number of iterations required.

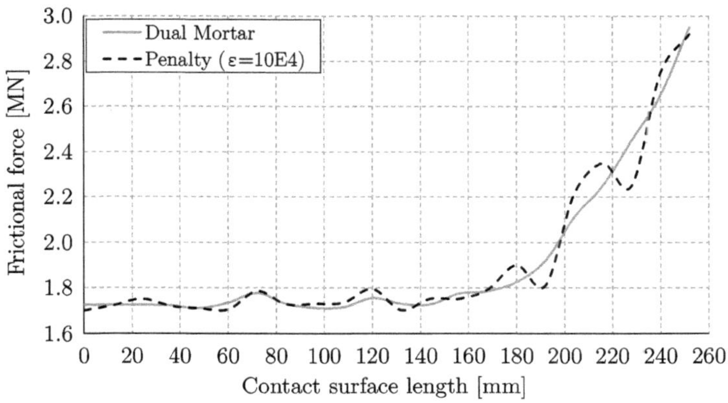

Fig. 13. Extrusion: evolution of the frictional force.

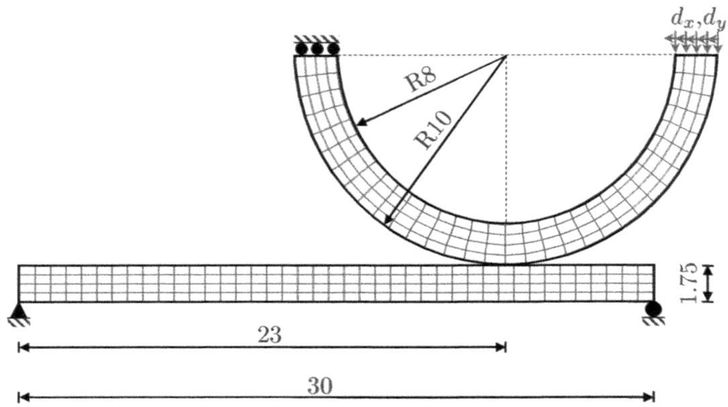

Fig. 14. Frictional beam: geometry and finite element discretization (dimensions in mm).

The problem is composed by two beams: a straight beam and a curved beam, which are represented in Fig. 14. The straight beam is simply supported and the curved beam is fixed on the left end in the yy direction and subjected to a vertical displacement $d_y = 1.2t$ at the right end. In addition, a prescribed horizontal displacement of $d_x = 2.0t$ is applied to both left and right ends. Due to the curvature and boundary conditions, the curved beam has a high structural stiffness, and therefore the straight beam is more prone to deform.

Both beams are assumed to be made of aluminium alloy, whose properties are listed in Table 1, with a coefficient of friction equal to $\mu = 0.3$. The total prescribed displacement, which is set to be equal to $d_x = 16$, was applied over 320 equally spaced pseudo-time increments of $\Delta t = 0.025$. The evolution of the effective plastic strain provided by the dual mortar method is shown in Fig. 15 for $t = 4$s and in Fig. 16 for the final configuration at $t = 8$s. There was no noticeable difference in the effective plastic strain contour between both methods. During the first 2 pseudo-seconds of the simulation, the majority of the displacement and deformation are undergone by the bottom beam. At $t = 4$s, the contact surface already presents significant levels of plastic strain. The presence of high frictional force causes shear distortion of the finite elements on the contact region and the bending process induces the growth of the contact area.

At the end of the simulation ($t = 8$s), the upper beam has considerable deformation levels at the right corner due to the prescribed displacement d_x, while the lower beam shows moderate levels of strain at the contact

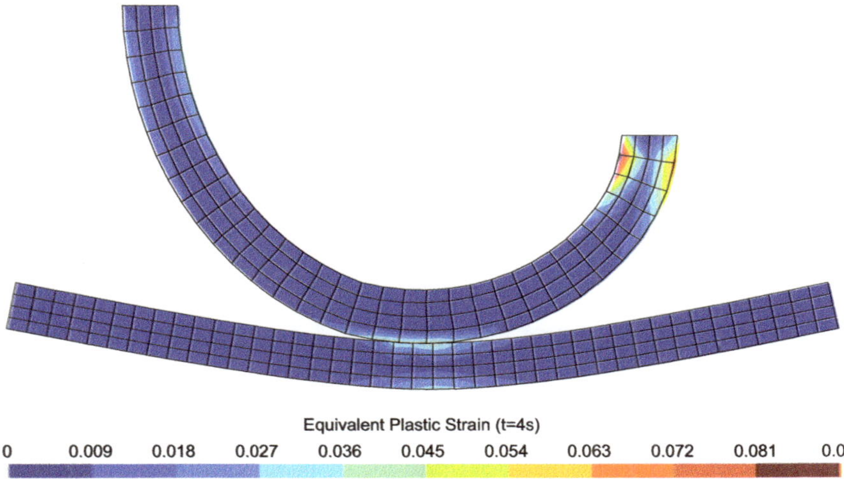

Fig. 15. Frictional beam: effective plastic strain at $t = 4$s.

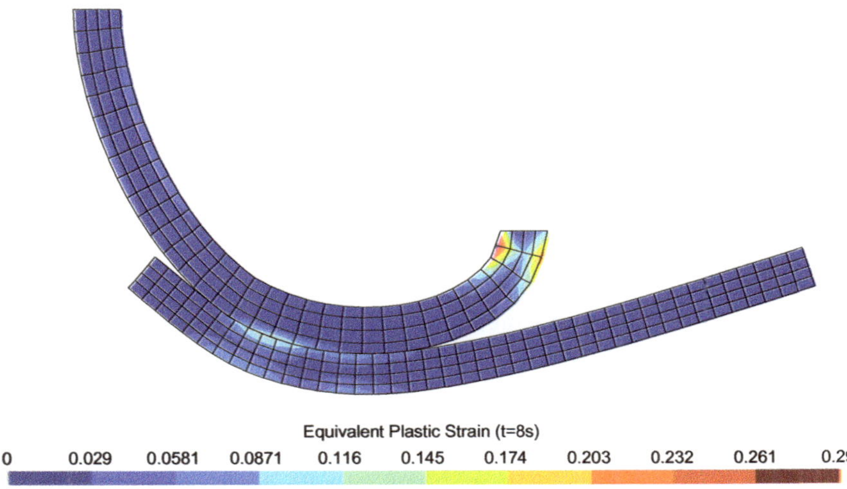

Fig. 16. Frictional beam: effective plastic strain at $t = 8$s.

surface. The spatial rate of convergence of the relative residual norm is shown in Table 4 for the representative steps of this analysis.

From Table 4, it is possible to observe that the Penalty method is able to solve the entire problem when employing $\epsilon_\eta = \epsilon_\tau = 10^4$. However when the curved beam starts to experience high levels of plastic deformation,

Table 4. Frictional beam: convergence behaviour.

t_n	$\Bbbk$	Penalty $\epsilon_\eta = \epsilon_\tau = 10^4$	Dual mortar $c_\eta = c_\tau = 10^0$
		Relative residual norm (%)	
001	1	0.457532E + 02	0.945446E + 01
	2	0.145766E + 00	0.298134E + 01
	3	0.303347E − 01	0.262611E − 04
	4	0.198423E − 02	0.444629E − 10
	5	0.435333E − 04	—
	6	0.254868E − 10	—
160	1	0.412056E + 02	0.145941E + 01[a]
	2	0.211690E + 00	0.112458E − 01
	3	0.125627E − 01	0.138134E − 02
	4	0.506982E − 02	0.138134E − 02
	5	0.022463E − 03	0.262611E − 04
	6	0.174102E − 04	0.444629E − 10
	7	0.512961E − 06	—
	8	0.512961E − 09	—
320	1	0.842507E + 01	0.226485E + 01[a]
	2	0.631476E − 01	0.286545E − 01
	3	0.400189E − 02	0.754524E − 04
	4	0.229077E − 03	0.917062E − 08
	5	0.990173E − 04	0.896271E − 12
	6	0.303569E − 06	—
	7	0.131313E − 09	—
$\sum(\Bbbk)$[b]		2281	1756

Notes: [a]Changes in the active set.
[b]Total number of iterations required.

the convergence rate of this method is compromised (see increment 160 in Table 4). On the other hand, the dual mortar method converges reasonably well during all the simulation. Nevertheless, the convergence of the single iteration PDASS algorithm is also affected when the upper beam undergoes plastic strains. In particular, multiple attempts to find the correct subset for the mortar segmentation occur after a given configuration update, which leads to changes on the contact pairs and, consequently, the active set. In addition, the finite elements at the most outer-left section of the upper beam get significantly distorted due to the boundary conditions, leading to the computation of negative Jacobian during the analysis.

With regard to the computational time required for carrying out the analysis, it is possible to conclude that the total number of iterations to attain convergence with the NTS–Penalty method, for each increment, is always higher than the dual mortar method. Although the computational cost of each iteration of the Penalty formulation is cheaper, the additional number of iterations makes the method more expensive. Furthermore, the NTS method also requires considerably smaller time steps in order to converge in both elastic and plastic domains when compared with the dual mortar method which, in an overall sense, is significantly more efficient.

Finally, with this example, we investigate the influence of the parameters c_η and c_τ introduced with the complementarity functions in Section 8. These parameters were defined within the context of small deformations in Ref. [31] and analysed at finite strains in Ref. [36]. The parameter c_η is a positive constant that relates the different physical units of z_η and $\tilde{g}$. Therefore, it seems logical to choose c_η such that the values of z_η and $\tilde{g}$ are balanced, and several authors[31,35,48,54] have suggested c_η to be of the order of the Elasticity modulus of the contact bodies. The same reasoning applies to the positive constant c_τ, i.e. it is logical to choose c_τ such that z_τ and $\tilde{u}_\tau$ are balanced. Nevertheless, the conditions on the tangential directions that result from the applied Coulomb's friction law are also related to the parameter c_η. Consequently, c_τ is also suggested to reflect the material parameters of the contacting bodies.[31,35] The convergence behaviour of the single-step PDASS algorithm for different values of c_η and c_τ is illustrated in Table 5. The number of Newton steps required to reach convergence for two representative steps, one within the elastic and other within the plastic

Table 5. Frictional beam: influence of the parameters c_η and c_τ.

	c_τ				
c_η	10^{-2}	10^{0}	10^{2}	10^{4}	10^{6}
Elastic domain: $t_n = 10$					
10^{-2}	5	4	4	—	*
10^{0}	4	4	4	6	*
10^{2}	4	4	5(1)	7(1)	*
10^{4}	5	5	6(1)	7(1)	*
10^{6}	*	*	8(1)	9(2)	*
Plastic domain: $t_n = 300$					
10^{-2}	6	6	6	—	*
10^{0}	6	7(1)	7(1)	7(1)	*
10^{2}	6	7(1)	7(1)	8(2)	*
10^{4}	7(1)	7(1)	9(3)	10(3)	*
10^{6}	*	*	*	*	*

Note: *More than 10 iterations and three changes in the active set.

Table 6. Frictional beam: influence of the parameters ϵ_η and ϵ_τ.

	ϵ_τ			
ϵ_η	10^2	10^4	10^6	10^8
Elastic domain: $t_n = 10$				
10^2	**	**	**	***
10^4	8	7	6	8
10^6	7	7	7	***
10^8	8	***	8	***
Plastic domain: $t_n = 300$				
10^2	**	**	**	**
10^4	**	9	10	10
10^6	8	9	9	***
10^8	8	10	10	***

Notes: **Penetration over than 10% of the finite element height.
***Locking/over-constraint.

domain, is investigated. It can be observed that, within the elastic domain, there is almost no degradation of the convergence rate for a wide range of c_η and c_τ. Therefore, the sensitivity of the algorithm, with regard to the choice of c_η and c_τ, is low within the elastic domain. For the plastic domain, there is still a reasonable range of c_η and c_τ that do not affect the convergence rate. However, the spectrum of complementarity parameters for which there is no deterioration of the convergence rate is noticeably more limited.

The additional source of nonlinearity caused by the material behaviour significantly affects the number of iterations needed to achieve convergence and the number of active set changes. These results suggest that the values of c_η and c_τ should reflect the plastic material parameters of the bodies in contact.

It is important to emphasize that the choice of c_η and c_τ only enhances or deteriorates convergence of the dual mortar method and not the accuracy of the results. On the other hand, the choice of penalty multipliers, ϵ_η and ϵ_τ, has a great influence on the accuracy of the method. In Table 6, the convergence behaviour of the NTS–Penalty formulation for different values of ϵ_η and ϵ_τ is illustrated, Again, the number of Newton steps needed to attain convergence for two representative steps, one elastic and one plastic, is investigated.

Within both the elastic and plastic domains, it is possible to observe the well-known behaviour of the NTS–Penalty method. For low values of

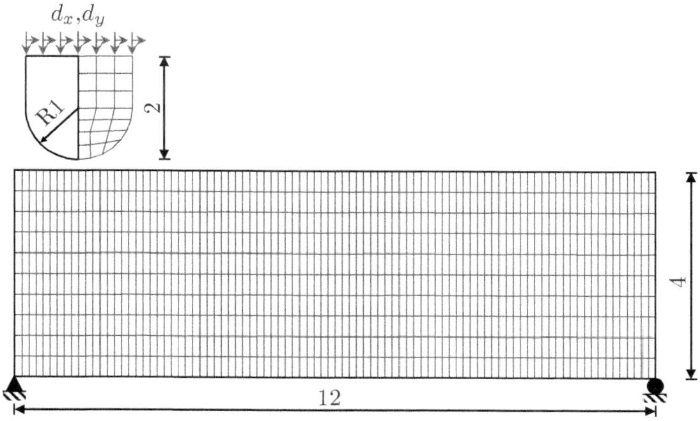

Fig. 17. Ironing: geometry and finite element discretization (dimensions in mm).

ϵ_η and ϵ_τ the penetration can be significant and the frictional constraints are not fulfilled. On the other hand, high values of these parameters lead to locking and/or over-constraint, which prevent the problem from being solved. Therefore, the selection of the penalty parameters is a compromise between the accuracy of the solution and finding a solution for the problem.

9.5. *Ironing*

The so-called Ironing problem[36,55] is a variation of the cylinder-to-flat contact problem, where after the indentation stage the tool is set to slide over the flat surface. This combination of loads generates high levels of deformation.

The geometry, finite element discretization and loads of this problem are depicted in Fig. 17. The simulation is divided in two stages. In the first stage, the semi-circular tool, already in contact with the base, is vertically pushed against the base $d_y = 0.3$, and this stage is set to take 30 increments. During the second stage, the tool is horizontally displaced by $d_x = 9.6$. This load is equally divided into 400 increments.

For this simulation, the values chosen for the penalty multipliers were $\epsilon_\eta = \epsilon_\tau = 10^9$ and for the complementarity parameters $c_\eta = c_\tau = 10^{-1}$. A static coefficient of friction equal to $\mu_{\mathcal{Q}} = 0.4$ and a kinetic coefficient of friction of $\mu_{\mathcal{L}} = 0.3$ are considered. The materials chosen for the tool and the block were, respectively, the tungsten alloy and the aluminium alloy (Table 1).

During the simulation, a layer of plastic deformation develops along the contact surface and the active set changes more than once for most increments. This happens due to the fact that once a given node at the left side of the contact surface reaches the yield stress, it no longer recovers from the deformation, leading to its deactivation. However, at the right side, material is being accumulated and new contact nodes become active. This constant change of the contact surface length and the number of contact nodes during a single increment poses additional challenges to the performance of the solution algorithms, as shown in Table 7.

Table 7. Ironing: convergence rate behaviour.

		Penalty	Single	Nested
t_n	k	$\epsilon_\eta = \epsilon_\tau = 10^9$	$c_\eta = c_\tau = 10^{-1}$	
		Relative residual norm (%)		
030	1	0.285466E + 02	0.772309E + 01	0.772308E + 01
	2	0.125468E + 00	0.392728E + 00	0.392728E + 00
	3	0.861073E − 01	0.337489E − 03	0.337488E − 03
	4	0.973016E − 02	0.229909E − 09	0.229908E − 09
	5	0.744196E − 03	0.111215E − 11	0.111214E − 11
	6	0.726018E − 04	—	—
	7	0.186181E − 09	—	—
230	1	0.626103E + 01	0.280271E + 01[a]	0.280271E + 01
	2	0.124845E + 00	0.158554E + 01	0.220156E + 01
	3	0.920084E − 01	0.436638E − 01	0.112644E + 00
	4	0.108697E − 02	0.125687E − 02	0.612545E − 02
	5	0.783435E − 03	0.274675E − 04	0.266750E − 04
	6	0.226584E − 06	0.246126E − 10	0.422109E − 07
	7	0.141980E − 09	—	0.199088E − 11
430	1	0.958624E + 02	0.706285E + 03[a]	0.706285E + 03
	2	0.111254E + 01	0.256844E + 02[a]	0.298211E + 01
	3	0.128455E + 00	0.862577E + 01	0.116925E − 01
	4	0.547556E − 01	0.150099E + 00	0.138852E − 02
	5	0.343761E − 02	0.546990E − 01	0.623577E − 04
	6	0.517235E − 03	0.734911E − 04	0.421596E − 07
	7	0.436580E − 04	0.149939E − 09	0.336952E − 09[b]
	8	0.140885E − 06	—	—
	9	0.269842E − 09	—	—
$\sum$(k)[c]		3452	2600	3258[d]

Notes: [a]Change in active set.
[b]Check of active contact set failed, Newton cycle has to be repeated.
[c]Total number of iterations required throughout the whole simulation.
[d]Considering changes in active set.

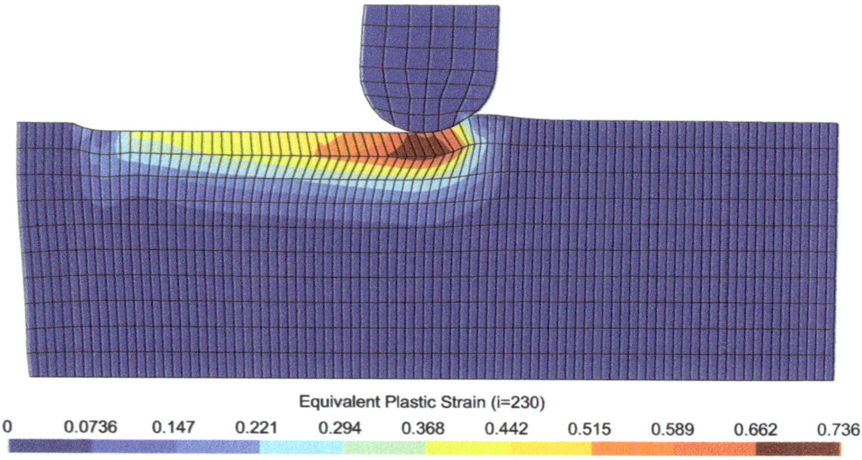

Fig. 18. Ironing contact: effective plastic strain at $d_x = 4.8$ and $d_y = 0.3$.

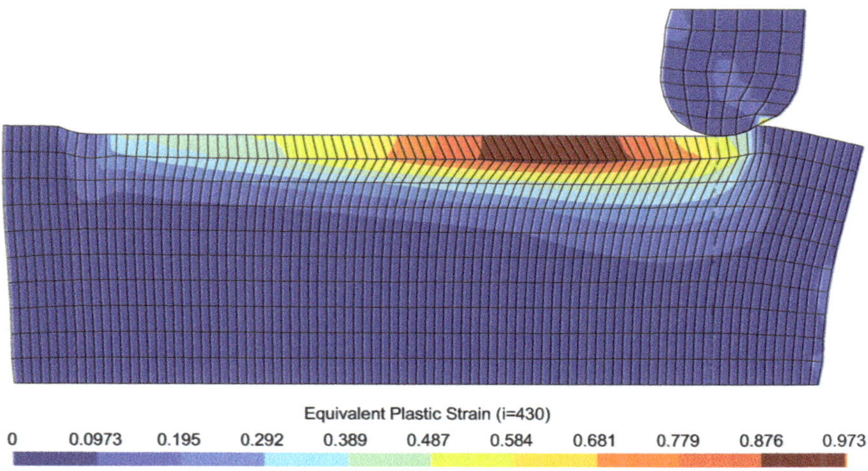

Fig. 19. Ironing: effective plastic strain at $d_x = 9.6$ and $d_y = 0.3$.

The accumulated material on the right-hand side also produces a high tangential load on the tool, which leads to its distortion and plastic deformation. Figures 18 and 19 show the effective strain at steps 230 and 430, respectively. Reducing the load of each increment helps to mitigate this phenomena; however, as soon the material reaches the plastic domain, even small loads incur in large distortions of the finite element mesh and, consequently, changes on the contact pairs occur.

The convergence evolution of the relative residual norm is listed in Table 7, for three representative steps and for three solution strategies: the Penalty method, the nested iterative strategy described in Algorithm 1 and single iterative strategy described in Algorithm 2.

From Table 7, it is possible to see that Penalty method is able to solve the problem when employing high values for the penalty multipliers ($\epsilon_\eta = \epsilon_\tau = 10^9$). Nevertheless, the convergence rate of the NTS method is poor when compared with the PDASSs. Furthermore, the level of penetration becomes erratic during the solution and it is difficult to know in advance whether a given set of penalty multipliers will lead to the solution of the problem or not. The convergence evolution provided by the PDASS algorithms was similar. This happens due to the very small size of the increments. Therefore, in order to enable a more detailed comparison between those methods, the number of increments is reduced by half and each increment is raised to twice their initial size (i.e. phase I: 15 increments with $\Delta d_y = 0.02$; phase II: 200 increments with $\Delta d_x = 0.096$).

Table 8 shows the convergence results provided by this new set of loading conditions. It's possible to observe that due to the increase of the load increment, the convergence rate of both strategies is deteriorated. In particular, at the plastic regime the nested strategy becomes more costly since the changes of active set require new Newton cycles to be performed. On the other hand, the single-step strategy also reveals a disadvantage: at the plastic regime, the contact pair changes more frequently during a single increment than that at the elastic regime. Therefore, due to sensitivity of this approach to changes of the active set, its convergence rate is severely deteriorated. Furthermore, depending on geometry discretization and increment loading size, the constant changes of active set may prevent the finding of a solution to the problem without significantly reducing the load increment.

Figure 20 displays the typical behaviour of the total reactions over the top surface of the tool. Forces increase during the indentation phase and stay relatively constant while the tool is sliding. However, a small oscillation of the horizontal force is expected since the amount of material accumulated at the right-hand side of the tool changes over the simulation. Both methods yield similar results for the vertical reactions. However, the horizontal reactions provided by the Penalty method became erratic at the last quarter of the simulation, when the highest levels of shear are attained.

Table 8. Ironing: comparison of PDASS algorithms.

		Nested strategy	Single strategy
t_n	$\Bbbk$	$c_\eta = c_\tau = 10^{-1}$	
		Relative residual norm (%)	
015	1	0.46320E + 02	0.39452E + 02
	2	0.21169E + 01	0.62015E + 00
	3	0.10101E + 00	0.30008E − 02
	4	0.36045E − 02	0.19961E − 04
	5	0.18604E − 04	0.85600E − 09
	6	0.45969E − 06	0.79367E − 11
	7	0.20781E − 11	—
115	1	0.10007E + 03	0.20508E + 02
	2	0.49089E + 02	0.95486E + 01[a]
	3	0.19065E + 01	0.89068E + 01[a]
	4	0.56002E + 01	0.78886E + 01
	5	0.27951E − 02	0.30219E + 00
	6	0.13355E − 04	0.11298E − 01
	7	0.80743E − 08	0.52594E − 04
	8	0.31092E − 11	0.61413E − 06
	9	—	0.46302E − 09
215	1	0.96452E + 02	0.99564E + 02[a]
	2	0.38491E + 01	0.49261E + 02[a]
	3	0.16482E + 00	0.22279E + 01[a]
	4	0.79050E − 01	0.31859E − 00
	5	0.31105E − 02	0.13692E − 01
	6	0.15924E − 04	0.39562E − 02
	7	0.36294E − 06	0.92615E − 04
	8	0.14820E − 08	0.41105E − 06
	9	0.46594E − 10[b]	0.26941E − 08
	10	—	0.12943E − 11

Notes: [a]Change in the active set.
[b]Check of active contact set failed, Newton cycle has to be repeated.

The influence of the complementarity parameters c_η and c_τ on the solution behaviour of this problem is illustrated in Table 9. Again, it is observed that changes on the deformation regime, from the elastic to the plastic domain, promote a disturbance in the convergence of the single-step PDASS. However, by choosing a suitable combination of these parameters for the different phases of the problems, one could circumvent this problem with a minimal effort.

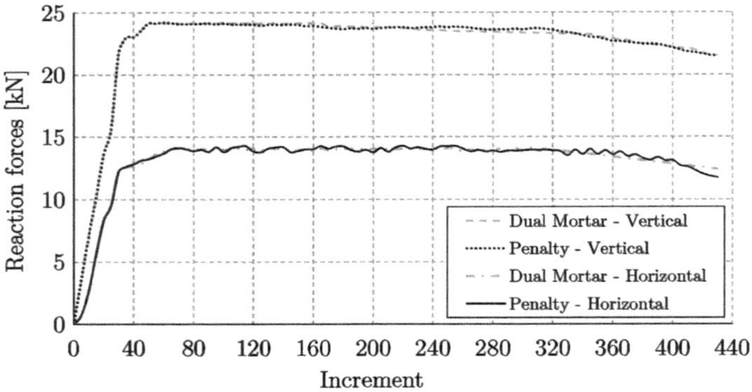

Fig. 20. Ironing: reaction forces.

Table 9. Ironing: influence of the parameters c_η and $c_\mathcal{T}$.

	$c_\mathcal{T}$			
c_η	10^{-2}	10^{-1}	10^{0}	10^{+1}
Elastic domain: $t_n = 10$				
10^{-2}	5	5	5	7
10^{-1}	5	5	5	7
10^{0}	6(1)	6	6	7
10^{+1}	6(1)	6(1)	7(1)	a
Plastic domain: $t_n = 200$				
10^{-2}	6	6	6	a
10^{-1}	6	6	7	8(2)
10^{0}	7(1)	7	8(1)	9(2)
10^{+1}	a	a	8	a

Note: [a]More than 10 iterations and three changes in the active set.

10. Outlook

This chapter presented a general overview of mortar formulations, a framework for the simulation of frictional contact problems with deformable solids undergoing finite elasto-plastic strains and several numerical examples. The numerical results demonstrate the robustness of the approach beyond the case of finite elasticity and endorse the applicability of the framework for the analysis of large inelastic deformations and metal forming problems. The accuracy and efficiency of the overall strategy is demonstrated against the well-known NTS penalty contact formulation. The results presented emphasize the benefits of using a mortar formulation with a dual basis for the Lagrange multipliers in contact problems involving inelastic material behaviour. In particular, the evaluation of the gap function as a semi-continuous weak integral yields an enhanced enforcement of constraints in the normal direction, which allows a more accurate evaluation of the contact forces not only in the normal direction but also in the tangential direction. Furthermore, it contributes to the correct fulfilment of additional sources of nonlinearity, which include finite plastic deformations and finite frictional sliding.

The superior correlation between the contacting surfaces provided by the mortar segmentation promotes a smaller initial value for the residual forces, which leads to a faster solution in every increment. Two PDASSs have been presented and applied for the solution of the class of problems addressed. While the single PDASS is generally more efficient than the nested iterative PDASS within the plastic domain, the robustness of the latter is usually greater than the former. The complementarity parameters employed in the definition of the nonlinear complementarity functions do not affect the accuracy of the solution as occurs with the penalty multipliers in the NTS formulation. Nevertheless, the selection of complementarity parameters has a significant impact in the efficiency of the solution whether the material behaviour is within the elastic or plastic domains. Due to the (possibly) strongly nonlinear nature of problem, the optimal convergence rate is only preserved when relatively low values for the complementarity parameters are used, particularly in the plastic domain, and small incremental steps are prescribed. An adaptive strategy should be devised to compute the ideal complementarity parameters during the analysis.

References

1. G. Weber and L. Anand, Finite deformation constitutive equations and a time integration procedure for isotropic, hyperelastic-viscoplastic solids, *Computer Methods in Applied Mechanics and Engineering*, **79**, 173–202 (1990).
2. A. Eterovic and K. Bathe, A hyperelastic-based large strain elasto-plastic constitutive formulation with combined isotropic-kinematic hardening using logarithmic stress and strain measures, *International Journal for Numerical Methods Engineering*, **30**, 1099–1114 (1985).
3. D. Perić, D. Owen, and M. Honnor, A model for large strain elastoplasticity based on logarithmic strain: Computational issues, *Computer Methods in Applied Mechanics and Engineering*, **94**, 35–61 (1992).
4. J. Simo, Algorithms for static and dynamic multiplicative plasticity that preserve the classical return mapping schemes of the infinitesimal theory, *Computer Methods in Applied Mechanics and Engineering*, **99**, 61–112 (1992).

5. A. Cuitiño and M. Ortiz, A material-independent method for extending stress update algorithms from small-strain plasticity to finite plasticity with multiplicative kinematics, *Engineering Computation*, **9**, 437–451 (1992).
6. M. Crisfield, *Nonlinear Finite Element Analysis of Solids and Structures, Vol. 2: Advanced Topics*, John Wiley & Sons, Chichester (1997).
7. J. Simo and T. Hughes, *Computational Inelasticity*, Springer, Berlin-Heidelberg (1998).
8. E. de Souza Neto, D. Perić, and D. Owen, *Computational Methods for Plasticity: Theory and Application*, John Wiley & Sons, Chichester (2008).
9. T. Laursen, *Computational Contact and Impact Mechanics: Fundamentals of Modeling Interfacial Phenomena in Nonlinear Finite Element Analysis*, Springer, Berlin-Heidelberg (2002).
10. P. Wriggers, *Computational Contact Mechanics*, John Wiley & Sons, Chichester (2006).
11. J. Hallquist, G. Goudreau, and D. Benson, Sliding interfaces with contact-impact in large-scale Lagrangian computations, *Computer Methods in Applied Mechanics and Engineering*, **51**, 107–137 (1985).
12. P. Wriggers, T. Vu Van, and E. Stein, Finite element formulation of large deformation impact-contact problems with friction, *Computers Structures*, **37**, 319–331 (1990).
13. T. Laursen and J. Simo, A continuum-based finite element formulation for the implicit solution of multi-body, large deformation frictional contact problems, *International Journal for Numerical Methods Engineering*, **36**, 3451–3485 (1993).
14. T. Erhart, W. Wall, and E. Ramm, Robust adaptive remeshing strategy for large deformation, transient impact simulations, *International Journal for Numerical Methods Engineering*, **65**, 2139–2166 (2006).
15. P. Papadopoulos and R. Taylor, A mixed formulation for the finite element solution of contact problems, *Computer Methods in Applied Mechanics and Engineering*, **94**, 373–389 (1992).
16. P. Hild, Numerical implementation of two nonconforming finite element methods for unilateral contact, *Computer Methods in Applied Mechanics and Engineering*, **184**, 99–123 (2009).
17. P. Wriggers, L. Krstulovic-Opara, and J. Korelc, Smooth c1-interpolations for two-dimensional frictional contact problems, *International Journal for Numerical Methods Engineering*, **51**, 1469–1495 (2001).
18. M. Puso and T. Laursen, A 3D contact smoothing method using gregory patches, *International Journal for Numerical Methods Engineering*, **54**, 1161–1194 (2002).
19. M. Stadler and G. Holzapfel, Subdivision schemes for smooth contact surfaces of arbitrary mesh topology in 3D, *International Journal for Numerical Methods Engineering*, **60**, 1161–1195 (2004).
20. C. Bernardi, Y. Maday, and A. Patera, A new nonconforming approach to domain decomposition: The mortar element method, In *Nonlinear Partial Differential Equations and Their Applications*, H. Brezis and J. Lions (eds.), Pitman/Wiley, London/New York (1994), pp. 13–51.

21. N. Kikuchi and J. Oden, Contact problems in elasticity: a study of variational inequalities and finite element methods, *Society for Industrial and Applied Mathematics Journal on Applied Mathematics*, Philadelphia (1988).
22. F. Facchinei and J. Pang, *Finite-Dimensional Variational Inequalities and Complementary Problems, Vol. 1*, Springer, Berlin-Heidelberg (1998).
23. B. Wohlmuth, A mortar finite element method using dual spaces for the lagrange multiplier, *Society for Industrial and Applied Mathematics Journal on Numerical Analysis*, **38**, 989–1012 (2000).
24. B. Wohlmuth, *Discretization Methods and Iterative Solvers Based on Domain Decomposition*, Springer, Berlin-Heidelberg (2001).
25. B. Flemisch and B. Wohlmuth, Stable Lagrange multipliers for quadrilateral meshes of curved interfaces in 3D, *Computer Methods in Applied Mechanics and Engineering*, **196**, 1589–1602 (2007).
26. P. Alart and A. Curnier, A mixed formulation for frictional contact problems prone to Newton like solution methods, *Computer Methods in Applied Mechanics and Engineering*, **92**, 353–375 (1991).
27. P. Christensen, A nonsmooth newton method for elastoplastic problems, *Computer Methods in Applied Mechanics and Engineering*, **191**, 1189–1219 (2002).
28. T. Koziara and N. Bicanic, Semismooth newton method for frictional contact between pseudo-rigid bodies, *Computer Methods in Applied Mechanics and Engineering*, **197**, 2763–2777 (2008).
29. M. Hintermüller, K. Ito, and K. Kunisch, The primal–dual active set strategy as a semi-smooth Newton method, *Society for Industrial and Applied Mathematics Journal on Optimization*, **13**, 865–888 (2003).
30. S. Hüeber and B. Wohlmuth, A primal–dual active set strategy for non-linear multibody contact problems, *Computer Methods in Applied Mechanics and Engineering*, **194**, 3147–3166 (2005).
31. S. Hüeber, G. Stadler, and B. Wohlmuth, A primal–dual active set algorithm for three-dimensional contact problems with Coulomb friction, *Society for Industrial and Applied Mathematics Journal on Scientific Computing*, **30**, 572–596 (2008).
32. S. Brunssen, F. Schmid, M. Schäfer, and B. Wohlmuth, A fast and robust iterative solver for nonlinear contact problems using a primal–dual active set strategy and algebraic multigrid, *International Journal for Numerical Methods Engineering*, **69**, 524–543 (2007).
33. C. Hager and B. Wohlmuth, Nonlinear complementarity functions for plasticity problems with frictional contact, *Computer Methods in Applied Mechanics and Engineering*, **198**, 3411–3427 (2003).
34. S. Hartmann and E. Ramm, A mortar based contact formulation for non-linear dynamics using dual Lagrange multipliers, *Finite Elements in Analysis and Design*, **44**, 245–258 (2004).
35. A. Popp, M. Gee, and W. Wall, A finite deformation mortar contact formulation using a primal–dual active set strategy, *Computer Methods in Applied Mechanics and Engineering*, **79**, 1354–1391 (2009).

36. M. Gitterle, A. Popp, M. Gee, and W. Wall, Finite deformation frictional mortar contact using a semi-smooth Newton method with consistent linearization, *International Journal for Numerical Methods Engineering*, **84**, 543–571 (2010).
37. A. Popp, M. Gitterle, M. Gee, and W. Wall, A dual mortar approach for 3D finite deformation contact with consistent linearization, *International Journal for Numerical Methods Engineering*, **83**, 1428–1465 (2010).
38. A. Popp, M. Gee, and W. Wall, Improved robustness and consistency of 3D contact algorithms based on a dual mortar approach, *Computer Methods in Applied Mechanics and Engineering*, **79**, 1354–1391 (2013).
39. T. Doca, F. Andrade Pires, and J. César de Sá, A frictional mortar contact approach for the analysis of large inelastic deformation problems, *International Journal of Solids and Structure*, **51**, 1697–1715 (2014).
40. S. Seitz, A. Popp, and W. Wolfgang, A semi-smooth newton method for orthotropic plasticity and frictional contact at finite strains, *Computer Methods in Applied Mechanics and Engineering*, **285**, 228–254 (2015).
41. E. de Souza Neto, D. Perić, M. Dutko, and D. Owen, Design of simple low order finite elements for large strain analysis of nearly incompressible solids, *International Journal of Solids and Structure*, **33**, 3277–3296 (1996).
42. E. de Souza Neto, F. Andrade Pires, and D. Owen, F-bar-based linear triangles and tetrahedra for finite strainanalysis of nearly incompressible solids. Part I: formulation and benchmarking, *International Journal of Numerical Methods Engineering*, **62**, 353–383 (2005).
43. E. Lee, Elastic–plastic deformation at finite strains, *Journal of Applied Mechanics*, **36**, 1–6 (1969).
44. M. Ortiz and E. Popov, Accuracy and stability of integration algorithms for elastoplastic constitutive relations, *International Journal for Numerical Methods Engineering*, **21**, 1561–1576 (1985).
45. M. Puso and T. Laursen, A mortar segment-to-segment frictional contact method for large deformations, *Computer Methods in Applied Mechanics and Engineering*, **193**, 601–629 (2004).
46. B. Yang, T. Laursen, and X. Meng, Two dimensional mortar contact methods for large deformation frictional sliding, *International Journal for Numerical Methods Engineering*, **62**, 1183–1225 (2005).
47. M. Puso, T. Laursen, and J. Solberg, A segment-to-segment mortar contact method for quadratic elements and large deformations, *Computer Methods in Applied Mechanics and Engineering*, **197**, 555–566 (2008).
48. S. Hüeber and B. Wohlmuth, Nonlinear complementarity functions for plasticity problems with frictional contact, *Society for Industrial and Applied Mathematics Journal on Numerical Analysis*, **43**, 157–173 (2005).
49. T. Laursen, M. Puso, and J. Sanders, Mortar contact formulations for deformable-deformable contact: Past contributions and new extensions for enriched and embedded interface formulations, *Computer Methods in Applied Mechanics and Engineering*, **3–15**, 205–208 (2012).
50. T. Doca and F. Andrade Pires, Analysis of a cylinder-to-flat contact problem at finite elasto-plastic strains, *Tribology International*, **79**, 92–98 (2014).

51. H. Hertz, Über die berührung fester elastischer körper (on contact between elastic bodies), *Journal für die Reine und Angewandte Mathematik*, **29**, 156–171 (1882).
52. K. Jonhson, *Contact Mechanics*, Cambridge University Press, Cambridge (1987).
53. J. Simo and T. Laursen, An augmented Lagrangian treatment of contact problems involving friction, *Computers & Structure*, **42**, 97–116 (1992).
54. S. Hüeber, *Discretization Techniques and Efficient Algorithms for Contact Problems*, PhD thesis, Universität Stuttgart (2008).
55. K. Fischer and P. Wriggers, Mortar based frictional contact formulation for higher order interpolations using the moving friction cone, *Computer Methods in Applied Mechanics and Engineering*, **195**, 5020–5036 (2006).

Chapter 2

Standard and Generalized High-order Mortar-based Finite Elements in Computational Contact Mechanics

A. P. C. Dias, S. P. B. Proenca and M. L. Bittencourt

In this chapter, we present high-order mortar-based contact elements using Lagrange multiplier and penalty methods. We seek approximation solutions for 2D small and large deformation contact problems, through the standard and generalized high-order finite element methods (labelled as *hp*-FEM and GFEM, respectively). We validate the elements for small and large deformation with known solutions in the literature. The *h*- and *p*-refinements assert that the high-order interpolation has superior performance in contact problems with better accuracy of results for stresses with lower time/cost processing. In the case of the GFEM contact element, we observe that the enrichment benefits goes beyond the smoothing of the stress fields. Our findings show that the contact elements proposed here can provide more accurate results than the ones obtained by the conventional low-order FEM.

1. Introduction

The efficiency and advantages of the *hp*-FEM when applied to nonlinear structural problems have been widely studied in the literature.[1–5] The advantages over classical methods are the exponential convergence rates for smooth problems,[6] flexibility in using elements with large aspect ratio, locking-free with respect to the thickness of plate and shell elements and incompressible materials.[7]

In the last years, the *hp*-FEM has been applied to frictionless contact problems considering small deformations in Refs. [8–10]. The results have showed that the use of *hp*-FEM in this class of problems increases the accuracy of results for contact stresses with less degrees of freedom (dofs).

The mortar and mortar-based linear elements are commonly used for the discretization of the contact terms.[11–18] Recent papers have shown gains in the solution accuracy with the use of quadratic contact elements.[11,12,19–23]

The mortar method was originally proposed for domain decomposition techniques[24] and later adapted for frictionless contact problems.[25] The method consists of elements which are built from nodes and edges of different bodies in contact. The formulation was extended to frictional contact problems considering large deformations and low-order interpolation.[26]

Comparative simulations in large deformation showed better accuracy with fewer dofs using quadratic contact elements.[11,12,21] The papers[8,9] applied high-order Legendre polynomials for interpolating the node-to-segment (NTS) and segment-to-segment (STS) contact elements and obtained significantly accurate results for small deformations. Isogeometric high-order interpolation with NURBS for the construction of the contact elements was used in Refs. [27–30]. The obtained results for the Hertz problem show that high-order interpolation with NURBS reduces contact stress oscillations in the on/off contact interface, however requiring very fine meshes on the contact area. Overviews about the use of high-order contact elements are presented in Refs. [11, 21, 31].

The present work is an extension of the fully high-order mortar-based contact element proposed in Ref. [31]. In that work, computational solutions of the high-order mortar-based contact element in 2D problems considered small and large deformations. The penalty method was used to impose the contact restriction on the mortar-based integrals. Furthermore, the Neo-Hookean isotropic compressible hyperelastic material model was considered for large deformations. Here, we extend such an approach to a formulation using the Lagrange multiplier method, then compare the Lagrange multiplier and penalty solutions. In general, the analyses of small and large deformation problems show that high-order interpolation has superior performance in contact problems in terms of accuracy of stresses with lower time/cost processing.

We also address the Generalized Finite Element Method (GFEM) applied to the development of a contact element (GFEM contact element).[32] GFEM is a Partition of Unity (PU) based Galerkin method, according to which the basic interpolation provided by the PU is modified using functions with good approximation properties, referred to as enrichment functions. In fact, this concept was primarily introduced in the Partition of Unity Method (PUM) framework.[33]

In PU-based methods such as the GFEM and the eXtended Finite Element Method (XFEM),[34] a mesh of finite elements is used to provide a PU, which is commonly defined using linear Lagrangian shape functions. The GFEM/XFEM enrichment framework, while dispensing with costly mesh refinements, has proved to be efficient in a variety of applications, mainly when the solution local *behaviour* is of major interest, and therefore, it must be properly reproduced.

Problems of linear fracture mechanics are likely among the most benefited by the method. For instance, crack tip stress concentrations are better reproduced by enriching the nearby nodal approximation with branch functions.[35–40] Furthermore, mesoscale cracking modelling of polycrystalline materials and solids are other classes of problems in which the GFEM/XFEM features are advantageous with respect to conventional FEM approaches.[40–43]

Since crack propagation problems including dynamic effects started to be addressed,[44–46] time integration schemes, lumped mass techniques, numerical stability and several other related issues also began to be considered.[47,48] Regarding the structural nonlinear dynamic behaviour, the GFEM was also successfully applied in combination with an implicit time-stepping technique.[49]

Even though the GFEM is being successfully applied to solve a large variety of mechanical problems, its extension to general nonlinear analysis purposes including contact problems still demands further developments, mainly in order to assess the effectiveness of the method compared to the conventional FEM approaches. Therefore, this work also investigate some key points regarding GFEM accuracy when applied to contact problems using mortar-based elements. A neo-Hookean material model is adopted for modelling the nonlinear elastic behaviour. Moreover, shifted polynomial functions are used for the enrichment of the approximation. Even though the results might be similar to the ones obtained by means of the FEM traditional polynomial refinement, we found that the topic has not been fully explored, and still some interesting studies remain worth exploring.

2. The weak form of the boundary value problem

The weak statement of the equilibrium of the body is provided by the principle of virtual work (PVW). Consider that we can write the spatial virtual work equation $\delta\Pi_s$ for the mechanical system as

$$\delta\Pi_s = \int_\Omega \boldsymbol{\sigma} : \delta\boldsymbol{\epsilon} d\Omega - \int_\Omega \mathbf{f} \cdot \delta\mathbf{u} d\Omega - \int_{\Gamma_\sigma} \bar{\mathbf{t}} \cdot \delta\mathbf{u} d\Gamma, \tag{1}$$

where Ω denotes the current domain occupied by the body, with boundary $\partial\Omega = \Gamma_\sigma \bigcup \Gamma_u$ and $\Gamma_\sigma \bigcap \Gamma_u = \emptyset$. Γ_σ and Γ_u denote the Neumann (with prescribed edge loads) and Dirichlet (with prescribed displacements) boundaries, respectively. $\delta\mathbf{u}$ is the virtual displacement vector from the current position, $\boldsymbol{\sigma}$ is the Cauchy stress tensor, $\delta\boldsymbol{\epsilon}$ is the virtual strain tensor, $\mathbf{f}$ is the body force vector and $\bar{\mathbf{t}}$ is the vector field of prescribed tractions at the Neumann part of the current solid boundary.

For problems in which large deformation kinematics cannot be neglected, it is possible to use different descriptions of the strain. A total Lagrangian approach is here chosen, for which the Green strain tensor $\mathbf{E}$ and the second Piolla–Kirchhoff stress tensor $\mathbf{S}$ are conjugated in the definition of the internal virtual work. Moreover, hyperelasticity is assumed for describing the nonlinear elastic constitutive model. Accordingly, the behaviour of the material is said to be path-independent and a stored strain energy function per unit of non-deformed volume (Π_{int}) can be established. Further derivations can be found in Refs. [50, 51].

Different hyperelastic models are available. Among the possibilities, the neo-Hookean compressible material model is chosen here. Therefore, following Ref. [21], the corresponding strain energy function is defined as

$$\Pi_{\text{int}} = \frac{1}{2}\mu\left[tr\left(\mathbf{C}\right) - 3\right] + \frac{\lambda}{4}\left(\mathbf{J}^2 - 1\right) - \left(\frac{\lambda}{2} + \mu\right)\ln\left(\mathbf{J}\right), \tag{2}$$

where $\mathbf{C}$ is the right Cauchy–Green deformation tensor, directly related to $\mathbf{E}$, $\mathbf{J}^2$ is the determinant of $\mathbf{C}$ and $ln(\mathbf{J})$ is the natural logarithm of $\mathbf{J}$. The Lamè parameters μ and λ are two material-independent quantities.[50]

The stress tensor work conjugated to the rate of $\mathbf{E}$ is the second Piolla–Kirchhoff tensor $\mathbf{S}$. In other words, the work conjugacy concept enables constructing a totally Lagrangian constitutive equation from those equations as follows:

$$\mathbf{S} = \frac{\partial\Pi_{\text{int}}}{\partial\mathbf{E}} = 2\frac{\partial\Pi_{\text{int}}}{\partial\mathbf{C}}. \tag{3}$$

Finally, we write the virtual work equation in terms of the second Piolla–Kirchhoff tensor $\mathbf{S}$. It is a material or total Lagrangian description[52] and can be written as: find $u \in V_t$ such that for $\forall\delta u \in V$

$$\delta\Pi_s = \int_{\Omega} \mathbf{S} \colon \delta\mathbf{E} d\Omega - \int_{\Omega_0} \mathbf{f}\cdot\delta\mathbf{u} d\Omega - \int_{\Gamma_\sigma} \bar{\mathbf{t}}\cdot\delta\mathbf{u} d\Gamma_\sigma + \delta\Pi_c, \tag{4}$$

where $\delta\Pi_c$ is the contact virtual work that will be defined in the following sections, V_t is the standard solution space for the elasticity problem and V

the test space, respectively defined as

$$V_t = \left\{\mathbf{u} \in H^1(\Omega) : \mathbf{u} = \bar{\mathbf{u}}(\mathbf{x}, t) \ \text{ for } \mathbf{x} \text{ on } \Gamma_u\right\}, \tag{5}$$

$$V = \left\{\mathbf{w} \in H^1(\Omega) : \mathbf{w}(\mathbf{x}) = 0 \text{ for } \mathbf{x} \text{ on } \Gamma_u\right\}. \tag{6}$$

$H^1(\Omega)$ is the Sobolev space that consists of all vector valued functions over Ω whose values and first derivatives are square integrable over the domain.

Similarly, we can write the virtual work equation considering updated Lagrangian formulation[52] in terms of the Kirchhoff stress tensor $\boldsymbol{\tau}$ as

$$\delta\Pi_s = \int_\Omega \boldsymbol{\tau}\text{:grad}(\delta\mathbf{u})\, d\Omega - \int_\Omega \mathbf{f} \cdot \delta\mathbf{u} d\Omega - \int_{\Gamma_\sigma} \bar{\mathbf{t}} \cdot \delta\mathbf{u} d\Gamma_\sigma + \delta\Pi_c. \tag{7}$$

The Kirchhoff stress tensor field $\boldsymbol{\tau}$ is defined in terms of the deformation gradient tensor $\mathbf{F}$ as[50,52]:

$$\boldsymbol{\tau} = 2\mathbf{F}\frac{\partial\Pi_{\text{int}}}{\partial\mathbf{C}}\mathbf{F}^T. \tag{8}$$

Further detailed description on the derivation of the weak form of the BVP, including hyperelasticity behaviour.[53]

3. The *hp*-FEM and GFEM main features

The finite element method (FEM)[54] provides a strategy for defining trial/test approximations for the displacement fields presented in the weak form of the boundary value problem (BVP). Using the FEM, the complete virtual work equation can be computed by the contributions of each of the elements of the discretization mesh. For instance, considering for the sake of simplicity the static linear elastic model, the internal virtual work restricted to one typical element, can be written as follows:

$$\delta\mathbf{d}^T\left(\int_{\Omega_e} \mathbf{B}^T\mathbf{D}\mathbf{B} d\Omega_e\right)\mathbf{d} = \delta\mathbf{d}^T\mathbf{K}_e\mathbf{d}, \tag{9}$$

where $\mathbf{d}$ and $\delta\mathbf{d}$ are, respectively, real and virtual nodal displacement vectors, while $\mathbf{B}$ is the matrix relating strain components to nodal displacement derivatives, and $\mathbf{D}$ the matrix representing the elastic constitutive stiffness tensor. As pointed out in the above relation, the integral computed in each element domain Ω_e gives rise to the element stiffness matrix $\mathbf{K}_e$. An analogous concept is applied to the external virtual work, then leading to the element nodal force vector $\mathbf{f}_e$. The main difference between the standard FEM, *hp*-FEM and GFEM formulations is in the definition of the shape functions that will interpolate the displacement field and geometry, as can be seen in the next sections.

3.1. *hp-FEM formulation*

In the *hp*-FEM, the displacement field and coordinates can be interpolated similarly to the standard FEM as

$$\mathbf{u}^e(\xi) = \sum_{i=1}^{N_{ne}} N_i(\xi)\mathbf{u}_i, \tag{10}$$

where N_{ne} is the number of nodes per element.

In the present work, the shape functions $N_i(\xi)$ are obtained from the Lagrange polynomials for any interpolation order. The local coordinate is $\xi \in [-1, 1]$.

The Lagrange polynomials of degree P, $h_i^{(P)}(\xi)$ are used as the polynomial basis for the construction of 1D nodal interpolation functions $N_i^{(P)}(\xi)$. These polynomials are defined in the standard element $\Omega_{st} = \{\xi | -1 \leq \xi \leq 1\}$ as

$$N_i^{(P)}(\xi) = h_i^{(P)}(\xi) = \frac{\prod_{j=1, j\neq i}^{N_{ne}} (\xi - \xi_j)}{\prod_{j=1, j\neq i}^{N_{ne}} (\xi_i - \xi_j)}, \quad i, j = 1, \ldots, N_{ne}, \tag{11}$$

where $N_{ne} = P + 1$ is the number of nodes of the finite element. One shape function is associated with each element node.

The shape functions of quadrilateral finite elements for the domain discretization are given by the tensor product of the Lagrange polynomials

$$N_{jk}^{(P)}(\xi, \eta) = h_j^{(P)}(\xi) h_k^{(P)}(\eta), \quad j, k = 1, \ldots, (N_{ne})^2, \tag{12}$$

with j and k indices of the nodes in directions ξ and η, respectively.

To interpolate the nodal Lagrange multipliers, we use a dual basis proposed in Ref. [55], which is biorthogonal to the Lagrange basis. References [12] and [55] discuss the details of the formulation and advantages when using a basis with this feature.

The dual basis functions $M_j(\xi)$ are built in such a way to satisfy the following orthogonality relation:

$$\mathbf{G} = \int_{-1}^{1} N_i(\xi) M_j(\xi)\, d\xi = w_i \delta_{ij}, \quad w_i \neq 0, \tag{13}$$

where w_i are the weights for the Gauss–Lobatto–Legendre integration quadrature and are defined as[6]

$$w_i = \int_{-1}^{1} N_i(\xi)d\xi. \tag{14}$$

It is important to define the mass or projection matrix for the one standard unidimensional element as[6]

$$\mathbf{M}_L = \int_{-1}^{1} N_i(\xi)N_j(\xi)d\xi = \mathbf{V}_L^T\mathbf{W}\mathbf{V}_L \tag{15}$$

and the matrix $\mathbf{W}$ is given by

$$\mathbf{W} = \begin{bmatrix} w_1 & 0 & 0 & 0 \\ 0 & w_2 & 0 & 0 \\ 0 & 0 & \ddots & 0 \\ 0 & 0 & 0 & w_{n_p} \end{bmatrix}_{(n_p,n_p)} . \tag{16}$$

where n_p is the number of the integration points used and $\mathbf{V}_L$ is the Vandermonde matrix of the Lagrange basis

$$\mathbf{V}_L = \begin{bmatrix} N_1(\xi_1) & N_2(\xi_1) & \cdots & N_{Nne}(\xi_1) \\ N_1(\xi_2) & N_2(\xi_2) & \cdots & N_{Nne}(\xi_2) \\ N_1(\xi_3) & N_2(\xi_3) & \cdots & N_{Nne}(\xi_3) \\ \vdots & \vdots & \ddots & \vdots \\ N_1(\xi_{n_p}) & N_2(\xi_{n_p}) & \cdots & N_{Nne}(\xi_{n_p}) \end{bmatrix}_{(n_p,Nne)} . \tag{17}$$

The Vandermonde matrix has in its columns the values of each shape function calculated on all the integration points applied.

The Vandermonde matrix for the polynomials of the dual basis $M_i(\xi)$ is

$$\mathbf{V}_M = \mathbf{G}\mathbf{M}_L^{-1}\mathbf{V}_L = \begin{bmatrix} M_1(\xi_1) & M_2(\xi_1) & \cdots & M_{Nne}(\xi_1) \\ M_1(\xi_2) & M_2(\xi_2) & \cdots & M_{Nne}(\xi_2) \\ M_1(\xi_3) & M_2(\xi_3) & \cdots & M_{Nne}(\xi_3) \\ \vdots & \vdots & \ddots & \vdots \\ M_1(\xi_{n_p}) & M_2(\xi_{n_p}) & \cdots & M_{Nne}(\xi_{n_p}) \end{bmatrix}_{(n_p,Nne)} . \tag{18}$$

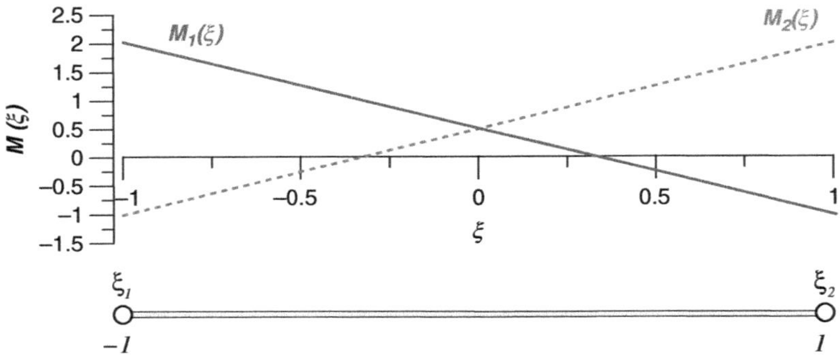

Fig. 1. Interpolation functions of the dual basis for $P = 1$.

For interpolation order $P = 1$, we have 2 nodes in the unidimensional element and the dual basis functions result in

$$M_1^1(\xi) = \frac{(1 - 3\xi)}{2}, \tag{19}$$

$$M_2^1(\xi) = \frac{(1 + 3\xi)}{2}. \tag{20}$$

Figure 1 shows the interpolation functions of the dual basis for $P = 1$.

For interpolation order $P = 2$, we have 3 nodes in the unidimensional element and the dual basis functions result in

$$M_1^2(\xi) = \frac{1}{4}\left(5\xi^2 - 2\xi - 1\right), \tag{21}$$

$$M_2^2(\xi) = \frac{1}{4}\left(5\xi^2 + 2\xi - 1\right), \tag{22}$$

$$M_3^2(\xi) = \frac{1}{2}\left(3 - 5\xi^2\right). \tag{23}$$

Figure 2 shows the interpolation functions of the dual basis for $P = 2$.

For interpolation order $P = 3$, we have 4 nodes in the unidimensional element (Fig. 3) and the basis functions are constructed in an analogous way.

3.2. *GFEM formulation*

Figure 4 shows a typical cloud ω_α. It is a set of elements that have a common node as vertex.

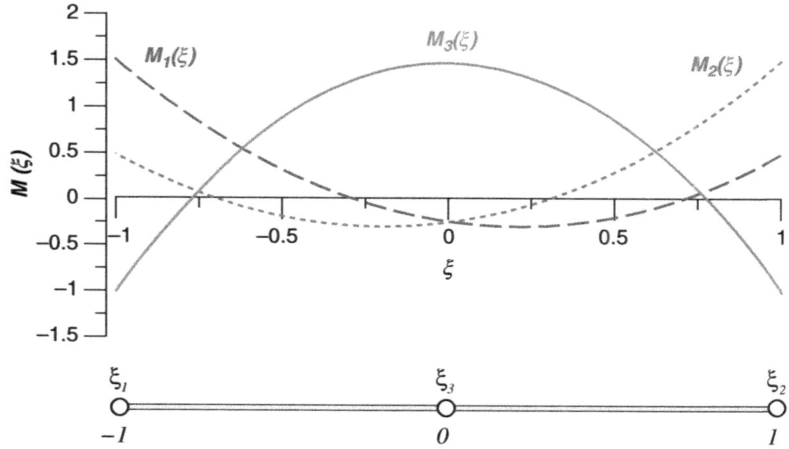

Fig. 2. Interpolation functions of the dual basis for $P = 2$.

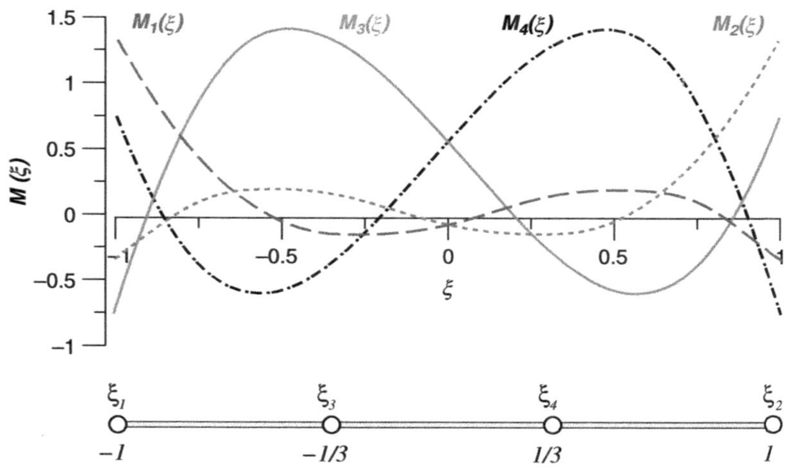

Fig. 3. Interpolation functions of the dual basis for $P = 3$.

Inside of each cloud, the shape functions are constructed by the product between the so-called enrichment functions $L_{\alpha j}$ and the linear *PU* functions belonging to the elements in the cloud and attached to the vertex α (N_α).

Specifically for the GFEM, the finite elements of a mesh and their attached nodal shape functions provide the *PU*. In GFEM, the shape

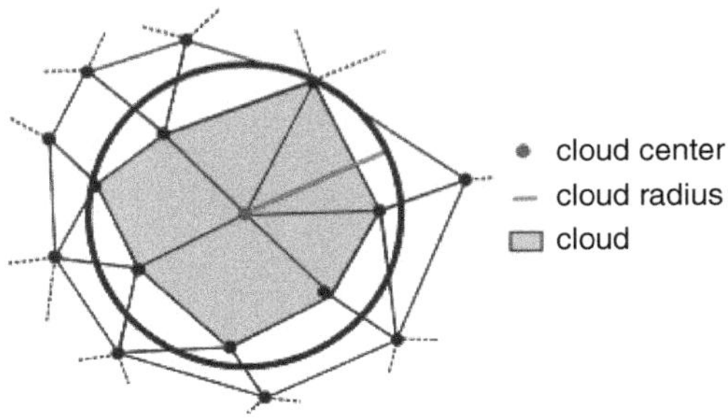

Fig. 4. The representation of the cloud, its center and radius.

functions ($\phi_{\alpha j}$) are defined locally in a cloud domain ω_α. Therefore,

$$\phi_{\alpha j} = N_\alpha L_{\alpha j}, \tag{24}$$

where α identifies the nodal cloud and $j = 1, \ldots, N_L$, and N_L is the total number of enrichment functions adopted for each cloud. The PU concept is explored again to paste together the cloud approximations, therefore building a conform global approximation. The GFEM global approximation for each component of the displacement field, restricting only to $2D$ problems, is given by the following relations:

$$\hat{\mathbf{u}} = \sum_{\alpha=1}^{N_{nc}} N_\alpha \mathbf{u}_\alpha + \sum_{\alpha=1}^{N_{nc}} \sum_{j=2}^{N_L} \phi_{\alpha j} \mathbf{b}_{\alpha j}, \tag{25}$$

$$\hat{\mathbf{v}} = \sum_{\alpha=1}^{N_{nc}} N_\alpha \mathbf{v}_\alpha + \sum_{\alpha=1}^{N_{nc}} \sum_{j=2}^{N_L} \phi_{\alpha j} \mathbf{c}_{\alpha j}, \tag{26}$$

where N_L is the total number of clouds, N_{nc} is the number of enriched clouds, $\mathbf{u}_\alpha$ and $\mathbf{v}_\alpha$ are parameters associated with usual dofs of finite elements, and $\mathbf{b}_{\alpha j}$ and $\mathbf{c}_{\alpha j}$ are additional vectors of the nodal parameters introduced by enrichment. Unlike the FEM, the nodal dofs in the GFEM, in general, do not have a direct physical meaning. However, there are no restrictions to the type and number of enrichment functions to be attached to a given node. Therefore, the number of dofs is totally flexible, and consequently the size of the resulting element stiffness matrix may change drastically, depending on the adopted enrichment scheme. Moreover, once a Galerkin approach is used, resulting element load force vectors also present variable sizes.

The enrichment functions aim to improve the quality of local approximation and can present special features or even be defined from previous knowledge of the solution. For instance, the branch functions can be used as enrichment functions in order to more accurately represent the singularity in the vicinity of the crack tip. Notwithstanding, the polynomial enrichment scheme can provide an efficient strategy to be used and worthy of exploring when applying GFEM to a broad class of problems involving smooth solutions.

As a rule, the enrichment functions are applied to each of the displacement components. It is also common to use an enrichment scheme aiming to achieve a certain complete polynomial degree. The general form of the polynomial enrichment function, for a 2D problem, can be expressed by the shifted approach as

$$L(m,n) = \frac{(X - X_\alpha)^m (Y - Y_\alpha)^n}{h^{m+n}}, \tag{27}$$

where X_α and Y_α are the coordinates of the node α to which the enrichment is applied. One of the most remarkable advantages of the shifted enrichment functions is that they are zero in the node where they are imposed. It follows that the physical meaning of the original dof in such a node is preserved. This fact enables, in principle, to directly enforce displacement boundary conditions in the same way as in the FEM. However, the nodal enforcement of the boundary condition (null displacement, for instance) does not guarantee that this condition is verified in the remaining regions of the boundary between nodes. This specific aspect is addressed in more detail in Ref. [56], but, in short, a penalty alternative is adopted in this work for the correct enforcement of displacement boundary conditions.

Notwithstanding the favourable aspects mentioned above, the polynomial enrichment approach may introduce linear dependencies in the resulting system of equations, therefore affecting the numerical stability of the model.[57] However, as followed in this work, once carefully considered the polynomial enrichment approach can be explored efficiently in nonlinear analyses.

4. Contact problem description and variational formulation

The contact physical phenomenon is characterized by the interaction between different parts of the solids, which, during the mechanical process, may tend to occupy the same position in space at the same time. Since this is not physically admissible, interaction forces between the solid parts

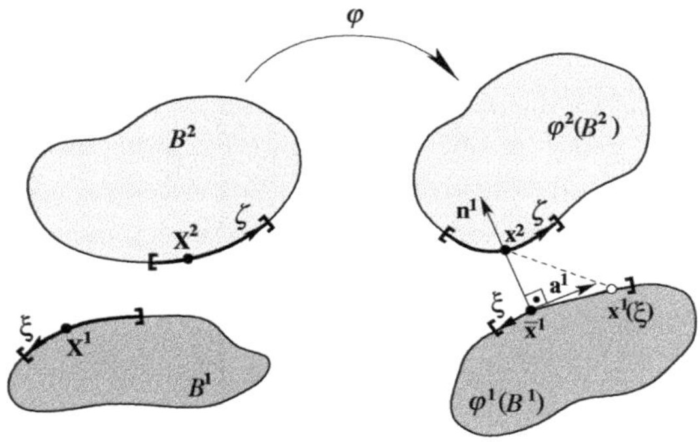

Fig. 5. Current configuration of the bodies B^{α}.[21]

arise and the solids must deform so that no penetration is observed. As the kinematical restrictions in some portions of the external boundaries are not known beforehand, the BVP automatically becomes nonlinear, regardless of the adopted kinematic and constitutive model.

In this section, we consider problems involving nonlinear effects of large deformations with two or more bodies in frictionless contact. The formulation is based on the Refs. [21, 31, 58]. A brief overview is presented next.

Consider two elastic bodies B^{α} with $\alpha = 1, 2$, each one occupying a delimited domain $\Omega^{\alpha} \subset R^2$, as shown in Fig. 5. The boundary Γ^{α} consists of three parts: Γ_{σ} with prescribed edge loads, Γ_u with prescribed displacements and Γ_c called contact surface where the two bodies B^1 and B^2 get in contact, as shown in Fig. 8. The following relationships must be satisfied: $\Gamma^{\alpha} = \Gamma^{\alpha}_{\sigma} \cup \Gamma^{\alpha}_{u} \cup \Gamma^{\alpha}_{c}$ and $\Gamma^{\alpha}_{\sigma} \cap \Gamma^{\alpha}_{u} \cap \Gamma^{\alpha}_{c} = \emptyset$. In this region, once contact occurs, an impenetrability condition must be verified. A pair of contacting points must be considered in order to express such a condition mathematically. It is then considered that two distinct points of the bodies in the initial configuration, with material coordinates, respectively, $\mathbf{X}^1$ and $\mathbf{X}^2$, can occupy the same position in the current configuration, $\mathbf{x}^2 = \varphi(\mathbf{X}^2, t) = \mathbf{x}^1 = \varphi(\mathbf{X}^1, t)$, where contact conditions have to be formulated.

The strategy commonly used to solve contact problems is to treat it as a constrained minimization problem.[21,58,59] The objective function is the total potential energy of the bodies in contact $\Pi(\mathbf{u})$, which is the sum of the potential energy of the mechanical system $(\Pi_s(\mathbf{u}))$ and of the contact region $(\Pi_c(\mathbf{u}))$. The solution of a frictionless contact problem can be obtained from the minimization of the total potential energy subject to inequality constraints of non-penetration between bodies. Mathematically, we can define the problem as

$$(P)\begin{cases}\min \Pi(\mathbf{u}) = \Pi_s(\mathbf{u}) + \Pi_c(\mathbf{u}), \\ \text{s.t.} \qquad\qquad g(\mathbf{u}) \geq 0.\end{cases} \tag{28}$$

The constraints are given by the penetration function or gap function $g(\mathbf{u})$, that represents the specific non-penetration condition of bodies. If $g = 0$, a perfect contact occurs; for penetration between bodies, $g < 0$; and for a non-penetration condition, $g > 0$. Once friction is taken into account, additional restrictions are necessary to add in the minimization problem, as well as a constitutive law for the friction. In this work, only frictionless contact effects are analysed. The potential energy of the contact is constructed according to the constrained minimization method adopted. Assuming that the total potential energy functional $\Pi(\mathbf{u})$ is convex, the solution of the constrained minimization problem (28) is equivalent to the solution by the variational formulation presented in Section 4.3 using the PVW. The variational formulation will be used in this work for the solution of the contact problems.

4.1. *Normal contact description*

We can establish the local contact kinematics required by the physics of the problem. Considering Fig. 5, the current configuration points on the bodies B^1 and B^2 are given by

$$\mathbf{x}^\alpha = \mathbf{X}^\alpha + \mathbf{u}^\alpha, \tag{29}$$

where $\mathbf{X}^\alpha$ is the vector position of the point in the initial configuration and $\mathbf{u}^\alpha$ is the displacement vector fields of bodies with $\alpha = 1, 2$.

The gap vector is defined as

$$\mathbf{g} = \left(\mathbf{x}^2 - \mathbf{x}^1\right), \tag{30}$$

and the mathematical condition to prevent a body from penetrating into another one is expressed by a usual normal penetration function g_N defined as

$$g_N = \begin{cases}\mathbf{g}\cdot\mathbf{n}^1 & \text{if } \mathbf{g}\cdot\mathbf{n}^1 < 0 \ \text{(with penetration)}, \\ 0 & \text{if } \mathbf{g}\cdot\mathbf{n}^1 \geq 0 \ \text{(without penetration)}.\end{cases} \tag{31}$$

Equation (31) defines the magnitude of numerical penetration between bodies. The special case of multiple projection coordinates is described in Ref. [21].

The normal unit vector $\mathbf{n}^1$ is associated to the body B^1 and defined as

$$\mathbf{n}^1(\boldsymbol{\xi}) = \frac{\mathbf{e}_3 \times \mathbf{a}^1(\boldsymbol{\xi})}{\|\mathbf{e}_3 \times \mathbf{a}^1(\boldsymbol{\xi})\|}, \tag{32}$$

where the tangent vector is given by $\mathbf{a}^1(\boldsymbol{\xi}) = \mathbf{x}^1_{,\xi}(\boldsymbol{\xi})$ and $\mathbf{e}_3$ is the unit vector perpendicular to the plane of analysis. Note that the tangent vector is not a normalized vector, because it will be necessary in both unitary and absolute forms.

These vectors are computed in $\bar{\mathbf{x}}^1(\boldsymbol{\xi})$, which is the orthogonal projection of the contactor point $\mathbf{x}^2$ on surface Γ_c^1 (Fig. 5). Considering that the contact boundary area is locally convex, we can associate with each point $\mathbf{x}^2$ on Γ_c^2 an orthogonal projection point, $\bar{\mathbf{x}}^1$ on Γ_c^1, given by the solution of the minimum distance problem[60]

$$\hat{d}^1(\boldsymbol{\xi}) = \|\mathbf{x}^2 - \bar{\mathbf{x}}^1\| = \min_{\mathbf{x}^1 \subset \Gamma_c^1} \|\mathbf{x}^2 - \mathbf{x}^1(\boldsymbol{\xi})\|. \tag{33}$$

In 3D analysis, the coordinate $\boldsymbol{\xi} = (\xi_1, \xi_2)$ denotes the parameterization of the boundary surface Γ_c^1 by convective coordinates[21,61] (Fig. 6). For 2D problems, it is possible to parameterize the contact boundary with a single convective coordinate $\boldsymbol{\xi} = \xi_1$. The metric and curvature tensors for a point on the target surface reduce to scalars and are defined, respectively, as[21]

$$a_{11} = \mathbf{a}^1(\xi_1) \cdot \mathbf{a}^1(\xi_1), \tag{34}$$

$$b_{11} = \mathbf{x}^1_{,\xi\xi}(\xi_1). \tag{35}$$

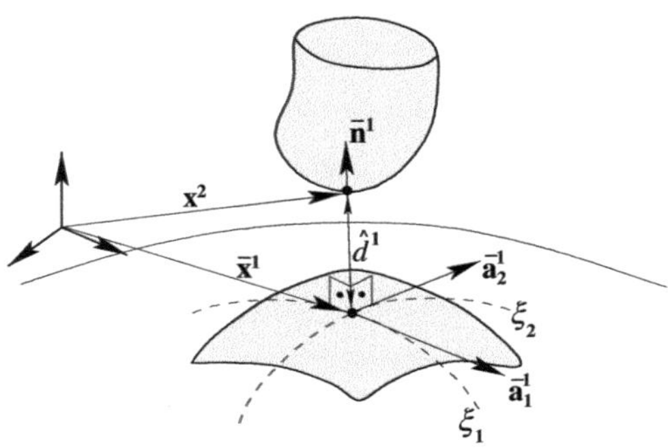

Fig. 6. Orthogonal projection for the minimum distance problem with convective coordinates.[21]

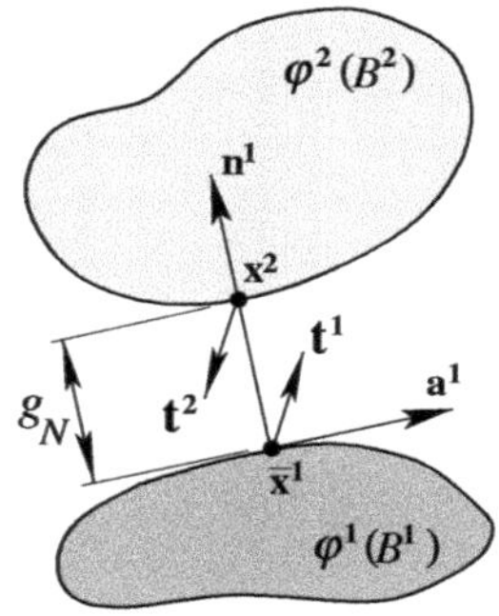

Fig. 7. Stress vector on the contact interface.[21]

4.2. *Treatment of contact restrictions*

A classic strategy of normal contact formulation is to treat normal contact as a unilateral restriction problem.[62,63] It is used when it is essential to satisfy the geometric contact constraint correctly. This type of strategy does not require a constitutive relation for the contact interface.

The necessary condition for no penetration between the bodies is $g_N \geq 0$. According to Fig. 7, when there is contact ($g_N < 0$), a stress vector $\mathbf{t}^1$, associated with the contactor, arises in order to prevent interpenetration and is given by

$$\mathbf{t}^1 = t_N \mathbf{n}^1 + t_T \bar{\mathbf{a}}^1, \tag{36}$$

where t_N and t_T are the normal and tangential contact stress components. The stress vector acts on both surfaces and $\mathbf{t}^1(\boldsymbol{\xi}) = -\mathbf{t}^2(\boldsymbol{\xi})$, according to the action and reaction principle at a given contact point. In the case of frictionless contact, $t_T = 0$.

When there is contact, $g_N = 0$ and the contact stress is $t_N < 0$; for no contact, $g_N > 0$ and $t_N = 0$. These conditions are represented by the following set of constraint equations:

$$\begin{aligned} g_N &\geq 0, \\ t_N &\leq 0, \\ t_N g_N &= 0. \end{aligned} \tag{37}$$

The previous equations define the Hertz–Signorini–Moreau conditions, which are the basis for the treatment of frictionless contact problems as a constrained minimization problem. This set of equations is known as the Karush–Kuhn–Tucker (KKT) conditions in optimization theory.[64,65] The

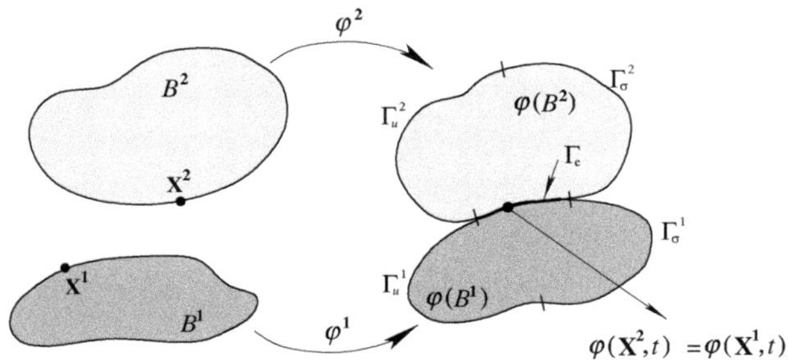

Fig. 8. Finite deformation contact problem representation.[21]

restrictions are imposed on the contact potential energy, as presented in the following section.

4.3. *Contact virtual work*

We can write the variation of the total potential energy for the contact problem (Eq. (28)) using the PVW[52] as

$$\delta\Pi(\mathbf{u}) = \delta\Pi_s(\mathbf{u}) + \delta\Pi_c(\mathbf{u}), \tag{38}$$

where $\delta\Pi_s$ is the system virtual work previously defined and $\delta\Pi_c$ the contact virtual work defined as

$$\delta\Pi_c(\mathbf{u}) = \int_{\Gamma_c} \mathbf{t}\cdot\delta\mathbf{g}d\Gamma_c, \tag{39}$$

where $\mathbf{t}$ is the contact traction vector field[11] and $\delta\mathbf{g}$ the variation of the gap vector that will be defined in the following sections.

4.3.1. *Lagrange multiplier method*

The classic frictionless contact potential energy contribution through the Lagrange multiplier method is defined as[21]

$$\Pi_c^{LM} = \int_{\Gamma_c} \lambda_N g_N d\Gamma, \tag{40}$$

where λ_N is the Lagrange multiplier that can be identified as the normal contact stress t_N.

The contact potential energy is only added to the total potential energy when the contact restriction is active. Therefore, the contact virtual work for 2D cases can be obtained by the first variation of the contact potential energy as

$$\delta \Pi_c^{LM} = \int_{\Gamma_c} \left(\lambda_N \delta g_N + \delta \lambda_N g_N \right) d\Gamma. \tag{41}$$

4.3.2. *Penalty method*

The contact potential energy through the penalty method is defined as[21]

$$\Pi_c = \frac{1}{2} \int_{\Gamma_c} t_N g_N d\Gamma. \tag{42}$$

Penalty parameters are added to the contact potential energy by considering normal contact stress proportional to the penetration function

$$t_N = \epsilon_N g_N, \tag{43}$$

where ϵ_N is the scalar penalty parameter in normal direction.

The contact potential energy is obtained by replacing Eq. (43) in (42)

$$\Pi_c^{PE} = \frac{1}{2} \int_{\Gamma_c} \epsilon_N g_N^2 d\Gamma. \tag{44}$$

Similarly to the Lagrange multiplier method, the contact virtual work can be obtained by the first variation of (44) as

$$\delta \Pi_c^{PE} = \int_{\Gamma_c} \epsilon_N g_N \delta g_N d\Gamma. \tag{45}$$

4.4. *Contact linearizations*

The linearization of the contact contributions is necessary for the solution of the nonlinear system of equations associated to the weak form of the proposed problem. The variation δg_N is defined as[21]

$$\delta g_N = \left[\boldsymbol{\eta}^2 - \boldsymbol{\eta}^1 \right] \cdot \mathbf{n}^1, \tag{46}$$

where $\boldsymbol{\eta}$ indicates the virtual displacements.

The linearizations of the contact virtual work for the Lagrange multiplier and penalty methods are obtained, respectively, as

$$\begin{aligned} \frac{\partial \delta \Pi_c^{LM}}{\partial \mathbf{u}} \Delta \mathbf{u} &= \frac{\partial \delta \Pi_c^{LM}}{\partial \lambda_N} \Delta \lambda_N + \frac{\partial \delta \Pi_c^{LM}}{\partial g_N} \Delta g_N + \frac{\partial \delta \Pi_c^{LM}}{\partial \delta g_N} \Delta \delta g_N \\ &= \int_{\Gamma_c} \left(\Delta \lambda_N \delta g_N + \delta \lambda_N \Delta g_N + \lambda_N \Delta \delta g_N \right) d\Gamma, \end{aligned} \tag{47}$$

$$\frac{\partial \delta \Pi_c^{PE}}{\partial \mathbf{u}}\Delta\mathbf{u} = \frac{\partial \delta \Pi_c^{PE}}{\partial g_N}\Delta g_N + \frac{\partial \delta \Pi_c^{PE}}{\partial \delta g_N}\Delta \delta g_N$$
$$= \int_{\Gamma_c} (\epsilon_N \delta g_N \Delta g_N + \epsilon_N g_N \Delta\delta g_N)\, d\Gamma. \quad (48)$$

The linearizations of the normal gap g_N and the variation δg_N are given, respectively, as[21]

$$\Delta g_N = \left[\Delta\mathbf{u}^2 - \Delta\mathbf{u}^1(\boldsymbol{\xi})\right] \cdot \mathbf{n}^1(\boldsymbol{\xi}), \quad (49)$$

$$\Delta\delta g_N = -\left(\boldsymbol{\eta}^1_{,\xi}\Delta\xi + \Delta\mathbf{u}^1_{,\xi}\delta\xi + \mathbf{x}^1_{,\xi\xi}\Delta\xi\delta\xi\right)\cdot\mathbf{n}^1$$
$$+\frac{g_N}{a_{11}}\left(\boldsymbol{\eta}^1_{,\xi} + \mathbf{x}^1_{,\xi\xi}\delta\xi\right)\cdot\left[\mathbf{n}^1\otimes\mathbf{n}^1\right]\left(\Delta\mathbf{u}^1_{,\xi} + \mathbf{x}^1_{,\xi\xi}\Delta\xi\right). \quad (50)$$

In the above equations, $\Delta\mathbf{u}^1$ is the increment in $\mathbf{x}^1 = \mathbf{X}^1 + \mathbf{u}^1$, and $\Delta\xi$ is the increment in $\boldsymbol{\xi}$. The increment $\Delta\xi$, the variation of the relative velocity $\delta\xi$ and the linearization of $\delta\xi$ are defined, respectively, as[21]

$$\Delta\xi = H^{11}\left[\left(\Delta\mathbf{u}^2 - \Delta\mathbf{u}^1(\boldsymbol{\xi})\right)\cdot\mathbf{a}^1 + g_N\mathbf{n}^1\cdot\Delta\mathbf{u}^1_{,\xi}\right], \quad (51)$$

$$\delta\xi = H^{11}\left[\left(\boldsymbol{\eta}^2 - \boldsymbol{\eta}^1\right)\cdot\mathbf{x}^1_{,\xi} + g_N\mathbf{n}^1\cdot\boldsymbol{\eta}^1_{,\xi}\right], \quad (52)$$

$$\Delta\delta\xi = \left[-\mathbf{a}^1\cdot\left(\delta\xi\Delta\mathbf{u}^1_{,\xi} + \boldsymbol{\eta}^1_{,\xi}\Delta\xi\right) - \left(\mathbf{a}^1\cdot\mathbf{a}^1_{,\xi} - g_N\mathbf{n}^1\cdot\mathbf{a}^1_{,\xi\xi}\right)\delta\xi\Delta\xi\right.$$
$$+g_N\left(\boldsymbol{\eta}^1_{,\xi\xi}\Delta\xi + \Delta\mathbf{u}^1_{,\xi\xi}\delta\xi\right)\cdot\mathbf{n}^1 - \left(\boldsymbol{\eta}^1_{,\xi} + \mathbf{a}^1_{,\xi}\delta\xi\right)\cdot\mathbf{a}^1\Delta\xi$$
$$-\left(\Delta\mathbf{u}^1_{,\xi} + \mathbf{a}^1_{,\xi}\Delta\xi\right)\cdot\mathbf{a}^1\delta\xi + \left(\boldsymbol{\eta}^2 - \boldsymbol{\eta}^1\right)\cdot\left(\Delta\mathbf{u}^1_{,\xi} + \mathbf{a}^1_{,\xi}\Delta\xi\right)$$
$$\left.+\left(\Delta\mathbf{u}^2 - \Delta\mathbf{u}^1\right)\cdot\left(\boldsymbol{\eta}^1_{,\xi} + \mathbf{a}^1_{,\xi}\delta\xi\right)\right]H^{11}, \quad (53)$$

where $H^{11} = \left[a_{11} - g_N b_{11}\right]^{-1}$.

5. Contact discretization

The *hp*-FEM and GFEM are used to discretize the domain terms and also the contact interfaces using a finite number of contact elements. These methods transform the integral equations resulting from the variational formulation on algebraic equations. The contact elements are built from entities of the two contact surfaces, such as nodes and edges. Each element has its own displacement and residual vectors and tangent matrix, which determines the transfer of the contact forces between the surfaces.

Basically, three types of contact elements are found in the literature: node-to-node (NTN), node-to-segment (NTS) and segment-to-segment (STS) contact elements (Fig. 9) can be applied only to small deformation problems. We have to apply NTS or STS contact elements to solve

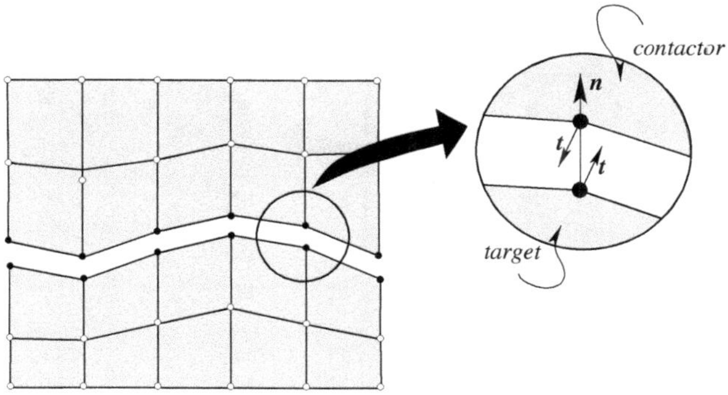

Fig. 9. NTN contact element representation.

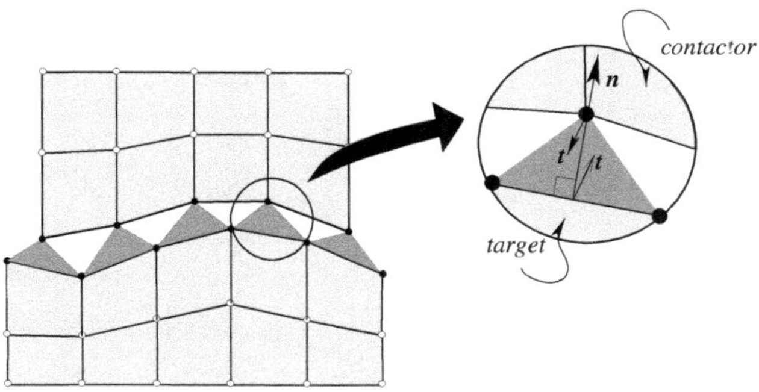

Fig. 10. NTS contact element representation.

problems involving large deformation and sliding effects. In NTS contact elements, the contact restrictions are applied on the nodes of the finite elements,[66] similar to the NTN, however the target is a segment (see Fig. 10). Although still widely used, this formulation has some disadvantages concerning the robustness in certain applications. For instance, in patch test problems with non-conforming meshes, the NTS contact element transmits incorrectly contact stresses through the contact interfaces.

In recent years, attention has been focused on the development of the STS contact elements[67] (Fig. 11). This formulation is currently coupled with the mortar domain decomposition method[68] and widely used to solve 2D and 3D structural contact problems.[13,17,26,31,69,70]

This section presents the high-order contact element proposed in Ref. [31] and a new GFEM contact element using enrichment interpolation functions. We also consider the assembly of the contact elements and the frictionless contact discretization.

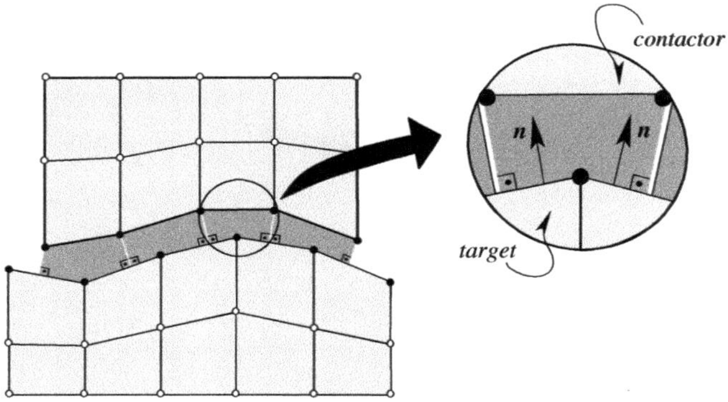

Fig. 11. STS contact element representation.

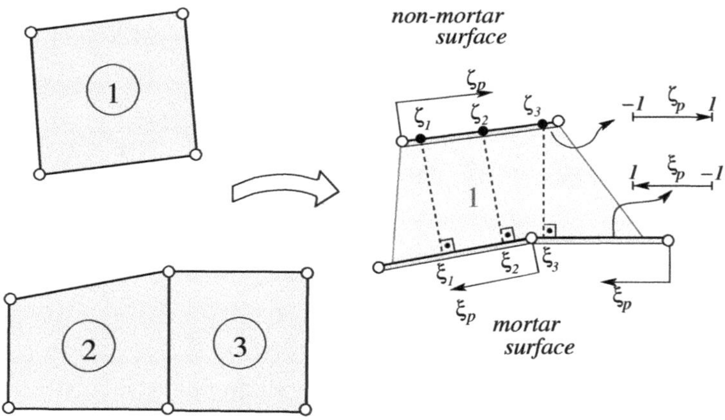

Fig. 12. Determination of the mortar-based contact element.

5.1. *Determination of the mortar-based contact element*

The contact surfaces can be discretized by non-conforming meshes with different number of segments such that

$$\Gamma_c^{nm} \approx \bigcup_{i=1}^{n_s} \Gamma_{c_i}^{nm} \text{ and } \Gamma_c^{m} \approx \bigcup_{j=1}^{n_m} \Gamma_{c_j}^{m}. \tag{54}$$

In the previous equations, n_s and n_m indicate the number of segments discretizing the non-mortar and mortar surfaces, respectively (see Fig. 12). A common contact interface is usually considered and defined here as $\Gamma_c^h = \Gamma_c^{nm}$. The non-mortar and mortar surfaces are associated to the edges of the elements in contact, as illustrated in Fig. 12. In this example, the finite element 1 is the contactor and elements 2 and 3 are the targets. The number of mortar-based contact elements is defined by the number of non-mortar segments $\Gamma_{c_s}^{nm}$. Therefore, in the considered example, we have one mortar based contact element constituted of one non-mortar segment $\Gamma_{c_s}^{nm}$ and two mortar segments $\Gamma_{c_m}^{m}$.

The current coordinates can be interpolated for each surface Γ_c^{nm} and Γ_c^{m} using (29), respectively, as

$$\mathbf{x}^{nm}(\zeta) = \sum_{i=1}^{N_{ns}} N_i^{nm}(\zeta)\left(\mathbf{X}_i^{nm} + \mathbf{u}_i^{nm}\right), \tag{55}$$

$$\mathbf{x}^{m}(\xi) = \sum_{i=1}^{N_{ns}} N_i^{m}(\xi)\left(\mathbf{X}_i^{m} + \mathbf{u}_i^{m}\right), \tag{56}$$

where N_{ns} is the number of nodes per segment. The nodes of the domain elements are used for the discretization of the contact element, and in Fig. 12, linear interpolation was used with $N_{ns} = 2$.

The shape functions N_i^{nm} and N_i^{m} are obtained from the Lagrange polynomials of any interpolation order. The local coordinates are $\zeta, \xi \in [-1, 1]$ for the non-mortar and mortar surfaces, respectively.

The current coordinates can be interpolated in GFEM similarly as

$$\mathbf{x}^{m}(\xi) = \sum_{i=1}^{N_{ns}} N_i(\xi)\mathbf{X}_i + \sum_{i=1}^{N_{enr}} N_i^{enr}(\xi)\mathbf{u}_i^{enr}, \tag{57}$$

where N_{ns} is the number of nodes in the target segment whose PU (L_2, L_3 or L_4) contains the point ξ; N_{enr} is the number of nodal-enriched shape functions containing that point expressed as

$$N_i^{enr}(\xi) = N_i(\xi)L(\mathbf{X}(\xi)). \tag{58}$$

The Lagrange multipliers are defined in the non-mortar surface as

$$\lambda_{N_s}(\zeta) = \sum_{i=1}^{Nns} M_i^{nm}(\zeta)\lambda_{N_i}, \tag{59}$$

where M_i^{nm} is the dual basis functions. Similarly for the GFEM,

$$\lambda_{N_s}(\zeta) = \sum_{i=1}^{Nns}\sum_{j=1}^{N_{enr}} M_i^{nm}(\zeta)\lambda_{N_j}L_j(\mathbf{X}(\zeta)). \tag{60}$$

The points ξ_p are projections of the non-mortar points ζ_p to the mortar surface obtained by solving the minimum distance problem given in (33). An algorithm to compute the projection points ξ_p is presented in.[31] The projections of the integration points ζ_p of a single non-mortar surface may be in different mortar surfaces, as can be seen in Fig. 12. Therefore, although the example has only one mortar-based contact element, it has projections in two distinct target surfaces.

The projections on different target surfaces also imply the discontinuity of the normal vector and, consequently, of the gap function. The strategy of averaged normal is used here to minimize this effect as presented in Ref. [11].

Given the above definitions, we can integrate numerically the virtual work expressions (41) or (45) in all contact segments n_s of the non-mortar surface Γ_c^{nm}, using the Gauss–Legendre quadrature.[6] The following sections present the discretization process for the frictionless contact element.

5.2. *Frictionless mortar-based contact element*

In the case of frictionless contact, we integrate the expressions for the virtual work (41) and (45), respectively, on the non-mortar surface of each element as

$$C_e^{LM} = \sum_{p=1}^{n_p} \left[\lambda_N(\zeta_p)\delta g_N(\zeta_p) + \delta\lambda_N(\zeta_p) g_N(\zeta_p)\right] ||\mathbf{a}_{0p}^{nm}||\, w_p, \tag{61}$$

$$C_e^{PE} = \sum_{p=1}^{n_p} \left[g_N(\zeta_p)\delta g_N(\zeta_p)\right] ||\mathbf{a}_{0p}^{nm}||\, w_p, \tag{62}$$

where ζ_p and w_p are the integration point coordinates and weights, respectively, n_p the number of integration points and $||\mathbf{a}_{0p}^{nm}||$ the Jacobian of the contact element.

We define next the terms to be evaluated on ζ_p and ξ_p in matrix format.

The normal gap and Lagrange multiplier for an integration point can be written as

$$g_{N_p} = [\mathbf{x}^{nm}(\zeta_p) - \mathbf{x}^m(\xi_p)] \cdot \mathbf{n}^m(\xi_p) = \mathbf{w}_p^T \mathbf{B}_p \mathbf{n}_{mp}, \tag{63}$$

$$\lambda_{N_p} = \mathbf{w}_p^T \mathbf{M}_p. \tag{64}$$

The Lagrange multipliers on the integration points N_p are equivalent to the contact stresses, i.e. $t_{N_p} = \lambda_{N_p}$. Matrices $\mathbf{B}_p$ and $\mathbf{M}_p$ and also vector

$\mathbf{w}_p$ are defined for any interpolation order for the Lagrange and penalty methods, respectively, as

$$\mathbf{B}_p = \begin{bmatrix} N_1^{nm}(\zeta_p)\mathbf{1} \\ \vdots \\ N_{Nns}^{nm}(\zeta_p)\mathbf{1} \\ -N_1^{m}(\xi_p)\mathbf{1} \\ \vdots \\ -N_{Nns}^{m}(\xi_p)\mathbf{1} \\ \mathbf{0}_{(nnod\times 2)} \end{bmatrix}, \tag{65}$$

$$\mathbf{M}_p = \begin{bmatrix} \mathbf{0}_{(4*Nns\times 1)} \\ M_1^{nm}(\zeta_p) \\ \vdots \\ M_{Nns}^{nm}(\zeta_p) \end{bmatrix}, \tag{66}$$

$$\mathbf{w}_p^T = (\mathbf{x}_1^{nm^T}, \ldots, \mathbf{x}_{Nns}^{nm^T}, \mathbf{x}_1^{m^T}, \ldots, \mathbf{x}_{Nns}^{m^T}, \lambda_{N_1}, \ldots, \lambda_{N_{Nns}}), \tag{67}$$

where $\mathbf{1}$ is the 2×2 *identity* matrix

$$\mathbf{1} = \begin{bmatrix} 1 & 0 \\ 0 & 1 \end{bmatrix}. \tag{68}$$

Vector $\mathbf{w}_p$ is composed of the contactor and the target current nodal coordinates and Lagrange multipliers. Normal vector $\mathbf{n}_{mp}$ is evaluated on ξ_p from the tangent vector $\mathbf{a}_{mp}$ given by

$$\mathbf{a}_{mp} = \mathbf{B}_{p,\xi}^T \mathbf{w}_p, \tag{69}$$

and therefore

$$\mathbf{n}_{mp} = \frac{\mathbf{e}_3 \times \mathbf{a}_{mp}}{||\mathbf{e}_3 \times \mathbf{a}_{mp}||}. \tag{70}$$

Matrix $\mathbf{B}_{p,\xi}$ is defined as

$$\mathbf{B}_{p,\xi} = \begin{bmatrix} \mathbf{0}_{(2Nns\times 2)} \\ N_{1,\xi}^{m}(\xi_p)\mathbf{1} \\ \vdots \\ N_{Nns,\xi}^{m}(\xi_p)\mathbf{1} \\ \mathbf{0}_{(Nns\times 2)} \end{bmatrix}. \tag{71}$$

To compute the tangent vector $\mathbf{a}_{0p}^{nm}$ on ζ_p and the curvature vector of the target surface $\mathbf{c}_{mp}$ on ξ_p, we define matrices $\mathbf{B}_{p,\zeta}^T$ and $\mathbf{B}_{p,\xi\xi}^T$ and the vector of the initial coordinates, respectively, as

$$\mathbf{B}_{p,\zeta} = \begin{bmatrix} N_{1,\zeta}^m(\zeta_p)\mathbf{1} \\ \vdots \\ N_{Nns,\zeta}^m(\zeta_p)\mathbf{1} \\ \mathbf{0}_{(2Nns\times 2)} \\ \mathbf{0}_{(Nns\times 2)} \end{bmatrix}, \tag{72}$$

$$\mathbf{B}_{p,\xi\xi} = \begin{bmatrix} \mathbf{0}_{(2Nns\times 2)} \\ N_{1,\xi\xi}^m(\xi_p)\mathbf{1} \\ \vdots \\ N_{Nns,\xi\xi}^m(\xi_p)\mathbf{1} \\ \mathbf{0}_{(Nns\times 2)} \end{bmatrix}, \tag{73}$$

$$\mathbf{W}_p^T = \left(\mathbf{X}_1^{nm^T}, \ldots, \mathbf{X}_{Nns}^{nm^T}, \mathbf{X}_1^{m^T}, \ldots, \mathbf{X}_{Nns}^{m^T}, 0, \ldots, 0_{N_{Nns}}\right). \tag{74}$$

Based on that, vectors $\mathbf{a}_{0p}^{nm}$ and $\mathbf{c}_{mp}$ are calculated as

$$\mathbf{a}_{0p}^{nm} = \mathbf{B}_{p,\zeta}^T \mathbf{W}_p, \tag{75}$$

$$\mathbf{c}_{mp} = \mathbf{B}_{p,\xi\xi}^T \mathbf{w}_p. \tag{76}$$

Equations (61) and (62), require the variation of the normal gap δg_{N_p} and Lagrange multipliers $\delta\lambda_{N_p}$

$$\delta g_{N_p} = \boldsymbol{\eta}_p^T \mathbf{B}_p \mathbf{n}_{mp}, \tag{77}$$

$$\delta\lambda_{N_p} = \boldsymbol{\eta}_p^T \mathbf{M}_p, \tag{78}$$

where $\boldsymbol{\eta}_p$ is the vector of virtual displacements

$$\boldsymbol{\eta}_p^T = \left(\boldsymbol{\eta}_1^{nm^T}, \ldots, \boldsymbol{\eta}_{Nns}^{nm^T}, \boldsymbol{\eta}_1^{m^T}, \ldots, \boldsymbol{\eta}_{Nns}^{m^T}, \delta\lambda_{N_1}, \ldots, \delta\lambda_{N_{Nns}}\right). \tag{79}$$

Vectors $\mathbf{w}_p^T$ and $\mathbf{W}_p^T$ are different for the penalty method and do not require the variables related to the Lagrange multipliers. Matrices $\mathbf{B}_p$, $\mathbf{B}_{p,\xi}$,

$\mathbf{B}_{p,\xi\xi}$ and $\mathbf{B}_{p,\zeta}$ are also rewritten as

$$\mathbf{w}_p^T = \left(\mathbf{x}_1^{nm^T}, \ldots, \mathbf{x}_{Nns}^{nm^T}, \mathbf{x}_1^{m^T}, \ldots, \mathbf{x}_{Nns}^{m^T}\right), \tag{80}$$

$$\mathbf{W}_p^T = \left(\mathbf{X}_1^{nm^T}, \ldots, \mathbf{X}_{Nns}^{nm^T}, \mathbf{X}_1^{m^T}, \ldots, \mathbf{X}_{Nns}^{m^T}\right), \tag{81}$$

$$\mathbf{B}_p = \begin{bmatrix} N_1^{nm}(\zeta_p)\mathbf{1} \\ \vdots \\ N_{Nns}^{nm}(\zeta_p)\mathbf{1} \\ -N_1^{m}(\xi_p)\mathbf{1} \\ \vdots \\ -N_{Nns}^{m}(\xi_p)\mathbf{1} \end{bmatrix}, \tag{82}$$

$$\mathbf{B}_{p,\xi} = \begin{bmatrix} \mathbf{0}_{(2Nns\times 2)} \\ N_{1,\xi}^{m}(\xi_p)\mathbf{1} \\ \vdots \\ N_{Nns,\xi}^{m}(\xi_p)\mathbf{1} \end{bmatrix}, \tag{83}$$

$$\mathbf{B}_{p,\xi\xi} = \begin{bmatrix} \mathbf{0}_{(2Nns\times 2)} \\ N_{1,\xi\xi}^{m}(\xi_p)\mathbf{1} \\ \vdots \\ N_{Nns,\xi\xi}^{m}(\xi_p)\mathbf{1} \end{bmatrix}, \tag{84}$$

$$\mathbf{B}_{p,\zeta} = \begin{bmatrix} N_{1,\zeta}^{m}(\zeta_p)\mathbf{1} \\ \vdots \\ N_{Nns,\zeta}^{m}(\zeta_p)\mathbf{1} \\ \mathbf{0}_{(2Nns\times 2)} \end{bmatrix}. \tag{85}$$

Similarity for the GFEM, matrices $\mathbf{B}_p$, $\mathbf{B}_{p,\xi}$, $\mathbf{B}_{p,\xi\xi}$ and $\mathbf{B}_{p,\zeta}$ use the enriched interpolation functions in place of the standard shape functions.

With these definitions, the contact virtual work expressions (61) and (62) can be written as

$$C_e^{LM} = \sum_{p=1}^{n_p} \left[\boldsymbol{\eta}_p^T \left(\mathbf{B}_p \mathbf{n}_{mp} \mathbf{M}_p^T + \mathbf{M}_p \mathbf{n}_{mp}^T \mathbf{B}_p^T \right) \mathbf{w}_p \right] ||\mathbf{a}_{0p}^{nm}|| \, w_p. \tag{86}$$

$$C_e^{PE} = \epsilon_N \sum_{p=1}^{n_p} \left[\boldsymbol{\eta}_p^T \left(\mathbf{B}_p \mathbf{n}_{mp} \mathbf{n}_{mp}^T \mathbf{B}_p^T \right) \mathbf{w}_p \right] ||\mathbf{a}_{0p}^{nm}|| \, w_p. \tag{87}$$

We can next linearize the contact virtual work in the direction of the increment displacement $\Delta\mathbf{u}$ from a current state of deformation φ_k using the Taylor's series expansion. By truncating in the first-order term and then replacing the discretized terms, we have

$$C_e \left(\varphi_k, \boldsymbol{\eta}_p \right) + DC_e \left(\varphi_k, \boldsymbol{\eta}_p \right) [\Delta\mathbf{u}] = \boldsymbol{\eta}_p^T \boldsymbol{\psi}_c + \boldsymbol{\eta}_p^T \mathbf{K}_c \Delta\mathbf{u} = \mathbf{0}, \tag{88}$$

where $\boldsymbol{\psi}_c$ is the contact residual force and $\mathbf{K}_c$ is the contact tangent matrix. Equation (88) must be satisfied for any kinematically compatible virtual displacement $\boldsymbol{\eta}_p$.

Given these definitions, we can express the contact residual force vector in element e for the Lagrange multiplier and penalty methods, respectively, as

$$\boldsymbol{\psi}_e^{LM} = \sum_{p=1}^{n_p} \left(\mathbf{B}_p \mathbf{n}_{mp} \mathbf{M}_p^T + \mathbf{M}_p \mathbf{n}_{mp}^T \mathbf{B}_p^T \right) \mathbf{w}_p \, ||\mathbf{a}_{0p}^{nm}|| \, w_p, \tag{89}$$

$$\boldsymbol{\psi}_{Ne}^{PE} = \sum_{p=1}^{n_p} \left(\mathbf{B}_p \mathbf{n}_{mp} \mathbf{n}_{mp}^T \mathbf{B}_p^T \right) \mathbf{w}_p \, ||\mathbf{a}_{0p}^{nm}|| \, w_p. \tag{90}$$

We also need the discrete linearization of the gap function variation $\Delta(\delta g_{N_p})$ to obtain the contact tangent matrix. From Eq. (50), we have

$$\begin{aligned} \Delta(\delta g_{N_p}) = \boldsymbol{\eta}_p^T \Bigg[& -\mathbf{B}_{p,\xi} \mathbf{n}_{mp} \mathbf{a}_{mp}^T \mathbf{B}_p^T - \mathbf{B}_p \mathbf{a}_{mp} \mathbf{n}_{mp}^T \mathbf{B}_{p,\xi}^T \\ & - g_{N_p} \mathbf{B}_{p,\xi} \mathbf{n}_{mp} \mathbf{n}_{mp}^T \mathbf{B}_{p,\xi}^T \frac{\mathbf{c}_{mp}^T \mathbf{n}_{mp}}{||\mathbf{a}_{mp}||^2} \mathbf{B}_p \mathbf{a}_{mp} \mathbf{a}_{mp}^T \mathbf{B}_p^T \Bigg] \\ & \times \frac{\Delta\mathbf{w}_p}{||\mathbf{a}_{mp}||^2 - g_{N_p} \mathbf{c}_{mp}^T \mathbf{n}_{mp}}. \end{aligned} \tag{91}$$

Based on that, the contact tangent matrix of one contact element for the Lagrange multiplier method is

$$\mathbf{K}_{Ne}^{LM} = \sum_{p=1}^{n_p} \left[\mathbf{K}_{N_p}^{LM1} - \left[\mathbf{K}_{N_p}^{LM2} + \mathbf{K}_{N_p}^{LM3} \right] \frac{\lambda_{N_p}}{||\mathbf{a}_{mp}||^2 - g_{N_p}\mathbf{c}_{mp}^T\mathbf{n}_{mp}} \right] ||\mathbf{a}_{0p}^{nm}||\, w_p, \tag{92}$$

where matrices $\mathbf{K}_{N_p}^{LM1}$, $\mathbf{K}_{N_p}^{LM2}$ and $\mathbf{K}_{N_p}^{LM3}$ are given, respectively, as

$$\mathbf{K}_{N_p}^{LM1} = \mathbf{B}_p\mathbf{n}_{mp}\mathbf{M}_p^T + \mathbf{M}_p\mathbf{n}_{mp}^T\mathbf{B}_p^T, \tag{93}$$

$$\mathbf{K}_{N_p}^{LM2} = \mathbf{B}_{p,\xi}\mathbf{n}_{mp}\mathbf{a}_{mp}^T\mathbf{B}_p^T + \mathbf{B}_p\mathbf{a}_{mp}\mathbf{n}_{mp}^T\mathbf{B}_{p,\xi}^T, \tag{94}$$

$$\mathbf{K}_{N_p}^{LM3} = g_{N_p}\mathbf{B}_{p,\xi}\mathbf{n}_{mp}\mathbf{n}_{mp}^T\mathbf{B}_{p,\xi}^T + \frac{\mathbf{c}_{mp}^T\mathbf{n}_{mp}}{||\mathbf{a}_{mp}||^2}\mathbf{B}_p\mathbf{a}_{mp}\mathbf{a}_{mp}^T\mathbf{B}_p^T. \tag{95}$$

Similarly for the penalty method, we obtain

$$\mathbf{K}_{Ne}^{PE} = \sum_{p=1}^{n_p} \left[\mathbf{K}_{N_p}^{PE1} - \left(\mathbf{K}_{N_p}^{PE2} + \mathbf{K}_{N_p}^{PE3} \right) \frac{g_{N_p}}{||\mathbf{a}_{mp}||^2 - g_{N_p}\mathbf{c}_{mp}^T\mathbf{n}_{mp}} \right] ||\mathbf{a}_{0p}^{nm}||\, w_p, \tag{96}$$

where matrices $\mathbf{K}_{N_p}^{PE1}$, $\mathbf{K}_{N_p}^{PE2}$ and $\mathbf{K}_{N_p}^{PE3}$ are given, respectively, as

$$\mathbf{K}_{N_p}^{PE1} = \mathbf{B}_p\mathbf{n}_{mp}\mathbf{n}_{mp}^T\mathbf{B}_p^T, \tag{97}$$

$$\mathbf{K}_{N_p}^{PE2} = \mathbf{B}_{p,\xi}\mathbf{n}_{mp}\mathbf{a}_{mp}^T\mathbf{B}_p^T + \mathbf{B}_p\mathbf{a}_{mp}\mathbf{n}_{mp}^T\mathbf{B}_{p,\xi}^T, \tag{98}$$

$$\mathbf{K}_{N_p}^{PE3} = g_{N_p}\mathbf{B}_{p,\xi}\mathbf{n}_{mp}\mathbf{n}_{mp}^T\mathbf{B}_{p,\xi}^T + \frac{\mathbf{c}_{mp}^T\mathbf{n}_{mp}}{||\mathbf{a}_{mp}||^2}\mathbf{B}_p\mathbf{a}_{mp}\mathbf{a}_{mp}^T\mathbf{B}_p^T. \tag{99}$$

The term $\mathbf{c}_{mp}$ is a null vector, $\mathbf{c}_{mp}^T = [0\ 0]^T$, in the case of interpolation order $P = 1$, because it carries the mortar segment curvature information.

The contact residual vector and the tangent matrix were obtained for one contact element. The respective global terms are constructed using the assembly procedure

$$\boldsymbol{\psi}_N^C = \sum_{e=1}^{N_{ce}} \boldsymbol{\psi}_{Ne}^{PE}, \tag{100}$$

$$\mathbf{K}_N^C = \sum_{e=1}^{N_{ce}} \mathbf{K}_{Ne}^{PE}, \tag{101}$$

where N_{ce} is the number of active contact elements. We calculate the contact elements on the integration points, generating residue and stiffness on the element nodes. As the integration points can be projected on different targets (see Fig. 12), the contact elements may generate residue and stiffness contributions in different dofs. This should be taken into account in the construction of the contact element connectivity.

5.3. *Contact element activation strategies*

The activation strategies of contact elements were originally developed for the Lagrange multiplier method and later applied to the penalty method.[71]

The basic idea of the mortar-based contact elements with Lagrange multiplier is the satisfaction of the contact restrictions on a given surface instead of in some points as in the classic contact formulation.

5.3.1. *Activation strategy CAS1*

The first contact activation strategy (*CAS1*) hereby adopted was discussed in Ref. [72].

We evaluate the signal of the area on the gap function graph, $g^A_{N_e}$, for one element as

$$g^A_{N_e} = \int_{\Gamma_e^{nm}} g_N d\Gamma_e = \sum_{p=1}^{n_p} \left[\mathbf{w}_p^T \mathbf{B}_p \mathbf{n}_{mp}\right] ||\mathbf{a}_{0p}^{nm}||\, w_p. \tag{102}$$

If $g^A_{N_e} \geq 0$, there is no normal contact stress, $t_N = 0$. The KKT conditions are fulfilled and the element is deactivated. If there is contact element penetration, $g^A_{N_e} < 0$ and the stress condition $t_N \leq 0$ are checked. In the present work, the verification for the stress condition is done in an average way as

$$t^A_{N_e} = \int_{\Gamma_e^{nm}} t_N d\Gamma_e. \tag{103}$$

To activate the element, Eqs. (102) and (103) must satisfy the following conditions:

$$g^A_{N_e} < 0, \tag{104}$$

$$t^A_{N_e} \leq 0. \tag{105}$$

Therefore, for elements in partial contact condition, the activation or deactivation can be evaluated using Eqs. (102) and (103).

5.3.2. *Activation strategy CAS2*

The second contact activation strategy (*CAS2*) is very simple and used only for the penalty method. The gap function g_{N_p} is evaluated at each integration point. If $g_{N_p} \geq 0$, there is no contact stress, $t_{N_p} = 0$, the KKT conditions are satisfied for this integration point and we turn off this contact integration point. On the other hand, if $g_{N_p} < 0$, the contact stress is zero and the contact integration point is activated.

6. Results

In this section, we present the results obtained using the *hp*-FEM and GFEM mortar-based contact elements.

6.1. *hp-FEM contact element*

Firstly, we verify the behaviour of the *hp*-FEM contact element in small deformation with the analytical solution of the Hertz problem. Then, we present studies of accuracy and processing time of the contact elements for h- and p-refinements for the Hertz problem. We finish presenting a study of the accuracy and processing time for the h- and p-refinements for a large deformation problem. The *hp*-FEM contact element passes the patch test proposed in Ref. [73] as can be seen in Ref. [31].

6.1.1. *Hertz contact problem*

As depicted in Fig. 13, the Hertz problem consists of a frictionless contact problem between two cylindrical bodies made of isotropic, linear elastic

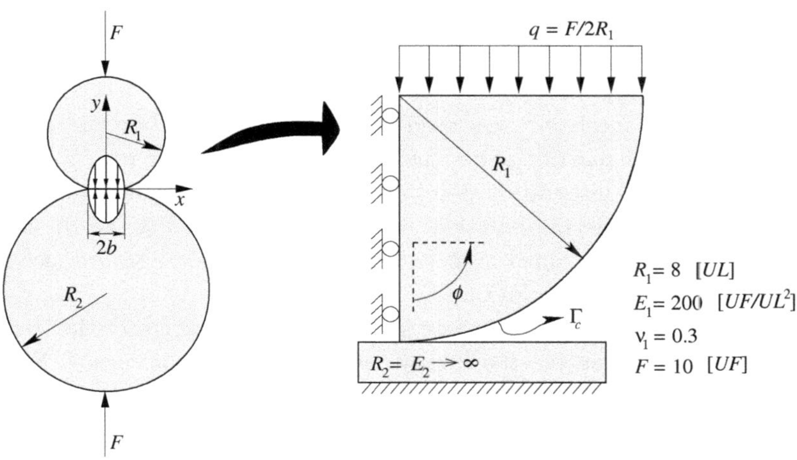

Fig. 13. 2D Hertz contact problem.

material under infinitesimal deformation for plane strain state. The problem is widely used in the contact literature to verify the behaviour of different formulations.

For the sake of briefness, a simplified version of the problem is hereby adopted, consisting of the contact between a cylinder and the rigid soil represented in Fig. 13. According to Shigley *et al.*,[74] the associated analytical solution for both the normal contact stress t_N^{Hertz} and the length b can be obtained from the analytical Hertz contact stress between two cylinders, by adopting very large values to the radius R_2 and the Young's modulus E_2 of one of the cylinders. Therefore, t_N^{Hertz} and b are given respectively as[74]:

$$t_N^{\text{Hertz}}(x) = \frac{4R_1 q}{\pi b^2}\sqrt{b^2 - x^2}, \tag{106}$$

$$b = 2\sqrt{\frac{2R_1^2 q\,(1-\nu_1^2)}{E_1 \pi}}. \tag{107}$$

In Fig. 13, the set of assumed data is shown. The resulting distributed load is $q = 0.625[UF/UL]$ and the contact region $b = 0.68077[UL]$.

The soil was discretized trough a single finite element with constrained edges. The initial contact condition is $g_N = 0$ at the symmetry point of the cylinder. P being the interpolation order, the Gauss–Legendre quadrature was used with $n_p = P + 1$ integration points per contact element. The contact surface Γ_c is mapped along the quarter of circumference $\phi \in [-\frac{\pi}{2}, 0]$. We used the activation strategy CAS1 for the Lagrange multiplier method and CAS2 for the penalty method. The sequence of meshes m^1 to m^5 used is illustrated in Fig. 14.

Firstly, the mesh m^4 was used in conjunction with an interpolation order $P = 1$ and penalty parameter $\epsilon_N = 1.0 \times 10^5$. The results for contact stress at the integration points provided by the penalty method and compared to the Hertz analytical stress are shown in Fig. 15. It can be observed that the obtained numerical results are in good accordance with the classical analytical solution.

Next, again using the penalty method, we performed the h- and p-refinements to analyse the improvement in solution accuracy. For the p-refinement, we fixed the mesh m^1 and the interpolation order was varied from $P = 1$ to $P = 9$ ($p1$ to $p9$ labels). We have only one element in the contact region. For the h-refinement, we fixed the interpolation order $P = 1$ in the meshes from m^1 to m^5 ($h1$ to $h5$ labels). The percentage

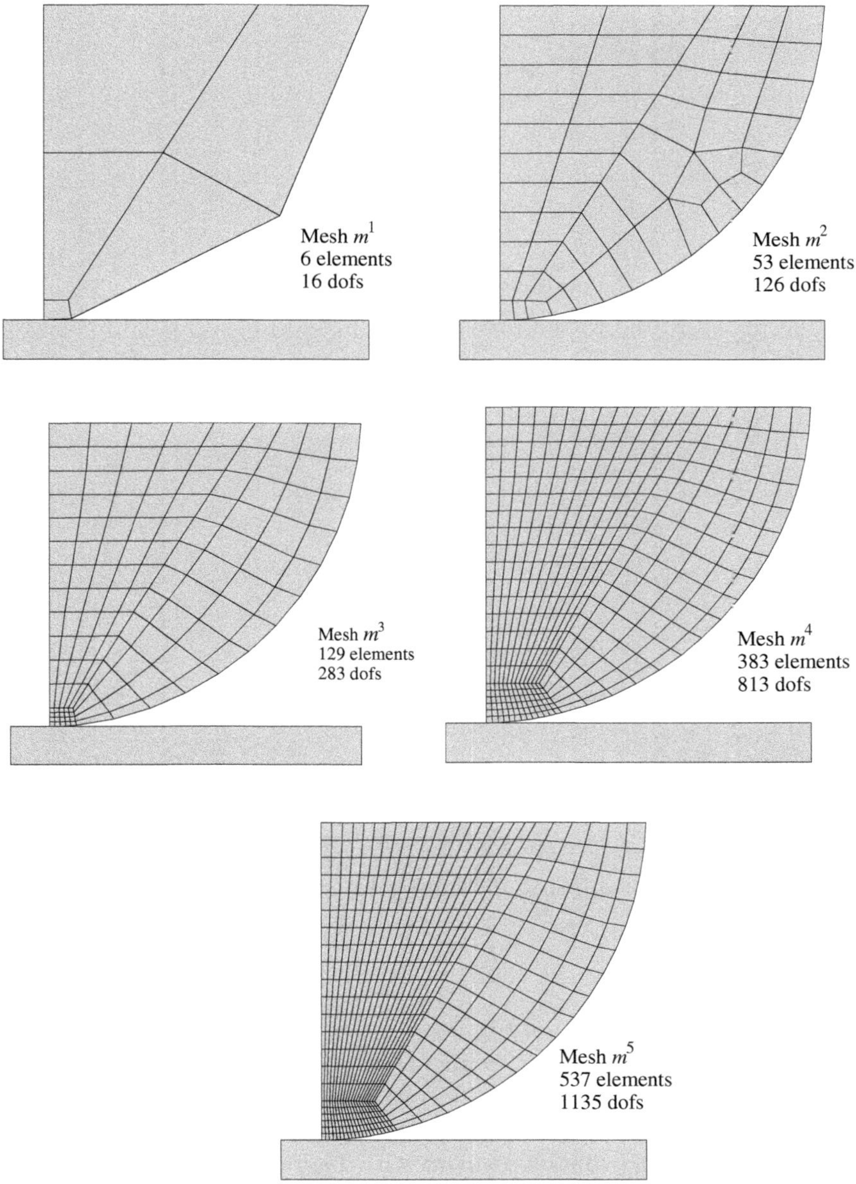

Fig. 14. Meshes for the Hertz problem.

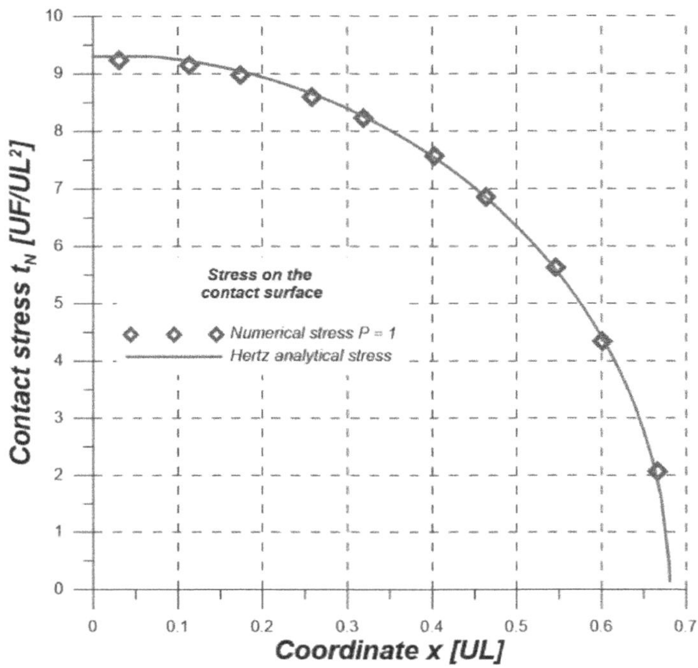

Fig. 15. Contact stress computed with mesh m^4 and interpolation order $P = 1$.

relative error η_σ for the contact stress using the L^2-norm is

$$\eta_\sigma = \frac{\|e_\sigma\|_{L^2}}{\sqrt{\int_{\Gamma_c} \left(t_N^{\text{Hertz}}\right)^2 d\Gamma_c}} \times 100\% = \frac{\sqrt{\int_{\Gamma_c} \left(t_N^{\text{Hertz}} - t_N\right)^2 d\Gamma_c}}{\sqrt{\int_{\Gamma_c} \left(t_N^{\text{Hertz}}\right)^2 d\Gamma_c}} \times 100\%. \tag{108}$$

The results are shown in Fig. 16. For the contact element with penalty method, the error reached a minimum value of 0.88% for the $p7$ refinement. The error norms did not decrease for the $p8$ and $p9$ refinements, and this may be due to the mesh resolution. A similar behaviour can be observed for the contact element with Lagrange multiplier above $p3$, however with a slight increase in the error from the $p5$ refinement due to numerical precision. Actually, the fact that the error does not decrease with the p-refinement and oscillates is also observed in the numerical solution of other problems.[6]

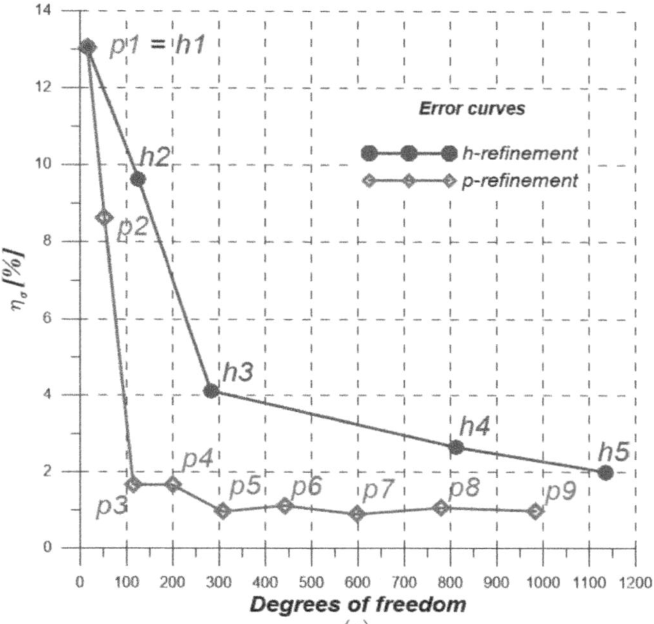

(a)

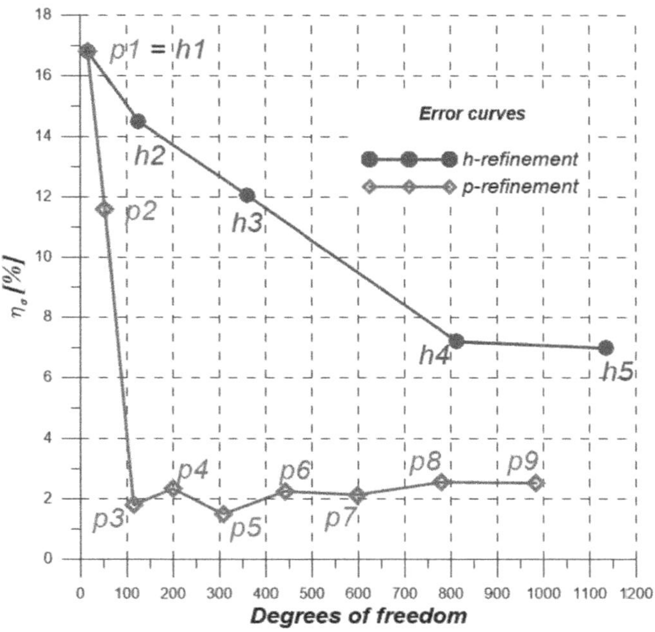

(b)

Fig. 16. Approximate error norms for the h- and p-refinements for penalty (a) and Lagrange multiplier (b) methods.

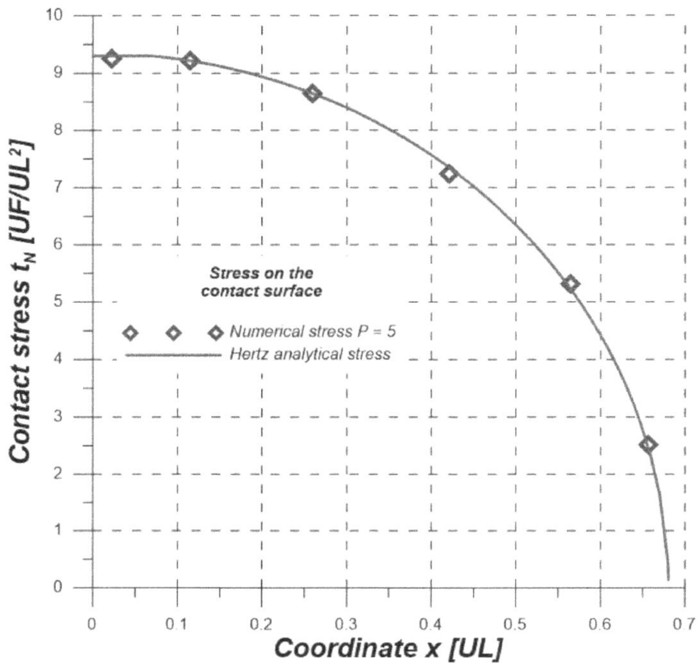

Fig. 17. Contact stress computed with mesh m^1 and interpolation order $P = 5$.

When comparing the h- and p-refinements, it is possible to obtain a more accurate contact stress using high-order contact elements with fewer dofs. In fact, for the $p5$ refinement, we obtained an error of 0.96% with 308 dofs, while for $h5$ refinement the error is 2.00% with 1135 dofs. The results show that the p-refinement provided the best solutions, with coarser meshes and fewer active contact elements. Figure 17 shows the contact stress for the $p5$ refinement.

We now present the results in terms of processing time performance calculating the time percentage as

$$\text{time percentage} = \frac{\text{processing time}}{\text{maximum processing time}} \times 100\%, \qquad (109)$$

where processing time is the overall runtime of the program considering the sum of element evaluation, contact evaluation and solver time of each refinement. The maximum processing time is the most expensive solution time, which was the $h5$ refinement for the penalty method.

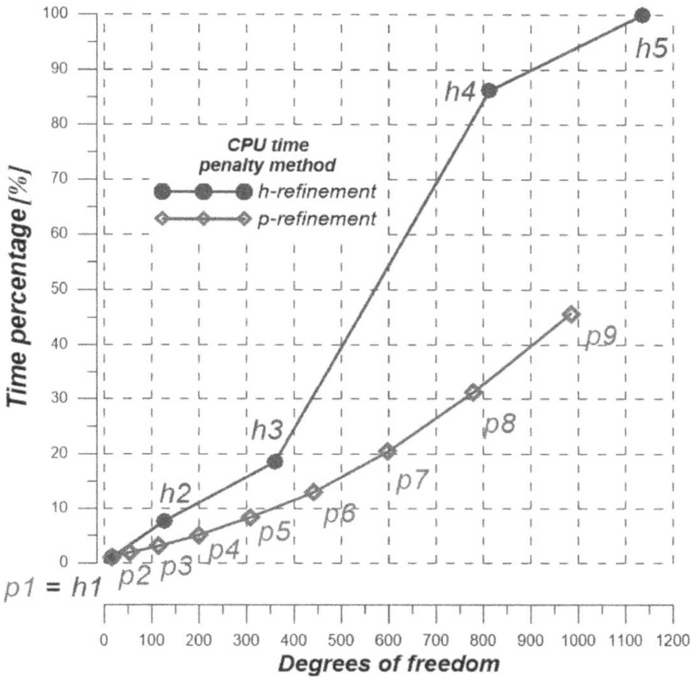

Fig. 18. Time percentage in terms of the number of dofs.

Figure 18 illustrates the results of the time percentage in terms of the number of dofs, obtained using Matlab direct solver. We observe that the best performance was obtained with the p-refinement. Comparing the cases $p9$ and $h5$, we can see that the $p9$ refinement took only 45.75% of the maximum time ($h5$ refinement) with better stresses and less dofs. A more detailed analysis for this problem is given in Ref. [31].

6.1.2. *Hyperelastic blocks*

In this section we perform the h- and p-refinements to analyse the improvement in solution accuracy for a large deformation contact problem between two hyperelastic blocks. The geometric and material parameters of the blocks are shown in Fig. 19 and the used meshes in Fig. 20. It were considered neo-Hookean compressible hyperelastic material for both blocks and plane strain state with updated Lagrangian formulation. Penalty parameter $\epsilon_N = 5.0 \times 10^4$ and $n_p = P + 1$ integration points per contact element were used. A total vertical displacement $w = -0.5\,[UL]$ was applied on the top

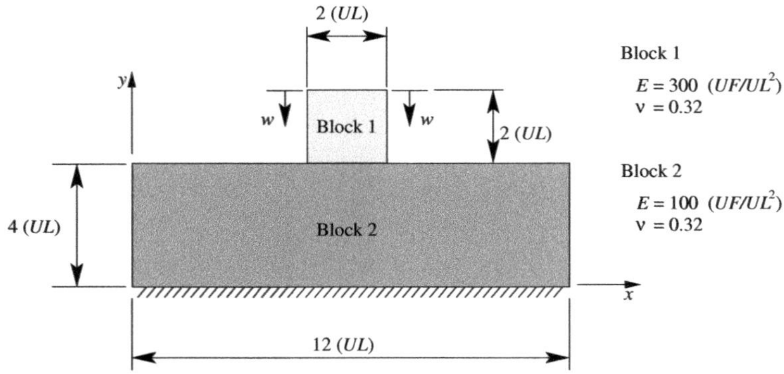

Fig. 19. Geometric and material parameters for the hyperelastic blocks problem.

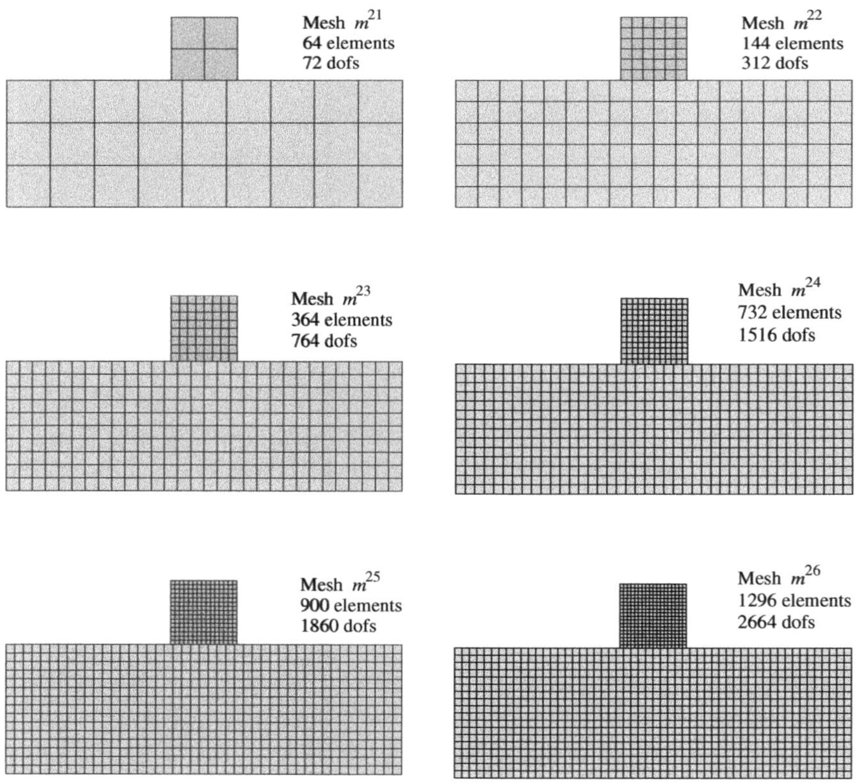

Fig. 20. Meshes for the hyperelastic blocks problem.

of block 1 with 5 steps. The bottom edge of block 1 was assumed as the non-mortar surface and the top edge of block 2 as the mortar surface. We used the activation strategy CAS1 for the Lagrange multiplier method and CAS2 for the penalty method.

Figure 21 shows the vertical displacement field u_y for the deformed domain at different steps of the solution for interpolation order $P = 1$ and mesh m^{24}.

For the p-refinement, we used mesh m^{21} and vary the interpolation order from $P = 1$ to $P = 5$ ($p1$ to $p5$ labels). For the h-refinement, we fixed the interpolation order $P = 1$ and change the meshes from m^{21} to m^{26} ($h1$ to $h6$ labels). The results are shown in Fig. 22 in terms of the total reaction force on top of block 1.

We observed that the force converged to the value 45.01 $[UF]$. In order to compute the reaction force error, we took this value as the converged or reference force value F_C and computed the percentage error η_F as

$$\eta_F = \frac{F_N - F_C}{F_C} \times 100\%. \tag{110}$$

The results in terms of the number of dofs can be seen in Fig. 23. The processing time, also in terms of the number of dofs, is shown in Fig. 24. The processing time was computed using Eq. (109), where the maximum processing time was for the $h6$ refinement with the penalty method.

We obtained the reaction force $F_{p3} = 45.4149[UF]$ for the $p3$ refinement with 588 dofs for the penalty method. The closest reaction force was obtained with the $h6$ refinement, $F_{h5} = 45.3917$, but with 1850 dofs and more processing time when compared to the $p3$ refinement.

If we compare the results of the $h4$ and $p5$ refinements for the penalty methods, we observe a significant gain in accuracy and cost with the $p5$ refinement. The $p5$ presented an error of $\eta_{F_{p5}} = 0.0367\%$ with 1600 dofs and time percentage $= 21.18\%$; $h4$ achieved an error of $\eta_{F_{h4}} = 1.5134\%$ with 1516 dofs and time percentage $= 56.07\%$.

We can now proceed with the same comparison for the Lagrange multiplier method. The $p5$ refinement presented an error of $\eta_{F_{p5}} = 0.1147\%$ with 1611 dofs and time percentage $= 18.11\%$; $h4$ presented an error of $\eta_{F_{h4}} = 1.4474\%$ with 1529 dofs and time percentage $= 49.54\%$. We can compare the time results in Fig. 24. All results also show the better performance of the p-refinement in terms of accuracy and time for large deformation problems.

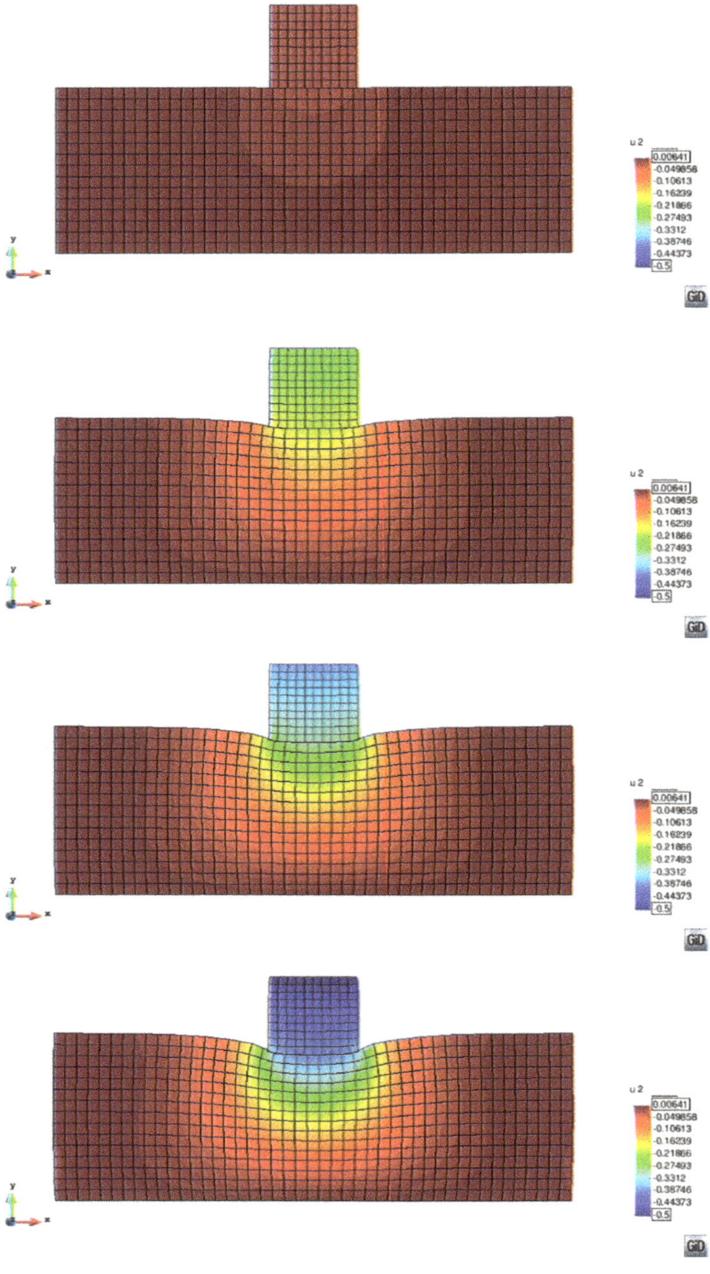

Fig. 21. Displacement field u_y for the deformed structure in load steps 1, 3, 4 and 5, respectively.

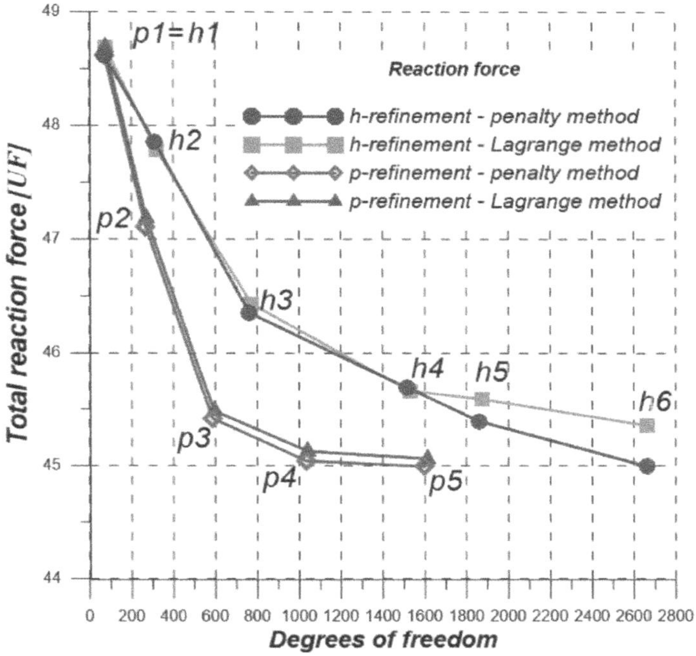

Fig. 22. Total reaction force convergence for h- and p-refinements.

6.2. *Generalized finite element method contact element results*

In what follows, two numerical examples are presented. In all of them, only enrichment schemes with shifted polynomial functions were investigated. The first example is related to a contact patch-test aiming essentially to evaluate accuracy of displacement and stress results that can be provided by the GFEM mortar-based contact element. The second example considers the contact between two beams.

6.2.1. *Contact patch test problem*

The example depicted in Fig. 25 consists of two rectangular solids composed of the same linear elastic material, one positioned on the top of the other in order that contact occurs between them. The upper solid is subjected to a uniform distributed pressure in the vertical direction. The adopted meshes do not conform at the contact interface. The main aim is to evaluate the

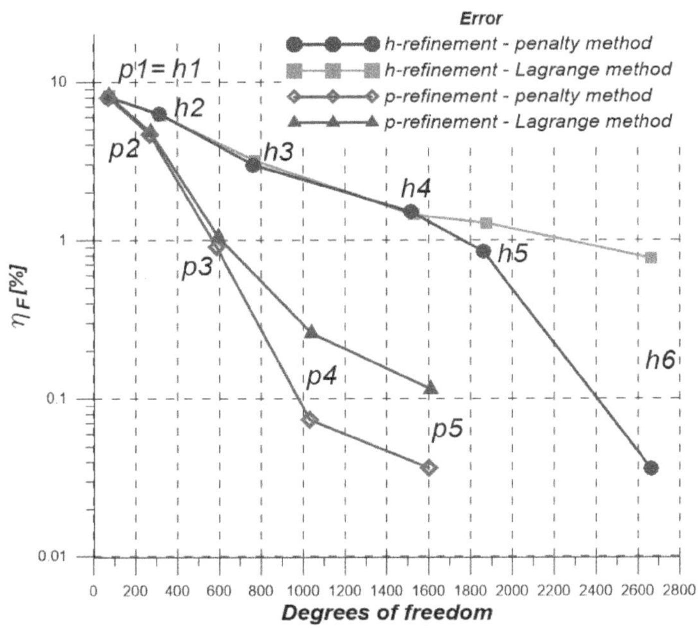

Fig. 23. Error curves for the h- and p-refinements.

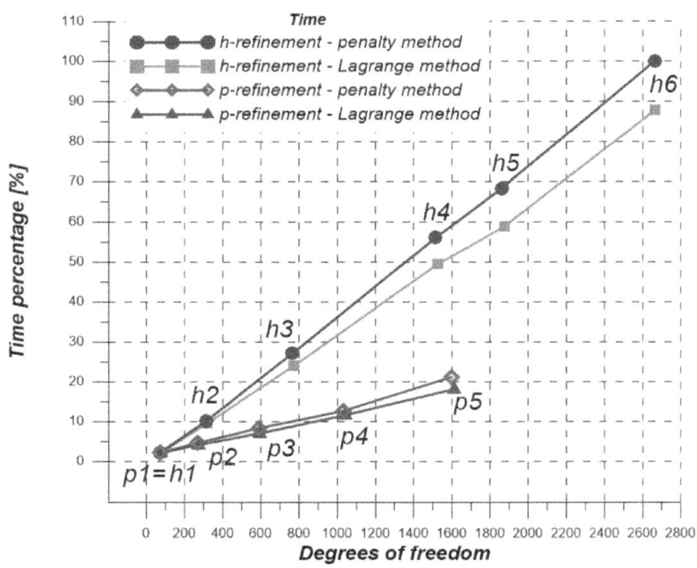

Fig. 24. Processing time in terms of the number of dofs.

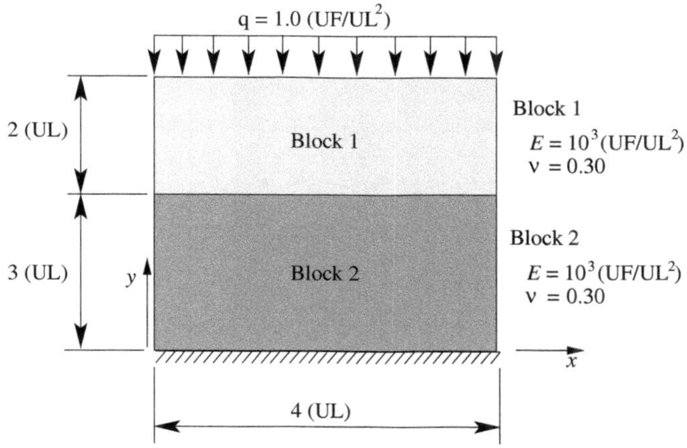

Fig. 25. Contact patch test using GFEM contact elements.

behaviour of the GFEM contact element. In this case, linear enrichment was adopted, as well a penalty parameter of $\epsilon_N = 1.0 \times 10^6$.

The patch test problem is modelled using meshes of bi-quadratic (Q8) finite elements. The vertical displacement and stress fields are presented in Fig. 26. As shown in Fig. 26, the vertical stress field, obtained by the enriched mortar contact elements, is sufficiently accurate and close to the expected value.

6.2.2. *Beams contact problem*

A large displacement contact problem of two cantilever beams, one positioned over the other covering a common span of 5 units and presenting an initial gap of 0.1, as depicted in Fig. 27, was considered in the present example. A large vertical displacement is imposed at the clamped edge of one of the beams, such that it pushes the other beam, resulting in a large sliding contact between the beams surfaces. We consider the Saint Venant–Kirchhoff constitutive model for the material, and self weight is neglected.

The reference solution for the proposed problem is a FEM model with a regular mesh of 160 (4×40) bi-quadratic (Q8) elements for each of the beams, resulting in 2274 dofs for the whole structural set. Adding the Lagrange multipliers, the total number of dofs is 2329.

The proposed mesh for the GFEM model is composed by 40 (2×20) bi-linear (Q4) elements for each of the beams, resulting in 252 dofs for the not-enriched model. Adding the Lagrange multipliers associated to the

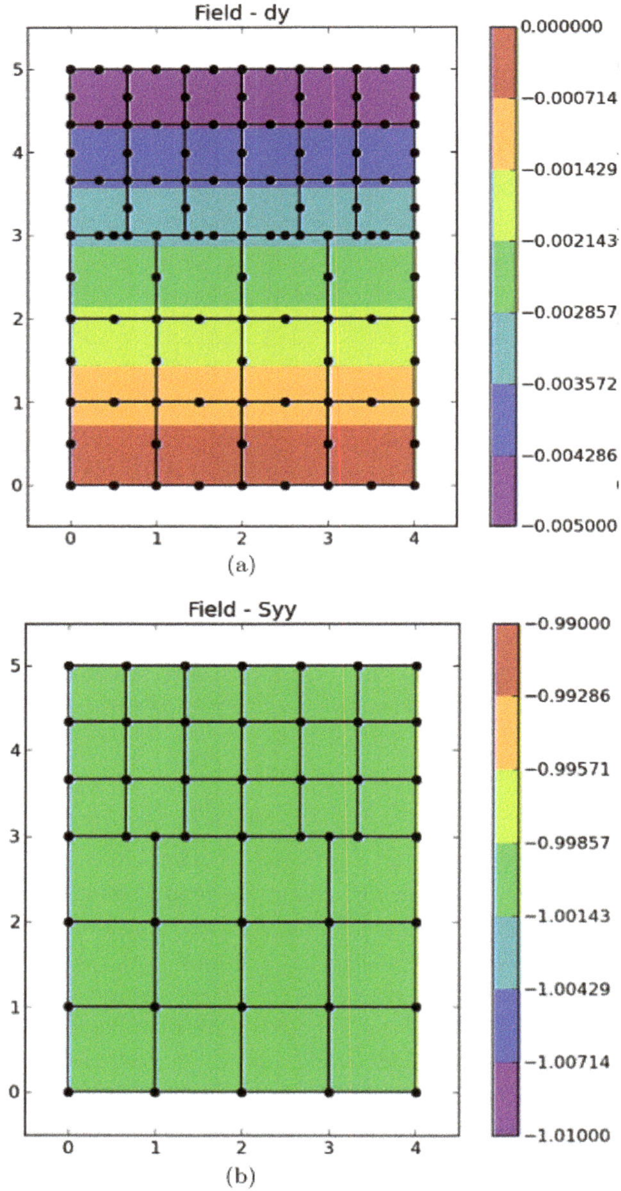

Fig. 26. Vertical displacement (a) and stress (b) components for the contact patch test using GFEM contact elements.

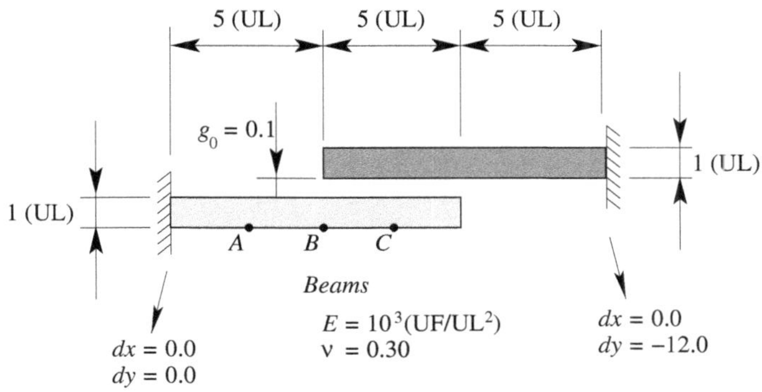

Fig. 27. Two beams contact problem.

mortar contact elements, the total number of dofs is 263. The GFEM model is enriched using shifted functions. First-degree enrichment is applied over the structural set nodes, except the ones in which Dirichlet boundary conditions are applied, resulting in 743 dofs.

The displacement is imposed downwards up to 12 [LU] at the right clamped edge. The nonlinear solution is computed in 50 time step increments. The adopted convergence criterion is the L_2 norm of the post-processed displacement increment on the integration points.

The final equilibrium configuration for the reference solution model is presented in Fig. 28. In particular, von Mises and horizontal stress component are shown. The same stress fields for the GFEM model are presented in Figs. 29 and 30, both for the not-enriched and the first-degree enrichment model. As it can be observed, the fields for the enriched model are smoother and more continuous across the elements.

The evolution of the von Mises stress (Svm) during the solution process, until the final time (1.0), for the points A, B and C indicated in Fig. 27 are depicted in Figs. 31, 32 and 33.

In order to advance in the results analysis, the gaps and real tractions at the contact interface are verified in a time step for which most of the beam surfaces remain in contact. Taking the contact gap results at time step 10 (vertical imposed displacement equal to 2.4 [LU]), we notice that all the models have similar results (Fig. 35).

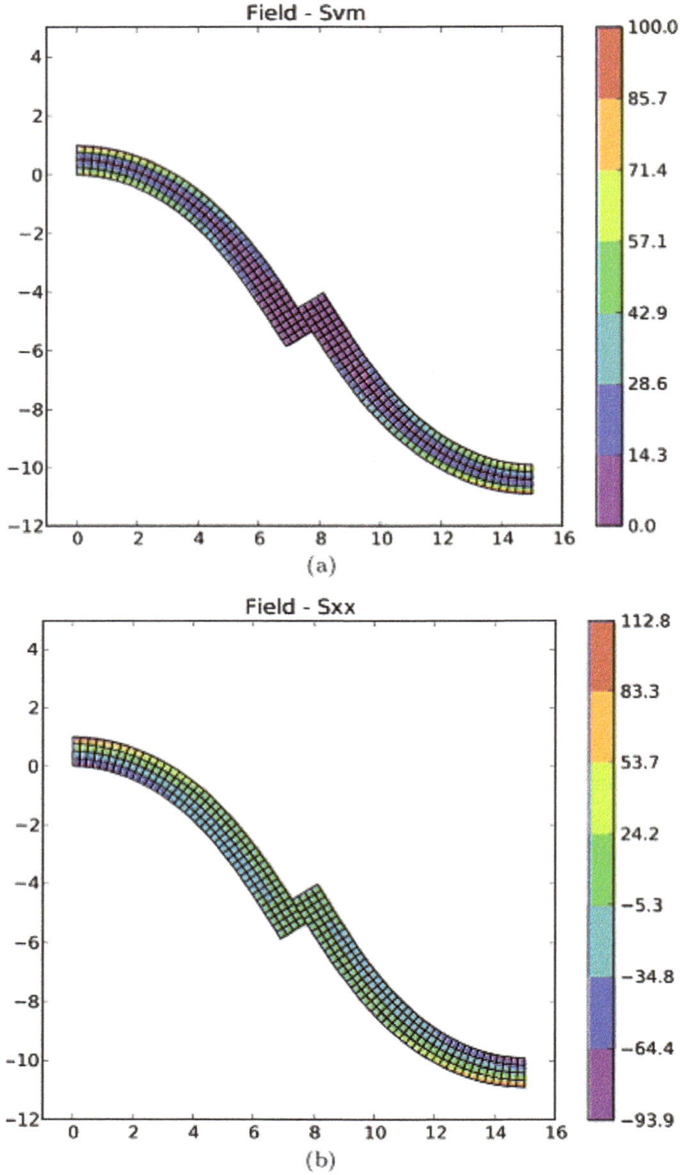

Fig. 28. Final equilibrium configuration achieved with the FEM reference solution model; von Mises stress (a) and horizontal stress component (b) fields.

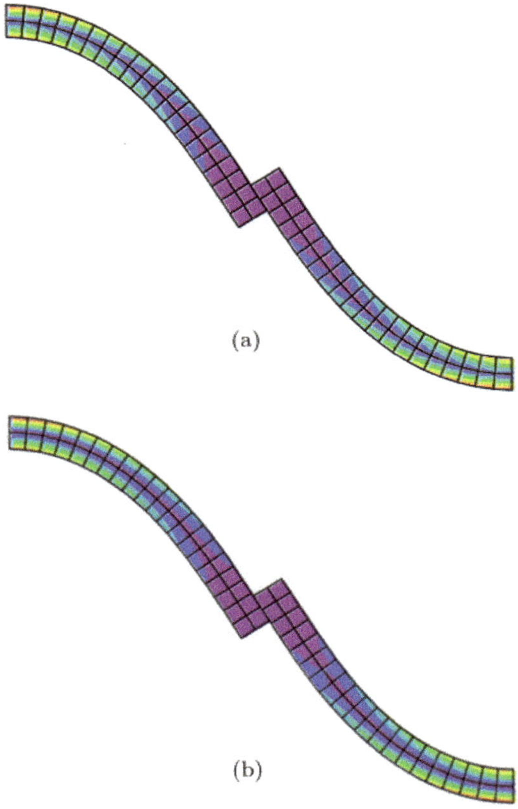

Fig. 29. von Mises stress (Svm) field for the not-enriched (a) and the first-degree enrichment GFEM model (b).

Even though the gaps results are similar (Fig. 34), the real tractions evaluated at the contact interface differ considerably for the not-enriched and the first-degree enrichment GFEM model, as depicted in Fig. 35.

By analysing those results, we observe that the enrichment benefits goes beyond the field smoothing presented in Figs. 29 and 30, since considerable traction error is observed for the not-enriched GFEM model, while the enriched model is much closer to the reference solution. In fact, even if one generates a refined mesh of bi-linear elements, the traction results remain in correct. For instance, for a bi-linear element mesh, similar to the one

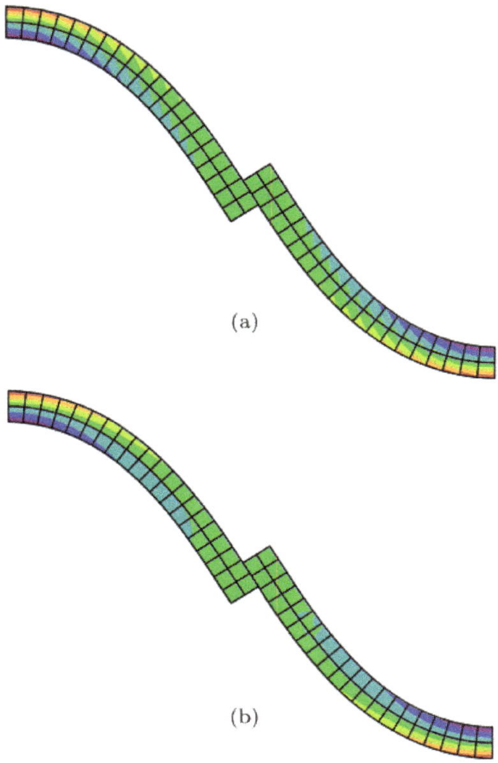

Fig. 30. Horizontal stress component (Sxx) field for the not-enriched (a) and the first-degree enrichment GFEM model (b).

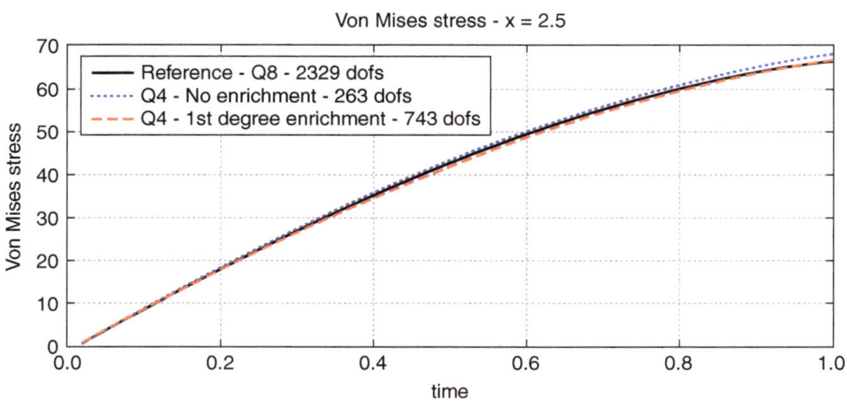

Fig. 31. von Mises stress value evolution during the nonlinear process, at point A.

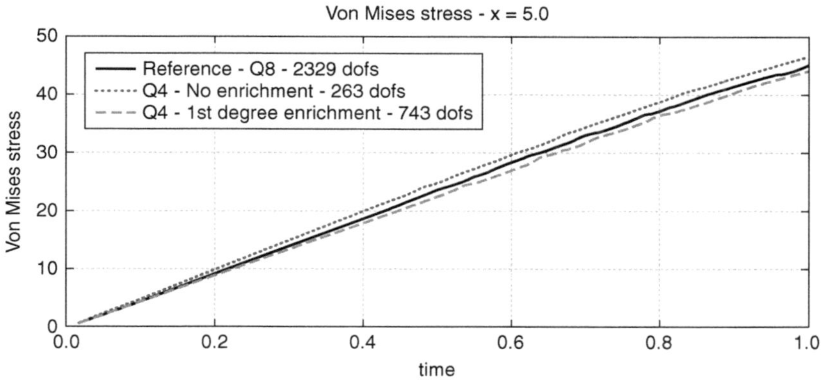

Fig. 32. von Mises stress value evolution during the nonlinear process, at point B.

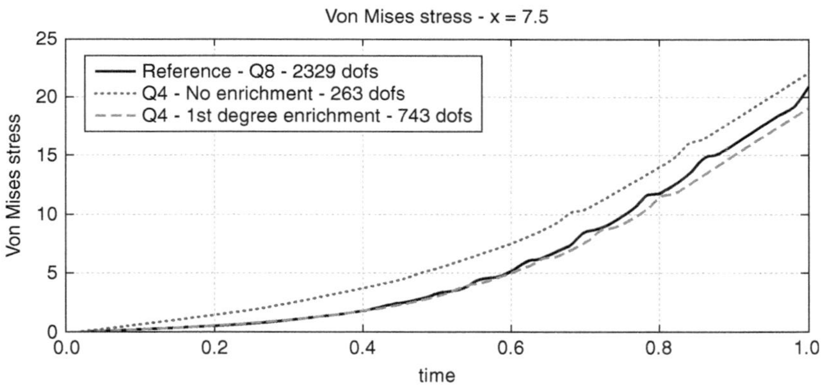

Fig. 33. von Mises stress value evolution during the nonlinear process, at point C.

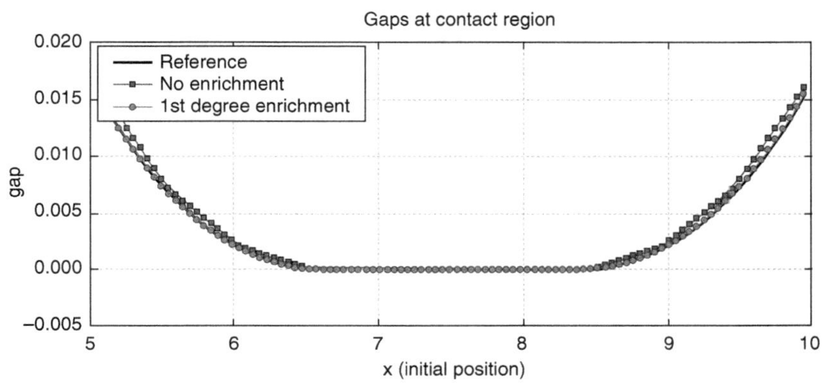

Fig. 34. Gaps at contact region for time step 10/50.

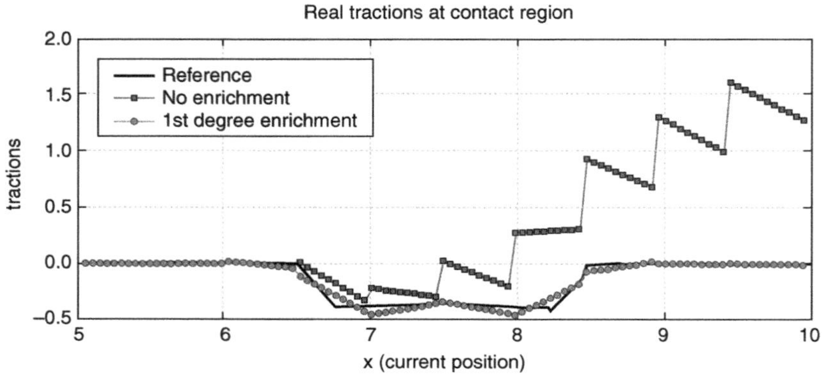

Fig. 35. Real tractions at contact region for time step 10/50.

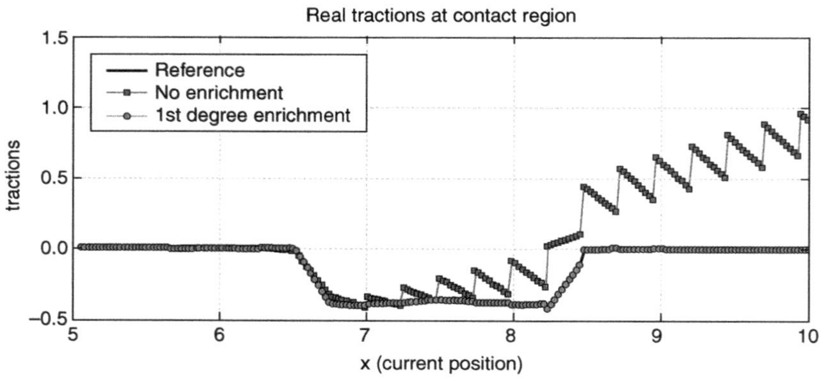

Fig. 36. Real tractions at contact region for time step 10/50 refined mesh GFEM model.

used for the reference solution (160 elements for each of the beams), the tractions results remain in accurate. The traction results for such refined GFEM model are presented in Fig. 36.

7. Concluding remarks

In this work, we presented a fully high-order mortar-based contact element with Lagrange multiplier method for the solution of 2D contact problems, considering small and large deformations. The proposed formulation was an extension of the work in Ref. [31] for higher-order interpolation functions constructed with Lagrange polynomials. We validated the contact element in small and large deformation with known solutions in the literature.

The h- and p-refinements and in small and large deformations revealed that the high-order interpolation has superior performance in contact problems, when the objective of the analysis is to increase the accuracy of results for stresses and forces generated by contact with a lower processing time.

We also considered the use of the GFEM to obtain a mortar-based contact element. The relevant contribution is the formulation, the computational implementation and the numerical assessment of the GFEM frictionless contact element. By analysing the GFEM contact element, one concludes that the enrichment benefits goes beyond the smoothing of the stress fields. The achieved results have shown that, in some cases, the contact elements can provide results even more accurate than the ones delivered by the standard FEM.

References

1. S. M. Holzer and Z. Yosibashi, The p-version of the finite element method in incremental elasto-plastic analysis, *International Journal for Numerical Methods in Engineering*, **39**, 1859–1878 (1996).
2. B. Jeremic and C. Xenophontos, Application of the p-version of the finite element method to elastoplasticity with localization of deformation, *Communications in Numerical Methods in Engineering*, **15**, 867–876 (1999).
3. A. Duster and E. Rank, The p-version of the finite element method compared to an adaptive h-version for the deformation theory of plasticity, *Computer Methods in Applied Mechanics and Engineering*, **190**, 1925–1935 (2001).
4. U. Heisserer, S. Hartmann, A. Duester, W. Bier, Z. Yosibash, and E. Rank, p-fem for finite deformation powder compaction, *Computer Methods in Applied Mechanics and Engineering*, **197**, 727–740 (2008).
5. M. L. Bittencourt, *Computational Solid Mechanics: Variational Formulation and High Order Approximation*, CRC Press, New York, US (2014).
6. G. Karniadakis and S. Spencer, *Spectral/hp Element Methods for Computional Fluid Dynamics*, 2nd edn., Oxford Science Publications, Oxford, UK (2004).
7. S. Dong and Z. Yosibashi, A parallel spectral element method for dynamic three-dimensional nonlinear elasticity problems, *Computers and Structures*, **87**, 59–72 (2009).
8. I. PáCzelt, B. A. Szabó, and T. Szabó, Solution of contact problem using the hp-version of the finite element method, *Computers and Mathematics with Applications*, **38**, 49–69 (1999).
9. D. F., A. Duster, V. Nubel, and E. Rank, A comparison of the h-, p-, hp-, and r p-version of the fem for the solution of the 2D hertzian contact problem, *Computational Mechanics*, **45**, 513–522 (2010).
10. A. Konyukhov and K. Schweizerhof, Incorporation of contact for high-order finite elements in covariant form, *Computer Methods in Applied Mechanics and Engineering*, **198**, 1213–1223 (2009).

11. K. A. Fischer and P. Wriggers, Mortar based frictional contact formulation for higher order interpolations using the moving friction cone, *Computer Methods in Applied Mechanics and Engineering*, **195**, 5020–5036 (2006).
12. A. Popp, W. G. Michael, and A. W. Wolfgang, A finite deformation mortar contact formulation using a primal–dual active set strategy, *International Journal for Numerical Methods in Engineering*, **79**, 1354–1391 (2009).
13. M. A. Puso, A 3D mortar method for solid mechanics, *International Journal for Numerical Methods in Engineering*, **59**, 315–336 (2004).
14. M. A. Puso and T. A. Laursen, A mortar segment-to-segment contact method for large deformation solid mechanics, *Computer Methods in Applied Mechanics and Engineering*, **193**, 601–629 (2004a).
15. M. A. Puso and T. A. Laursen, A mortar segment-to-segment contact method for large deformation, *Computer Methods in Applied Mechanics and Engineering*, **193**, 4891–4913 (2004b).
16. B. Yang, T. A. Laursen, and X. Meng, A finite deformation mortar contact formulation using a primal–dual active set strategy, *International Journal for Numerical Methods in Engineering*, **79**, 1354–1391 (2009).
17. B. Yang and T. A. Laursen, A large deformation mortar formulation of self contact with finite sliding, *Computer Methods in Applied Mechanics and Engineering*, **197**, 756–772 (2008).
18. B. Yang, *Mortar Finite Element Method for Large Deformation Contact Mechanics*, VDM Verlag Dr. Muller Aktiengesellschaft and Co. KG (2009).
19. M. A. Puso, T. A. Laursen, and J. Solberg, A segment-to-segment mortar contact method for quadratic elements and large deformations, *Computer Methods in Applied Mechanics and Engineering*, **197**, 555–566 (2008).
20. M. Gitterle, A. Popp, W. G. Michael, and A. W. Wolfgang, Finite deformation frictional mortar contact using a semi-smooth newton with consistent linearization, *International Journal for Numerical Methods in Engineering*, **84**, 543–571 (2010).
21. P. Wriggers, *Computational Contact Mechanics*, 2nd edn., Springer, Berlin, DE (2006).
22. B. I. Wohlmuth, A. Popp, M. W. Gee, and W. A. Wall, An abstract framework for a priori estimates for contact problems in 3D with quadratic finite elements, *Computational Mechanics*, **49**, 735–747 (2012).
23. A. Popp, B. I. Wohlmuth, M. W. Gee, and W. A. Wall, Dual quadratic mortar finite element methods for 3D finite deformation contact, *SIAM Journal on Scientific Computing*, **34**(4), B421–B446 (2012).
24. C. Bernardi, N. Debit, and Y. Maday, Coupling finite element and spectral methods: First results, *Mathematics of Computation*, **54**, 21–39 (1990).
25. F. B. Belgacem, P. Hild, and P. Laborde, The mortar finite element method for contact problems, *Mathematical and Computer Modelling*, **28**, 263–271 (1998).
26. T. W. Mcdevitt and T. A. Laursen, A mortar-finite element formulation for frictional contact problems, *International Journal for Numerical Methods in Engineering*, **48**, 1525–1547 (2000).

27. Temizer, P. Wriggers, and T. J. R. Hughes, Contact treatment in isogeometric analysis with nurbs, *Computer Methods in Applied Mechanics and Engineering*, **200**, 1100–1112 (2011).
28. L. Lorenzis, I. Temizer, P. Wriggers, and G. Zavarise, A large deformation frictional contact formulation using nurbs-based isogeometric analysis, *International Journal for Numerical Methods in Engineering*, **87**, 1278–1300 (2011).
29. L. Lorenzis, P. Wriggers, and G. Zavarise, A mortar formulation for 3D large deformation contact using nurbs-based isogeometric analysis and the augmented lagragioan method, *Computational Mechanics*, **49**, 1–20 (2012).
30. L. Lorenzis, J. A. Evans, T. J. R. Hughes, and A. Reali, Isogeometric collocation: Neumann boundary conditions and contact, *Computer Methods in Applied Mechanics and Engineering*, **284**, 21–54 (2015).
31. A. P. C. Dias, A. L. Serpa, and M. L. Bittencourt, High-order mortar-based element applied to nonlinear analysis of structural contact mechanics, *Computer Methods in Applied Mechanics and Engineering*, **294**, 19–55 (2015).
32. D. Piedade Neto, *On the Generalized Finite Element Method in Nonlinear Solid Mechanics Analyses*, PhD thesis, University of São Paulo, USP (2013).
33. J. M. Melenk and I. Babuska, The partition of unity finite element method: Basic theory and applications, *Computer Methods in Applied Mechanics and Engineering*, **39**, 289–314 (1996).
34. T. Belytschko, R. Gracie, and G. Ventura, A review of the extended/ generalized finite element methods for material modeling, *Modelling and Simulation in Materials Science and Engineering*, **17**, 24 (2009).
35. T. Belytschko and T. Black, Elastic crack growth in finite elements with minimal remeshing, *International Journal for Numerical Methods in Engineering*, **45**, 601–620 (1999).
36. N. Moes, J. Dolbow, and T. Belytschko, A finite element method for crack growth without remeshing, *International Journal for Numerical Methods in Engineering*, **46**, 131–150 (1999).
37. C. A. Duarte, O. N. Hamzeh, T. J. Liszka, and W. W. Tworzydlo, A generalized finite element method for the simulation of three-dimensional dynamic crack propagation, *Computer Methods in Applied Mechanics and Engineering*. **15**, 2227–2262 (2001).
38. J. P. Pereira, C. A. Duarte, D. Guoy, and X. Jiao, Hp-generalized fem and crack surface representation for non-planar 3-d cracks, *International Journal for Numerical Methods in Engineering*, **77**, 601–633 (2009).
39. N. Sukumar, B. Moes, B. Moran, and T. Belytschko, Extended finite element method for three-dimensional crack modeling, *International Journal for Numerical Methods in Engineering*, **48**, 1549–1570 (2000).
40. N. Sukumar, D. J. Srolovitz, T. J. Baker, and J. H. Prevost, Brittle fracture in polycrystalline microstructures with the extended finite element method, *International Journal for Numerical Methods in Engineering*, **56**, 2015–2037 (2003).
41. A. Simone, C. A. Duarte, and E. Van der Giessen, A generalized finite element method for polycrystals with discontinuous grain boundaries, *International Journal for Numerical Methods in Engineering*, **67**, 1122–1145 (2006).

42. J. Alfaiate, A. Simone, and L. J. Sluys, Non-homogeneous displacement jumps in strong embedded discontinuities, *International Journal for Numerical Methods in Engineering*, **40**, 5799–5817 (2003).
43. S. Mariani and U. Perego, Extended finite element method for quasi-brittle fracture, *International Journal for Numerical Methods in Engineering*, **58**, 103–126 (2003).
44. T. Menouillard and T. Belytschko, Dynamic fracture with meshfree enriched xfem, *ActaMechanica*, **21**, 53–69 (2010).
45. R. Larsson, J. Mediavilla, and M. Fagerstrom, Dynamic fracture modeling in shell structures based on xfem, *International Journal for Numerical Methods in Engineering*, **86**, 499–527 (2011).
46. Z. L. Liu, T. Menouillard, and T. Belytschko, An xfem/spectral element method for dynamic crack propagation, *International Journal of Fracture*, **169**, 183–198 (2011).
47. H. Talebi, C. Samaniego, E. Samaniego, and T. T. Rabczuk, On the numerical stability and mass-lumping schemes for explicit enriched meshfree methods, *International Journal for Numerical Methods in Engineering*, **89**, 1009–1027 (2012).
48. T. Elguerdj, A. Gravouil, and H. Maigre, An explicit dynamics extended finite element method. Part 1: Mass lumping for arbitrary enrichment functions, *Computer Methods in Applied Mechanics and Engineering*, **198**, 2297–2317 (2009).
49. D. Piedade Neto and S. P. B. Proenca, Generalized finite element method in linear and nonlinear structural dynamic analyses, *Engineering Computations*, **33**, 806–830 (2016).
50. J. Bonet and R. D. Wood, *Nonlinear Continuum Mechanics for Finite Element Analysis*, 2nd edn., Cambridge, Cambridge, UK (2008).
51. G. A. Holzapfel, *Nonlinear Solid Mechanics: A Continuum Approach for Engineering*, John Wiley and Sons. Chichester, West Sussex (2000).
52. K. J. Bathe, *Finite Element Procedures in Engineering Analysis*, Prentice-Hall (1982).
53. T. A. Laursen, *Computational Contact and Impact Mechanics*, 1st edn., Springer (2002).
54. T. J. R. Hughes, *The Finite Element Method: Linear Static and Dynamic Finite Element Analysis*, Prentice-Hall, Inc., New Jersey (2000).
55. B. P. Lamichhane, *Higher Order Mortar Finite Elements with Dual Lagrange Multiplier Spaces and Applications*, PhD thesis, Institute for Applied Analysis and Numerical Simulation, University of Stuttgart (2006).
56. D. Piedade Neto, M. D. C. Ferreira, and S. P. B. Proenca, An object-oriented class design for the generalized finite element method programming, *Latin American Journal of Solids and Structures*, **10**, 1267–1291 (2013).
57. I. Babuska and U. Banerjee, *Stable Generalized Fnite Element Method (SGFEM)*, technical report ices report 11-07 edn., The Institute for Computational Engineering and Sciences, The University of Texas at Austin (2011).

58. T. A. Laursen, *Computational Contact and Impact Mechanics: Fundamentals of Modeling Interfacial Phenomena in Nonlinear Finite Element Analysis*, Springer-Verlag, Berlin, DE (2002).
59. H. J. C. Barbosa and J. Ghaboussi, Discrete finite element method for multiple deformable bodies, *Finite Elements in Analysis and Desing*, **7**, 145–158 (1990).
60. A. Curnier, Unilateral contact: Mechanical modelling. in p. wriggers and p. panagiotopoulos, editors, *New Developments in Contact Problems*, Springer, pp. 1–54 (1999).
61. M. A. Crisfield, *Non-linear Finite Element Analysis of Solids and Structures*, 1st edn., Vol. 2, John Wiley and Sons, Chichester, UK (2000).
62. N. Kikuchi and J. T. Oden, *Contact Problems in Elasticity: A Study of Variational Inequalities and Finite Element Methods. Society for Industrial Applied Mathematics*, Vol. 8, pp. 19–54, Philadelphia, US (1988).
63. K. L. Johnson, *Contact Mechanics*, Cambridge University Press (1985).
64. M. S. Bazaraa and C. M. Shetty, *Nonlinear Programming: Theory and Algorithms*, John Willey and Sons, New York, US (1979).
65. D. G. Luenberger, *Linear and Non Linear Programming*, Addison–Wesley Publishing Company-Inc., Second edn., Boston, US (1984).
66. T. R. J. Hughes, R. L. Taylor, and W. A. Kanoknukulchai, Finite element method for large displacement contact and impact problems, In K.J. Bathe, J.T. Oden, W.Wunderlich, and E.L. Wilson (eds.), *Formulations and Computational Algorithms in FE Analysis*, MIT Press, Berkeley, US, pp. 468–495 (1977).
67. J. C. Simo, P. Wriggers, and R. L. Taylor, A perturbed lagrangian formulation for the finite element solution of contact problems, *Computer Methods in Applied Mechanics and Engineering*, **50**, 163–180 (1985).
68. B. I. Wohlmuth, *Discretization Methods and Iterative Solvers Based on Domain Decomposition*, Springer (2001).
69. M. A. Puso and T. A. Laursen, Mesh tying on curved interfaces in 3D, *Engineering Computations*, **20**, 305–319 (2003).
70. B. Yang, T. A. Laursen, and X. Meng, Two dimensional mortar contact methods for large deformation frictional sliding, *International Journal for Numerical Methods in Engineering*, **62**, 1183–1225 (2005).
71. K. A. Fischer and P. Wriggers, Frictionless 2D contact formulations for finite deformations based on the mortar method, *Computational Mechanics*, **36**, 226–244 (2005).
72. H. Patrick, Numerical implementation of two nonconforming finite element methods for unilateral contact, *Computer Methods in Applied Mechanics and Engineering*, **184**, 99–123 (2000).
73. K. J. Bathe and N. El-abbasi, Stability and path test performance of contact discretizations and a new solution algorithm, *Computers and Structures*, **79**, 1473–1486 (2001).
74. J. Shigley, C. Mischke and R. Budynas, *Mechanical Engineering Design*, 7th edn., McGraw-Hill (2004).

Chapter 3

A Large Deformation Frictionless Contact Treatment in NURBS-based Isogeometric Analysis

J. Kopačka, D. Gabriel, R. Kolman and J. Plešek

This chapter deals with numerical treatment of a large deformation frictionless contact problem by isogeometric analysis. The symmetry-preserving formulation, also known as two-half-pass formulation is presented together with the Gauss-point-to-segment discretization. The contact constraints regularization is performed by the penalty method. In the chapter, all key aspects of the numerical solution of contact problems are mentioned, including contact kinematics, spatial discretization by the finite element method, linearization, contact detection and solution of the nonlinear equation system. The performances of the resulting algorithm are showed by means of several contact benchmark problems including contact patch test, Hertz problem and ironing problem.

1. Introduction

The main difficulty encountered in computational contact analysis is non-smoothness. It naturally arises from inequality constraints as well as the geometric discontinuities inducted by spatial discretization. Contact analysis based on traditional finite elements utilizes element facets to describe a contact surface. The facets are C^0-continuous so that surface normals can experience jump across facets boundaries leading to artificial oscillations in normal and tangent contact forces. A global nonlinear equation solving scheme is also affected by non-smoothness. It was pointed out by Eterovic and Bathe[1] that the Newton–Raphson method admits, in general, only super-linear convergence even for C^2-continuous contact boundary, unless Lipschitz continuity is enforced.

Non-smoothness is usually attacked by contact smoothing techniques, where contact interface is approximated by various species of splines. In 2D, the first attempts applied the Hermite cubic interpolation.[2–4] Another option represents C^1-continuous Overhauser (Catmull–Rom) splines,[5] or non-uniform rational B-splines (NURBS).[6] Also in 3D, a lot of different options of contact surface smoothing were presented in the literature: local diffuse approximation with quadratic basis,[7] cubic B-Splines,[8] Bézier patch,[9,10] NURBS,[11] Gregory patches,[10,12] or recently Nagata patches.[13] All these contact smoothing techniques concentrate on the better approximation of the contact boundary. A different approach was proposed by Belytschko *et al.*,[14] where rather than contact boundary the signed distance function was smoothed by a moving least-squares approximation with a polynomial basis.

Contact smoothing techniques introduce an additional geometry on the top of the existing finite element mesh. This adds an extra layer of data management and increases computational overhead. On the other hand, a hierarchical enrichment[15] enriches the shape function space of the contact layer finite elements. Contact interface geometry is exactly represented, but at the same time the bulk of the contacting bodies can be described by lower-order shape functions. As an example, it can be mention the local enrichment with high-order Lagrange and Hermite interpolation,[16] NURBS,[17] or isogeometrically enriched finite elements.[18]

Another remedy to the geometric discontinuity provides isogeometric analysis (IGA).[19,20] The fundamental idea is to accurately describe a physical domain of interest by a proper mathematical representation (e.g. B-spline, NURBS, etc.) and then utilize the same basis for the analysis. This is in contrast with the classical finite element method where the basis is given in advance by the element type and the geometry of physical domain is only then approximated with the aid of isoparametric mapping.

Isogeometric NURBS-based contact analysis has some additional advantages: preserving geometric continuity, facilitating patch-wise contact search, supporting a variationally consistent formulation and having a uniform data structure for the contact surface and the underlying volumes.

The geometric basis and formulation for frictionless isogeometric contact were given by Lu.[21] Herein, the contact constraints were regularized by the penalty method and contact virtual work was discretized by finite strain segment-to-segment contact element.

Soon after, Temizer *et al.*[22] considered finite deformation frictionless quasi-static thermomechanical contact problems. Two penalty-based contact algorithms were studied herein. The former was called knot-to-segment (KTS) algorithm. It is the straightforward extension of the classical node-to-segment (NTS) algorithm.[23] Since NURBS control points are not interpolatory, contact constraints were enforce directly at the physical points of the quadrature points. It was shown in this reference that this approach is over-constrained. The latter was called mortar–KTS algorithm. In this algorithm, a mortar projection to control pressures was employed to obtain the correct number of constraints. This penalty-based mortar–KTS algorithm was later extended to frictional contact,[24,25] studied in conjunction with augmented Lagrangian method,[26] and the interior point method.[27]

At the same time as Temizer *et al.*,[22] Kim and Youn[28] published their 2D isogeometric frictionless contact treatment utilizing non-conforming mortar method in 2D linear elasticity regime. A new matching algorithm was proposed to combine the mortar method with the isogeometric NURBS-based analysis.

Although Temizer *et al.*,[22] called their algorithm KTS, the more concise name is Gauss-point-to-surface (GPTS)[29] because of the evaluation of the contact constraints at the Legendre–Gauss quadrature points. Only Matzen *et al.*,[30] came up with a real isogeometric analogy of the NTS discretization called point-to-segment (PTS) contact formulation. To collocate the contact integrals, it was proposed to use Greville and Botella points. Also two post-process methods to obtain contact stress distribution from discrete values of the Lagrange multipliers at collocation points were investigated and compared. This concept was recently enhanced by weighting.[31]

Dittmann *et al.*,[32] proposed an isogeometric thermomechanical mortar contact algorithm. In comparison to Ref. [22] where mortar-based formulation was also used, in this work a series of simplifications were considered so that established numerical techniques for mortar methods such as segmentation[33,34] could be transferred to IGA without modification. Recently, a dual mortar formulation for IGA was derived and investigated by Seitz *et al.*[35]

Isogeometric collocation methods (IGA-C)[36,37] have been recently proposed as an alternative to standard Galerkin approaches. They provide a significant reduction in computational cost especially for higher-order discretizations. They can be interpreted as a one-point quadrature rule in the IGA context. The imposition of Neumann boundary conditions and the enforcement of contact constraints in IGA-C context was studied by de Lorenzis *et al.*[38] The application of IGA-C to large deformation frictional contact problems with consistent linearization was presented in the work of Kruse *et al.*[39]

It is well known that accurate evaluation of the contact stresses at the contact interface requires appropriate refinement. As the NURBS multivariate basis functions are defined as tensor product, the number of unknowns increase rapidly as the knot insertion requires control point addition along the whole parametric dimension. The local basis function refinement provides T-spline-based IGA.[40,41] It was first applied to frictionless large deformation contact problem in the work of Dimitri *et al.*[42] and recently in the work of Dimitri and Zavarise,[43] where debonding problem was also studied. Here, the cubic T-splines and cubic NURBS were considered. The GPTS contact discretization and the penalty method were used to treat the contact constraints. An alternative to T-splines is the NURBS-based hierarchical refinement scheme[44] applied to contact problems by Temizer and Hesch[45] where large deformation frictionless contact was considered, or in the work of Hesch *et al.*,[46] where a mortar contact algorithms applied to a phase-field approach to brittle fracture were studied.

This chapter presents a description of all key aspects of the numerical solution of contact problems. The goal is to provide the reader with the simple framework to be able to implement the isogeometric contact algorithm on his own. Therefore, a simple yet powerful symmetry-preserving contact formulation[47] is considered. It is based on the Gauss-point-to-segment discretization where contact constraints are enforced at the Gaussian quadrature point level by the penalty method. The resulting system of nonlinear equations is linearized in order to use the Newton–Raphson method. Only an incomplete linearization is made to keep an implementation simple. The performance of the overall algorithm is presented by means of several numerical examples including patch test, Hertz problem, and ironing problem.

2. B-spline and NURBS

This section gives a brief overview of the main concerns of B-spline and NURBS. For more detailed description as well as efficient algorithms see, e.g. Ref. [48]. Throughout this chapter, we use p to indicate the polynomial degree, n to indicate the number of basis functions, d_{p} to indicate the number of parametric dimensions, and d_{s} to indicate the number of spatial dimensions.

A B-spline or NURBS object is called patch. An example of the NURBS patch is in Fig. 1(a). The patch is parameterized by the linear combination

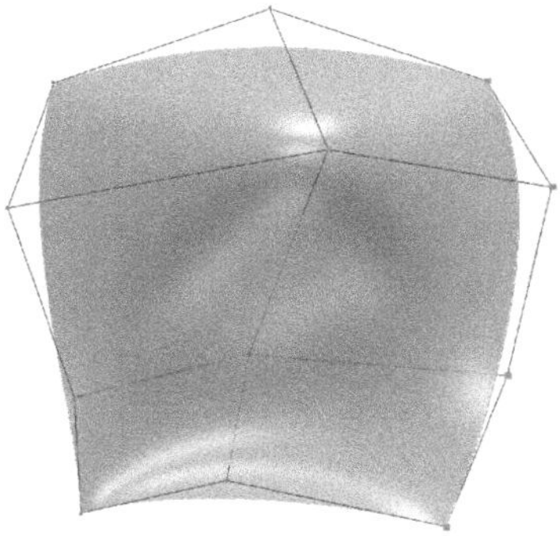

(a) NURBS surface

(b) NURBS curve

Fig. 1. An example of the NURBS patch.

of control points $\mathbf{P}_A \in \mathbb{R}^{d_s}$ and basis functions $N_A(\boldsymbol{\xi}) : \mathbb{R}^{d_p} \mapsto [0,1]$. A particular B-spline/NURBS basis function is defined with the aid of the so-called knot vector

$$\Xi^i = \left\{ \underbrace{\xi_1^i, \ldots, \xi_{p_i+1}^i}_{p_i+1 \text{ equal terms}}, \xi_{p_i+2}^i, \ldots, \xi_{n_i}^i, \underbrace{\xi_{n_i+1}^i, \ldots, \xi_{n_i+p_i+1}^i}_{p_i+1 \text{ equal terms}} \right\}, \tag{1}$$

where $i = 1, \ldots, d_\mathrm{p}$. The knot vector is a non-decreasing sequence of parametric coordinates. The knot vector is said to be uniform if the knots are unequally spaced in the parametric space. If the first and the last knot value appears $p_i + 1$ times, the knot vector is called open. B-spline/NURBS object with open knot vectors are interpolatory at the corner of a patch. It means that the boundary of a B-spline/NURBS object with d_p parametric dimensions is itself a B-spline/NURBS object with $d_\mathrm{p} - 1$ dimensions (see Fig. 1(b)).

The B-spline basis functions are defined by the Cox–de Boor recursion formula. For $p = 0$ it is

$$N_{j,0}(\xi) = \begin{cases} 1 & \xi \in [\xi_j, \xi_{j+1}), \quad j = 1 \ldots n \\ 0 & \text{otherwise} \end{cases} \tag{2}$$

and for $p > 0$

$$N_{j,p}(\xi) = \frac{\xi - \xi_j}{\xi_{j+p} - \xi_j} N_{j,p-1}(\xi) + \frac{\xi_{j+1+p} - \xi}{\xi_{j+1+p} - \xi_{j+1}} N_{j+1,p-1}(\xi). \tag{3}$$

B-splines are known to be unable to exactly describe some curves and surfaces. For instance, circles or cylinders. NURBS was developed to extend interpolatory capability of the B-splines. While the B-spline basis functions are ordinary polynomials, the NURBS basis functions are rational polynomials. The extension originates from projection geometry of conic sections. More details can be found in Ref. [48]. The pth degree NURBS basis function is defined as

$$R_j^p(\xi) = \frac{N_{j,p}(\xi)w_j}{\sum_{\hat{j}=1}^{n} N_{\hat{j},p}(\xi)w_{\hat{j}}}, \tag{4}$$

where w_j is referred to as the jth weight, which together with the cartesian coordinates of the $\mathbf{P}_j$th control point form the so-called homogeneous coordinates.

Multivariate NURBS objects can be constructed simply by tensor product of univariate NURBS basis functions (4). For example, for $d_\mathrm{p} = 2$

$$R_{j_1,j_2}^{p_1,p_2}(\xi^1,\xi^2) = \frac{N_{j_1,p_1}(\xi^1)N_{j_2,p_2}(\xi^2)w_{j_1,j_2}}{\sum_{\hat{j}_1=1}^{n_1}\sum_{\hat{j}_2=1}^{n_2} N_{\hat{j}_1,p_1}(\xi^1)N_{\hat{j}_2,p_2}(\xi^2)w_{\hat{j}_1,\hat{j}_2}} \tag{5}$$

and similarly for higher parametric dimensions.

With multivariate NURBS basis functions at hand, we can introduce surface discretization by

$$\mathbf{x}(\xi^1,\xi^2) = \sum_{j_1=1}^{n_1}\sum_{j_2=1}^{n_2} R_{j_1,j_2}^{p_1,p_2}(\xi^1,\xi^2)\mathbf{P}_{j_1,j_2}, \tag{6}$$

where $\mathbf{P}_{j_1,j_2} \in \mathbb{R}^{d_\mathrm{s}}$ is the control net, i.e. array of coordinates of control points.

3. Contact kinematics

Without loss of generality, let us consider two deformable bodies $\mathcal{B}^{(i)}$, $i = \{1,2\}$ in d_s dimensional Euclidean space $\mathbb{R}^{d_\mathrm{s}}$ coming into contact, which are depicted in Fig. 2. The position of a particle $P^{(i)}$ of the body $\mathcal{B}^{(i)}$ at time $t = 0$ is described by the position vector $\mathbf{X}^{(i)} \in \bar{\Omega}_0^{(i)} \subset \mathbb{R}^{d_\mathrm{s}}$. Further, let us consider existence of a regular mapping $\boldsymbol{\chi}^{(i)} : \bar{\Omega}_0^{(i)} \times \mathbb{T} \mapsto \bar{\Omega}_\mathrm{t}^{(i)}$, which will be called motion. Its image is the position vector $\mathbf{x}^{(i)} \in \bar{\Omega}_\mathrm{t}^{(i)} \subset \mathbb{R}^{d_\mathrm{s}}$ of the particle $P^{(i)}$ at time $t \in \mathbb{T} = [0,T]$. Now the positions of the particle $P^{(i)}$ in initial and current configuration can be linked through the displacement

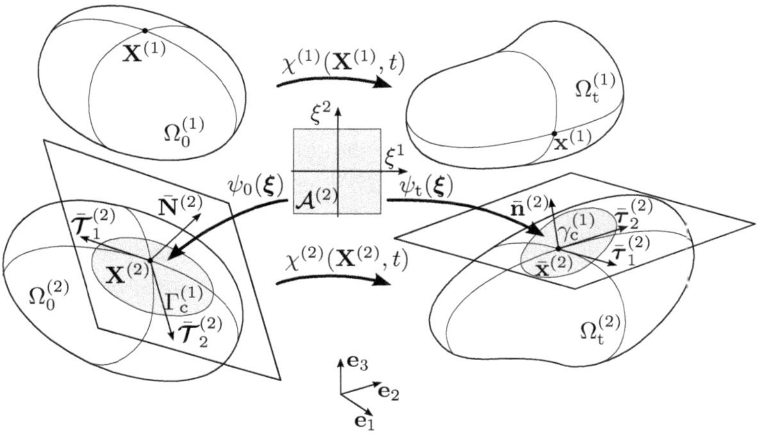

Fig. 2. Two deformable bodies coming into contact depicted in the initial and current configuration.

field defined as

$$\mathbf{u}^{(i)}(\mathbf{X}^{(i)}, t) := \mathbf{x}^{(i)}(\mathbf{X}^{(i)}, t) - \mathbf{X}^{(i)}. \tag{7}$$

The boundaries $\partial\Omega^{(i)}$ of domains $\Omega^{(i)}$ in the initial and current position are covered by disjoint sets

$$\partial\Omega_0^{(i)} = \Gamma_\mathrm{u}^{(i)} \cup \Gamma_\sigma^{(i)} \cup \Gamma_\mathrm{c}^{(i)}, \tag{8}$$

$$\partial\Omega_\mathrm{t}^{(i)} = \gamma_\mathrm{u}^{(i)} \cup \gamma_\sigma^{(i)} \cup \gamma_\mathrm{c}^{(i)}, \tag{9}$$

where $\Gamma_\mathrm{u}^{(i)}, \gamma_\mathrm{u}^{(i)}$ and $\Gamma_\sigma^{(i)}, \gamma_\sigma^{(i)}$ are those portions of the boundary in the initial and current configurations where displacements and tractions are prescribed, respectively. The sets $\Gamma_\mathrm{c}^{(i)}, \gamma_\mathrm{c}^{(i)}$ represent the potential contact interface in the initial and current configuration where contact can possibly take place.

The physical fact that two bodies cannot penetrate each other during the motion can be expressed by the non-penetration condition

$$\Omega_\mathrm{t}^{(1)} \cap \Omega_\mathrm{t}^{(2)} = \emptyset, \quad \forall\, t \in \mathbb{T}. \tag{10}$$

This certainly elegant term, however, is not suitable for numerical purposes, and therefore it will be useful to reformulate it slightly. To generalize the notation, let us introduce another body index as $(k) := \{1, 2\} \setminus (i)$, i.e. (k) is the complement body to the body (i). The closest point $\bar{\mathbf{X}}^{(k)}$ is assigned to each point $\mathbf{X}^{(i)} \in \Gamma_\mathrm{c}^{(i)}$ in the sense

$$\bar{\mathbf{X}}^{(k)}(\mathbf{X}^{(i)}, t) = \arg \min_{\mathbf{X}^{(k)} \in \Gamma_\mathrm{c}^{(k)}} \left\| \mathbf{x}^{(i)}(\mathbf{X}^{(i)}, t) - \mathbf{x}^{(k)}(\mathbf{X}^{(k)}, t) \right\|, \tag{11}$$

where $\| \bullet \|$ is the Euclidean norm. It should be noted that the body $\mathcal{B}^{(i)}$ is usually termed as the slave body and $\mathcal{B}^{(k)}$ as the master body. It is probably due to the dependence implied by the closest point projection. From now on, the variables which depend on the closest point projection will be denoted with the bar $\bar{\bullet}$. To find the position vector $\bar{\mathbf{X}}^{(k)}$, the potential contact boundary $\Gamma_\mathrm{c}^{(k)}$ is parameterized by convective coordinates $\boldsymbol{\xi} \in \mathcal{A} \subset \mathbb{R}^{d_\mathrm{s}-1}$ and described by the mapping $\psi_0^{(k)} : \mathcal{A} \mapsto \Gamma_\mathrm{c}^{(k)}$, see Fig. 2. Further, a squared distance function is introduced

$$d(\boldsymbol{\xi}) := \frac{1}{2} \left\| \mathbf{x}^{(i)}(\mathbf{X}^{(i)}, t) - \mathbf{x}^{(k)}(\mathbf{X}^{(k)}(\boldsymbol{\xi}), t) \right\|^2. \tag{12}$$

Minimization of this function yields system on nonlinear algebraic equations for unknown convective coordinates $\bar{\boldsymbol{\xi}}$ of the nearest point $\bar{\mathbf{X}}^{(k)}$

$$\left(\mathbf{x}^{(i)} - \bar{\mathbf{x}}^{(k)} \right) \cdot \frac{\partial \bar{\mathbf{x}}^{(k)}}{\partial \mathbf{X}^{(k)}} \frac{\partial \bar{\mathbf{X}}^{(k)}}{\partial \xi^\alpha} = 0, \tag{13}$$

where $\alpha = 1, \ldots, d_\mathrm{s} - 1$ is the index of the convective coordinate component. The solution of this system of nonlinear algebraic equations, $\bar{\boldsymbol{\xi}}$, determines the desired position vector $\bar{\mathbf{X}}^{(k)} = \psi_0^{(k)}(\bar{\boldsymbol{\xi}})$.

It is worth pointing out that convective coordinates implicitly depend not only on the initial position vector $\mathbf{X}^{(i)}$ but also on the spatial position vectors $\mathbf{x}^{(i)}(\mathbf{X}^{(i)}, t)$. It will be crucial to take this fact into consideration for the calculation of directional derivatives of the contact kinematic quantities.

Equation (13) can be further recast considering the definition of the deformation gradient

$$\bar{\mathbf{F}}^{(k)} := \frac{\partial \bar{\mathbf{x}}^{(k)}}{\partial \mathbf{X}^{(k)}} \tag{14}$$

and the vectors tangent to the initial contact boundary $\Gamma_\mathrm{c}^{(k)}$ at point $\mathbf{X}^{(k)}(\bar{\boldsymbol{\xi}})$

$$\bar{\boldsymbol{\mathcal{T}}}_\alpha^{(k)} := \frac{\partial \bar{\mathbf{X}}^{(k)}}{\partial \xi^\alpha}. \tag{15}$$

The tangent vector $\bar{\boldsymbol{\mathcal{T}}}_\alpha^{(k)}$ can be transformed to the current configuration by the push forward operation

$$\bar{\boldsymbol{\tau}}_\alpha^{(k)} = \bar{\mathbf{F}}^{(k)} \bar{\boldsymbol{\mathcal{T}}}_\alpha^{(k)}. \tag{16}$$

Equation (13) now gains a particularly simple form

$$\left(\mathbf{x}^{(i)} - \bar{\mathbf{x}}^{(k)}\right) \cdot \bar{\boldsymbol{\tau}}_\alpha^{(k)} = 0. \tag{17}$$

It can be interpreted as the orthogonality condition in the current configuration.

Usually, rather then the initial position vector $\bar{\mathbf{X}}^{(k)}$ its spatial counterpart $\mathbf{x}^{(k)}(\mathbf{X}^{(k)}(\bar{\boldsymbol{\xi}}), t)$ will be in question. Therefore it is useful to replace the mapping $\psi_0^{(k)}$ by the composite mapping $\psi_t^{(k)} : \mathcal{A} \mapsto \gamma_\mathrm{c}^{(k)}$ defined as

$$\psi_t^{(k)} := \boldsymbol{\chi}^{(k)} \circ \psi_0^{(k)}. \tag{18}$$

The spatial position vector $\mathbf{x}^{(k)}(\bar{\boldsymbol{\xi}}, t)$ now directly depends on the convective coordinates, and the vector tangent to the current contact boundary $\gamma_\mathrm{c}^{(k)}$ at point $\mathbf{x}^{(k)}(\bar{\boldsymbol{\xi}}, t)$ can be defined as

$$\bar{\boldsymbol{\tau}}_\alpha^{(k)} := \frac{\partial \bar{\mathbf{x}}^{(k)}}{\partial \xi^\alpha}. \tag{19}$$

With the aid of tangent vectors, the normal vector to the initial contact boundary $\Gamma_\mathrm{c}^{(k)}$ is constructed as

$$\bar{\mathbf{N}}^{(k)} := \frac{\mathbf{e}_3 \times \bar{\boldsymbol{\mathcal{T}}}_1^{(k)}}{\left\| \mathbf{e}_3 \times \bar{\boldsymbol{\mathcal{T}}}_1^{(k)} \right\|}, \qquad \bar{\mathbf{N}}^{(k)} := \frac{\bar{\boldsymbol{\mathcal{T}}}_1^{(k)} \times \bar{\boldsymbol{\mathcal{T}}}_2^{(k)}}{\left\| \bar{\boldsymbol{\mathcal{T}}}_1^{(k)} \times \bar{\boldsymbol{\mathcal{T}}}_2^{(k)} \right\|}, \tag{20}$$

for $d_\mathrm{s} = 2$ and $d_\mathrm{s} = 3$, respectively, and its spatial counterparts

$$\bar{\mathbf{n}}^{(k)} := \frac{\mathbf{e}_3 \times \bar{\boldsymbol{\tau}}_1^{(k)}}{\left\| \mathbf{e}_3 \times \bar{\boldsymbol{\tau}}_1^{(k)} \right\|}, \qquad \bar{\mathbf{n}}^{(k)} := \frac{\bar{\boldsymbol{\tau}}_1^{(k)} \times \bar{\boldsymbol{\tau}}_2^{(k)}}{\left\| \bar{\boldsymbol{\tau}}_1^{(k)} \times \bar{\boldsymbol{\tau}}_2^{(k)} \right\|}. \tag{21}$$

The normal vector allows to define an essential contact kinematic quantity, the normal gap function, in the form

$$g_\mathrm{N}^{(i)}(\mathbf{X}^{(i)}, t) := -\left(\mathbf{x}^{(i)} - \bar{\mathbf{x}}^{(k)}\right) \cdot \bar{\mathbf{n}}^{(k)}, \tag{22}$$

which can be understood as the signed distance function. According to our definition (22), the gap is open, i.e. bodies are not in contact, when the normal gap function is negative. Conversely, a positive value of the normal gap function indicates penetration of the bodies. With the normal gap function at hand, the non-penetration condition (10) can be reformulated as

$$g_N^{(i)} \le 0, \qquad \forall\, \mathbf{X}^{(i)} \in \Gamma_c^{(i)}, t \in \mathbb{T}. \tag{23}$$

This condition is essential for the contact problem formulation, which will be presented in Section 6. It is the inequality character of the non-penetration condition why the numerical solution of contact problems is so challenging.

4. Contact traction vector

The traction vector $\mathbf{t}_{\mathrm{c}}^{(k)}$ acting on the current master contact boundary $\gamma_c^{(k)}$ is given by Cauchy's stress theorem[49]

$$\mathbf{t}_c^{(k)} = \mathbf{n}^{(k)} \cdot \boldsymbol{\sigma}^{(k)}, \tag{24}$$

where $\boldsymbol{\sigma}^{(k)}$ is the Cauchy stress tensor (also known as the true stress tensor). Note that the bar indicating the quantities evaluated at the projection $\bar{\boldsymbol{\xi}}$ is omitted in this section. For the differential of contact force in the current configuration it holds

$$d\mathbf{f}_c^{(k)} = \mathbf{t}_c^{(k)} d\gamma^{(k)} = \mathbf{n}^{(k)} \cdot \boldsymbol{\sigma}^{(k)} d\gamma^{(k)}. \tag{25}$$

To express the counterpart of $\mathbf{t}_c^{(k)}$ in the initial configuration, Nanson's relation[50]

$$\mathbf{n}^{(k)} = J^{(k)} \frac{d\Gamma^{(k)}}{d\gamma^{(k)}} \mathbf{N}^{(k)} \cdot \left(\mathbf{F}^{(k)}\right)^{-1}, \tag{26}$$

will be exploited. Here, $J^{(k)}$ denotes the Jacobian of mapping from initial to current configuration, i.e. $J^{(k)} = \det \mathbf{F}^{(k)}$. Nanson's relation establishes relation between normal vectors in the initial and current configurations. Substituting from (26) into (25) yields

$$d\mathbf{f}_c^{(k)} = J^{(k)} \frac{d\Gamma^{(k)}}{d\gamma^{(k)}} \mathbf{N}^{(k)} \cdot \left(\mathbf{F}^{(k)}\right)^{-1} \cdot \boldsymbol{\sigma}^{(k)} d\gamma^{(k)}, \tag{27}$$

$$= \mathbf{N}^{(k)} \cdot J^{(k)} \left(\mathbf{F}^{(k)}\right)^{-1} \cdot \boldsymbol{\sigma}^{(k)} d\Gamma^{(k)}, \tag{28}$$

$$= \mathbf{N}^{(k)} \cdot \mathbf{P}^{(k)} d\Gamma^{(k)}, \tag{29}$$

where the first Piola–Kirchhoff stress tensor has been introduced as

$$\mathbf{P}^{(k)} = J^{(k)} (\mathbf{F}^{(i)})^{-1} \cdot \boldsymbol{\sigma}^{(k)}. \tag{30}$$

The first Piola–Kirchhoff traction vector $\mathbf{T}_{\mathrm{c}}^{(k)}(\mathbf{X}^{(k)})$ acting on the initial contact boundary $\Gamma_{\mathrm{c}}^{(k)}$ is then given as

$$\mathbf{T}_{\mathrm{c}}^{(k)} = \mathbf{N}^{(k)} \cdot \mathbf{P}^{(k)}. \tag{31}$$

Now, for the differential of contact force it holds

$$\mathrm{d}\mathbf{f}_{\mathrm{c}}^{(k)} = \mathbf{t}_{\mathrm{c}}^{(k)} \mathrm{d}\gamma^{(k)} = \mathbf{T}_{\mathrm{c}}^{(k)} \mathrm{d}\Gamma^{(k)}. \tag{32}$$

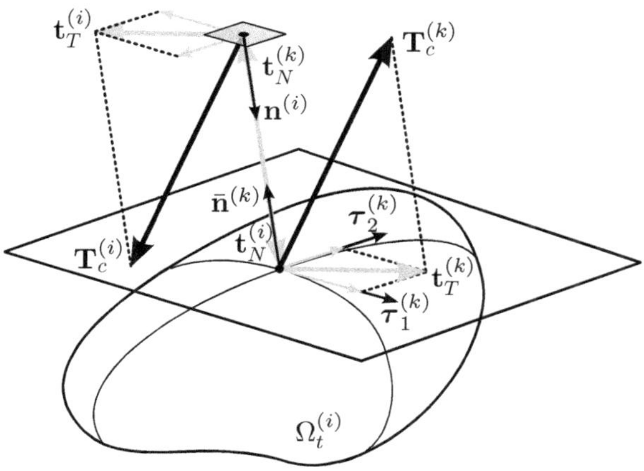

Fig. 3. Resolution of the contact traction vector into the normal and tangent components in the current configuration.

Thus, the vectors $\mathbf{t}_c^{(k)}$ and $\mathbf{T}_c^{(k)}$ have different length but they are pointing in the same direction. The components of the first Piola–Kirchhoff stress tensor can be interpreted as the stresses that result from the force that acts on a surface in the initial configuration. Both contact traction vectors $\mathbf{t}_c^{(k)}$, $\mathbf{T}_c^{(k)}$ at some particular point $\mathbf{X}^{(k)}$ can be resolved into the normal and tangential direction in the current (spatial) tangent basis.[51] This vector resolution is depicted in Fig. 3. The resolution of the contact traction vector $\mathbf{T}_c^{(k)}$ into the normal and tangential direction in the current basis yields

$$\mathbf{T}_c^{(k)} = \mathbf{t}_N^{(k)}\mathbf{n}^{(k)} + \mathbf{t}_T^{(k)}. \tag{33}$$

The normal component of the contact traction vector is

$$\mathbf{t}_N = t_N^{(k)}\mathbf{n}^{(k)} = -p_c^{(k)}\mathbf{n}^{(k)}, \tag{34}$$

where the contact pressure, $p_c^{(k)}$, has been introduced. The contact pressure is *positive* when the normal contact traction is negative, otherwise it is equal to zero. Since the frictionless contact is considered, the tangential component $\mathbf{t}_T^{(k)}$ is equal to zero vector

$$\mathbf{t}_T^{(k)} = t_T^{\alpha(k)}\boldsymbol{\tau}_\alpha^{(k)} = \mathbf{0}. \tag{35}$$

According to the law of action and reaction, it must hold

$$p_c^{(i)} = p_c^{(k)}. \tag{36}$$

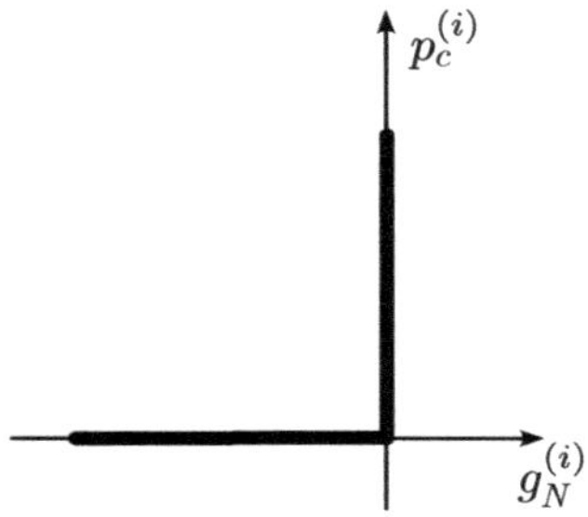

Fig. 4. A scheme of the normal contact conditions.

In what follows, it is considered that on the contact interface it holds

$$\mathbf{n}^{(i)} = -\bar{\mathbf{n}}^{(k)}. \tag{37}$$

Consequently,

$$\mathbf{T}_c^{(i)} = \mathbf{t}_N^{(i)} = t_N^{(i)}\mathbf{n}^{(i)} = -p_c^{(i)}\mathbf{n}^{(i)} = p_c^{(i)}\bar{\mathbf{n}}^{(k)}. \tag{38}$$

5. Normal contact conditions

The physical behaviour at the contact boundary in the normal direction can be described using the Hertz–Signoriniho–Moreau conditions

$$p_{\mathrm{c}}^{(i)} \geq 0, \qquad g_{\mathrm{N}}^{(i)} \leq 0, \qquad p_{\mathrm{c}}^{(i)} g_{\mathrm{N}}^{(i)} = 0, \qquad \forall\, \mathbf{X}^{(i)} \in \Gamma_{\mathrm{c}}^{(i)}, \quad \forall\, t \in \mathbb{T}. \tag{39}$$

These conditions are known in mathematical optimization as the Karush–Kuhn–Tucker (KKT) conditions formulating the first-order necessary conditions for a solution in nonlinear programming to be optimal (Fig. 4). The first inequality $(39)_1$ indicates that only non-negative contact pressure is admissible at the contact interface. The second inequality $(39)_2$ represents the non-penetration condition. The third equality $(39)_3$, is the so-called complementarity condition which assures that compression is admissible only when the gap is closed and, conversely, the potential contact interface, $\Gamma_{\mathrm{c}}^{(i)}$, is stress-free only when the gap is open.

6. Contact problem formulation

In this section, the large deformation frictionless contact problem is first formulated in the strong sense. Then, the problem is reformulated into the weak form which will serve as the basis for the finite element discretization presented in Section 9.

6.1. *Strong form*

After all the necessary variables have been defined, one can proceed to formulate the contact initial-boundary value problem in the strong sense. The Lagrangian description in the initial configuration is chosen having regard to the linearization because integration regions will not depend on the displacement field.

At each time instance $t \in \mathbb{T}$, the conservation of momentum[50] must be met

$$\operatorname{Div} \mathbf{P}^{(i)} + \rho_0^{(i)} \mathbf{b}_0^{(i)} = \rho_0^{(i)} \mathbf{A}^{(i)} \quad \text{in } \Omega_0^{(i)}, \tag{40}$$

where Div denotes the divergence with respect to initial configuration, $\mathbf{P}$ is the first Piola–Kirchhoff stress tensor, $\mathbf{b}_0$ is the prescribed body force per unit reference mass, ρ_0 is the reference density and $\mathbf{A}$ is the material acceleration field. The equation of momentum conservation is further supplemented by initial conditions

$$\mathbf{u}^{(i)}\Big|_{t=0} = \mathbf{u}_0^{(i)} \quad \text{in } \bar{\Omega}_0^{(i)}, \tag{41}$$

$$\mathbf{V}^{(i)}\Big|_{t=0} = \mathbf{V}_0^{(i)} \quad \text{in } \bar{\Omega}_0^{(i)}, \tag{42}$$

boundary conditions

$$\mathbf{u}^{(i)} = \hat{\mathbf{u}}^{(i)} \quad \text{on } \Gamma_u^{(i)}, \tag{43}$$

$$\mathbf{N}^{(i)} \cdot \mathbf{P}^{(i)} = \hat{\mathbf{T}}^{(i)} \quad \text{on } \Gamma_\sigma^{(i)}, \tag{44}$$

and contact conditions on $\Gamma_c^{(i)}$

$$(\mathbf{N}^{(i)} \cdot \mathbf{P}^{(i)}) \cdot \bar{\mathbf{n}}^{(k)} = p_c^{(i)}, \tag{45}$$

$$p_c^{(i)} = p_c^{(k)}, \tag{46}$$

$$p_c^{(i)} \geq 0, \qquad g_N^{(i)} \leq 0, \qquad p_c^{(i)} g_N^{(i)} = 0, \tag{47}$$

where $\mathbf{u}_0^{(i)}$ and $\mathbf{V}_0^{(i)}$ are prescribed initial displacement and initial material velocity fields, respectively. $\hat{\mathbf{u}}^{(i)}$ and $\hat{\mathbf{T}}^{(i)}$ are prescribed displacement and traction fields, respectively.

This system of equations is necessary to is necessary to supplement by the constitutive relation between stress and strain, such as Hooke's law,[52] but in this work it will not be further specified. The solution of the problem (40)–(47) is the displacement field, $\mathbf{u}^{(i)}$, which is required to be C^2-continuous, and the contact pressure field, $p_c^{(i)}$. The solution must satisfy (40)–(47) in every single point of $\bar{\Omega}_0^{(i)}$, hence the name strong form.

As will be shown in the next section, this requirement can be weakened by considering the so-called weak form.

6.2. *Weak form*

As is well known, to address a particular problem by the finite element method (FEM), it is first necessary to convert the strong formulation of the problem to the so-called weak formulation which has lower demands on the smoothness of the solution. In this section, the weak formulation of the contact initial-boundary values problem will be derived using the method of weighted residuals.

First of all, appropriate function spaces will be introduced. The weighting space, $\mathcal{W}^{(i)}$, is defined as a class of all kinematically admissible displacement fields, which are zero on the boundary $\Gamma_{\mathrm{u}}^{(i)}$

$$\mathcal{W}^{(i)} := \left\{\delta\mathbf{u}^{(i)}(\mathbf{X}^{(i)}) \in H^1(\Omega_0^{(i)})|\delta\mathbf{u}^{(i)}(\mathbf{X}^{(i)}) = \mathbf{0} \text{ on } \Gamma_u^{(i)}\right\} \tag{48}$$

and the solution space of the trial functions, $\mathcal{S}_t^{(i)}$, will be defined as the same class of functions but which, in addition, fulfils Dirichlet's boundary conditions on the boundary $\Gamma_u^{(i)}$

$$\mathcal{S}_t^{(i)} := \left\{\mathbf{u}^{(i)}(\mathbf{X}^{(i)}, t) \in H^1(\Omega_0^{(i)})|\mathbf{u}^{(i)}(\mathbf{X}^{(i)}, t) = \hat{\mathbf{u}}^{(i)} \text{ on } \Gamma_u^{(i)}\right\}. \tag{49}$$

Here, H^1 denotes the Sobolev space[53] of all functions defined on $\Omega_0^{(i)}$ that are, as well as all their first partial derivatives, square integrable. The space of contact pressure is defined as

$$\mathcal{M}_{\mathrm{t}}^{(i)} := \left\{p_{\mathrm{c}}^{(i)}(\mathbf{X}^{(i)}, t) \in H^{1/2}(\Gamma_{\mathrm{c}}^{(i)})|p_{\mathrm{c}}^{(i)} \geq 0\right\}, \tag{50}$$

where $H^{1/2}$ is the Sobolev trace space on the boundary $\Gamma_{\mathrm{c}}^{(i)}$ of all functions defined on $\Omega_0^{(i)}$ that are, as well as all their first partial derivatives, square integrable.

The Bubnov–Galerkin method of weighted residuals[54] yields the weak form of the equation of momentum conservation (40) as

$$\int_{\Omega_0^{(i)}} \delta\mathbf{u}^{(i)} \cdot \left(\rho_0^{(i)}\mathbf{A}^{(i)} - \operatorname{Div}\mathbf{P}^{(i)} - \rho_0^{(i)}\mathbf{b}_0^{(i)}\right) \mathrm{d}\Omega = 0. \tag{51}$$

The exploitation of Green's theorem (integration per partes), Gauss–Ostrogradsky theorem, and properties of the weighting space $\mathcal{W}^{(i)}$ finally yields

$$\int_{\Omega_0^{(i)}} \left(\delta\mathbf{u}^{(i)} \cdot \rho_0^{(i)}\mathbf{A}^{(i)} + \mathrm{Grad}\,\delta\mathbf{u}^{(i)} : \mathbf{P}^{(i)} - \delta\mathbf{u}^{(i)} \cdot \rho_0^{(i)}\mathbf{b}_0^{(i)}\right) \mathrm{d}\Omega^{(i)}$$

$$- \int_{\Gamma_\sigma^{(i)}} \delta\mathbf{u}^{(i)} \cdot \hat{\mathbf{T}}^{(i)}\,\mathrm{d}\Gamma^{(i)} - \int_{\Gamma_\mathrm{c}^{(i)}} \delta\mathbf{u}^{(i)} \cdot p_\mathrm{c}^{(i)}\bar{\mathbf{n}}^{(k)}\,\mathrm{d}\Gamma^{(i)} = 0, \tag{52}$$

where Grad denotes the gradient operation with respect to the initial coordinates. In this complex expression, one can identify

$$\delta\Pi_{\mathrm{int}}^{(i)} = \int_{\Omega_0^{(i)}} \delta\mathbf{u}^{(i)} \cdot \rho_0^{(i)}\mathbf{A}^{(i)}\,\mathrm{d}\Omega^{(i)} + \int_{\Omega_0^{(i)}} \mathrm{Grad}\,\delta\mathbf{u}^{(i)} : \mathbf{P}^{(i)}\,\mathrm{d}\Omega^{(i)}, \tag{53}$$

$$\delta\Pi_{\mathrm{ext}}^{(i)} = -\int_{\Omega_0^{(i)}} \delta\mathbf{u}^{(i)} \cdot \rho_0^{(i)}\mathbf{b}_0^{(i)}\,\mathrm{d}\Omega^{(i)} - \int_{\Gamma_\sigma^{(i)}} \delta\mathbf{u}^{(i)} \cdot \hat{\mathbf{T}}^{(i)}\,\mathrm{d}\Gamma^{(i)}, \tag{54}$$

$$\delta\Pi_\mathrm{c}^{(i)} = -\int_{\Gamma_\mathrm{c}^{(i)}} \delta\mathbf{u}^{(i)} \cdot p_\mathrm{c}^{(i)}\bar{\mathbf{n}}^{(k)}\,\mathrm{d}\Gamma^{(i)}, \tag{55}$$

where $\delta\Pi_{\mathrm{int}}^{(i)}$ consists of inertial term and stress divergence term, $\delta\Pi_{\mathrm{ext}}^{(i)}$ comprises of terms expressing virtual work due to volume and surface forces, respectively, and $\delta\Pi_\mathrm{c}^{(i)}$ denotes virtual work done by contact forces i.e. the contact residual term. With these definitions at hand, one can finally write

$$\delta\Pi = \sum_{i=1}^{2} \left(\delta\Pi_{\mathrm{int}}^{(i)} + \delta\Pi_{\mathrm{ext}}^{(i)} + \delta\Pi_\mathrm{c}^{(i)}\right) = 0. \tag{56}$$

In addition to the balance of momentum (40), the contact constraints $(47)_2$ are also enforced in the weak sense

$$\delta\Pi_\mathrm{N}^{(i)} = \int_{\Gamma_\mathrm{c}^{(i)}} \delta p_\mathrm{c}^{(i)} g_\mathrm{N}^{(i)}\,\mathrm{d}\Gamma^{(i)} \leq 0, \tag{57}$$

where $\delta p_\mathrm{c}^{(i)} \in \mathcal{M}_\mathrm{t}^{(i)}$ is the weighting contact pressure field.

Now we can proceed to formulate the contact initial-boundary value contact problem in the weak form: Let $\rho_0^{(i)}$, $\mathbf{b}_0^{(i)}$, $\mathbf{u}_0^{(i)}$, $\mathbf{v}_0^{(i)}$ defined on $\bar{\Omega}^{(i)}$ and $\hat{\mathbf{T}}^{(i)}$, $\hat{\mathbf{u}}^{(i)}$ defined on $\Gamma_\sigma^{(i)}$, $\Gamma_\mathrm{u}^{(i)}$, respectively, are given square-integrable functions. In every time instance $t \in \mathbb{T}$, we seek for a displacement field

$\mathbf{u}^{(i)} \in \mathcal{S}_{\mathrm{t}}^{(i)}$ and a contact pressure field $p_{\mathrm{c}}^{(i)} \in \mathcal{M}_{\mathrm{t}}^{(i)}$, which for all weight functions $\delta\mathbf{u}^{(i)} \in \mathcal{W}^{(i)}$ and $\delta p_{\mathrm{c}}^{(i)} \in \mathcal{M}_{\mathrm{t}}^{(i)}$ satisfies

$$\sum_{i=1}^{2} \left(\delta\Pi_{\mathrm{int}}^{(i)} + \delta\Pi_{\mathrm{ext}}^{(i)} + \delta\Pi_{\mathrm{c}}^{(i)} \right) = 0, \tag{58}$$

$$\sum_{i=1}^{2} \delta\Pi_{\mathrm{N}}^{(i)} \leq 0. \tag{59}$$

7. Enforcement of contact constraints

In the just-described weak formulation of the initial-boundary contact problem, the displacement field and the contact pressure field have been introduced. They are bonded through KKT conditions. One can recognize the method of Lagrange multipliers and put $p_{\mathrm{c}}^{(i)} = \lambda_{\mathrm{N}}^{(i)}$. In this section, a much simpler method — the method of penalty function — will be presented.

The penalty method is the most commonly used method of regularization of contact traction vector components. The popularity of the penalty method comes from the fact that it is easy to implement and it has a vivid physical interpretation. The penalty function may, in fact, be seen as the spring, which is active only in that part of the contact boundary, where is a mutual penetration of the contacting bodies. The higher the value of penalty function, the less the resultant penetration. Hence the main disadvantage of this method, namely, that the contact conditions are fulfilled only approximately for the final value of the penalty function.

The principle of regularization of the contact pressure by the penalty method is shown in Fig. 5. It consists in replacing the unknown contact pressure field with a new one which is dependent on the displacement field.

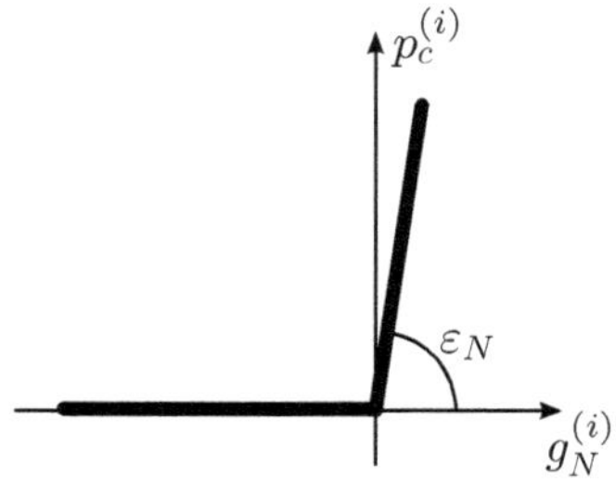

Fig. 5. A scheme of regularization of the normal contact constraints by the penalty method.

Thereby, the system of Eqs. (58)–(59) reduces to Eq. (58), where the only unknown is the displacement field, $\mathbf{u}$. The contact pressure is prescribed by the function

$$p_c^{(i)} = \epsilon_N \langle g_N^{(i)} \rangle, \tag{60}$$

where ϵ_N is the normal penalty function, most commonly a constant, and $\langle \bullet \rangle$ are so-called Macaulay's brackets defined as

$$\langle \bullet \rangle := \frac{|\bullet| + \bullet}{2}. \tag{61}$$

It is thus an operator that for a positive argument returns the argument itself and for a negative argument returns zero.

The contact constraints are theoretically fulfilled exactly for $\epsilon_N \to \infty$, but in practice, however, the penalty function significantly higher than the stiffness of the contacting bodies results in the ill-conditioning of the linear system of equations. This problem can be quite simply eliminated by Uzawa's augmentation method.[23,51]

8. Linearization

After contact constraints regularization, the resulting system of equations (56) nonlinearly depends on the displacements field, $\mathbf{u}^{(i)}$. An efficient way to find a solution of system of nonlinear equations is the Newton–Raphson method. It is based on iterative solving of the linearized problem. To linearize a nonlinear problem, the concept of directional derivative has to be introduced. The directional derivative $D_{\Delta\mathbf{u}}[F](\mathbf{u})$ of a functional F at a point $\mathbf{u}$ and in the direction $\Delta\mathbf{u}$ is defined as

$$D_{\Delta\mathbf{u}}\left[F\right]\left(\mathbf{u}\right) = \frac{d}{d\alpha} F(\mathbf{u} + \alpha\Delta\mathbf{u})\bigg|_{\alpha=0}. \tag{62}$$

With the concept of directional derivatives at hand, one can proceed to construct the linearization of the weak form (56) at the point $\mathbf{u}$

$$\delta\Pi_{\text{int,ext}}^{(i)}\Big|_{\mathbf{u}} + \delta\Pi_{\text{c}}^{(i)}\Big|_{\mathbf{u}} + D_{\Delta\mathbf{u}}\left[\delta\Pi_{\text{int,ext}}^{(i)}\right] + D_{\Delta\mathbf{u}}\left[\delta\Pi_{\text{c}}^{(i)}\right] = 0. \tag{63}$$

The linearization of the virtual work due to internal and external forces will be not specified in this work. The interested reader is referred to Refs. [50, 55]. After substitution from (60), the directional derivative of the contact residual term (55) in the direction of $\Delta\mathbf{u}$ is

$$D_{\Delta\mathbf{u}}\left[\delta\Pi_{\mathrm{c}}^{(i)}\right] = -\int_{\Gamma_{\mathrm{c}}^{(i)}} \delta\mathbf{u}^{(i)} \cdot \epsilon_{\mathrm{N}} D_{\Delta\mathbf{u}}\left[\langle g_{\mathrm{N}}^{(i)}\rangle\right] \bar{\mathbf{n}}^{(k)}\,\mathrm{d}\Gamma^{(i)}, \tag{64}$$

where the derivative of normal vector has been neglected as a term of higher order. The directional derivatives of the normal gap function (22) are

$$D_{\Delta\mathbf{u}}\left[\langle g_{\mathrm{N}}^{(i)}\rangle\right] = -H(g_{\mathrm{N}}^{(i)})(\Delta\mathbf{u}^{(i)} - \Delta\bar{\mathbf{u}}^{(k)}) \cdot \bar{\mathbf{n}}^{(k)}, \tag{65}$$

where $H(\bullet)$ denotes Heaviside step function, which results from the derivative of Macaulay's brackets. The final expression for directional derivative of the contact residual term is

$$D_{\Delta\mathbf{u}}\left[\delta\Pi_{\mathrm{c}}^{(i)}\right] = -\int_{\Gamma_{\mathrm{c}}^{(i)}} \delta\mathbf{u}^{(i)} \cdot \bar{\mathbf{n}}^{(k)}[\epsilon_{\mathrm{N}} H(g_{\mathrm{N}}^{(i)})(\Delta\mathbf{u}^{(i)} - \Delta\bar{\mathbf{u}}^{(k)}) \cdot \bar{\mathbf{n}}^{(k)}]\,\mathrm{d}\Gamma^{(i)}. \tag{66}$$

This certainly compact expression, however, represents only incomplete linearization due to the negligence of the derivative of the normal vector in (64). Therefore, one cannot expect the quadratic rate of convergence of the Newton–Raphson method. Nevertheless, this inaccuracy is outweighed by the simplicity of implementation.

9. Finite element discretization

The aim of the spatial discretization is to construct finite-dimensional subspaces $\mathcal{W}^{h(i)}$ and $\mathcal{S}_{\mathrm{t}}^{h(i)}$ of the infinite-dimensional spaces $\mathcal{W}^{(i)}$ and $\mathcal{S}_{\mathrm{t}}^{(i)}$ defined above. In the isogeometric finite element analysis, the geometry of the whole region $\Omega_0^{(i)} \subset \mathbb{R}^{d_{\mathrm{s}}}$ is accurately described by the mapping $\mathbf{X}(\boldsymbol{\xi}) : \Omega_\square \mapsto \Omega_0^{(i)}$, where $\Omega_\square \subset \mathbb{R}^{d_{\mathrm{p}}}$ indicates the parametric domain (In $\mathbb{R}^1$, the parametric domain is usually the interval $\Omega_\square = [0,1]$, in $\mathbb{R}^2$ it is the square $\Omega_\square = [0,1] \times [0,1]$, etc.). Here, for the sake of simplicity, it is assumed that the domain $\Omega_0^{(i)}$ is represented by a single patch. In the NURBS-based IGA, this mapping is given by linear combination of NURBS basis functions and coordinates of control points

$$\mathbf{X}^{(i)}(\boldsymbol{\xi}) = \sum_{A=1}^{n_{\mathrm{cp}}} R_A(\boldsymbol{\xi})\mathbf{X}_A^{(i)}, \tag{67}$$

where $\mathbf{X}_A^{(i)} \in \Omega_0^{(i)}$ is the initial position vector of the Ath control point, $R_A(\boldsymbol{\xi})$ is the NURBS basis function associated with the Ath control point, and n_{cp} is the number of control points of the patch. The index of global basis function, A, is related to indices of univariate basis function by the connectivity array $A = \texttt{INC}(j_1, j_2, \ldots, j_{d_p})$, cf. Eq. (6).

The fundamental idea of the spatial discretization by the finite element method is the decomposition of the domain $\Omega_0^{(i)}$ into the so-called finite elements $\Omega_e^{(i)} \subset \mathbb{R}^{d_s}$

$$\Omega_0^{(i)} = \bigcup_{e}^{n_{\mathrm{el}}} \Omega_0^{e(i)}, \tag{68}$$

where n_{el} is the number of elements. Therefore, it is also useful to consider local mapping defined over one individual knot span, which can be interpreted as a finite element, as $\mathbf{X}^{e(i)}(\tilde{\boldsymbol{\xi}}) : \tilde{\Omega}_{\square} \mapsto \Omega_0^{e(i)}$, where $\tilde{\Omega}_{\square} \subset \mathbb{R}^{d_{\mathrm{P}}}$ is the parent domain (In $\mathbb{R}^1$ the parent domain is usually the interval $\tilde{\Omega}_{\square} = [-1, 1]$, in $\mathbb{R}^2$ it is the bi-unit square $\Omega_{\square} = [-1, 1] \times [-1, 1]$, etc.). Then element mapping is given as

$$\mathbf{X}^{e(i)}(\tilde{\boldsymbol{\xi}}) = \sum_{a=1}^{n_{\mathrm{ecp}}} R_a(\tilde{\boldsymbol{\xi}}) \mathbf{X}_a^{e(i)}, \tag{69}$$

where $\mathbf{X}_a^{e(i)} \in \Omega_0^{e(i)}$ is the initial position vector of the ath control point, $R_a(\boldsymbol{\xi})$ is the NURBS basis function associated with the ath control point, and n_{ecp} indicates number of element control points. The index of local and global control point are joined by connectivity array $A = \texttt{IEN}(e, a)$, where e is the index of element.

Further, the so-called isoparametric concept, first introduced by Irons,[56] will be employed. The fundamental idea consists of the fact that any field being discretized on the region Ω_0 uses the same basis functions as the domain Ω_0 itself in the mapping $\mathbf{X}^{(i)}(\boldsymbol{\xi})$. Consequently, the discretized displacement field, displacement test functions and the field of spatial coordinates, defined on the domain $\Omega_0^{(i)}$, can be written as

$$\mathbf{u}^{h(i)}(\boldsymbol{\xi}, t) = \sum_{A=1}^{n_{\mathrm{cp}}} R_A(\boldsymbol{\xi}) \mathbf{d}_A^{(i)}(\mathbf{X}_A^{(i)}, t), \tag{70}$$

$$\delta\mathbf{u}^{h(i)}(\boldsymbol{\xi}) = \sum_{A=1}^{n_{\mathrm{cp}}} R_A(\boldsymbol{\xi}) \mathbf{c}_A^{(i)}(\mathbf{X}_A^{(i)}), \tag{71}$$

$$\mathbf{x}^{(i)}(\boldsymbol{\xi}, t) = \sum_{A=1}^{n_{\mathrm{cp}}} R_A(\boldsymbol{\xi}) \mathbf{x}_A^{(i)}(\mathbf{X}_A^{(i)}, t), \tag{72}$$

where $\mathbf{d}_A^{(i)} \in \mathbb{R}^{d_s}$ are control point-displacements, $\mathbf{c}_A^{(i)} \in \mathbb{R}^{d_s}$ are control point-weighted displacements, and $\mathbf{x}_A^{(i)} \in \mathbb{R}^{d_s}$ are control point position vectors in the current configuration.

The set of all functions defined in this way form the finite-dimensional subspaces $\mathcal{W}^{h(i)} \subset \mathcal{W}^{(i)}$ and $\mathcal{S}_t^{h(i)} \subset \mathcal{S}_t^{(i)}$

$$\mathcal{W}^{h(i)} = \left\{ \delta\mathbf{u}^{h(i)} = \sum_{A=1}^{n_{\mathrm{cp}}} R_A \mathbf{c}_A^{(i)} | \delta\mathbf{u}^{h(i)} = \mathbf{0} \text{ on } \Gamma_u^{h(i)} \right\}, \tag{73}$$

$$\mathcal{S}_t^{h(i)} = \left\{ \mathbf{u}^{h(i)} = \sum_{A=1}^{n_{\mathrm{cp}}} R_A \mathbf{d}_A^{(i)} | \mathbf{u}^{h(i)} = \hat{\mathbf{u}}^{(i)} \text{ on } \Gamma_u^{h(i)} \right\}. \tag{74}$$

Now we can proceed to discretize the weak form introduced in Section 6.2 simply by replacing the infinite-dimensional spaces $\mathcal{W}^{(i)}$ and $\mathcal{S}_t^{(i)}$ with their finite-dimensional subspaces $\mathcal{W}^{h(i)}$ and $\mathcal{S}_t^{h(i)}$. After substituting into Eq. (56) from the definition of spaces $\mathcal{W}^h$ and $\mathcal{S}_t^h$, and after some manipulations which can be found, e.g. in Refs. [51, 55], one can write discrete form of (56) in the matrix notation

$$\delta\Pi^h = \mathbf{c}^T \left[\mathbf{M}\ddot{\mathbf{d}}(t) + \mathbf{F}_{\mathrm{int}}(\mathbf{d}(t)) - \mathbf{F}_{\mathrm{ext}}(t) + \mathbf{F}_{\mathrm{c}}(\mathbf{d}(t)) \right], \tag{75}$$

$$= \mathbf{c}^T \mathbf{R}(\mathbf{d}(t)) = 0, \tag{76}$$

where $\mathbf{c}$ is the vector of control point weighting coefficients of displacement, $\mathbf{R}$ is the residual vector, also known as out-of-balance vector,[55] $\mathbf{M}$ is the mass matrix, $\ddot{\mathbf{d}}$ is the vector of control point accelerations, $\mathbf{F}_{\mathrm{int}}$ is the vector of internal control point forces, $\mathbf{F}_{\mathrm{ext}}$ is the vector of external control point forces and $\mathbf{F}_{\mathrm{c}}$ is the vector of control point contact forces.

Equation (75) must hold for all displacement weighting functions from the space $\mathcal{W}^{h(i)}$. But this means that the vector $\mathbf{c}$ of control point weighting coefficients can be arbitrary. Therefore, the term in square brackets has to be equal to zero vectors

$$\mathbf{M}\ddot{\mathbf{d}}(t) + \mathbf{F}_{\mathrm{int}}(\mathbf{d}(t)) - \mathbf{F}_{\mathrm{ext}}(t) + \mathbf{F}_{\mathrm{c}}(\mathbf{d}(t)) = \mathbf{0}, \tag{77}$$

This is the final semi-discrete system of ordinary differential equations (ODE). The word semi-discrete means that the weak formulation was discretized in space but remains continuous in time. Temporal discretization, however, is not covered in this chapter.

9.1. Gauss-point-to-segment contact discretization

Further in this section, the discretization of the contact residual term $\delta\Pi_{\mathrm{c}}$ will be presented. Because this term is integrated only over the contact boundary $\Gamma_{\mathrm{c}}^{h(i)}$, to establish a global vector of control point contact forces, $\mathbf{F}_{\mathrm{c}}^{(i)}$, it is sufficient to consider the restriction of functions $\mathbf{u}^{h(i)}$ and $\delta\mathbf{u}^{h(i)}$ on $\Gamma_{\mathrm{c}}^{h(i)}$. By the substitution of (71) into (55)

$$\delta\Pi_c^{h(i)} = -\int_{\Gamma_c^{(i)}} \sum_{A=1}^{n_{\mathrm{cp}}} R_A \mathbf{c}_A^{(i)} \cdot \epsilon_N \langle g_N^{(i)} \rangle \mathbf{n}^{(i)}\, d\Gamma^{(i)}, \tag{78}$$

Note that the superscript h indicating discretized quantities has been omitted to simplify the notation. Similarly to the bulk elements, $\Omega_0^{e(i)}$, also the boundary $\Gamma_c^{(i)}$ is decomposed into finite elements, here called segments

$$\Gamma_c^{(i)} = \bigcup_{s=1}^{n_s} \Gamma_c^{s(i)}. \tag{79}$$

Consequently, the contact residual can be assemble segment-wise

$$\delta\Pi_c^{h(i)} = \bigcup_{s=1}^{n_{\mathrm{s}}} -\left(\mathbf{c}^{s(i)}\right)^{\mathrm{T}} \int_{\Gamma_c^{s(i)}} \epsilon_N \langle g_N^{(i)} \rangle \mathbf{N}^{s(i)}\, d\Gamma^{s(i)}, \tag{80}$$

where $\mathbf{c}^{s(i)}$ is the segment matrix of control point weighted-displacements, and $\mathbf{N}^{s(i)}$ is the segment matrix defined as

$$\mathbf{N}^{s(i)} = \begin{bmatrix} R_1 \bar{\mathbf{n}}^{(k)} \\ \vdots \\ R_{n_{\mathrm{scp}}} \bar{\mathbf{n}}^{(k)} \end{bmatrix}, \tag{81}$$

where n_{scp} is the number of segment control points. Finally, one can introduce the vector of equivalent control point contact forces through the identity

$$\delta\Pi_c^{h(i)} = \bigcup_{s=1}^{n_s} (\mathbf{c}^{s(i)})^T \mathbf{F}_c^{s(i)}, \tag{82}$$

where for $\mathbf{F}_c^{s(i)}$ holds

$$\mathbf{F}_c^{s(i)} = -\int_{\Gamma_c^{s(i)}} \epsilon_N \langle g_N^{(i)} \rangle \mathbf{N}^{s(i)}\, d\Gamma^{s(i)}. \tag{83}$$

The global vector of control point contact forces is then assembled in a conventional manner

$$\mathbf{F}_{\mathrm{c}} = \sum_{i=1}^{2} \bigcup_{s=1}^{n_s} \mathbf{F}_c^{s(i)}. \tag{84}$$

What remains is to discretize the linearization of the contact residual (66). By following the same steps as in the case of contact residual vector, one comes to the expression

$$D_{\Delta\mathbf{u}}[\delta\Pi_c^{(i)}] = \bigcup_{s=1}^{n_{\mathrm{s}}} (\mathbf{c}^{s(i)})^{\mathrm{T}} \mathbf{K}_c^{s(i)} \begin{bmatrix} \Delta\mathbf{d}^{s(i)} \\ \Delta\mathbf{d}^{s(k)} \end{bmatrix}, \tag{85}$$

where the contact tangent matrix is

$$\mathbf{K}_c^{s(i)} = -\int_{\Gamma_c^{s(i)}} \epsilon_N H(g_N) \left[\mathbf{N}^{s(i)} \left(\mathbf{N}^{s(i)}\right)^T - \mathbf{N}^{s(i)} \left(\mathbf{N}^{s(k)}\right)^T\right]. \tag{86}$$

The global contact tangent matrix stemming from the conventional finite element assembly procedure is

$$\mathbf{K}_c = \sum_{i=1}^{2} \bigcup_{s=1}^{n_s} \mathbf{K}_c^{s(i)}. \tag{87}$$

Notice that the expression for the contact tangent matrix arises from the discretization of the continuous term for the directional derivative of the contact residual (66).

10. Solution procedures

After finite element discretization one arrives to the system of nonlinear ODE. The source of nonlinearity in contact is the normal gap function. Its evaluation implements the contact detection procedure, which is the subject of this section. The solution of the system of nonlinear equations is mentioned at the end of this section.

10.1. *Contact detection*

By the contact detection it is understood the evaluation of the normal gap function, (22), in all Gaussian quadrature points of the potential contact boundary Γ_{c}. The importance of this procedure results from the fact that this evaluation must be performed in each iteration of the nonlinear solver. Computational time of this procedure determines the resulting time demands of the whole contact algorithm. Indeed, the simplest contact detection procedure consists in solving (13) by a nonlinear solver, such as the Newton–Raphson method, in the loop over slave quadrature points and master contact patches. The ability to perform loop over contact patches, instead of contact segments, is one of the most important benefits of the isogeometric contact analysis.

It is obvious that searching for the closest point in a cycle over all quadrature points is not optimal, because if the bodies are not in contact, the exact value of the gap function is not necessary. Instead, only logical value is sufficient to indicate that no contact occurs. Therefore, the contact detection is usually divided into two phases: a global search, which consists of effectively detecting and sorting all potential candidate slave points and their corresponding candidate master segments, and a local search for closest point projections of slave points onto master segments.

The closest point projection of a point onto a free-form-surface (FFS), as spline surfaces are sometimes called, also poses a common problem to computer graphics. Therefore, these methods can be directly employed in the isogeometric contact analysis. Computational methods for the orthogonal projection of points in CAD/CAM applications were presented in a review paper.[57] In order to guarantee robustness of the calculation, it is recommended to initially employ a subdivision based global scheme, followed by a Newton-type iteration method or the second-order algorithms with small sensitivity to the choice of initial estimation such as the sphere approximation method[58] and the torus approximation method.[59]

10.1.1. *Global contact search*

One of the first subdivision schemes was presented by Piegel and Tiller.[60] The so-called base surface is subdivided into quadrilaterals and a slave point is projected onto the closest quadrilateral. The parametric coordinates of the projection are recovered from the parametric coordinates of the corner point of the closest quadrilateral. A similar concept was proposed by Ma and Hewitt[61] where the base surface is recursively subdivided into Bézier surfaces and their control polygons are utilized to find the projection. An improved version of this method was proposed in the work of Selimovic,[62] where a large number of subdivided parts are excluded from the computation by introducing new elimination criteria, namely the end-point interpolation, convex hull property and tangent cones. An even more efficient method is the circular clipping algorithm.[63] A sphere is used as an elimination region. After subdivision, the surface segments located outside the sphere are repeatedly eliminated until the termination condition is satisfied. The concept of the clipping sphere method was further improved in the work of Oh *et al.*,[64] by using the separating axis and k-DOP (Discrete Orientation Polytopes) type of bounding scheme.

10.1.2. *Local contact search*

When the global search is successfully done, a local one is performed. It consists in the solution of the closest point projection problem (11). The existence and uniqueness of this problem were investigated for C^0, C^1 and C^2 continuous master segment by Konyukhov and Schweizerhof.[65] An analytical solution exists only for linear triangular segments, but bilinear quadrilateral, as well as higher order, elements have to be treated numerically. The Newton–Raphson method is usually applied to this end.

Interesting alternatives to the Newton-type iteration method are the geometric iteration methods.[66] In particular, the torus approximation method[59] has proven to be a very powerful technique, especially for NURBS surfaces of higher order where the costly evaluation of base functions and their derivations by recurrent formulas must be carried out.

10.2. *Solution of nonlinear equations*

Newton–Raphson method is one of the effective techniques for solving systems of nonlinear algebraic equations. Its principle consists in converting the system of equations into the residual form

$$\mathbf{R}(\mathbf{d}) = \mathbf{0}, \tag{88}$$

and its subsequent linearization according to iterative scheme

$$\mathbf{R}(\mathbf{d}^{j+1}) \approx \mathbf{R}(\mathbf{d}^j) + \left.\frac{\partial \mathbf{R}}{\partial \mathbf{d}}\right|_{\mathbf{d}=\mathbf{d}^j} \Delta\mathbf{d}^j = \mathbf{0}, \tag{89}$$

where j indicates iteration counter. For the increment $\Delta\mathbf{d}^j$, one gets from the last equation

$$\Delta\mathbf{d}^j = \left.\left[\frac{\partial \mathbf{R}}{\partial \mathbf{d}}\right]^{-1}\right|_{\mathbf{d}=\mathbf{d}^j} (-\mathbf{R}(\mathbf{d}^j)). \tag{90}$$

Then a new estimate of the solution can be calculated from the increment $\Delta\mathbf{d}^j$ as

$$\mathbf{d}^{j+1} = \mathbf{d}^j + \Delta\mathbf{d}^j. \tag{91}$$

The whole process is repeated until $\mathbf{R}(\mathbf{d}^{j+1})$ and $\Delta\mathbf{d}^j$ are smaller than the selected tolerances. It can be shown that the Newton–Raphson method converges quadratically[67] to the nearest root of the nonlinear system, provided that appropriate initial guess, $\mathbf{d}^0$, is chosen so that the tangent matrix is positive definite during the whole iteration process.

If the inertial forces in (75) can be neglected and external loads do not depend on time, the residual vector for quasi-static problems is obtained as

$$\mathbf{R}(\mathbf{d}^j) = \mathbf{F}_{\text{int}}(\mathbf{d}^j) + \mathbf{F}_{\text{c}}(\mathbf{d}^j) - \mathbf{F}_{\text{ext}} = \mathbf{0} \tag{92}$$

and the tangent matrix is

$$\frac{\partial \mathbf{R}}{\partial \mathbf{d}} = \left.\frac{\partial \mathbf{F}_{\text{int}}}{\partial \mathbf{d}}\right|_{\mathbf{d}=\mathbf{d}^j} + \left.\frac{\partial \mathbf{F}_{\text{c}}}{\partial \mathbf{d}}\right|_{\mathbf{d}=\mathbf{d}^j} \tag{93}$$

$$= \mathbf{K}(\mathbf{d}^j) + \mathbf{K}_{\text{c}}(\mathbf{d}^j), \tag{94}$$

$$(\mathbf{K} + \mathbf{K}_{\text{c}})\,\Delta\mathbf{d}^j = \mathbf{F}_{\text{ext}} - \mathbf{F}_{\text{int}} - \mathbf{F}_{\text{c}}. \tag{95}$$

This is the final form of the Newton–Raphson scheme where the vector of equivalent contact forces $\mathbf{F}_{\text{c}}$ is given by (84) and the contact tangent matrix $\mathbf{K}_{\text{c}}$ is defined by (87). The remaining matrices and vectors are given by standard finite element method procedures and the reader can find them, for example, monographs in Refs. [50, 54 and 55].

11. Numerical examples

Three numerical examples are presented in this section to demonstrate properties of the presented contact algorithm, namely, the contact patch test, the Hertz problem and and contact ironing problem.

11.1. *Contact patch test*

In the first example, the contact patch test according to Taylor and Papadopoulos[68] is presented. The purpose of this example is to test the ability of contact discretization to accurately transfer constant pressure loads from one body to another. Two blocks of different size are pressed against each other. Dimensions are depicted in the Fig. 6(a), where $a = 1$ m. Both blocks are subjected to a pressure $p = 1$Pa. The same material with Poisson's ratio $\nu = 0.3$ and Young's modulus $E = 1000$Pa is used for both blocks. The analytical solution is $\sigma_y = -1$Pa.

Either of the blocks is discretized by one bivariate NURBS patch of order $p^{(i)} = 1$ in each parametric dimension. The lower block is discretized

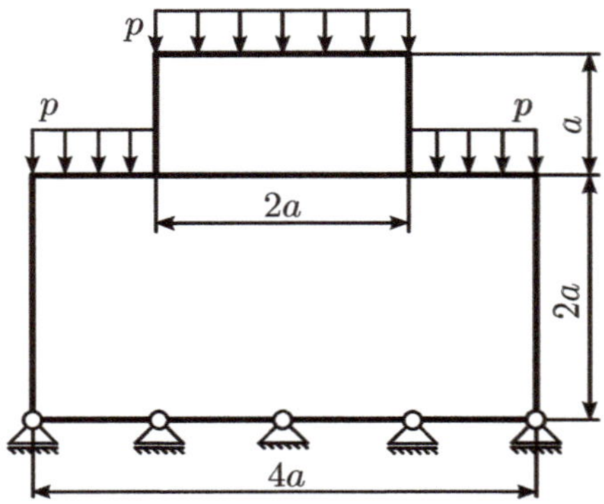

(a) Dimensions and boundary conditions.

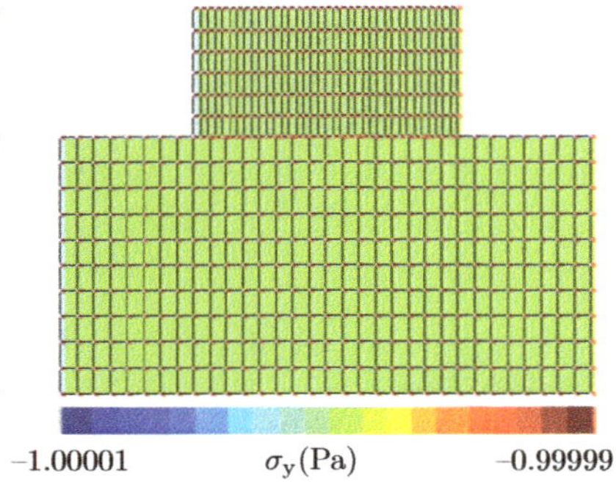

(b) Contour of stress component σ_y.

Fig. 6. 2D contact patch test.

by $n_1^{(1)} = 33$ times $n_2^{(1)} = 11$ control points and the upper block by $n_1^{(2)} = 37$ times $n_2^{(2)} = 7$ control points. Corresponding uniform knot vectors were considered.

In Fig. 6(b), there are contours of σ_y stress component, which perfectly match the analytical solution. In fact, the maximum error of the numerical solution is in the order of machine precision. Note that the same results were obtained for bi-quadratic and bi-cubic NURBS patches.

11.2. *Hertzian contact problem*

In the second example, the classic Hertzian problem is investigated. In particular, the plane strain contact of cylinder with rigid plane is considered. The material parameters, geometry, and boundary conditions as well as IGA mesh parameterization are taken from the work of Temizer *et al.*[22]

The cylinder with the outer radius $r_o = 1\,\mathrm{m}$ is made of a linear elastic material characterized by the Poisson's ratio $\nu = 0.3$ and Young's modulus $E = 1\,\mathrm{Pa}$. The vertical load per unit length of the cylinder is $p_d = 10^{-3}\,\mathrm{Pa}$, so that the total vertical force is $F_c = 2r_o p_d = 2 \times 10^{-3}\,\mathrm{N}$. In the work of Temizer *et al.*,[22] one can find analytical solution for the contact area radius $a = \sqrt{4F_c r_o/(\pi E^*)}$, where $E^* = E/(1-\nu^2)$, and maximum contact pressure $p_o = 2F_c/(\pi a)$. The contact pressure distribution is $p_c(X) = p_0\sqrt{1-(X/a)^2}$. As in Ref. [22], the cylinder is modelled with a non-zero inner radius $r_i = 0.1\,\mathrm{m}$. The corresponding vertical load on the inner radius is $p_i(\phi) = p_d \cos(\phi)$.

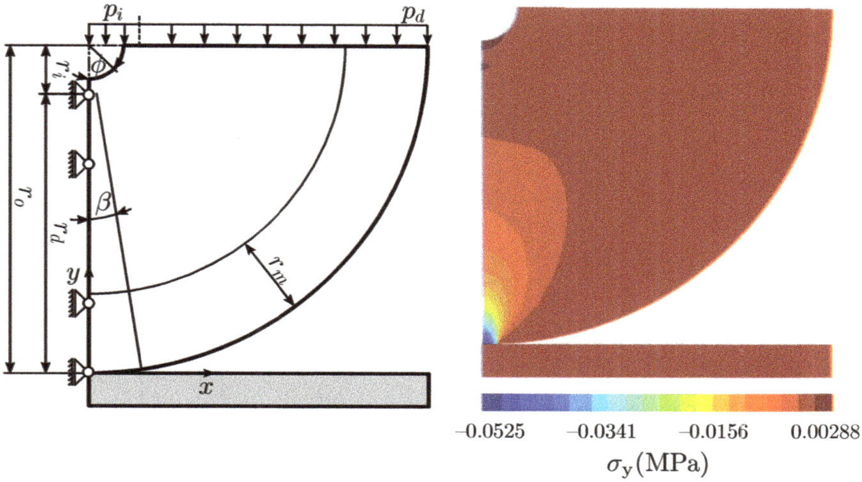

(a) (b)

Fig. 7. Hertzian contact problem. (a) Parameters of the IGA mesh and (b) Contours of σ_y^* stress component.

The entries of the knot vector are located in such a way that approximately 75% of the elements in the radial direction are distributed in the outer circular ring (see Fig. 7(a)) of radius $r_{\rm m} = 0.1r_d$, where $r_d = r_{\rm o} - r_{\rm i}$. Likewise, approximately 75% of the elements in the angular direction are distributed in the region defined by the angle $\beta = 0.1\pi/2$. Particularly, 48 linear elements along the radial direction and 81 linear elements along the angular direction are considered. The r-refinement is employed during increasing the polynomial order of the basis functions, i.e. the control points are relocated but their number is kept constant.

For illustration, the contours of the normalized vertical stress component $\sigma_{\rm y}^* = \sigma_{\rm y}/p_{\rm o}$ are shown in Fig. 7(b). The detailed study of contact pressure distribution in dependence on the contact area radius a is presented in Fig. 8. In this figure, the plots in the left column were obtained for the penalty parameter $\epsilon_{\rm N} = 10^4\,{\rm N/m^3}$, whereas the plots in the right column were obtained for $\epsilon_N = 10^6\,{\rm N/m^3}$. Rows of Fig. 8 differs in the polynomial order p of the basis functions. In each subplot of Fig. 8 there is analytical solution, the contact pressure at Gauss points ($p_{c_g}^{(i)} = \epsilon_N g_{N_g}^{(i)}$), and a post-processed contact pressure denoted in the legend as NURBS.

Here, we employed post-processing scheme for the contact pressure that was originally proposed by Sauer.[16] The idea is to calculate control point

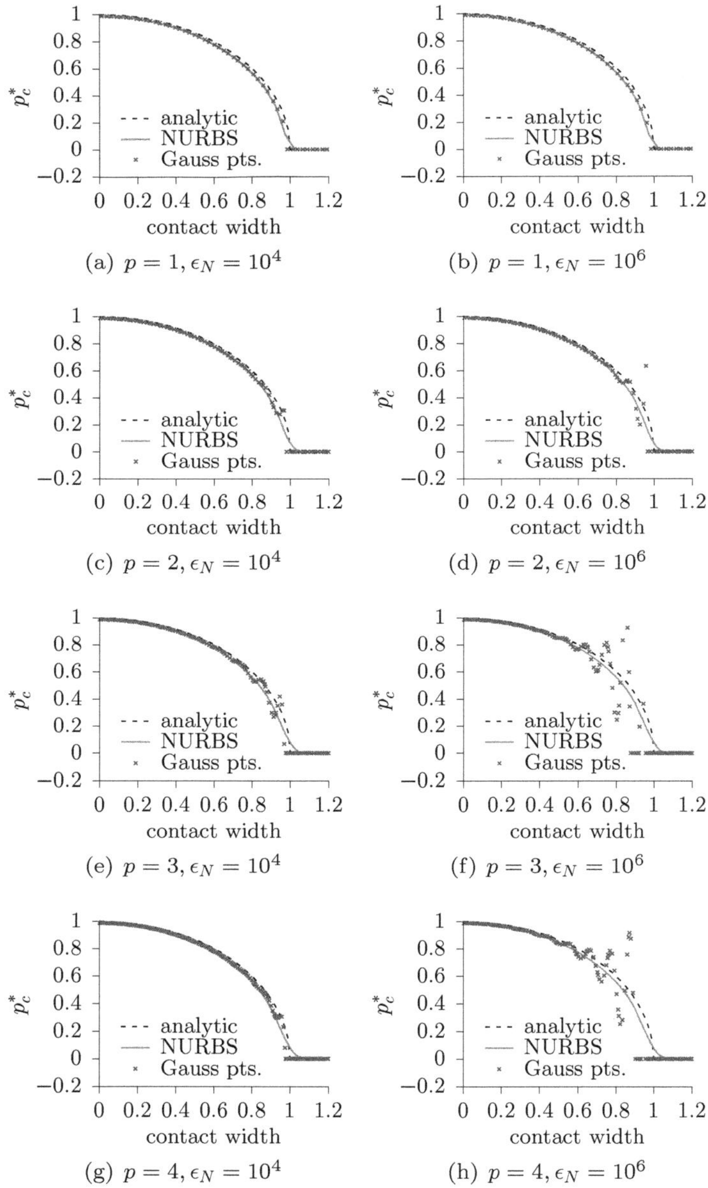

Fig. 8. Contact pressure distribution.

contact pressures $p_{cA}^{(i)}$ to get post-processed contact pressure in the form

$$p_c^{p(i)} = \sum_{A=1}^{n_{\mathrm{cp}}} R_A p_{cA}^{(i)}. \tag{96}$$

The calculation of the control point values of the contact pressure is inspirited by the mortar method.[22] In particular, control point pressures are evaluated as a weighted average

$$p_{cA}^{(i)} = \frac{\int_{\Gamma_c^{(i)}} R_A \epsilon_N \langle g_N^{(i)} \rangle \, d\Gamma^{(i)}}{\int_{\Gamma_c^{(i)}} R_A \, \mathrm{d}\Gamma^{(i)}}. \tag{97}$$

One can see in Fig. 8 that post-processed contact pressure shows very good agreement with the analytical solution. On the other hand, the contact pressure at Gauss points exhibits oscillations that increase with the polynomial order of the basis functions. This is the expected result because the GPTS formulation leads to the over-constrained system of equations. Note that calculation of the weighted average quantities is the bottom line of the mortar–KTS algorithm.[22] While Eq. (97) was used only once in a post-processing solution, the mortar FE algorithm uses this formula in each iteration of the nonlinear solution to obtain the correct number of contact constraints.

11.3. *3D ironing problem*

In the last numerical example, the finite deformation contact of a cylindrical elastic die with a rubber block is analysed. This problem was first formulated in the work of Puso and Laursen[69] and similar investigations were made by Popp *et al.*[70]

The Poisson's ratio and Young's modulus of the elastic die are $E = 1000\mathrm{Pa}$, $\nu = 0.3$ and of the rubber block $E = 1\mathrm{Pa}$, $\nu = 0.3$. The conventional incompressible neo-Hookean material model is considered.[50] The dimensions of the block are $a_{\mathrm{x}} = 9\,\mathrm{m}$, $a_{\mathrm{y}} = 4\,\mathrm{m}$, and $a_{\mathrm{z}} = 3\,\mathrm{m}$. The cylindrical die has outer radius $r_{\mathrm{o}} = 3\,\mathrm{m}$, height $h = 5.2\,\mathrm{m}$ and wall thickness $t = 0.2\,\mathrm{m}$. Note that only 45° sector of the cylinder is considered, in order to represent the die only by single patch.

Both the die and the block are discretized by one trivariate NURBS patch of the first order in each parametric dimension. The exception is

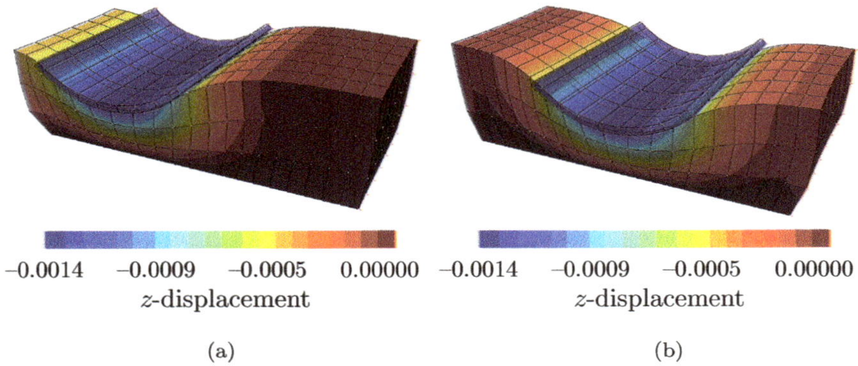

Fig. 9. Two characteristic stages of deformation of the ironing problem. (a) $u_1 = 0.0, u_3 = -1.4$ and (b) $u_1 = 2.0, u_3 = -1.4$.

the rounded direction where second order is considered in order to exactly describe the geometry. The die consists of $n_1^{(1)} = 10$, $n_2^{(1)} = 6$ and $n_3^{(1)} = 2$ control points in each parametric dimension. Similarly, the block is discretized by $n_1^{(2)} = 21$, $n_2^{(2)} = 7$ and $n_3^{(2)} = 4$ control points. Again, uniform knot vectors are considered.

The initial position of the die centre is 2.5 m from the left edge of the block and 3.0m from the top surface of the block. The die is first pressed into the block by prescribing a vertical displacement $u_3 = 1.4$ m. This stage is recorded in Fig. 9(a). Then it slides along the block until a horizontal displacement $u_1 = 4.0$m in the x-direction is reached. The moment when the die is in the middle of the block is illustrated in Fig. 9(b). It can be concluded that presented contact formulation is capable to solve finite sliding contact problems.

12. Conclusions

This chapter dealt with the treatment of contact in IGA. From the review of the literature mentioned in the first section, it is clear that this topic has undergone rapid development. In general, it is not possible to describe all the details and contexts of different isogeometric contact formulations. Instead, authors have intended, on the example of one simple contact formulation, to systematically present all aspects of the numerical solution of contact problems and pointed out the specific properties of the GSA. An emphasis was put on maximum simplicity, so that even an inexperienced reader would be able to implement the contact algorithm into an isogeometric finite element code. To this end, a simple large deformation frictionless contact formulation[47] was chosen. An essential feature of this formulation is an unbiased treatment of the contact residual, which means that there is no need to distinguish between master and slave bodies. Moreover, a complicated projection[33] from one contact body to the other is not required. Further simplification was achieved by using the penalty method to regularize contact constraints and to convert the nonlinear programming problem to the unconstrained optimization problem. Then, the obtained nonlinear equation system was linearized before discretization, making the linearization independent of discretization. Note that only incomplete linearization was considered, again with an emphasis on the simplicity of the resulting implementation.

So far, described steps to treat contact problems do not depend on discretization and can therefore be used for both classic FEA and IGA. A custom discretization by the finite element method consists in the choice of the base of the finite dimensional functional spaces in which the solution is sought. For classic FEA, basis functions are only C^0 continuous at the interface of elements, whereas in IGA, they are designed to achieve higher continuity at element interfaces thanks to the use of spline basis functions. Using the spline basis functions itself implies better properties of the solution of contact problems. This is mainly due to an unambiguous definition of the normal vector field at the contact interface.

The most significant difference between contact algorithms for classic FEA and IGA lies in the detection of contact. While in the classic FEA nodal points interpolate the elements, i.e. lie directly on their surfaces, in the IGA the control points generally do not lie on the surface of the represented geometry. On the other hand, the properties of the spline representation, such as the convex hull, allow for the construction of highly efficient contact detection algorithms. Another advantage of the IGA representation is the fact that local contact search is possible in the loop over patches rather than over elements.

The performance of the resulting IGA contact algorithm was tested by means of three numerical examples. The contact patch test proved that the symmetry contact formulation was able to accurately transfer load from one body to another. On the Hertz problem, it was demonstrated that oscillations of the Gauss point solution can be efficiently eliminated in post-processing. Finally, the 3D ironing problem showed that the used contact formulation is also capable of solving problems with large deformations and sliding of contact interface.

Acknowledgements

The work of D. Gabriel, R. Kolman and J. Plešek was supported by the European Regional Development Fund under Grant No. CZ.02.1.01/0.0/0.0/15_003/0000493 (Centre of Excellence for Nonlinear Dynamic Behaviour of Advanced Materials in Engineering). The work of J. Kopačka was supported by the Czech Science Foundation under grant number 16-03823S within institutional support RVO:61388998.

References

1. A. L. Eterovic and K. J. Bathe, An interface interpolation scheme for quadratic convergence in the finite element analysis of contact problems, In P. Wriggers and W. Wagner (eds.), *Nonlinear Computational Mechanics: State of the Art*, 1st edn., pp. 703–715, Springer-Verlag Berlin, Heidelberg (1991).
2. V. Padmanabhan and T. A. Laursen, A framework for development of surface smoothing procedures in large deformation frictional contact analysis, *Finite Elements in Analysis and Design*, **37**(3), 173–198 (2001).
3. P. Wriggers, L. Krstulović-Opara, and J. Korelc, Smooth c^1-interpolations for two-dimensional frictional contact problems, *International Journal for Numerical Methods in Engineering*, **51**(12), 1469–1495 (2001).
4. M. Tur, E. Giner, F. J. Fuenmayor, and P. Wriggers, 2D contact smooth formulation based on the mortar method, *Computational Methods in Applied Mechanics and Engineering*, **247–248**, 1–14 (2012).
5. N. El-Abbasi, S. A. Meguid, and A. Czekanski, On the modelling of smooth contact surfaces using cubic splines, *International Journal for Numerical Methods in Engineering*, **50**(4), 953–967 (2001).
6. M. Stadlar, G. A. Holzapfel, and J. Korelc, c^n continuous modelling of smooth contact surfaces using nurbs and application to 2D problems, *International Journal for Numerical Methods in Engineering*, **57**(15), 2177–2203 (2003).
7. D. Chamoreta, P. Saillardb, A. Rassineuxc, and J. M. Bergheaua, New smoothing procedures in contact mechanics, *Journal of Computational and Applied Mathematics*, **168**(1–2), 107–116 (2004).
8. J. J. M. Noz, Modelling unilateral frictionless contact using the null-space method and cubic b-spline interpolation, *Computer Methods in Applied Mechanics and Engineering*, **197**(9–12), 979–993 (2008).
9. S. Stupkiewicz, J. Lengiewicz, and J. Korelc, Sensitivity analysis for frictional contact problems in the augmented lagrangian formulation, *Computer Methods in Applied Mechanics and Engineering*, **199**(33–36), 2165–2176 (2010).
10. J. Lengiewicz, J. Korelc, and S. Stupkiewicz, Automation of finite element formulations for large deformation contact problems, *International Journal for Numerical Methods in Engineering*, **85**(10), 1252–1279 (2011).
11. M. Stadlar and G. A. Holzapfel, Subdivision schemes for smooth contact surfaces of arbitrary mesh topology in 3D, *International Journal for Numerical Methods in Engineering*, **60**(7), 1161–1195 (2004).

12. M. A. Puso and T. A. Laursen, A 3D contact smoothing method using gregory patches, *International Journal for Numerical Methods in Engineering*, **54**(8), 1161–1194 (2002).
13. D. M. Neto, M. C. Oliveira, L. F. Menezes, and J. L. Alves, A contact smoothing method for arbitrary surface meshes using nagata patches, *Computer Methods in Applied Mechanics and Engineering*, **299**, 283–315 (2016).
14. T. Belytschko, W. J. T. Daniel, and G. Ventura, A monolithic smoothing-gap algorithm for contact-impact based on the signed distance function, *International Journal for Numerical Methods in Engineering*, **55**(1), 101–125 (2002).
15. A. Konyukhov and K. Schweizerhof, Incorporation of contact for high-order finite elements in covariant form, *Computer Methods in Applied Mechanics and Engineering*, **198**(13–14), 1213–1223 (2009).
16. R. A. Sauer, Local finite element enrichment strategies for 2D contact computations and a corresponding post-processing scheme, *Computational Mechanics*, **52**, 301–319 (2013).
17. C. J. Corbett and R. A. Sauer, Nurbs-enriched contact finite elements, *Computer Methods in Applied Mechanics and Engineering*, **275**, 55–75 (2014).
18. C. J. Corbett and R. A. Sauer, Three-dimensional isogeometrically enriched finite elements for frictional contact and mixed-mode debonding, *Computer Methods in Applied Mechanics and Engineering*, **284**, 781–806 (2015).
19. T. J. R. Hughes, J. A. Cottrell, and Y. Bazilevs, Isogeometric analysis: CAD, finite elements, NURBS, exact geometry and mesh refinement, *Computer Methods in Applied Mechanics and Engineering*, **194**(39–41), 4135–4195 (2005).
20. J. A. Cottrell, T. J. R. Hughes, and Y. Bazilevs, *Isogeometric Analysis: Toward Integration of CAD and FEA*, John Wiley & Sons (2009).
21. J. Lu, Isogeometric contact analysis: Geometric basis and formulation for frictionless contact, *Computational Methods in Applied Mechanics and Engineering*, **200**, 726–741 (2010).
22. İlker Temizer, P. Wriggers, and T. J. R. Hughes, Contact treatment in isogeometric analysis with nurbs, *Computational Methods in Applied Mechanics and Engineering*, **200**, 1100–1112 (2011).
23. P. Wriggers, *Computational Contact Mechanics*, 2nd edn., Springer (2006).
24. İlker Temizer and P. Wriggers, Inelastic analysis of granular interfaces via computational contact homogenization, *International Journal for Numerical Methods in Engineering*, **84**, 883–915 (2010).
25. L. De Lorenzis, I. Temizer, P. Wriggers, and G. Zavarise, A large deformation frictional contact formulation using nurbs-based isogeometric analysis, *International Journal for Numerical Methods in Engineering*, **87**, 1278–1300 (2011).
26. L. De Lorenzis, P. Wriggers, and G. Zavarise, A mortar formulation for 3D large deformation contact using nurbs-based isogeometric analysis and the augmented lagrangian method, *Computational Mechanics*, **49**, 1–20 (2012).
27. İlker Temizer, M. M. Abdalla, and Z. Gürdal, An interior point method for isogeometric contact, *Computer Methods in Applied Mechanics and Engineering*, **276**, 589–611 (2014).

28. J.-Y. Kim and S.-K. Youn, Isogeometric contact analysis using mortar method, *International Journal for Numerical Methods in Engineering*, **89**, 1559–1581 (2011).
29. L. De Lorenzis, P. Wriggers, and T. J. R. Hughes, Isogeometric contact: A review, *GAMM-Mitteilungen*, **37**(1), 85–123 (2014).
30. M. E. Matzen, T. Cichosz, and M. Bischoff, A point to segment contact formulation for isogeometric, nurbs based finite elements, *Computer Methods in Applied Mechanics and Engineering*, **255**, 27–39 (2013).
31. M. E. Matzen and M. Bischoff, A weighted point-based formulation for isogeometric contact, *Computer Methods in Applied Mechanics and Engineering*, **308**, 73–95 (2016).
32. M. Dittmann, M. Franke, İlker Temizer, and C. Hesch, Isogeometric analysis and thermomechanical mortar contact problems, *Computer Methods in Applied Mechanics and Engineering*, **274**, 192–212 (2014).
33. M. A. Puso and T. A. Laursen, Mesh tying on curved interfaces in 3D, *Engineering Computations*, **20**(3), 305–321 (2003).
34. M. A. Puso, A 3D mortar method for solid mechanics, *International Journal for Numerical Methods in Engineering*, **59**, 315–336 (2004).
35. A. Seitz, P. Farah, J. Kremheller, B. I. Wohlmuth, W. A. Wall, and A. Popp, Isogeometric dual mortar methods for computational contact mechanics, *Computer Methods in Applied Mechanics and Engineering*, **301**, 259–280 (2016).
36. F. Auricchio, L. B. ao da Veiga, T. J. R. Hughes, A. Reali, and G. Sangalli, Isogeometric collocation methods, *Mathematical Models and Methods in Applied Sciences*, **20**(11), 2075–2107 (2010).
37. F. Auricchio, L. Beirão da Veiga, T. J. R. Hughes, A. Reali, and G. Sangalli, Isogeometric collocation for elastostatics and explicit dynamics, *Computational Methods in Applied Mechanics and Engineering*, **249–252**, 2–14 (2012).
38. L. De Lorenzis, J. A. Evans, T. J. R. Hughes, and A. Rea, Isogeometric collocation: Neumann boundary conditions and contact, *Computer Methods in Applied Mechanics and Engineering*, **284**, 21–54 (2015).
39. R. Kruse, N. Nguyen-Thanh, L. De Lorenzis, and T. J. R. Hughes, Isogeometric collocation for large deformation elasticity and frictional contact problems, *Computer Methods in Applied Mechanics and Engineering*, **296**, 73–112 (2015).
40. T. W. Sederberg, J. Zheng, A. Bakenov, and A. Nasri, T-splines and t-nurccs, *ACM Transactions on Graphics (TOG) — Proceedings of ACM SIGGRAPH 2003*, **22**(3), 477–484 (2003).
41. T. W. Sederberg, D. L. Cardon, G. T. Finnigan, N. S. North, J. Zheng, and T. Lyche, T-spline simplification and local refinement, *ACM Transactions on Graphics (TOG) — Proceedings of ACM SIGGRAPH 2004*, **23**(3), 276–283 (2004).
42. R. Dimitri, L. De Lorenzis, M. A. Scott, P. Wriggers, R. L. Taylor, and G. Zavarise, Isogeometric large deformation frictionless contact using

t-splines, *Computer Methods in Applied Mechanics and Engineering*, **269**, 394–414 (2014).
43. R. Dimitri and G. Zavarise, Isogeometric treatment of frictional contact and mixed mode debonding problems, *Computational Mechanics*, **60**(2), 315–332 (2017).
44. A. V. Vuong, C. Giannelli, B. Jüttler, and B. Simeon, A hierarchical approach to adaptive local refinement in isogeometric analysis, *Computer Methods in Applied Mechanics and Engineering*, **200**(49–52), 3554–3567 (2011).
45. İlker Temizer and C. Hesch, Hierarchical nurbs in frictionless contact, *Computer Methods in Applied Mechanics and Engineering*, **299**, 161–186 (2016).
46. C. Hesch, M. Franke, M. Dittmann, and İlker Temizer, Hierarchical nurbs and a higher-order phase-field approach to fracture for finite-deformation contact problems, *Computer Methods in Applied Mechanics and Engineering*, **301**, 242–258 (2016).
47. D. Gabriel, J. Plešek, and M. Ulbin, Symmetry preserving algorithm for large displacement frictionless contact by the pre-discetization penalty method, *International Journal for Numerical Methods in Engineering*, **61**(15), 2615–2638 (2004).
48. L. A. Piegel and W. Tiller, *The NURBS Book (Monographs in Visual Communication)*, 2 edn., Springer-Verlag (1997).
49. M. E. Gurtin, E. Fried, and L. Anand, *The Mechanics and Thermomechanics of Continua*, Cambridge University Press (2010).
50. R. de Borst, M. A. Crisfield, J. J. C. Remmers, and C. V. Verhoosel, *Nonlinear Finite Element Analysis of Solids and Structures*, 2 edn., John Wiley & Sons (2012).
51. T. A. Laursen, *Computational Contact and Impact Mechanics*, Springer (2002).
52. S. Timoshenko and J. N. Goodier, *Theory of elasticity*, McGraw-Hill (1969).
53. L. Debnath and P. Mikusinski, *Introduction to Hilbert Spaces with Applications*. Elsevier (2005).
54. O. C. Zienkiewicz, R. L. Taylor, and J. Z. Zhu, *The Finite Element Method: Its Basis and Fundamentals*, 6 edn., vol. 1, Butterworth-Heinemann (2005).
55. K. J. Bathe, *Finite Element Procedures*, Prentice-Hall (1996).
56. B. M. Irons, Engineering applications of numerical integration in stiffness methods, *The American Institude of Aeronatics and Astronatics*, **4**(11), 2035–2037 (1966).
57. K. Ko and T. Sakkalis, Orthogonal projection of points in cad/cam applications: An overview, *Journal of Computational Design and Engineering*, **1**(2), 116–127 (2014).
58. S.-M. Hu and J. Wallner, A second order algorithm for orthogonal projection onto curves and surfaces, *Computer Aided Geometric Design*, **22**(3), 251–260 (2005).
59. X.-M. Liu, L. Yang, J.-H. Yong, H.-J. Gu, and J.-G. Sun, A torus patch approximation approach for point projection on surfaces, *Computer Aided Geometric Design*, **26**(5), 593–598 (2009).

60. L. A. Piegel and W. Tiller, Parametrization for surface fitting in reverse engineering, *Computer-Aided Design*, **33**(8), 593–603 (2001).
61. Y. L. Ma and W. Hewitt, Point inversion and projection for nurbs curve and surface: Control polygon approach, *Computer Aided Geometric Design*, **20**(2), 79–99 (2003).
62. I. Selimovic, Improved algorithms for the projection of points on nurbs curves and surfaces, *Computer Aided Geometric Design*, **23**, 439–445 (2006).
63. X.-D. Chen, J.-H. Yong, G. Wang, J.-C. Paul, and G. Xu, Computing the minimum distance between a point and a nurbs curve, *Computer-Aided Design*. **40**, 1051–1054 (2008).
64. Y.-T. Oh, Y.-J. Kim, J. Lee, M.-S. Kim, and G. Elber, Efficient point-projection to freeform curves and surfaces, *Computer Aided Geometric Design*, **29**, 242–254 (2012).
65. A. Konyukhov and K. Schweizerhof, On the solvability of closest point projection procedures in contact analysis: Analysis andsolution strategy for surfaces of arbitrary geometry, *Computer Methods of Applied Mechanics Engineering*, **197**(33–40), 3045–3056 (2008).
66. J. Kopačka, D. Gabriel, J. Plešek, and M. Ulbin, Assessment of methods for computing the closest point projection, penetration, and gap functions in contact searching problems, *International Journal for Numerical Methods in Engineering*, **105**(11), 803–833 (2016).
67. J. Nocedal and S. J. Wright, *Numerical Optimization*, Springer (1999).
68. R. L. Taylor and P. Papadopoulos, On a patch test for contact problems in two dimensions, In P. Wriggers and W. Wagner (eds.), *Nonlinear Computational Mechanics*, pp. 690–702. Springer-Verlag, Berlin (1991).
69. M. A. Puso and T. A. Laursen, A mortar segment-to-segment contact method for large deformation solid mechanics, *Computer Methods in Applied Mechanics and Engineering*, **193**(6–8), 601–629 (2004).
70. A. Popp, M. Gitterle, M. W. Gee, and W. A. Wall, A dual mortar approach for 3D finite deformation contact with consistent linearization, *International Journal for Numerical Methods in Engineering*, **83**(11), 1428–1465 (2010).

Chapter 4

Treatment of Non-matching Interfaces in Partitioned Fluid–Structure Interaction Problems

J. A. González and K. C. Park

In this work, we consider the dynamic interaction of incompressible fluids with flexible structures. Due to the particular characteristics of fluid and solid systems, different formulations are usually applied to each medium producing non-matching interfaces. Also the movements of the fluid on the free surface due to sloshing motions are usually bigger than the elastic displacements of the structure and is also a source of interface mismatching. To investigate this problem, a variational formulation for the linear-vibration of bounded fluid-structure systems taking into account the combined effect of sloshing, acoustic waves and initial stress due to gravity, is introduced. The systems are modelled using finite elements and coupled with localized Lagrange multipliers to study the particular case of an inviscid homogeneous liquid in a closed container. A key approach of the present localized formulation is to treat the non-matching interfaces by the method of LLMs using the FEM with a Lagrangian description of motion for the fluid and the structure.

1. Introduction

Vibrating structures inducing waves in a fluid and the opposite case of acoustic or gravity waves inducing structural vibrations constitute an active area of research. The two interconnected systems, the flexible structure and the enclosed fluid, can be strongly coupled, and in that case the fluid-structure system must be studied using specific techniques.

Fluid structure interaction (FSI) problems involving only internal acoustics or only sloshing are relatively well understood. The concurrent interaction of both intense acoustics and sloshing with flexible containers, however, is not well understood, and presents challenge for the design of high-pressure liquid and gaseous containers.

Normally, in the structural domain the primary variable to describe the motion is the displacement. For the fluid domain, several different approaches can be used but normally include velocity and pressure as variables. Using velocity and pressure variables on the fluid requires additional efforts to satisfy the equilibrium and compatibility conditions between the fluid and the structure. A displacement formulation for the fluid domain allows to impose directly the compatibility and equilibrium conditions along the interface, simplifying the treatment of the interface. It also simplifies the application of interface conditions.

Treatment of the interface in this work is based on the *Localized Lagrange Multiplier* (LLM) method,[1–3] a general variational framework used to solve partitioned systems that introduces a *kinematic frame* in the fluid-structure interface. Two multiplier fields separately connect the interface frame to the fluid on one side and to the structure *wet surface* on the other. Both multiplier spaces are discretized as delta functions collocated at the fluid interface and wet-surface structural nodes. These multipliers can be physically interpreted as interaction point forces. For an inviscid fluid, only multipliers normal-to-the-interface are needed. The main goal of this interface treatment is *complete decoupling* of fluid and structure models, in the sense that those can be constructed separately by different teams, or different mesh generators. Consequently, finite element meshes will not necessarily match over the interface.

This localized approach, initially dedicated to structural substructuring and contact-impact problems,[4] allows the treatment of non-matching meshes and can be constructed in order to preserve the constant-stress interface patch test, as demonstrated by Park *et al.*[5]

The LLM method proposed by Park and Felippa[1–3] is a general variational framework used to solve partitioned systems in mechanics that also introduces between the substructures an intermediate surface called *frame*, that is endowed with independent degrees of freedom (dof) and treated with a FEM discretization to approximate the interface variables. Finally, this *frame* is connected to the coupled systems using classical Lagrange multipliers defined on the substructure interface nodes, obtaining an approximation that naturally allows the treatment of non-matching meshes and can be constructed in order to preserve the constant-stress interface patch test, as demonstrated by Park *et al.*[5] It is this treatment of the contact conditions that will facilitate the consideration of large motions of the fluid at the interface surface.

In this work, we apply this LLM formulation for the simulation of structures experiencing significant FSI effects with coupled acoustic and gravity waves. Many theoretical and implementation issues behind the techniques proposed[6] are reviewed, extended and explained in detail as appropriate to understand its possibilities. Additional theory topics are also included, as treatment of fluid quasi-incompressibility, procedures for interfacing non-matching meshes in the case of inviscid fluids and treatment of the interfaces using LLMs. This is especially so as the sloshing fluid should be allowed to move tangentially along the container walls. Different example problems are used to demonstrate the capabilities of the proposed methodology, showing the generality of the formulation for modeling sloshing and elastic structure-fluid interaction problems.

2. Statement of the problem

We consider the problem of determining the motions of an *inviscid*, *homogeneous* and *compressible* liquid with a free surface Σ_f and contained in a linear elastic structure, as the case represented in Fig. 1. The internal fluid is considered to be initially at rest occupying a volume V_f under hydrostatic equilibrium due to the action of the gravity field $\mathbf{g}$. At the same moment, the elastic structure containing the liquid occupies a volume V_s

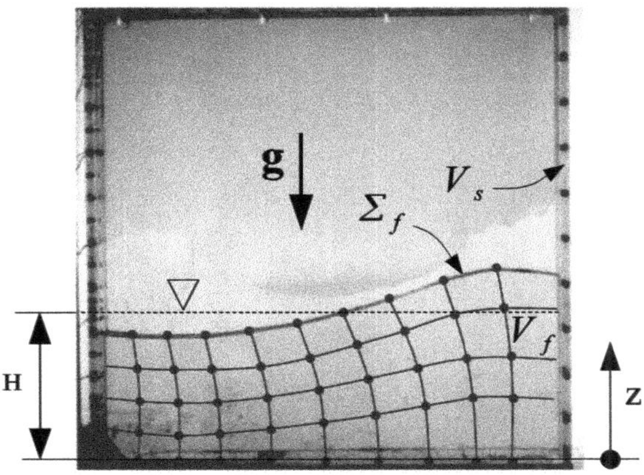

Fig. 1. Container with a liquid in motion presenting large runs up and down of the fluid on the walls of the structure. Finite elements are used to describe the motion of the fluid using the Lagrangian approach.

and presents a *wet surface*, common interface with the fluid denoted Γ_b. The structure presents external forces and/or displacements will produce a coupled oscillation of the two systems.

A global reference system $\mathbf{X}(X, Y, Z)$ is introduced with the vertical direction Z and its unit vector $\mathbf{k}$ measured from the bottom of the tank, defining the undisturbed fluid level inside the tank as H. In order to describe the motion of the two systems, we use the displacement fields of the fluid $\mathbf{u}_f$ and the structure $\mathbf{u}_s$ defined on V_f and V_s, respectively, that added to the reference configurations $\mathbf{X}_f$ and $\mathbf{X}_s$ provide the current positions $\mathbf{x}_f$ and $\mathbf{x}_s$

$$\begin{aligned} \mathbf{x}_f &= \mathbf{X}_f + \mathbf{u}_f, \\ \mathbf{x}_s &= \mathbf{X}_s + \mathbf{u}_s, \end{aligned} \tag{1}$$

magnitudes expressed in the global system.

3. Localized Lagrange multiplier method for FSI

The LLM treatment of the problem is schematized in Fig. 2 where the total FSI system is divided into three subsystems: the fluid, the *interface frame* and the structure.

To derive the equations of motion for the FSI system, we use the variational formulation proposed by Park and Felippa[1–3] and later particularized to FSI problems.[6,7] Under this variational framework, the problem is treated like if all systems were entirely independent, formulating the virtual work of the system by summing up the contributions of each subsystem and adding finally the interface contribution obtained multiplying the constraint equations by the LLMs. The variational functional that represents the stationary condition of the total energy of the system $\delta\Pi$ is then composed of three terms

$$\delta\Pi = \delta\Pi_f + \delta\Pi_s + \delta\pi^{\text{int}} = 0, \tag{2}$$

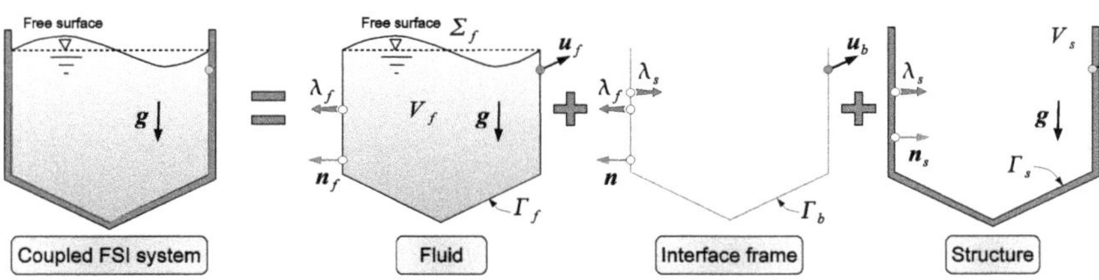

Fig. 2. Partitioning of an internal fluid–structure coupled system into three different subsystems: fluid, interface frame and structure.

where $\delta\Pi_f$ is the virtual work for the fluid; $\delta\Pi_s$ is the virtual work for the flexible structure; and $\delta\pi^{\text{int}}$ is the partition interface constraint that groups contributions from the fluid and the structure. It should be noted that $\delta\Pi_f = 0$ yields the governing equations of motion for the fluid. Similarly, $\delta\Pi_s = 0$ yields the equations of motion for the structure. This implies that the present formulation may utilize two independently developed fluid and flexible structural dynamics analysis software modules to perform FSI analysis, with the addition of an interface treatment module whose main function is to enforce the interface constraint $\delta\pi^{\text{int}} = 0$.

In the following sections, the approximations adopted for each one of these three virtual work expressions are described.

4. Fluid formulation

In this section, we derive the approximation used for the fluid.

In the absence of gravity effects, a fluid with no shear stiffness presents rotational zero energy modes. These free rotational modes in the vertical plane disappear when the gravity effects are introduced.[8,9]

Because our interest is in fluid–structure interaction, a pure displacement formulation is preferred, where the absence of the pressure will facilitate the connection with the structure.

4.1. *Lagrangian description of fluid motion*

Let us consider an inviscid and homogeneous fluid under the action of gravity field $\mathbf{g}$, initially at rest with density ρ_f^0 and hydrostatic pressure due to gravity $p_f^0 = p_{\text{int}} + \rho_f^0 g(H - z)$. Next, we suppose small deviations from this equilibrium position and introduce a Lagrangian displacement field $\mathbf{u}_f$ following the fluid particle, as represented in Fig. 3.

Conservation of mass equation states

$$\rho_f^0 = \rho_f J, \tag{3}$$

with $J = \det \mathbf{F}$ and where $\mathbf{F} = \nabla \mathbf{u}_f$ is the deformation gradient tensor.

Linearizing the kinematics, the inverse of the dilatation J can be approximated as $J^{-1} \approx (1 - \nabla \cdot \mathbf{u}_f)$ and substituted back in (3) to provide a linearized version of mass conservation

$$\rho_f = \rho_f^0(1 - \nabla \cdot \mathbf{u}_f), \tag{4}$$

On the other hand, for general equations of state, the speed of sound (c_f) relates variation of pressure with variation of density in adiabatic

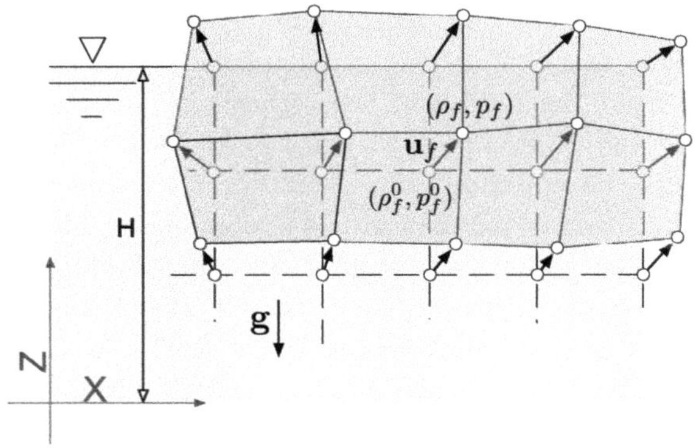

Fig. 3. Initial state of the fluid under the action of gravity. Discretization and Lagrangian description of the fluid motion.

conditions, i.e.

$$p_f = p_f^0 + c_f^2(\rho_f - \rho_f^0) + \cdots , \tag{5}$$

and neglecting all terms involving squares and higher powers of $(\rho_f - \rho_f^0)$, this relation can be substituted in (4) to express the fluid pressure in Lagrangian form as

$$p_f = p_f^0 - \rho_f^0 c^2 \boldsymbol{\nabla} \cdot \mathbf{u}_f, \tag{6}$$

an *equation of state* for the fluid that relates Lagrangian pressure with displacements. This approximation is well known and has been extensively used to model fluid motion in Lagrangian frameworks.[8–13]

4.2. *Principle of virtual work*

We can write the equation for the principle of virtual work in terms of the first Piola–Kirchhoff tensor as

$$\begin{aligned} \delta\Pi_f = & \int_{V_f} \mathbf{T} \colon \delta\mathbf{F}\, dV - \int_{V_f} \mathbf{b}_f \cdot \delta\mathbf{u}_f\, dV \\ & - \int_{\partial V_f} \mathbf{t}_f \cdot \delta\mathbf{u}_f\, dA + \int_{V_f} \rho_f^0 \ddot{\mathbf{u}}_f \cdot \delta\mathbf{u}_f\, dV, \end{aligned} \tag{7}$$

where $\mathbf{b}_f = \rho_f^0 \mathbf{g}$ is the body load per unit undeformed volume due to gravity and $\mathbf{t}_f$ is the surface traction vector per unit initial area.

We observe from this equality that the stress tensor conjugate to the deformation gradient $\mathbf{F}$ is the so-called *first Piola–Kirchhoff stress tensor* given as

$$\mathbf{T} = J\boldsymbol{\sigma}\mathbf{F}^{-\mathsf{T}}, \tag{8}$$

note that like $\mathbf{F}$, the first Piola–Kirchhoff stress tensor is an unsymmetric tensor.

For an inviscid fluid, the Cauchy stress tensor $\boldsymbol{\sigma}$ presents only a hydrostatic component and can be expressed as

$$\boldsymbol{\sigma} = -p_f\mathbf{I}, \tag{9}$$

a symmetric tensor in the deformed configuration.

A linear expansion of the deformation gradient tensor gives

$$J \approx (1 + \boldsymbol{\nabla} \cdot \mathbf{u}_f), \tag{10}$$

$$\mathbf{F}^{-\mathsf{T}} \approx (\mathbf{I} - \boldsymbol{\nabla}^{\mathsf{T}}\mathbf{u}_f), \tag{11}$$

first-order approximation of the deformations.

Substituting this linearization in (8) gives us an approximation of the first Piola–Kirchhoff stress tensor

$$\mathbf{T} \approx -(1 + \boldsymbol{\nabla} \cdot \mathbf{u}_f)(p_f\mathbf{I})(\mathbf{I} - \boldsymbol{\nabla}^{\mathsf{T}}\mathbf{u}_f), \tag{12}$$

and substituting in the general principle of virtual work (7) gives

$$\begin{aligned}\delta\Pi_f = &-\int_{V_f} p_f(\boldsymbol{\nabla} \cdot \delta\mathbf{u}_f)\, dV - \int_{V_f} p_f(\boldsymbol{\nabla} \cdot \mathbf{u}_f)(\boldsymbol{\nabla} \cdot \delta\mathbf{u}_f)\, dV \\ &+ \int_{V_f} p_f(1 + \boldsymbol{\nabla} \cdot \mathbf{u}_f)\boldsymbol{\nabla}^{\mathsf{T}}\mathbf{u}_f : \boldsymbol{\nabla}\delta\mathbf{u}_f\, dV \\ &- \int_{V_f} \rho_f^0\mathbf{g} \cdot \delta\mathbf{u}_f\, dV - \int_{\partial V_f} \mathbf{t}_f \cdot \delta\mathbf{u}_f\, dA \\ &+ \int_{V_f} \rho_f^0\ddot{\mathbf{u}}_f \cdot \delta\mathbf{u}_f\, dV.\end{aligned} \tag{13}$$

Considering that the hydrostatic pressure equation writes

$$\boldsymbol{\nabla}p_f^0 = \rho_f^0\mathbf{g}, \tag{14}$$

and that using Green–Gauss theorem the following equality holds

$$\int_{V_f} \rho_f^0\mathbf{g} \cdot \delta\mathbf{u}_f\, dV = \int_{\partial V_f} p_f^0\mathbf{n}_f \cdot \delta\mathbf{u}_f\, dA - \int_{V_f} p_f^0(\boldsymbol{\nabla} \cdot \delta\mathbf{u}_f)\, dV, \tag{15}$$

let us now modify the gravitational body force term in (13) to obtain

$$\delta\Pi_f = -\int_{V_f} (p_f - p_f^0)(\nabla \cdot \delta\mathbf{u}_f)\, dV - \int_{V_f} p_f(\nabla \cdot \mathbf{u}_f)(\nabla \cdot \delta\mathbf{u}_f)\, dV,$$
$$+ \int_{V_f} p_f(1 + \nabla \cdot \mathbf{u}_f)\nabla^{\mathsf{T}}\mathbf{u}_f \colon \nabla\delta\mathbf{u}_f\, dV,$$
$$- \int_{\partial V_f} (\mathbf{t}_f + p_f^0\mathbf{n}_f) \cdot \delta\mathbf{u}_f\, dA + \int_{V_f} \rho_f^0 \ddot{\mathbf{u}}_f \cdot \delta\mathbf{u}_f\, dV, \tag{16}$$

where we can appreciate that the second volume integral and the second term of the sum in the third volume integral present second-order terms.

The boundary integral of (16) extended to the fluid surface ∂V_f can be divided in two different regions: the free surface Σ_f and the surface in contact with the structure Γ_f with $\partial V_f = \Gamma_f \cup \Sigma_f$. When integrating on Σ_f, integrand $(\mathbf{t}_f + p_f^0\mathbf{n}_f)$ represents the pressure jump through the free surface. We permit the Lagrangian pressure of the fluid to take the external pressure value p_{int} on the free surface, making

$$\int_{\Sigma_f} (\mathbf{t}_f + p_f^0\mathbf{n}_f) \cdot \delta\mathbf{u}_f\, dA = 0, \tag{17}$$

equality that can also be extended to the complete fluid boundary ∂V_f because the normal displacement of the fluid is zero in Γ_f. In applying previous condition, we are taking gravity as the sole force restoring the flatness of the free surface: thus, we ignore any pressure discontinuity across the air–liquid interface due to surface tension. This approximation is valid for all waves except ripples with very short wave length.[14]

Now we substitute in (16) the approximation of Lagrangian pressure derived in (6) to obtain the final form from where we will derive the equations of motion

$$\delta\Pi_f = \int_{V_f} (\rho_f^0 c_f^2 - p_f)(\nabla \cdot \mathbf{u}_f)(\nabla \cdot \delta\mathbf{u}_f)\, dV$$
$$+ \int_{V_f} p_f(1 + \nabla \cdot \mathbf{u}_f)\nabla^{\mathsf{T}}\mathbf{u}_f \colon \nabla\delta\mathbf{u}_f\, dV$$
$$+ \int_{V_f} \rho_f^0 \ddot{\mathbf{u}}_f \cdot \delta\mathbf{u}_f\, dV, \tag{18}$$

where we can identify an acoustic term first, an initial stress term due to confinement pressure second and finally the term due to inertia forces.

After elimination of second-order terms and regrouping, we rewrite (18) as

$$\delta\Pi_f = \int_{V_f} \rho_f^0 c_f^2 (\nabla \cdot \mathbf{u}_f)(\nabla \cdot \delta\mathbf{u}_f)\, dV$$

$$+ \int_{V_f} p_f^0 (\nabla^\mathsf{T} \mathbf{u}_f : \nabla \delta\mathbf{u}_f)\, dV + \int_{V_f} \rho_f^0 \ddot{\mathbf{u}}_f \cdot \delta\mathbf{u}_f\, dV, \qquad (19)$$

a linearized approximation of the virtual work where the first acoustic term is pure volumetric and the second pure deviatoric. Note that in the deviatoric part appears the total Lagrangian pressure, function of the acoustic pressure and displacement dependent, see Eq. (6). During a time integration process, this Lagrangian pressure can be predicted or, as an approximation, taken constant and equal to the initial pressure of the fluid p_f^0.

4.3. *Mean dilatation method for incompressibility*

A standard finite element discretization is unfortunately not applicable to simulations involving incompressible or quasi-incompressible behaviour. It is well known that without further development such a formulation is kinematically over-constrained, resulting in the over-stiff phenomenon known as *volumetric locking*. Well-known solutions to this problem are: to impose the incompressibility condition by penalization or by using reduced integration methods,[15–18] to adopt an augmented Lagrangian formulation enforcing quasi-incompressibility condition[19,20] or use u–p elements that satisfy the inf-sup condition.[21,22] Among all these possibilities, a total displacement formulation with variational treatment of near incompressibility is preferred in order to facilitate the treatment of the interface without sacrifying theoretical support.

The three-field canonical functional of linear elastostatics identified as the *Hu–Washizu* (HW) functional in the mechanics literature[23,24] will be applied to treat nearly-incompressibility of the fluid. In this functional, three different interior fields, displacements, stresses and strains, are defined and independently varied.

In our case, the HW functional for the fluid uses independent kinematic descriptions for the volumetric and distortional deformation, introducing a new kinematic variable ($\bar{J}$) representing the dilatation independent of $\mathbf{u}_f$ and a Lagrange multiplier ($\bar{p}_v$) to enforce the condition $\bar{J} = J(\mathbf{u}_f)$.

The HW principle states,[25,26]

$$\Pi_f^{\mathrm{HW}}(\mathbf{u}_f, \bar{J}, \bar{p}_v) = \int_{V_f} U_f(\bar{J})\, dV + \int_{V_f} \hat{\Psi}_f(\mathbf{F})\, dV + \int_{V_f} \bar{p}_v(J - \bar{J})\, dV - \Pi_f^{\mathrm{ext}}(\mathbf{u}_f), \tag{20}$$

where $U_f(\bar{J})$ represents the volumetric strain energy and $\hat{\Psi}_f(\mathbf{F}) = \Psi_f(\hat{\mathbf{F}})$ the deviatoric part of the strain energy function.

The stationary conditions of (20) are

$$\int_{V_f} \mathbf{T} \colon \delta\mathbf{F}\, dV - \delta\Pi_f^{\mathrm{ext}} = 0, \tag{21}$$

$$\int_{V_f} \left(\frac{\partial U_f(\bar{J})}{\partial \bar{J}} - \bar{p}_v \right) \delta\bar{J}\, dV = 0, \tag{22}$$

$$\int_{V_f} (J - \bar{J})\delta\bar{p}_v\, dV = 0. \tag{23}$$

Mean dilatation procedure assumes that $(\bar{p}_v,\ \bar{J})$ and $(\delta\bar{p}_v,\ \delta\bar{J})$ are constant over the integration volume V_f. Introducing the volumetric strain energy function as

$$U_f(\bar{J}) = \frac{1}{2}\rho_f^0 c_f^2 (\bar{J} - 1)^2, \tag{24}$$

where $\rho_f^0 c_f^2$ is the volumetric stiffness, Eqs. (22) and (23) reduce to

$$\bar{J} = \frac{1}{V_f} \int_{V_f} J\, dV, \tag{25}$$

$$\bar{p}_v = \rho_f^0 c_f^2 (\bar{J} - 1), \tag{26}$$

conditions that will be incorporated in our finite element formulation for the fluid.

4.4. *Discrete approximation of fluid functional*

The mean dilatation approach for a given volume V_f leads to a constant pressure over that volume, as indicated by Eqs. (25) and (26). When this formulation is applied to an individual fluid element identified by superscript (e) in a finite element mesh, the acoustic pressure becomes constant over the element volume. In particular, considering (25) in an element with volume $V_f^{(e)}$ together with (10), an average of the element dilatation is obtained

$$(\bar{\boldsymbol{\nabla}} \cdot \mathbf{u}_f)^{(e)} = \frac{1}{V_f^{(e)}} \int_{V_f^{(e)}} \boldsymbol{\nabla} \cdot \mathbf{u}_f \, dV, \tag{27}$$

and using the expression of uniform acoustic pressure given by Eq. (26) at the element level, the uniform element acoustic pressure can be expressed

$$\bar{p}_v^{(e)} = \rho_f^0 c_f^2 (\bar{\boldsymbol{\nabla}} \cdot \mathbf{u}_f)^{(e)}, \tag{28}$$

where acoustic pressure $\bar{p}_v^{(e)}$ is constant inside the element.

Continuing with the discretization, recall the linearized virtual work Eq. (19) which for an element (e) is

$$\delta\Pi_f^{(e)} = \int_{V_f^{(e)}} \rho_f^0 c_f^2 (\boldsymbol{\nabla} \cdot \mathbf{u}_f)(\boldsymbol{\nabla} \cdot \delta\mathbf{u}_f) \, dV + \int_{V_f^{(e)}} p_f^0 (\boldsymbol{\nabla}^{\mathsf{T}} \mathbf{u}_f \colon \boldsymbol{\nabla} \delta\mathbf{u}_f) \, dV + \int_{V_f^{(e)}} \rho_f^0 \ddot{\mathbf{u}}_f \cdot \delta\mathbf{u}_f \, dV, \tag{29}$$

and apply the mean dilatation method. Considering that in the first volume integral the acoustic pressure $\bar{p}_v = \rho_f^0 c_f^2 (\bar{\boldsymbol{\nabla}} \cdot \mathbf{u}_f)$ is constant over the element volume, the average dilatations are introduced for an element (e) as

$$\delta\Pi_f^{(e)} = \rho_f^0 c_f^2 V_f^{(e)} (\bar{\boldsymbol{\nabla}} \cdot \mathbf{u}_f)^{(e)} (\bar{\boldsymbol{\nabla}} \cdot \delta\mathbf{u}_f)^{(e)} + \int_{V_f^{(e)}} p_f^0 (\boldsymbol{\nabla}^{\mathsf{T}} \mathbf{u}_f \colon \boldsymbol{\nabla} \delta\mathbf{u}_f) \, dV + \int_{V_f^{(e)}} \rho_f^0 \ddot{\mathbf{u}}_f \cdot \delta\mathbf{u}_f \, dV, \tag{30}$$

where we observe that the element stiffness matrix is composed of a pure volumetric acoustic term plus a pure deviatoric geometric term.

When element displacements are discretized and expressed as $\mathbf{u}_f = \mathbf{N}\mathbf{u}_f^{(e)}, \mathbf{N}_f$ collects the element shape functions and $\mathbf{u}_f^e$ collects nodal values of the element. Applied forces include body forces collected in $\mathbf{B}_f$ and surface tractions collected in $\mathbf{t}_f$. The fluid element matrices are given by

The stiffness matrix is finally composed of two terms,

$$\mathbf{K}^{(e)} = \mathbf{K}_{ac}^{(e)} + \mathbf{K}_{geo}^{(e)}, \tag{31}$$

$$\mathbf{K}_{ac}^{(e)} = \frac{\rho_f^0 c_f^2}{V_f^{(e)}} \int_{V_f^{(e)}} (\boldsymbol{\nabla} \cdot \mathbf{N}_f)^{\mathsf{T}} \, dV \int_{V_f^{(e)}} (\boldsymbol{\nabla} \cdot \mathbf{N}_f) \, dV, \tag{32}$$

$$\mathbf{K}_{geo}^{(e)} = \int_{V_f^{(e)}} p_f^0 (\boldsymbol{\nabla} \mathbf{N}_f)^{\mathsf{T}} (\boldsymbol{\nabla} \mathbf{N}_f) \, dV, \tag{33}$$

where $\mathbf{K}_{ac}^{(e)}$ is the volumetric stiffness matrix and $\mathbf{K}_{geo}^{(e)}$ is the geometrical stiffness matrix.

Spectral analysis of the stiffness matrix (31) for an 8-node, 24 dof, brick element with a regular (cubic-like) geometry under exact $2 \times 2 \times 2$ integration reveals a full rank with six zero-energy modes corresponding to rigid-body motions. The acoustic stiffness matrix (32) contributes with one pure-volumetric mode and the geometrical stiffness matrix (33) introduces the remaining 17 non-zero energy modes that include shear, torsion and filling modes.

After assembling the finite element matrices, we arrive to the matrix form of the discretized variational

$$\delta \Pi_f = \delta \mathbf{u}_f^{\mathsf{T}} \{ \mathbf{M}_f \ddot{\mathbf{u}}_f + \mathbf{K}_f \mathbf{u}_f - \mathbf{f}_f \}, \tag{34}$$

in which $\mathbf{M}_f$ and $\mathbf{K}_f$ are the mass and stiffness matrices of the fluid, $\mathbf{u}_f$ the vector of fluid node displacements and $\mathbf{f}_f$ the applied node force vector. As we can see, the second-order ODE representing the fluid equations of motion is only function of displacements and its time derivatives.

5. Structure formulation

The structure is described by the equation of motion for a continuum body. Because in the variational formulation previously derived for the fluid, liquid motion is considered to be small, similar consideration has to be taken for the structure. To this end, we write the the variational formulation of the structural partition with the following expression for the virtual work

$$\delta \Pi_s = \int_{V_s} \boldsymbol{\sigma}_s : \boldsymbol{\nabla} \delta \mathbf{u}_s \, dV - \int_{\partial V_s} \mathbf{t}_s \cdot \delta \mathbf{u}_s \, dA + \int_{V_s} \rho_s \ddot{\mathbf{u}}_s \cdot \delta \mathbf{u}_s \, dV, \tag{35}$$

where $\mathbf{u}_s$ are the structural displacements, $\boldsymbol{\sigma}_s$ the stress tensor and $\mathbf{t}_s$ the surface tractions.

5.1. *Discrete approximation of structure functional*

The structure is discretized using the classical FEM approximation, where the assembly of element contributions by the direct stiffness method leads to the semi-discrete equations of motion

$$\delta\Pi_s = \delta\mathbf{u}_s^{\mathsf{T}}\{\mathbf{M}_s\ddot{\mathbf{u}}_s + \mathbf{K}_s\mathbf{u}_s - \mathbf{f}_s\}, \tag{36}$$

in which $\mathbf{M}_s$ and $\mathbf{K}_s$ are the mass and stiffness matrices, $\mathbf{u}_s$ the vector of node displacements and $\mathbf{f}_s$ the applied node force vector.

6. Interface frame

To formulate the interface problem, instead of considering the direct interaction between the fluid and the structure, we insert between them a deforming non-physical surface or *interface frame* denoted as Γ_b and reformulate the FSI problem in terms of interaction of the two systems with this auxiliary surface using LLMs collocated on each side of the frame.

The interaction forces acting on the frame are represented in Fig. 4 where the LLMs connecting the fluid domain V_f with the frame are named using the vector quantity $\boldsymbol{\lambda}_f = \{\lambda_{f_n}, \lambda_{f_{t_1}}, \lambda_{f_{t_2}}\}^{\mathsf{T}}$ and the multipliers connecting the structure V_s are named $\boldsymbol{\lambda}_s = \{\lambda_{s_n}, \lambda_{s_{t_1}}, \lambda_{s_{t_2}}\}^{\mathsf{T}}$. These forces are expressed using a local orthonormal base system attached to the frame, $\mathbf{B} = [\mathbf{n}, \mathbf{t}_1, \mathbf{t}_2]$ that depends of the position occupied by the node on the

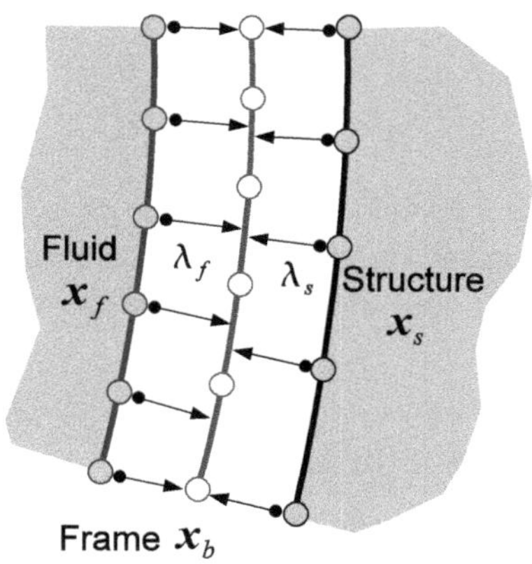

Fig. 4. Treatment of non-matching interfaces using LLM. The fields of multipliers on the fluid side $\boldsymbol{\lambda}_f$ and the structural side $\boldsymbol{\lambda}_s$ are discretized using Dirac delta functions collocated at the nodes. If the fluid is inviscid, only normal displacements of the frame and normal components of these multipliers are considered.

frame and that is used to describe $\boldsymbol{\lambda}_f$ and $\boldsymbol{\lambda}_s$. This frame local system is defined by convention in the following way; $\mathbf{t}_1$ and $\mathbf{t}_2$ are the orthogonal vectors contained in the frame tangent plane at the considered point $\boldsymbol{\xi}_p$ and vector $\mathbf{n}$ points towards structure V_s.

6.1. *Frame kinematics*

The motion of the interface frame is described using a displacement field $\mathbf{u}_b = \{u_{b_n}, u_{b_{t_1}}, u_{b_{t_2}}\}^\mathsf{T}$ and its initial configuration $\mathbf{X}_b$, providing a current position

$$\mathbf{x}_b = \mathbf{X}_b + \mathbf{u}_b, \tag{37}$$

that is restricted to follow the fluid–structure interface permanently. To do so, the frame-fluid and frame-structure relative positions are forced to be zero, conditions expressed in the frame local system as

$$\begin{aligned} \mathbf{B}^\mathsf{T}(\mathbf{x}_f - \mathbf{x}_b) &= \mathbf{0}, \\ \mathbf{B}^\mathsf{T}(\mathbf{x}_s - \mathbf{x}_b) &= \mathbf{0}, \end{aligned} \tag{38}$$

and extended to the fluid interface Γ_f and the structure interface Γ_s.

If the fluid is inviscid, tangential stress on the wall is zero and tangential displacements of the fluid and the structure at the interface become independent. In this case, only the kinematic condition of (38) in the normal direction applies, i.e.

$$\begin{aligned} \mathbf{n}^\mathsf{T} \cdot (\mathbf{x}_f - \mathbf{x}_b) &= 0, \\ \mathbf{n}^\mathsf{T} \cdot (\mathbf{x}_s - \mathbf{x}_b) &= 0, \end{aligned} \tag{39}$$

reducing the displacement field on the frame to $\mathbf{u}_b = \{u_b\}$ with only one component in the normal direction and transforming the LLMs of the fluid side to $\boldsymbol{\lambda}_f = \{\lambda_n\}$ and the structural side to $\boldsymbol{\lambda}_s = \{\lambda_s\}$ as normal components.

7. Variationally based interface coupling

A variational derivation of the fluid-structure interface equations, has the important advantage of preserving the symmetry of the equations of motion. In the present section, the variational framework proposed by Park *et al.*,[6] is used to obtain the coupling equations associated to the interface.

7.1. *Interface constraint functional*

Consider the fluid–structure boundary Γ_b shown in Fig. 4 where the interface frame presents an independent field of displacements ($\mathbf{u}_b$). The fluid boundary displacements ($\mathbf{u}_f$), the structure boundary displacements ($\mathbf{u}_s$), as well as their associated multipliers on the fluid ($\boldsymbol{\lambda}_f$) and the structure ($\boldsymbol{\lambda}_s$) will appear in the variational formulation of the interaction conditions. This variational is divided in the following way:

$$\delta\pi^{\text{int}} = \delta\pi_f^{\text{int}} + \delta\pi_s^{\text{int}}, \tag{40}$$

grouping contributions from the fluid and the structure.

Let us obtain first the interface functional for the fluid. It can be expressed by multiplication of the LLMs and the constraint equation related with the kinematic positioning of the frame (38) that is enforced in a weak sense using the variational form

$$\delta\pi_f^{\text{int}} = \int_{\Gamma_b} \delta\{\boldsymbol{\lambda}_f \cdot [\mathbf{B}^{\mathsf{T}}(\mathbf{x}_f - \mathbf{x}_b)]\}\, dA, \tag{41}$$

that is differentiated by parts to provide the expression for the total variation of the interface potential at the fluid side

$$\delta\pi_f^{\text{int}} = \int_{\Gamma_b} (\boldsymbol{\lambda}_f \cdot \delta\{\mathbf{B}^{\mathsf{T}}(\mathbf{x}_f - \mathbf{x}_b)\} + \delta\boldsymbol{\lambda}_f \cdot \{\mathbf{B}^{\mathsf{T}}(\mathbf{x}_f - \mathbf{x}_b)\})\, dA, \tag{42}$$

and similarly

$$\delta\pi_s^{\text{int}} = \int_{\Gamma_b} (\boldsymbol{\lambda}_s \cdot \delta\{\mathbf{B}^{\mathsf{T}}(\mathbf{x}_s - \mathbf{x}_b)\} + \delta\boldsymbol{\lambda}_s \cdot \{\mathbf{B}^{\mathsf{T}}(\mathbf{x}_s - \mathbf{x}_b)\})\, dA, \tag{43}$$

for the structural side.

To deal with the first term of Eqs. (42) and (43), we decompose them in the following way

$$\boldsymbol{\lambda}_f \cdot \delta\{\mathbf{B}^{\mathsf{T}}(\mathbf{x}_f - \mathbf{x}_b)\} = (\delta\mathbf{u}_f - \delta\mathbf{u}_b) \cdot \{\mathbf{B}\boldsymbol{\lambda}_f\}, \tag{44}$$

$$\boldsymbol{\lambda}_s \cdot \delta\{\mathbf{B}^{\mathsf{T}}(\mathbf{x}_s - \mathbf{x}_b)\} = (\delta\mathbf{u}_s - \delta\mathbf{u}_b) \cdot \{\mathbf{B}\boldsymbol{\lambda}_s\}, \tag{45}$$

a first-order approximation where the variations of the normal and tangential frame unitary vectors are considered to be very small. Substitution into

(42) and (43) leads the interface potential for the fluid to be

$$\delta\pi_f^{\text{int}} = \int_{\Gamma_b} (\delta\boldsymbol{\lambda}_f \cdot \{\mathbf{B}^{\mathsf{T}}(\mathbf{x}_f - \mathbf{x}_b)\} + (\delta\mathbf{u}_f - \delta\mathbf{u}_b) \cdot \{\mathbf{B}\boldsymbol{\lambda}_f\})\, dA, \tag{46}$$

and

$$\delta\pi_s^{\text{int}} = \int_{\Gamma_b} (\delta\boldsymbol{\lambda}_s \cdot \{\mathbf{B}^{\mathsf{T}}(\mathbf{x}_s - \mathbf{x}_b)\} + (\delta\mathbf{u}_s - \delta\mathbf{u}_b) \cdot \{\mathbf{B}\boldsymbol{\lambda}_s\})\, dA, \tag{47}$$

for the structure.

Grouping contributions from both sides (40), the final expression for the interface variation is obtained

$$\begin{aligned}\delta\pi^{\text{int}}(\mathbf{x}_f, \mathbf{x}_s, \boldsymbol{\lambda}_f, \boldsymbol{\lambda}_s, \mathbf{x}_b) = &\int_{\Gamma_b} \delta\mathbf{u}_f \cdot \{\mathbf{B}\boldsymbol{\lambda}_f\}\, dA + \int_{\Gamma_b} \delta\mathbf{u}_s \cdot \{\mathbf{B}\boldsymbol{\lambda}_s\}\, dA \\ &+ \int_{\Gamma_b} \delta\boldsymbol{\lambda}_f \cdot \{\mathbf{B}^{\mathsf{T}}(\mathbf{x}_f - \mathbf{x}_b)\}\, dA \\ &+ \int_{\Gamma_b} \delta\boldsymbol{\lambda}_s \cdot \{\mathbf{B}^{\mathsf{T}}(\mathbf{x}_s - \mathbf{x}_b)\}\, dA \\ &- \int_{\Gamma_b} \delta\mathbf{u}_b \cdot \{\mathbf{B}(\boldsymbol{\lambda}_f + \boldsymbol{\lambda}_s)\}\, dA,\end{aligned} \tag{48}$$

where independently varied fields are identified in parenthesis on the left-hand side.

Note that in Eq. (48), the following terms can be identified from the fluid side

$$G_{u_f}(\boldsymbol{\lambda}_f; \delta\mathbf{u}_f) = \int_{\Gamma_b} \delta\mathbf{u}_f \cdot \{\mathbf{B}\boldsymbol{\lambda}_f\}\, dA, \tag{49}$$

$$G_{\lambda_f}(\mathbf{x}_f, \mathbf{x}_b; \delta\boldsymbol{\lambda}_f) = \int_{\Gamma_b} (\delta\boldsymbol{\lambda}_f \cdot \{\mathbf{B}^{\mathsf{T}}(\mathbf{x}_f - \mathbf{x}_b)\})\, dA, \tag{50}$$

$$G_{u_b}(\boldsymbol{\lambda}_f; \delta\mathbf{u}_b) = -\int_{\Gamma_b} \delta\mathbf{u}_b \cdot \{\mathbf{B}\boldsymbol{\lambda}_f\}\, dA, \tag{51}$$

obtaining similar expressions for the structure; where $G_{u_f}(\boldsymbol{\lambda}_f; \delta\mathbf{u}_f)$ represents the virtual work of interface forces acting on the fluid, $G_{\lambda_f}(\mathbf{x}_f, \mathbf{x}_b; \delta\boldsymbol{\lambda}_f)$ constitutes the kinematic constraint between the fluid and the frame and $G_{u_b}(\boldsymbol{\lambda}_f; \delta\mathbf{u}_b)$ evaluates the virtual work done by the interface forces on the frame. Once these terms are combined with their counterparts from the structure, the final conditions that will have to be satisfied in the frame are

$$G_{u_b}(\boldsymbol{\lambda}_f; \delta\mathbf{u}_b) + G_{u_b}(\boldsymbol{\lambda}_s; \delta\mathbf{u}_b) = 0, \tag{52}$$

or frame equilibrium due to forces coming from both sides.

If the fluid model is inviscid, interaction between the fluid and the structure takes place only in the normal direction to the wet-surface $\mathbf{n}$, conventionally selected to be exterior to the fluid. In this case, the LLM (λ_f, λ_s) and the frame displacements (u_b) will present only the normal component and the appropriate form of the interface functional becomes

$$\begin{aligned}\delta\pi^{\text{int}}(\mathbf{x}_f, \mathbf{x}_s, \lambda_f, \lambda_s, x_b) &= \int_{\Gamma_b} \delta\mathbf{u}_f \cdot \{\lambda_f \mathbf{n}\}\, dA + \int_{\Gamma_b} \delta\mathbf{u}_s \cdot \{\lambda_s \mathbf{n}\}\, dA \\ &+ \int_{\Gamma_b} \delta\lambda_f(\mathbf{x}_f \cdot \mathbf{n} - x_b)\, dA + \int_{\Gamma_b} \delta\lambda_s(\mathbf{x}_s \cdot \mathbf{n} - x_b)\, dA \\ &- \int_{\Gamma_b} (\lambda_f + \lambda_s)\delta u_b\, dA, \end{aligned} \tag{53}$$

where λ_f and λ_s are LLM functions in the normal direction that link the independently varied normal displacement function of the frame u_b with the displacement fields of the fluid $\mathbf{u}_f$ and the structure $\mathbf{u}_s$.

Finally, it is important to mention that the preceding interface constraint functional is a linear approximation that will be valid for small variations of local system $\mathbf{B}$. For a complete nonlinear formulation of the moving frame, the reader is referred to the works.[4,27,28]

7.2. *Discretization of interface constraint functional*

To produce the matrix form of the interface equations, functional (53) is discretized by assuming shape functions for the independently varied fields: λ_f, λ_s and u_b.

Approximation for boundary displacements $\mathbf{u}_f$ and $\mathbf{u}_s$ come from the finite elements used for the fluid and structure models, respectively, and are expressed in the global reference system. The normal displacements $\mathbf{u}_b(\xi_1, \xi_2)$ of the interface frame are interpolated using isoparametric finite elements in the classical form

$$\mathbf{u}_b(\xi_1, \xi_2) = \mathbf{N}_b(\xi_1, \xi_2) \begin{Bmatrix} u_{b_1} \\ \vdots \\ u_{b_{n_b}} \end{Bmatrix} = N_{bi}(\boldsymbol{\xi}) u_{bi}, \tag{54}$$

where $\boldsymbol{\xi} = (\xi_1, \xi_2)$ is the frame coordinate, n_b is the number of nodes in the frame element and the shape functions for $\mathbf{N}_b$ will be piecewise-linear, C^0-continuous.[5,29–32]

If we assume discretized fields of normal multipliers $\lambda_f = \mathbf{N}_{\lambda_f} \boldsymbol{\lambda}_f$, $\lambda_s = \mathbf{N}_{\lambda_s} \boldsymbol{\lambda}_s$, insert these interpolations in $\delta\pi^{\text{int}}$, and integrate over Γ_b, the final expression of the discretized interface functional is obtained

$$\delta\pi^{\text{int}} = \delta\mathbf{u}_f^{\mathsf{T}}\mathbf{B}_f\mathbf{u}_f + \delta\mathbf{u}_s^{\mathsf{T}}\mathbf{B}_s\mathbf{u}_s + \delta\boldsymbol{\lambda}_f^{\mathsf{T}}\left(\mathbf{B}_f^{\mathsf{T}}\mathbf{x}_f - \mathbf{L}_f\mathbf{x}_b\right) + \delta\boldsymbol{\lambda}_s^{\mathsf{T}}\left(\mathbf{B}_s^{\mathsf{T}}\mathbf{x}_s - \mathbf{L}_s\mathbf{x}_b\right) - \delta\mathbf{u}_b^{\mathsf{T}}\left(\mathbf{L}_f^{\mathsf{T}}\boldsymbol{\lambda}_f + \mathbf{L}_s^{\mathsf{T}}\boldsymbol{\lambda}_s\right), \tag{55}$$

in which

$$\mathbf{B}_f = \int_{\Gamma_b} \mathbf{N}_f^{\mathsf{T}}\mathbf{n}\mathbf{N}_{\lambda_f}\, dA, \tag{56}$$

$$\mathbf{B}_s = \int_{\Gamma_b} \mathbf{N}_s^{\mathsf{T}}\mathbf{n}\mathbf{N}_{\lambda_s}\, dA, \tag{57}$$

$$\mathbf{L}_f = \int_{\Gamma_b} \mathbf{N}_{\lambda_f}^{\mathsf{T}}\mathbf{N}_b\, dA, \tag{58}$$

$$\mathbf{L}_s = \int_{\Gamma_b} \mathbf{N}_{\lambda_s}^{\mathsf{T}}\mathbf{N}_b\, dA, \tag{59}$$

and where $\mathbf{N}_f$ and $\mathbf{N}_s$ are the shape functions of the fluid and structure displacement elements, respectively, evaluated on the discretized interface surface Γ_b. In some cases, as in curved geometries, the discretized interface may differ slightly from the original one.[6]

The integral evaluations in (56–59) are greatly simplified by assuming that $\mathbf{N}_{\lambda_f}$ and $\mathbf{N}_{\lambda_s}$ are *Dirac* delta functions collocated at the fluid and structure interface nodes, respectively. If so, $\mathbf{B}_f$ and $\mathbf{B}_s$ become Boolean matrices that extract the boundary degrees of freedom from the complete state vectors $\mathbf{u}_f$ and $\mathbf{u}_s$, while $\mathbf{L}_f$ and $\mathbf{L}_s$ simply become the evaluation on the frame of boundary node positions.

Our choice for multipliers discretization, represented in Fig. 5, is to model them as concentrated forces, that is, support functions of the multipliers are *Dirac* delta functions

$$\lambda_f(\boldsymbol{\xi}) = N_{\lambda_f i}(\boldsymbol{\xi})\lambda_{fi}; \;\; N_{\lambda_f i}(\boldsymbol{\xi}) = \delta(\boldsymbol{\xi} - \boldsymbol{\xi}_p), \tag{60}$$

$$\lambda_s(\boldsymbol{\xi}) = N_{\lambda_s i}(\boldsymbol{\xi})\lambda_{si}; \;\; N_{\lambda_s i}(\boldsymbol{\xi}) = \delta(\boldsymbol{\xi} - \boldsymbol{\xi}_p), \tag{61}$$

where $\boldsymbol{\xi}_p$ is the frame coordinate of the fluid or structural interface node projected on the frame. If the interface is defined in terms of *pairs* or

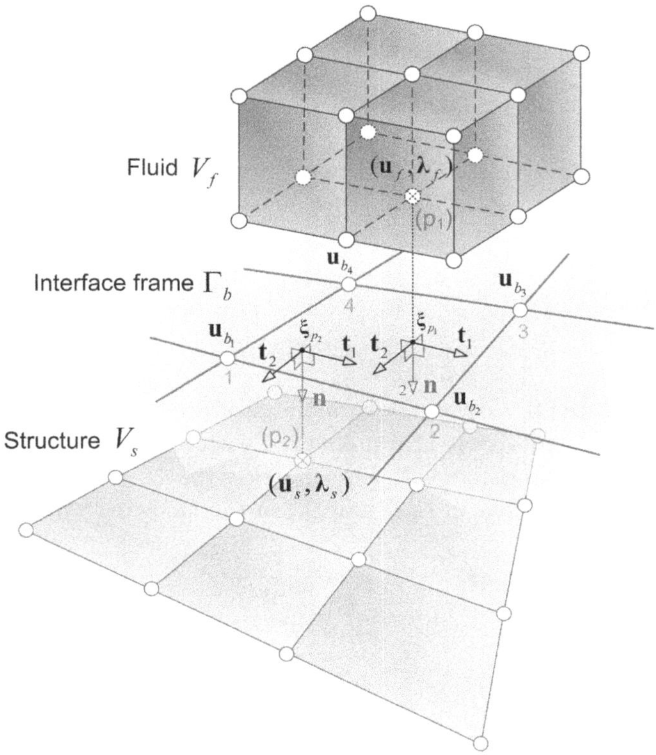

Fig. 5. Exploded view of the fluid–structure interface discretization. Interface nodes of the fluid and the structure are projected on the frame constituting *pairs* or couples formed by the hitting node and its closest frame element.

couples, formed by an interface node and its projection on the frame, then integrations of the multiplier field over the contact zone are reduced to summations over the pairs, i.e.

$$\int_{\Gamma_b} \boldsymbol{\lambda}_f(\boldsymbol{\xi}) \cdot \mathbf{f}(\boldsymbol{\xi})\, d\Gamma_b = \sum_{p=1}^{n_p} \boldsymbol{\lambda}_f^{(p)} \cdot \mathbf{f}(\boldsymbol{\xi}_p), \tag{62}$$

with $\mathbf{f}(\boldsymbol{\xi})$ a general function and n_p the total number of pairs on the fluid side. Expression (62) is the key in order to maintain the interface treatment generic, leading to modular coupling software since the frame needs to know very little information about the system it is interfacing.

To manage the discrete variables we have to introduce the interface nodal indicator $\boldsymbol{\mathcal{B}}$, the well known boolean finite element assembling operator defined in the following way

$$\mathbf{u}_f^{(p)} = \boldsymbol{\mathcal{B}}_{u_f}^{(p)} \mathbf{u}_f, \tag{63}$$

$$\mathbf{u}_s^{(p)} = \boldsymbol{\mathcal{B}}_{u_s}^{(p)} \mathbf{u}_s, \tag{64}$$

$$\mathbf{u}_b^{(p)} = \boldsymbol{\mathcal{B}}_{u_b}^{(p)} \mathbf{u}_b, \tag{65}$$

and

$$\boldsymbol{\lambda}_f^{(p)} = \boldsymbol{\mathcal{B}}_{\lambda_f}^{(p)} \boldsymbol{\lambda}_f, \tag{66}$$

$$\boldsymbol{\lambda}_s^{(p)} = \boldsymbol{\mathcal{B}}_{\lambda_s}^{(p)} \boldsymbol{\lambda}_s, \tag{67}$$

where $\boldsymbol{\mathcal{B}}_{\square}^{(p)}$ is used to extract the variable associated with a boundary node (p) from the global vector of $\square$ unknowns with $\square = u_f, u_s, \lambda_f, \lambda_s, u_b$. By using this operator together with (62), matrices (56–59) are simply obtained assembling contributions of each pair (p) in the following way

$$\mathbf{B}_f = \sum_{p=1}^{n_p} \boldsymbol{\mathcal{B}}_{u_f}^{(p)\mathsf{T}} \mathbf{n}(\boldsymbol{\xi}_p) \boldsymbol{\mathcal{B}}_{\lambda_f}^{(p)}, \tag{68}$$

$$\mathbf{B}_s = \sum_{p=1}^{n_p} \boldsymbol{\mathcal{B}}_{u_s}^{(p)\mathsf{T}} \mathbf{n}(\boldsymbol{\xi}_p) \boldsymbol{\mathcal{B}}_{\lambda_s}^{(p)}, \tag{69}$$

$$\mathbf{L}_f = \sum_{p=1}^{n_p} \boldsymbol{\mathcal{B}}_{\lambda_f}^{(p)\mathsf{T}} \mathbf{N}_b(\boldsymbol{\xi}_p) \boldsymbol{\mathcal{B}}_{u_b}^{(p)}, \tag{70}$$

$$\mathbf{L}_s = \sum_{p=1}^{n_p} \boldsymbol{\mathcal{B}}_{\lambda_s}^{(p)\mathsf{T}} \mathbf{N}_b(\boldsymbol{\xi}_p) \boldsymbol{\mathcal{B}}_{u_b}^{(p)}, \tag{71}$$

with sums extended to the n_p active pairs of the corresponding side: fluid or structure.

8. Frame construction

The interface formulation presented so far relies on the assumption that there exists a discretization for the frame representing the fluid–structure interface. But the problem of defining a mesh for the interface frame is not unique.[32] For example, it would be perfectly valid solutions: constructing a mesh matching the fluid discretization, making it coincident with the structural discretization, or defining a completely new mesh between the

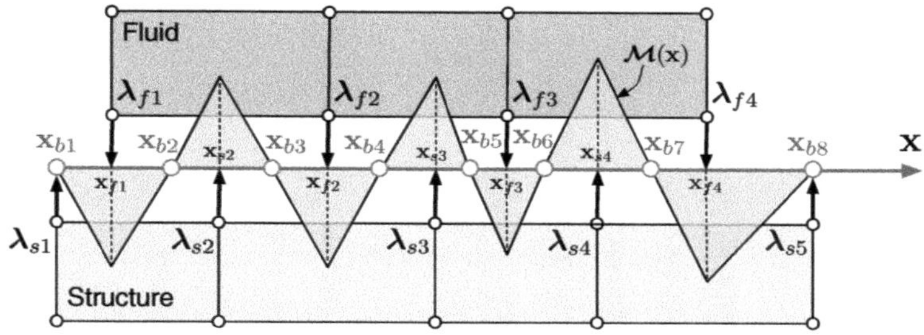

Fig. 6. Frame construction using a defined position of the fluid $\mathbf{x}_f$ and the structure $\mathbf{x}_s$. Frame nodes are located at $\mathbf{x}_b$, zero moment roots of the bending moment diagram $\mathcal{M}(\mathbf{x})$ produced by the forces acting on the frame.

two domains. However, these arbitrary discretizations in general will not satisfy the patch test through the interface.[33,34]

For constructing an interface mesh that satisfies the requirement of unicity and passes the patch test, the node locations of the interface frame are decided using the Zero Moment Rule (ZMR) originally proposed for contact-impact problems.[4] The main concept behind the ZMR is to assure that the LLMs can transmit a constant stress state through the interface frame exactly, passing the patch test. This condition can be satisfied if the interface node locations are determined by the roots of the frame moment-equilibrium condition.[5,29]

The process of constructing a frame is schematized in Fig. 6, where nodal positions of the fluid interface $\mathbf{x}_f$ and the structure interface $\mathbf{x}_s$ are considered known and a bending moment diagram $\mathcal{M}(\mathbf{x})$, due to the nodal forces corresponding to a constant stress state, is constructed. ZMR states that, for preservation of constant-stress states, the location of the frame nodes $\mathbf{x}_b$ must coincide with the roots of $\mathcal{M}(\mathbf{x})$. This rule can be expressed as

$$\mathbf{x}_b = \mathcal{M}_0(\mathbf{x}_f, \mathbf{x}_s), \tag{72}$$

where $\mathcal{M}_0$ represents the roots of the bending moment function for certain position of the fluid and the structure.

In the case of a viscous fluid, imposing the kinematic condition (38) fixes the relative position of fluid and structure meshes at the interface, therefore the interface frame mesh is constant. However, the case of an inviscid fluid requires free relative tangential motion between the meshes of the fluid and the structure at the interface, as dictated by condition (39), making the position of the fluid and structure interface nodes on the frame

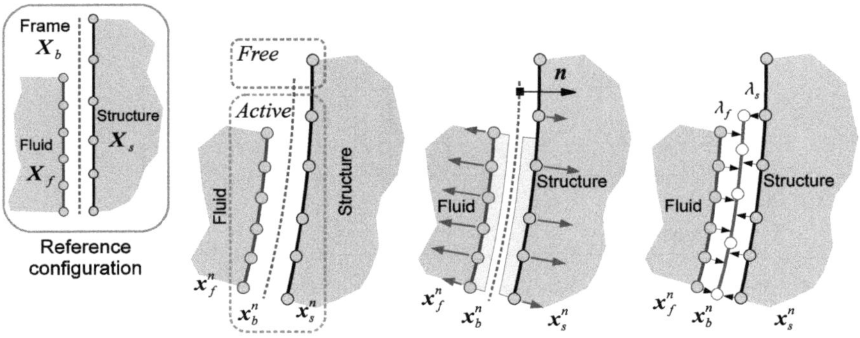

Fig. 7. Steps in the definition of the frame based on position $\mathbf{x}_b^n$: (a) Detection of active interface nodes. (b) Computation of nodal forces corresponding to a constant pressure transmission and (c) Construction of a new frame discretization using the ZMR. Separation between fluid and structure is exaggerated for better visualization.

time dependent. This situation requires the frame mesh to be recalculated every time step using a similar procedure to that employed by Park *et al.* in contact-impact problems.[5]

Let us suppose that we have arrived to time step t^n and we know the total positions of the fluid ($\mathbf{x}_f^n$), frame ($\mathbf{x}_b^n$) and structure ($\mathbf{x}_s^n$). The procedure used to construct a frame satisfying the ZMR for time step t^{n+1} is graphically explained in Fig. 7 and can be summarized in the following steps:

(1) Define the active boundary for t^{n+1}. Search for active interface nodes based on the total position of the fluid $\mathbf{x}_f^n$ relative to the position of the structure $\mathbf{x}_s^n$. Find free nodes and disconnect them from the interface region.
(2) Suppose the transmission of a constant normal stress through the active boundary. Compute the contribution of every interface element to that constant stress by evaluating its equivalent nodal forces.
(3) Using previous nodal forces together with their total position ($\mathbf{x}_f^n$, $\mathbf{x}_s^n$) apply the ZMR to obtain the location of the new frame nodes $\mathbf{x}_b^n = \boldsymbol{\mathcal{M}}_0(\mathbf{x}_f^n, \mathbf{x}_s^n)$.
(4) Construct a frame mesh with isoparametric elements using the previous frame nodes.
(5) Project the current position of the fluid and the structure active nodes on the frame elements forming a new group of pairs. For each pair,

calculate the position $\boldsymbol{\xi}_p$ of the interface node in the frame element, the normal in that position $\mathbf{n}(\boldsymbol{\xi}_p)$ and shape functions $\mathbf{N}_b(\boldsymbol{\xi}_p)$.

Once the frame is defined, the FSI problem is solved for time step t^{n+1} obtaining new positions $\mathbf{x}_f^{n+1}$ and $\mathbf{x}_s^{n+1}$. The frame definition algorithm can then be repeated for a new time step.

9. Partitioned fluid–structure formulation

The coupled partitioned fluid–structure interaction model can now be constructed following variational statement given in (2) for the discrete form of the total energy functional variation $\delta\Pi$. Inserting the discrete variational internal constraint (55), the discrete variational fluid Eq. (34) and the discrete variational structural Eq. (36) into the foregoing equation, the stationarity of the resulting expression yields the partitioned equations of motion:

$$\begin{aligned}\delta\Pi &= \delta\mathbf{u}_f^{\mathsf{T}}\{\mathbf{M}_f\ddot{\mathbf{u}}_f + \mathbf{K}_f\mathbf{u}_f + \mathbf{B}_f\boldsymbol{\lambda}_f - \mathbf{f}_f\}\\ &\quad + \delta\mathbf{u}_s^{\mathsf{T}}\{\mathbf{M}_s\ddot{\mathbf{u}}_s + \mathbf{K}_s\mathbf{u}_s + \mathbf{B}_s\boldsymbol{\lambda}_s - \mathbf{f}_s\}\\ &\quad + \delta\boldsymbol{\lambda}_f^{\mathsf{T}}\{\mathbf{B}_f^{\mathsf{T}}\mathbf{u}_f - \mathbf{L}_f\mathbf{u}_b - \mathbf{h}_f\} + \delta\boldsymbol{\lambda}_s^{\mathsf{T}}\{\mathbf{B}_s^{\mathsf{T}}\mathbf{u}_s - \mathbf{L}_s\mathbf{u}_b - \mathbf{h}_s\}\\ &\quad - \delta\mathbf{u}_b^{\mathsf{T}}\{\mathbf{L}_f^{\mathsf{T}}\boldsymbol{\lambda}_f + \mathbf{L}_s^{\mathsf{T}}\boldsymbol{\lambda}_s\},\end{aligned} \tag{73}$$

with vectors $\mathbf{h}_f = (\mathbf{B}_f^{\mathsf{T}}\mathbf{X}_f - \mathbf{L}_f\mathbf{X}_b)$ and $\mathbf{h}_s = (\mathbf{B}_s^{\mathsf{T}}\mathbf{X}_s - \mathbf{L}_s\mathbf{X}_b)$ function of the initial configuration.

The partitioned equations of motion for the frame-based FSI problem are obtained from the stationary condition of variational (73), that can be expressed as

$$\begin{bmatrix}\mathbf{K}_f + \mathbf{M}_f\frac{d^2}{dt^2} & \mathbf{0} & \mathbf{B}_f & \mathbf{0} & \mathbf{0}\\ \mathbf{0} & \mathbf{K}_s + \mathbf{M}_s\frac{d^2}{dt^2} & \mathbf{0} & \mathbf{B}_s & \mathbf{0}\\ \mathbf{B}_f^{\mathsf{T}} & \mathbf{0} & \mathbf{0} & \mathbf{0} & -\mathbf{L}_f\\ \mathbf{0} & \mathbf{B}_s^{\mathsf{T}} & \mathbf{0} & \mathbf{0} & -\mathbf{L}_s\\ \mathbf{0} & \mathbf{0} & -\mathbf{L}_f^{\mathsf{T}} & -\mathbf{L}_s^{\mathsf{T}} & \mathbf{0}\end{bmatrix}\begin{Bmatrix}\mathbf{u}_f\\ \mathbf{u}_s\\ \boldsymbol{\lambda}_f\\ \boldsymbol{\lambda}_s\\ \mathbf{u}_b\end{Bmatrix} = \begin{Bmatrix}\mathbf{f}_f\\ \mathbf{f}_s\\ \mathbf{h}_f\\ \mathbf{h}_s\\ \mathbf{0}\end{Bmatrix}, \tag{74}$$

where first two rows are the equilibrium equations of each system with terms $\mathbf{B}_f\boldsymbol{\lambda}_f$ and $\mathbf{B}_s\boldsymbol{\lambda}_s$ representing the interface forces exerted on the fluid and structure expressed in the global system, the third and fourth equations impose fluid-frame and structure-frame displacement compatibility, respectively, and the last equation enforces the frame equilibrium.

9.1. *Transient analysis of FSI systems*

The semi-discrete equations of motion (74) are directly integrated in time using Newmark's method, where displacements, velocities and accelerations at current time step t^{n+1} are related with known quantities of the previous time-step t^n by the expressions

$$\mathbf{u}^{n+1} = \mathbf{u}^n + \Delta t\dot{\mathbf{u}}^n + \Delta t^2\left[\left(\frac{1}{2}-\beta\right)\ddot{\mathbf{u}}^n + \beta\ddot{\mathbf{u}}^{n+1}\right], \tag{75}$$

$$\dot{\mathbf{u}}^{n+1} = \dot{\mathbf{u}}^n + \Delta t[(1-\gamma)\ddot{\mathbf{u}}^n + \gamma\ddot{\mathbf{u}}^{n+1}], \tag{76}$$

where $\Delta t = t^{n+1} - t^n$ is the time step and (β, γ) are time integration parameters that determine stability and accuracy characteristics. These parameters are taken to be the same for both fluid ($\mathbf{u} = \mathbf{u}_f$) and structure ($\mathbf{u} = \mathbf{u}_s$).

Suppose now that computations have proceeded until t^n. Inserting (75, 76) into the matrix equation (74) and moving information from the previous time-step to the right side yields

$$\begin{bmatrix} \mathbf{M}_f + \beta\Delta t^2\mathbf{K}_f & \mathbf{0} & \mathbf{B}_f & \mathbf{0} & \mathbf{0} \\ \mathbf{0} & \mathbf{K}_s + \beta\Delta t^2\mathbf{M}_s & \mathbf{0} & \mathbf{B}_s & \mathbf{0} \\ \mathbf{B}_f^\mathsf{T} & \mathbf{0} & \mathbf{0} & \mathbf{0} & -\mathbf{L}_f \\ \mathbf{0} & \mathbf{B}_s^\mathsf{T} & \mathbf{0} & \mathbf{0} & -\mathbf{L}_s \\ \mathbf{0} & \mathbf{0} & -\mathbf{L}_f^\mathsf{T} & -\mathbf{L}_s^\mathsf{T} & \mathbf{0} \end{bmatrix} \begin{Bmatrix} \ddot{\mathbf{u}}_f \\ \ddot{\mathbf{u}}_s \\ \boldsymbol{\lambda}_f \\ \boldsymbol{\lambda}_s \\ \ddot{\mathbf{u}}_b \end{Bmatrix}^{n+1}$$

$$= \begin{Bmatrix} \mathbf{g}_f \\ \mathbf{g}_s \\ \mathbf{h}_f \\ \mathbf{h}_s \\ \mathbf{0} \end{Bmatrix}^{n+1}, \tag{77}$$

where the free-term vector groups the dynamic forces acting on the fluid and the structure

$$\mathbf{g}_f^{n+1} = \mathbf{f}_f^{n+1} - \mathbf{K}_f\left[\mathbf{u}_f^n + \Delta t\dot{\mathbf{u}}_f^n + \Delta t^2\left(\frac{1}{2}-\beta\right)\ddot{\mathbf{u}}_f^n\right], \tag{78}$$

$$\mathbf{g}_s^{n+1} = \mathbf{f}_s^{n+1} - \mathbf{K}_s\left\{\mathbf{u}_s^n + \Delta t\dot{\mathbf{u}}_s^n + \Delta t^2\left(\frac{1}{2}-\beta\right)\ddot{\mathbf{u}}_s^n\right\}, \tag{79}$$

and the dynamic stiffness matrices appear on the diagonal of the system.

It has been demonstrated by Ross *et al.*,[31] that in partitioned systems connected with LLM, as in monolithic ones, Newmark's method with ($\beta = \frac{1}{4}, \gamma = \frac{1}{2}$) provides A-stability and global second-order accuracy. This selection of parameters, known as the Trapezoidal Rule, is used for time integration in all our numerical applications.

10. Numerical applications

The following applications serve to verify and illustrate the capability of the proposed formulation in the study of FSI systems. Examples are selected to verify a correct approximation of the three different effects considered in the fluid, i.e. initial hydrostatic state and sloshing, together with strong fluid–structure interaction effects.

10.1. *Sloshing in flexible containers*

When the structure containing the fluid is flexible, the hydrodynamic pressure on the walls arising due to the free surface oscillation causes the wall to deflect and move, which in turn alters the free surface oscillation and the hydrodynamic forces on the wall. This coupled effect is studied in the next two examples for the square and cubic tanks of the previous section.

In the 2D case, represented in Fig. 8, the elastic tank is modelled using a regular mesh with 30 beam elements with 3-dof per node, restricting displacements and rotation of a node located at the centre of the bottom plate. The material properties of the container are $E_s = 2.1 \cdot 10^{11}$Pa, $\nu_s = 0.3$ with density $\rho_s = 2500$ kg/m^3 and wall thickness $t_s = 2$ mm. The mode

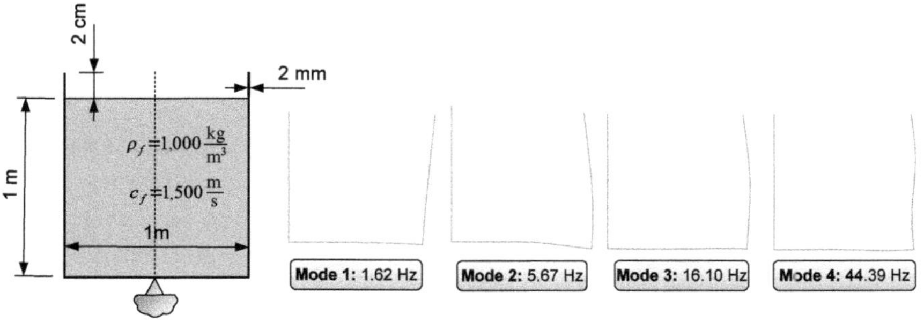

Fig. 8. Dimensions of the flexible container, boundary conditions and first four mode shapes of the structure vibrating *in vacuo*.

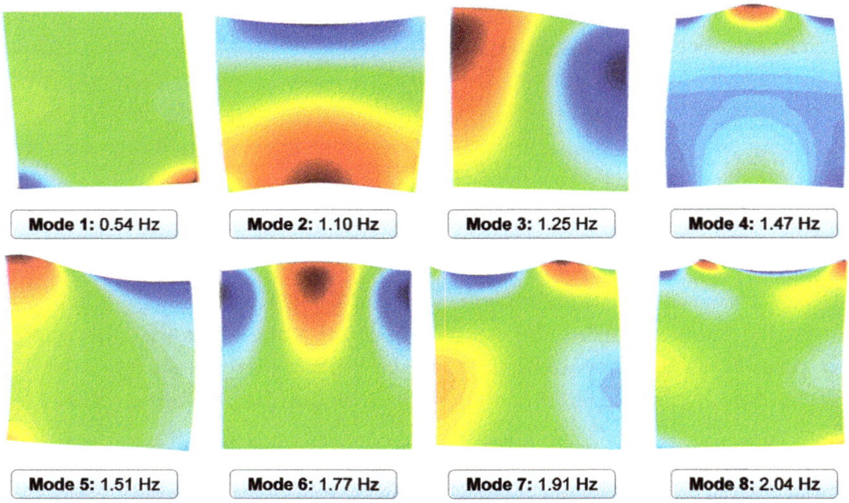

Fig. 9. Coupled mode shapes of the 2D results of the 3D sloshing problem with flexible walls. First natural frequencies of the fluid container (dry structure) are: 1.62 Hz, 5.67 Hz, 16.1 Hz and 44.39 Hz.

shapes and natural frequencies of the empty structure are also represented in Fig. 8 up to the fourth mode.

The results of performing a vibration analysis of the coupled system are presented in Fig. 9 using for the fluid exactly the same mesh as that in the rigid case. Added mass effect can be observed in the first two coupled modes, structure-driven coupled modes presents a lower frequency than their counterparts in the dry structure. Note that in the coupled case, the sloshing modes are appearing at higher frequencies: the third coupled mode with frequency 1.25 Hz corresponds to the first analytical sloshing mode that appears at 0.88 Hz in a rigid container and the sixth coupled mode with frequency 1.77 Hz corresponds to the second sloshing mode of 1.25 Hz. More complicated patterns for the fluid are observed at higher frequencies.

In the 3D case, coupled vibration analysis of the cubic tank is revisited introducing six flexible walls for the container. The mesh used for the fluid is exactly the same than in the rigid-wall case and the mesh for the structure is composed of 125 quadrilateral plate elements with 3-dof per node; the normal displacement and two in-plane rotations. Boundary conditions are somewhat different than in the previous example, each face of the tank is

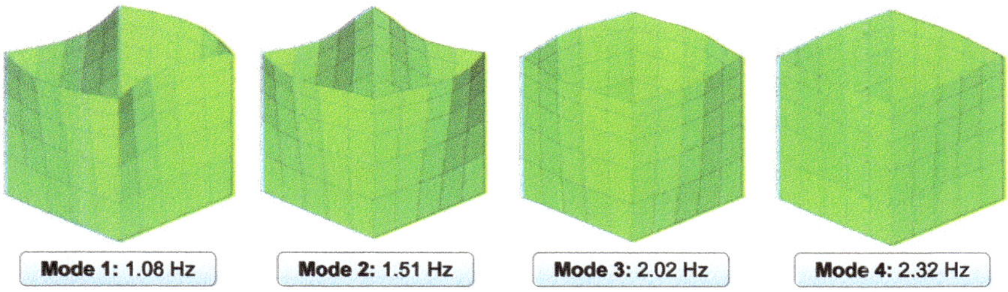

Fig. 10. Mode shapes and natural frequencies of the cubic container vibrating *in vacuo*.

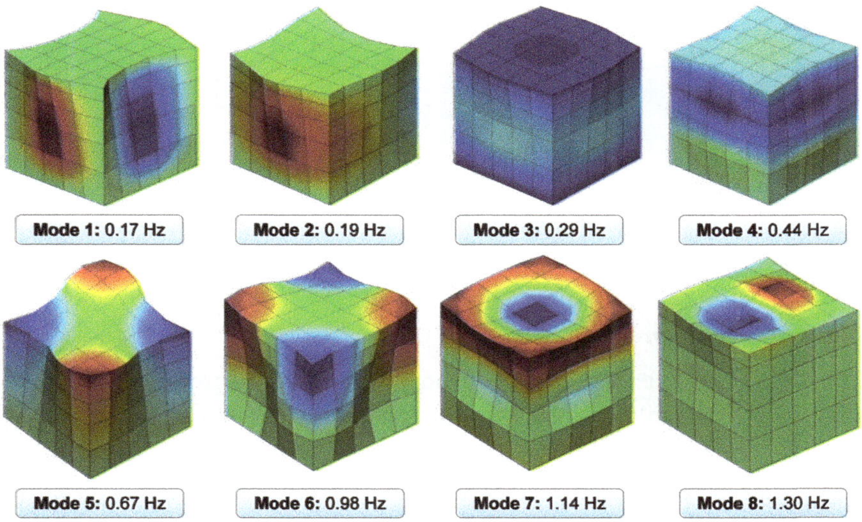

Fig. 11. Coupled mode shapes for the fluid contained in a cubic tank with corresponding natural frequencies.

considered as a simply supported plate. Properties used for the structure are $E_s = 2.1 \cdot 10^9$Pa, $\nu_s = 0.3$, density $\rho_s = 2500$ kg/m^3 and plate thickness $t_s = 2$ mm.

The natural frequencies and mode shapes of the dry structure can be found in Fig. 10 and the coupled mode shapes with their corresponding frequencies in Fig. 11. The first four coupled modes obtained are structure-driven modes with added mass effect, the third one is produced by the vibration of the bottom plate. Sloshing appears in the form of coupled modes after the fourth frequency.

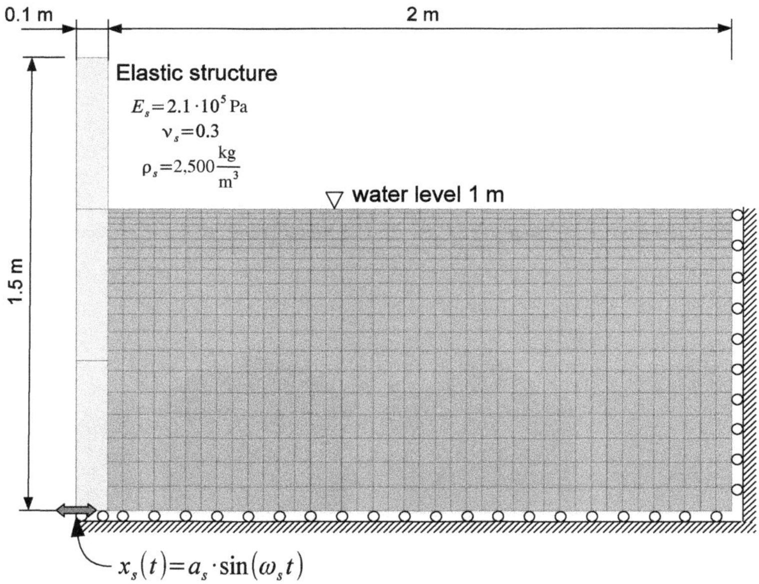

Fig. 12. Wave generator: geometry, boundary conditions and material properties.

10.2. *Wave generator*

This example is devised just to demonstrate that the proposed interfacing strategy can handle scenarios with drastic mesh transitions from structure to fluid without any further modification. A flexible structure modelled using only three quadrilateral elements, as represented in Fig. 12, is used to generate waves on a rectangular basin by imposing a harmonic horizontal motion $x_s(t) = a_s \sin(\omega_s t)$ at the left bottom node of the structure with frequency $\omega_s = 6.28$ rad/s and amplitude $a_s = 5$ cm. Fluid and structure present elastic properties, boundary conditions and dimensions given in Fig. 12 with meshes composed of three plane-strain quadrilateral elements for the structure and 800 quadrilateral fluid elements with 861 nodes for the fluid.

The dynamic calculation is performed integrating in time the equations of motion (77) using a time-step size $\Delta t = 1 \cdot 10^{-2}$ s. In Fig. 13, the waves generated on the fluid at different time steps are depicted, showing that the dynamic interface-frame definition algorithm can handle with no difficulty the contact of sharp edges. This underlines the powerful interface strategy behind the combination of LLMs and ZMR techniques, validating the interface construction algorithm presented in Section 8.

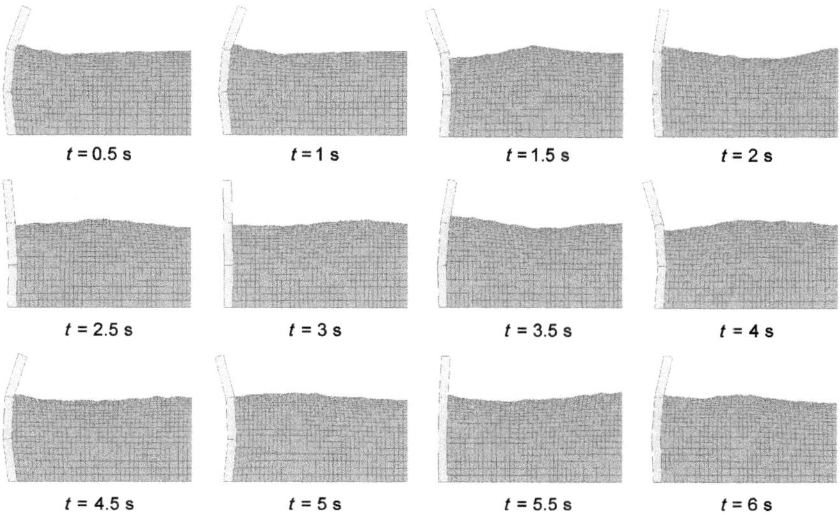

Fig. 13. Wave generator: deformed configurations at different time steps.

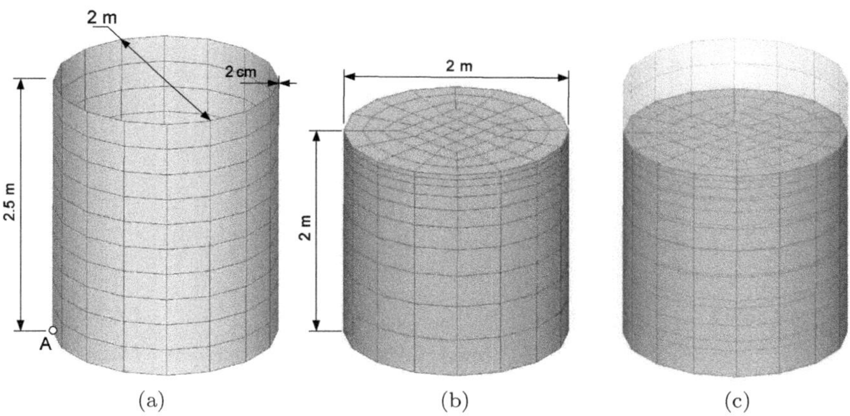

Fig. 14. Dimensions and meshes of the elastic cylindrical container (a), internal fluid (b) and coupled system presenting non-matching interfaces (c).

10.3. *Elastic cylindrical container*

In our last example, an elastic cylindrical container partially filled with water up to a height $H = 2\,\mathrm{m}$ is subject to a forced motion in the horizontal plane. The container is modelled using a regular mesh of 160 4-node structural shell elements with 176 nodes and 6 dof per node. The structural

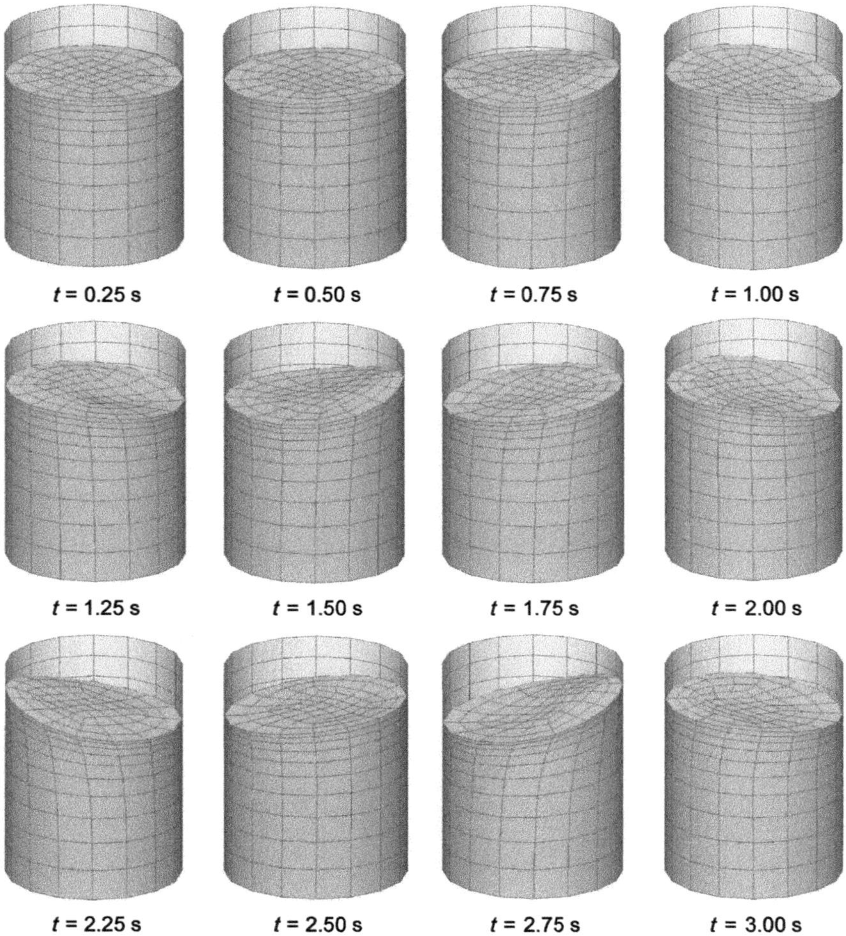

Fig. 15. Cylindrical container: deformed configurations for different time steps. Half cylinder is removed for a clear visualization of the free surface evolution.

properties are $E_s = 2.1 \cdot 10^9$ Pa, $\nu_s = 0.3$, density $\rho_s = 2500$ kg/m^3 and shell thickness $t_s = 2$ mm. All the nodes of the shell structure laying on the floor are allowed to move freely in the X–Y plane except point A, see Fig. 14, subjected to a sinusoidal forced motion in the X-direction of amplitude $a_s = 2$ cm and frequency of $\omega_s = 1$Hz.

The water inside the cylinder, with properties $\rho_f = 2500$ kg/m^3 and $c_f = 1500$ m/s, is modelled with 640 hexahedral finite elements and 803 nodes with 3 dof per node, exactly the same mesh used for the rigid case.

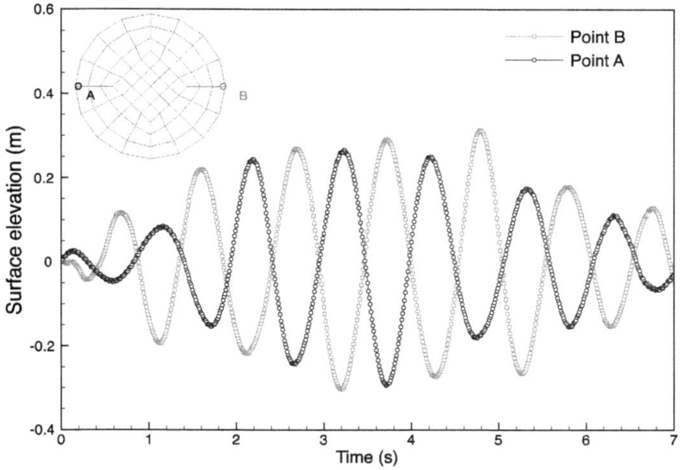

Fig. 16. Cylindrical container: evolution of the free-surface elevation at opposite points near the wall aligned with the direction of imposed motion.

A detail of the spatial discretization is shown in Fig. 14, emphasizing the presence of non-matching between the fluid and the structure. Transient simulation is performed integrating in time system (77) using the trapezoidal rule with time step $\Delta t = 1 \cdot 10^{-2}$s.

The frames in Fig. 15 show the evolution in time of the free surface due to the prescribed swaying motion imposed to the tank. Solution combines stretching of the container walls in the horizontal direction and large amplitude waves in the free surface with a swirl effect. Note that, when the fluid free-surface runs up and down the walls of the container, activation and deactivation of structural interface nodes takes place so they are not affected by the fluid.

Time history of free-surface elevation is given in Fig. 16 for two points near the wall. Points A and B are aligned with the direction of external excitation.

11. Conclusions

The main objective of the paper is to demonstrate applicability of the partitioned FSI formulation proposed by Park *et al.*,[6] for the simulation of structures experiencing significant FSI effects with coupled acoustic and gravity waves, using 2D and 3D models. Theoretical and implementation issues behind the techniques proposed in that reference are reviewed, extended and explained in detail as appropriate to understand its possibilities. Also additional theory topics are included, as treatment of fluid quasi-incompressibility, procedures for interfacing non-matching meshes in the case of inviscid fluids and treatment of the interface using LLMs.

A computational, fully implicit strategy for fluid–structure interaction has been presented that is able to represent correctly acoustic and gravitational effects on the fluid. Numerical examples have been provided, which give evidence of its accuracy, robustness and efficiency. The adopted computational framework allows the independent discretization of fluid and structure, introducing a unique discretization for the interface that passes the patch test following the fluid and structure interfaces during their relative motion. The strategy is very well suited for the modelling of internal fluids with a free-surface interacting with flexible structures.

References

1. K. C. Park and C. A. Felippa, A variational framework for solution method developments in structural mechanics, *Journal of Applied Mechanics*, **65**(1), 242–249 (1998).
2. K. C. Park, C. A. Felippa, and U. Gumaste, A localized version of the method of lagrange multipliers and its applications, *Computational Mechanics*, **24**, 476–490 (2000).
3. K. C. Park and C. A. Felippa, A variational principle for the formulation of partitioned structural systems, *International Journal for Numerical Methods in Engineering*, **47**, 395–418 (2000).
4. G. Rebel, K. C. Park, and C. A. Felippa, A contact formulation based on localized lagrange multipliers: formulation and application to two-dimensional problems, *International Journal for Numerical Methods in Engineering*, **54**(2), 263–297 (2002).
5. K. C. Park, C. A. Felippa, and G. Rebel, A simple algorithm for localized construction of non-matching structural interfaces, *International Journal for Numerical Methods in Engineering*, **53**(9), 2117–2142 (2002).
6. K. C. Park, R. Ohayon, C. A. Felippa, and J. A. González, Partitioned formulation of internal and gravity waves interacting with flexible structures, *Computer Methods in Applied Mechanics and Engineering*, **199**(9–12), 723–733 (2010).
7. K. C. Park, C. A. Felippa, and R. Ohayon, Partitioned formulation of internal fluid-structure interaction problems by localized lagrange multipliers, *Computer Methods in Applied Mechanics and Engineering*, **190**(24–25), 2989–3007 (2001).
8. O. Andrianarison and R. Ohayon, Compressibility and gravity effects in internal fluid-structure vibrations: Basic equations and appropriate variational formulations, *Computer Methods in Applied Mechanics and Engineering*, **195**(17–18), 1958–1972 (2006).

9. O. Andrianarison and R. Ohayon, Reduced models for modal analysis of fluid-structure systems taking into account compressibility and gravity effects, *Computer Methods in Applied Mechanics and Engineering*, **195**(41–43), 5656–5672 (2006).
10. A. Bermúdez and R. Rodríguez, Finite element computation of the vibration modes of a fluid–solid system, *Computer Methods in Applied Mechanics and Engineering*, **119**(3–4), 355–370 (1994).
11. A. Bermúdez, L. Hervella-Nieto, and R. Rodriguez, Finite element computation of three-dimensional elastoacoustic vibrations, *Journal of Sound and Vibration*, **219**(2), 279–306 (1999).
12. A. Bermúdez, L. Hervella-Nieto, and R. Rodríguez, Finite element computation of the vibrations of a plate-fluid system with interface damping, *Computer Methods in Applied Mechanics and Engineering*, **190**(24–25), 3021–3038 (2001).
13. R. Ohayon, Reduced symmetric models for modal analysis of internal structural-acoustic and hydroelastic-sloshing systems, *Computer Methods in Applied Mechanics and Engineering*, **190**(24-25), 3009–3019 (2001).
14. J. Lighthill, *Waves in Fluids*, Cambridge University Press (1978).
15. W. S. Park, C.-B. Yun, and C. Pyun, Infinite elements for 3-dimensional wave-structure interaction problems, *Engineering Structures*, **14**(5), 335–346 (1992).
16. W. Kanok-Nukulchai and B. Tam, Structure-fluid interaction model of tuned liquid dampers, *International Journal for Numerical Methods in Engineering*, **46**(9), 1541–1558 (1999).
17. F. Parrinello and G. Borino, Lagrangian finite element modelling of dam-fluid interaction: Accurate absorbing boundary conditions, *Computers & Structures*, **85**(11–14), 932–943 (2007).
18. D. K. Kishor, S. Gopalakrishnan, and R. Ganguli, Three-dimensional sloshing: A consistent finite element approach, *International Journal for Numerical Methods in Fluids*, **66**(3), 345–376 (2011).
19. J. Simo, R. Taylor, and K. Pister, Variational and projection methods for the volume constraint in finite deformation elasto-plasticity, *Computer Methods in Applied Mechanics and Engineering*, **51**(1–3), 177–208 (1985).
20. J. Simo and R. Taylor, Quasi-incompressible finite elasticity in principal stretches. continuum basis and numerical algorithms, *Computer Methods in Applied Mechanics and Engineering*, **85**(3), 273–310 (1991).
21. K. Bathe, The inf-sup condition and its evaluation for mixed finite element methods, *Computers & structures*, **79**(2), 243–252 (2001).
22. D. Chapelle and K. Bathe, The inf-sup test, *Computers & structures*, **47**(4–5), 537–545 (1993).
23. B. M. F. de Veubeke, Diffusion des inconnues hyperstatiques dans les voilures à longeron couplés, *Bull. Serv. Technique de L'Aéronautique*, **24**, 56 (1951).
24. K. Washizu, *Variational Methods in Elasticity and Plasticity*, Pergamon Press, New York (1972).

25. C. A. Felippa, A survey of parametrized variational principles and applications to computational mechanics, *Computer Methods in Applied Mechanics and Engineering*, **113**(1–2), 109–139 (1994).
26. J. Bonet and R. D. Wood, *Nonlinear Continuum Mechanics for Finite Element Analysis*, 2nd edn., Cambridge University Press (2008).
27. J. A. González, K. C. Park, and C. A. Felippa, Partitioned formulation of frictional contact problems using localized Lagrange multipliers, *Communications in Numerical Methods in Engineering*, **22**, 319–333 (2005).
28. J. A. González, K. C. Park, and C. A. Felippa, A formulation based on localized lagrange multipliers for bem-fem coupling in contact problems, *Computer Methods in Applied Mechanics and Engineering*, **197**, 623–640 (2008).
29. K. Park, C. Felippa, and G. Rebel, Interfacing nonmatching FEM meshes: The zero moment rule. In W. Wall, K. Bletzinger, and K. Schweizer (eds.), *Trends in Computational Structural Mechanics*, Barcelona, Spain (2001).
30. M. R. Ross. *Coupling and Simulation of Acoustic Fluid-Structure Interaction Systems Using Localized Lagrange Multipliers*, PhD thesis, University of Colorado (2006).
31. M. R. Ross, C. A. Felippa, K. C. Park, and M. A. Sprague, Treatment of acoustic fluid-structure interaction by localized lagrange multipliers: formulation, *Computer Methods in Applied Mechanics and Engineering*, **197**(33–40), 3057–3079 (2008).
32. M. R. Ross, M. A. Sprague, C. A. Felippa, and K. Park, Treatment of acoustic fluid-structure interaction by localized lagrange multipliers and comparison to alternative interface-coupling methods, *Computer Methods in Applied Mechanics and Engineering*, **198**(9–12), 986–1005 (2009).
33. O. Zienkiewicz and R. Taylor, *The Finite Element Method: Basic Formulation and Linear Problems*, 4th edn., Vol. 1, McGraw-Hill Book Company, London (1989).
34. M. A. Crisfield, Re-visiting the contact patch test, *International Journal for Numerical Methods in Engineering*, **48**, 435–449 (2000).

Chapter 5

An Eulerian-based Thermo-flexible Multi-body Approach for Simulating Rig Testing of Disc Brakes

N. Strömberg

In this chapter, we propose an Eulerian-based thermo-flexible multi-body approach in order to simulate rig testing of disc brake systems accurately and efficiently. A multi-body model of the disc, an assembly of flywheels and the shaft connecting the disc and flywheels is coupled to an Eulerian-based thermo-mechanical finite element model of the disc-pad system. By utilizing the Eulerian framework of the disc, the contact interface is modelled most accurately with Signorini contact, Coulomb frictional heating with a new temperature-dependent friction model that includes fading at high temperatures and Archard wear with a temperature-dependent wear coefficient. The governing equations are treated with a sequential approach, where first the mechanical contact problem is solved using the augmented Lagrangian approach with a non-smooth Newton method. Then, the multi-body model is solved for the brake moment obtained from the mechanical contact analysis using the average acceleration method. Finally, heat balance for the system including the frictional power from the two previous steps is obtained with the trapezoidal rule and formulating the nonlinear equations as a system of linear equations. Here, the non-symmetric convection matrix is stabilized by adding artificial conduction according to the streamline-upwind approach. The proposed sequential approach is implemented in an in-house code and utilized to study a vented disc-pad system to a heavy truck. This is discussed at the end of the chapter, showing the development of heat bands, hot spots and corresponding residual stresses.

1. Introduction

A brake dynamometer is an important test equipment for automotive industry when developing disc brake systems. It consists of a motor that accelerates an assembly of fly wheels up to a given angular velocity; the motor is then disengaged and the flywheels are decelerated with the disc brake system. In this work, a thermo-flexible multi-body approach for simulating such rig tests of disc brakes is presented and implemented in an in-house toolbox. The first step towards this toolbox was taken already in 1993 when a generalized standard model of contact, friction and wear was developed in Strömberg *et al.*[1] This model was then solved in one of the earliest implementations of the augmented Lagrangian approach in Strömberg.[2] Due to the success of this implementation, it was compared to an interior point method in Christensen *et al.*,[3] where the augmented Lagrangian approach outperformed an interior point method. The implementation was then expanded to 3D in Strömberg,[4] where the development of fretting wear was studied numerically for different examples. Shortly after, a complete thermomechanical model of contact, friction and wear including frictional heat and contact conductance was implemented using the augmented Lagrangian approach and it was numerically studied how frictional hot spots are influenced by wear conditions.[5] This work was then continued in Ireman *et al.*[6] A new generalized standard model for fretting fatigue was derived and solved numerically in Refs. [7, 8], where continuum damage was coupled to wear using gradient-based state variables. In Lundvall *et al.*,[9] a flexible multi-body approach for studying the transmission error in gears was suggested and implemented. Similar ideas are utilized in this chapter to develop the Eulerian-based thermo-flexible multi-body approach. Other relevant implicit approaches for structural dynamic contact problems with friction and wear can be found in Refs. [10–13]. Most of the approaches discussed so far are for problems with small displacements. Of course, such approaches are not applicable for our rotating disc. The straightforward approach would be to apply a total Lagrangian formulation. However, experience has shown that such a formulation is extremely computational costly, mainly due to the need of updating the contact interface between pad and disc. Therefore, an Eulerian framework for these type of problems was developed in Strömberg.[14] The disc is formulated in an Eulerian frame such that the material flows through the mesh like a fluid. A new convective term appears that so far has been stabilized using the standard streamline-upwind approach. This works well since the streamlines are well-defined by the rotation of the disc. A major advantage of the proposed formulation is that the contact interface is always geometrically well defined, we don't have to update the contact geometry and can also have a fine mesh of the contact interface at low extra computational cost without adopting any adaptive strategy.

The temperature history generated by the in-house toolbox can be exported to Abaqus or LS-Dyna where the corresponding residual stresses can be analyzed by using temperature dependent plasticity models. This has been done in a number of works in different projects, and some of the results are presented in the Ph.D. thesis by Rashid.[15] The temperature-dependent plasticity models have been developed when we studied residual stresses in castings from the solidification process in other projects, see e.g. Ref. [16]. The residual stresses developed in hard braking down a hill with a heavy truck were studied in Ref. [17] by using temperature-dependent J2-plasticity with isotropic hardening. Cyclic braking was studied in Ref. [18] by replacing the nonlinear isotropic hardening with linear kinematic hardening. Not only frictional heating is included in the toolbox but also the effect of wear is implemented. In Ref. [19], it was studied how the development of heat bands are influenced by wear and the evolution of the contact pressure. After repeated breaking with wear included in the model, not only one band is developed but instead two bands appear on the disc. This corresponds to what is found experimentally; it has been shown that experiment and numerical results correlate well during a previous collaboration with Volvo Truck.

The outline of the chapter is as follows: in the next two sections, the governing equations of the Eulerian-based thermo-flexible multi-body approach are presented. A temperature-dependent friction model with fading properties at high temperatures is also suggested, then we devote two sections on summarizing the corresponding finite element equations as well as presenting the sequential approach in detail. Finally, the numerical implementation is discussed and some concluding remarks are given.

2. Kinematics and balance in momentum

Let us consider a rigid multi-body system of two rotors and a torsional spring as depicted in Fig. 1 with moment of inertia J_1 (inertia of disc) and J_2 (inertia of flywheels), and a torsional spring stiffness k (stiffness of axis connecting disc and flywheels). The upper rotor is subjected to a brake torque $M_b = M_b(t)$ as a function of time t. The brake torque is in turn governed by a nonlinear thermo-mechanical frictional contact model of a pad–disc system including wear, see Figs. 1 and 4, where the large rotations of the disc $\theta_1 = \theta_1(t)$ are formulated in an Eulerian framework. By adopting

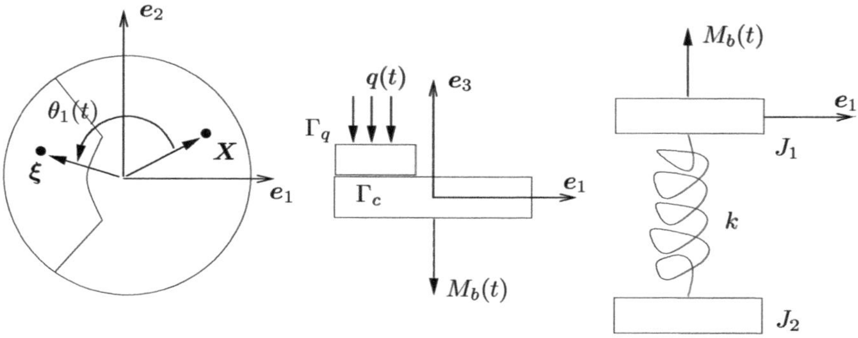

Fig. 1. An Eulerian-based thermo-flexible multi-body model of a brake dynamometer.

this formulation, unique contact pairs can be identified at the initiation of the simulation without any need of a contact search algorithm during the simulation. This is a major advantaged of the suggested approach compared to a standard Lagrangian formulation of the disc.

The rigid body dynamics of the rotors can be formulated as

$$\boldsymbol{J}\ddot{\boldsymbol{\theta}} + \boldsymbol{K}\boldsymbol{\theta} = \boldsymbol{M}, \tag{1}$$

where

$$\boldsymbol{J} = \begin{bmatrix} J_1 & 0 \\ 0 & J_2 \end{bmatrix},$$
$$\boldsymbol{K} = \begin{bmatrix} k & -k \\ -k & k \end{bmatrix}, \tag{2}$$
$$\boldsymbol{M} = \begin{Bmatrix} M_b = M_b(t) \\ 0 \end{Bmatrix}$$

and

$$\boldsymbol{\theta} = \begin{Bmatrix} \theta_1 = \theta_1(t) \\ \theta_2 = \theta_2(t) \end{Bmatrix}, \tag{3}$$

is a vector containing the angular positions $\boldsymbol{\theta}_i = \theta_i \boldsymbol{e}_3$ of the two rotors.

The spatial position $\boldsymbol{x}^1$ of a material point $\boldsymbol{X}^1$ of the disc is governed by the rigid body rotation of the upper rotor $\boldsymbol{\xi}^1 = \boldsymbol{Q}\boldsymbol{X}^1$ and superimposed small displacements $\boldsymbol{u}^1 = \boldsymbol{u}^1(\boldsymbol{\xi}^1, t)$, where

$$\boldsymbol{Q} = \boldsymbol{Q}(t) = \begin{bmatrix} \cos(\theta_1(t)) & -\sin(\theta_1(t)) & 0 \\ \sin(\theta_1(t)) & \cos(\theta_1(t)) & 0 \\ 0 & 0 & 1 \end{bmatrix}. \tag{4}$$

Thus,

$$\boldsymbol{x}^1 = \boldsymbol{Q}(t)\boldsymbol{X}^1 + \boldsymbol{u}^1(\boldsymbol{Q}(t)\boldsymbol{X}^1, t). \tag{5}$$

The kinematics of the pad is defined by small displacements $\boldsymbol{u}^2 = \boldsymbol{u}^2(\boldsymbol{X}^2, t)$.

By neglecting the rates of the small displacements $\boldsymbol{u}^1$ and $\boldsymbol{u}^2$, the velocity of a material point $\boldsymbol{X}^1$ passing a spatial point $\boldsymbol{\xi}^1$ can be written as

$$\boldsymbol{v}^1 = \boldsymbol{\omega} \times \boldsymbol{\xi}^1, \tag{6}$$

where $\boldsymbol{\omega} = \boldsymbol{\omega}(t) = \dot{\theta}_1(t)\boldsymbol{e}_3$ is the axial vector of $\dot{\boldsymbol{Q}}\boldsymbol{Q}^T$. The corresponding acceleration becomes

$$\boldsymbol{a}^1 = \dot{\boldsymbol{\omega}} \times \boldsymbol{\xi}^1 + \boldsymbol{\omega} \times \boldsymbol{\omega} \times \boldsymbol{\xi}^1. \tag{7}$$

At the contact interface Γ_c between the pad and the disc, see Figs. 1 and 4, we identify unique contact pairs of material points $\boldsymbol{X}_c^2$ of the pad and spatial points $\boldsymbol{\xi}_c^1$ of the disc such that $\boldsymbol{\xi}_c^1 = \boldsymbol{X}_c^2$. By using such contact pairs, we set up contact conditions for a material point $\boldsymbol{X}^1$ passing $\boldsymbol{\xi}_c^1$ at time t and $\boldsymbol{X}_c^2$. The relative displacement of these points are given by

$$\boldsymbol{\delta} = \boldsymbol{\delta}(t) = \boldsymbol{Q}(t)\boldsymbol{X}^1 + \boldsymbol{u}^1(\boldsymbol{\xi}_c^1, t) - (\boldsymbol{X}_c^2 + \boldsymbol{u}^2(\boldsymbol{X}_c^2, t)), \tag{8}$$

which reduces to $\boldsymbol{\delta} = \boldsymbol{u}^1(\boldsymbol{\xi}_c^1, t) - \boldsymbol{u}^2(\boldsymbol{X}_c^2, t)$ since $\boldsymbol{\xi}_c^1 = \boldsymbol{X}_c^2$. By neglecting the rates of the small displacements, the material time derivative of $\boldsymbol{\delta}$ becomes

$$\dot{\boldsymbol{\delta}} = \boldsymbol{\omega} \times \boldsymbol{\xi}_c^1. \tag{9}$$

By neglecting the body forces and introducing the Cauchy stress $\boldsymbol{\sigma}$ and the mass density ρ, the balance in momentum for the disc Ω^1 and the pad Ω^2 are formulated as

$$\begin{aligned} \mathrm{div}\boldsymbol{\sigma} &= \rho\left(\dot{\boldsymbol{\omega}} \times \boldsymbol{\xi}^1 + \boldsymbol{\omega} \times \boldsymbol{\omega} \times \boldsymbol{\xi}^1\right) \quad \text{in } \Omega^1, \\ \mathrm{div}\boldsymbol{\sigma} &= \boldsymbol{0} \quad \text{in } \Omega^2. \end{aligned} \tag{10}$$

Furthermore, at the contact interface Γ_c, the principle of action and reaction is formulated using a contact traction vector $\boldsymbol{p}$ defined by

$$\boldsymbol{p} = -\boldsymbol{\sigma}\boldsymbol{e}_3 \quad \text{on } \Gamma_c \text{ of the disc and pad.} \tag{11}$$

In addition, a pad pressure $q = q(t)$ is applied on Γ_q, see Fig. 1, i.e.

$$\boldsymbol{\sigma}\boldsymbol{e}_3 = -q\boldsymbol{e}_3 \quad \text{on } \Gamma_q. \tag{12}$$

Finally, notice that the brake torque is obtained by

$$\boldsymbol{M}_b = -\int_{\Gamma_c} \boldsymbol{\xi}_c^1 \times \boldsymbol{p}\, \mathrm{d}A. \tag{13}$$

3. Constitutive assumptions and balance in energy

By introducing the temperature fields T of the disc $T^1 = T^1(\xi^1, t)$ and pad $T^2 = T^2(X^2, t)$, respectively, and the corresponding engineering strains as

$$\epsilon_{ij}^1 = \frac{1}{2}\left(\frac{\partial u_i^1}{\partial \xi_j} + \frac{\partial u_j^1}{\partial \xi_i}\right) \quad \text{and} \quad \epsilon_{ij}^2 = \frac{1}{2}\left(\frac{\partial u_i^2}{\partial X_j} + \frac{\partial u_j^2}{\partial X_i}\right), \tag{14}$$

we formulate linear thermoelasticity for both the disc and the pad by

$$\sigma_{ij}^i = \lambda^i \epsilon_{kk}^i \delta_{ij} + 2G^i \epsilon_{ij}^i - (3\lambda^i + 2G^i)\alpha^i T^i \delta_{ij}, \tag{15}$$

where λ^i and G^i are Lame's coefficients and α^i is the thermal dilatation coefficient for the disc ($i = 1$) and pad ($i = 2$), respectively.[a] δ_{ij} is of course Kronecker's delta. The constitutive assumption in (15) could be expanded in different directions. For instance, one might add a yield criterion for the disc and orthotropic properties for the pad including continuum damage. The latter might also be coupled to wear at the contact interface following the ideas of Ireman *et al.*[7]

The contact traction vector is governed by Signorini's contact conditions including a wear gap $\delta_w = \delta_w(t)$ and Coulomb's law with a temperature-dependent friction $\mu = \mu(\mathcal{T})$, where $\mathcal{T}$ is the temperature at the contact interface. These assumptions are formulated by decomposing $\boldsymbol{p}$ into a normal component $p_n = \boldsymbol{p} \cdot \boldsymbol{e}_3$ and a tangential part $\boldsymbol{p}_t = \boldsymbol{p} - p_n\boldsymbol{e}_3$.

[a]The reference temperature is here assumed to be zero.

The normal contact condition is formulated using Signorini's contact conditions including wear, which can be formulated as

$$p_n \in \partial I_C(\delta_n), \tag{16}$$

where $\delta_n = \boldsymbol{\delta} \cdot \boldsymbol{e}_3$ and

$$I_C = I_C(x) = \begin{cases} 0 & \text{if } x \in C = \{x : x \leq \delta_w\}, \\ \infty & \text{otherwise,} \end{cases} \tag{17}$$

is the indicator function of the convex set C. The subdifferential of the indicator function in (16) is given by the following set:

$$\partial I_C(\delta_n) = \begin{cases} N_C(\delta_n) & \text{if } \delta_n \in C, \\ 0 & \text{otherwise,} \end{cases} \tag{18}$$

where

$$N_C(x) = \{X : X(x' - x) \leq 0 \ \forall x' \in C\}, \tag{19}$$

is the normal cone of C. This is equivalent to

$$\begin{cases} \min\limits_{\delta_n} \ - p_n \delta_n, \\ \text{s.t. } \delta_n \leq \delta_w. \end{cases} \tag{20}$$

The corresponding Karush–Kuhn–Tucker conditions read

$$p_n \geq 0, \qquad \delta_n \leq \delta_w, \qquad p_n(\delta_n - \delta_w) = 0, \tag{21}$$

which of course are recognized as Signorini's contact conditions with zero initial gap and an extra wear gap δ_w included. By using the augmented Lagrangian approach, these conditions can also be rephrased to

$$p_n = \frac{p_n + r(\delta_n - \delta_w) + |p_n + r(\delta_n - \delta_w)|}{2}, \tag{22}$$

where $r > 0$ is a penalty coefficient.

The wear gap δ_w is governed by the following representation of Arhard's assumption:

$$\dot{\delta}_w = k_w p_n |\boldsymbol{\omega} \times \boldsymbol{\xi}_c^1|, \tag{23}$$

where $k_w = k_w(\mathcal{T})$ is a temperature-dependent wear coefficient. Typically, the wear rate increases dramatically when the temperature increases to high levels.

The tangential part of the contact traction develops according to

$$\boldsymbol{p}_t = p_t\boldsymbol{e}_t = \mu p_n \frac{\boldsymbol{\omega} \times \boldsymbol{\xi}_c^1}{|\boldsymbol{\omega} \times \boldsymbol{\xi}_c^1|}, \tag{24}$$

during sliding. Thus, in this work, only fully developed sliding between the disc and pad is considered. Stick-slip phenomenon between pad and disc is a topic for future work. The friction coefficient $\mu = \mu(\mathcal{T})$ depends typically strongly on the temperature. From experiments, one can usually identify three characteristic values: the friction coefficient μ_{room} at room temperature $\mathcal{T}_{\text{room}}$, a peak value μ_{peak} at a certain temperature $\mathcal{T}_{\text{peak}}$ and a lower bound μ_{bound} at temperatures over a certain limit temperature $\mathcal{T}_{\text{bound}}$. We suggest to adopt a quadratic interpolation using these values according to

$$\mu(\mathcal{T}) = \mu_0 + \mu_1\mathcal{T} + \mu_2\mathcal{T}^2, \tag{25}$$

where μ_i are obtained from

$$\begin{Bmatrix} \mu_0 \\ \mu_1 \\ \mu_2 \end{Bmatrix} = \begin{bmatrix} 1 & \mathcal{T}_{\text{room}} & \mathcal{T}_{\text{room}}^2 \\ 1 & \mathcal{T}_{\text{peak}} & \mathcal{T}_{\text{peak}}^2 \\ 1 & \mathcal{T}_{\text{bound}} & \mathcal{T}_{\text{bound}}^2 \end{bmatrix}^{-1} \begin{Bmatrix} \mu_{\text{room}} \\ \mu_{\text{peak}} \\ \mu_{\text{bound}} \end{Bmatrix}. \tag{26}$$

This model is illustrated in Fig. 2.

The frictional power developed in the contact interface is

$$\mathcal{D}_f = \boldsymbol{p}_t \cdot (\boldsymbol{\omega} \times \boldsymbol{\xi}_c^1) = \mu p_n |\boldsymbol{\omega} \times \boldsymbol{\xi}_c^1| \tag{27}$$

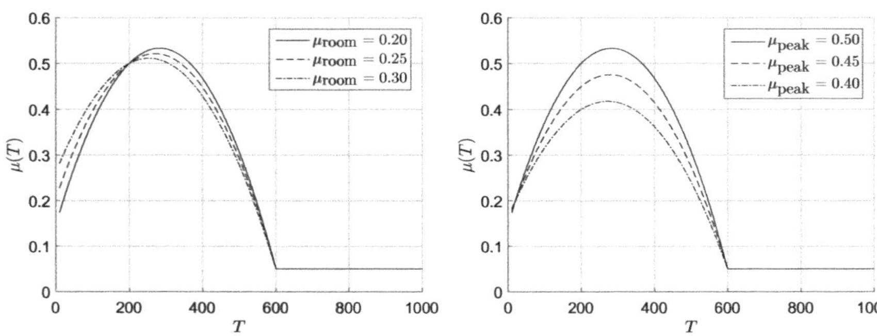

Fig. 2. Temperature-dependent friction coefficient.

and the total frictional power is consequently obtained by

$$\int_{\Gamma_c} \mathcal{D}_f \, \mathrm{d}A = \boldsymbol{\omega} \cdot \int_{\Gamma_c} \boldsymbol{\xi}_c^1 \times \boldsymbol{p}_t \, \mathrm{d}A = -\boldsymbol{\omega} \cdot \boldsymbol{M}_b, \tag{28}$$

where (13) and $\boldsymbol{p} = p_n \boldsymbol{e}_3 + \boldsymbol{p}_t$ also have been utilized.

By introducing Fourier's law of heat diffusion, i.e.

$$q_i = -k \frac{\partial T}{\partial x_i}, \tag{29}$$

where $\boldsymbol{q}^1 = \boldsymbol{q}^2 = \boldsymbol{q} = q_i \boldsymbol{e}_i$ represents the heat flux in the bodies and k is the thermal conductivity, we can formulate balance in energy for the bodies as

$$\rho c \dot{T} = k \sum_{i=1}^{3} \frac{\partial^2 T}{\partial x_i^2}, \tag{30}$$

where c is the specific heat capacity. For the disc, a convective term appears when the material time derivative is calculated, i.e.

$$\dot{T} = \frac{\partial T}{\partial \xi_i} \frac{\partial \xi_i}{\partial t} + \frac{\partial T}{\partial t} = (\boldsymbol{\omega} \times \boldsymbol{\xi})_i \frac{\partial T}{\partial \xi_i} + \frac{\partial T}{\partial t}. \tag{31}$$

Thus, the energy balance for the disc can also be formulated as

$$\rho c \frac{\partial T}{\partial t} + \rho c (\boldsymbol{\omega} \times \boldsymbol{\xi})_i \frac{\partial T}{\partial \xi_i} = k \sum_{i=1}^{3} \frac{\partial^2 T}{\partial x_i^2}. \tag{32}$$

The energy balance for the contact interface reads

$$\mathcal{D}_f + \boldsymbol{q}^1 \cdot \boldsymbol{e}_3 - \boldsymbol{q}^2 \cdot \boldsymbol{e}_3 = 0. \tag{33}$$

In addition, we assume that the heat flowing from the bodies to the common interface follow

$$\begin{aligned} \boldsymbol{q}^1 \cdot \boldsymbol{e}_3 &= \vartheta^1 p_n (T^1 - \mathcal{T}), \\ -\boldsymbol{q}^2 \cdot \boldsymbol{e}_3 &= \vartheta^2 p_n (T^2 - \mathcal{T}), \end{aligned} \tag{34}$$

where ϑ^i represents contact conductances for the disc and pad, respectively. By eliminating $\mathcal{T}$ from (33) and (34), we can write the energy balance at the contact interface as

$$\begin{aligned} \boldsymbol{q}^1 \cdot \boldsymbol{e}_3 &= \vartheta^0 p_n (T^1 - T^2) + \varsigma \mathcal{D}_f, \\ -\boldsymbol{q}^2 \cdot \boldsymbol{e}_3 &= \vartheta^0 p_n (T^2 - T^1) + (1 - \varsigma) \mathcal{D}_f, \end{aligned} \tag{35}$$

where

$$\vartheta^0 = \frac{\vartheta^1 \vartheta^2}{\vartheta^1 + \vartheta^2} \qquad \text{and} \qquad \varsigma = \frac{\vartheta^1}{\vartheta^1 + \vartheta^2}. \tag{36}$$

4. Finite element equations

The governing equations of the Eulerian-based thermo-mechanical model of the disc–pad system presented in the two previous sections are discretized in space using standard linear finite elements. Letting $\boldsymbol{d}_m$ and $\boldsymbol{T}_m$ containing nodal displacements and nodal temperatures of the disc ($m = 1$) and the pad ($m = 2$), respectively, the finite element formulation of the balance in momentum (2), together with boundary conditions (11)–(12) and thermoelasticity (15) inserted, becomes

$$\begin{aligned} \boldsymbol{K}_1\boldsymbol{d}_1 - \hat{\boldsymbol{K}}_1\boldsymbol{T}_1 + \left(\boldsymbol{C}_{N1} + \boldsymbol{C}_{T1}(\tilde{\boldsymbol{T}})\right)\boldsymbol{P}_N &= \boldsymbol{F}_a(\dot{\theta}_1, \ddot{\theta}_1), \\ \boldsymbol{K}_2\boldsymbol{d}_2 - \hat{\boldsymbol{K}}_2\boldsymbol{T}_2 + \left(\boldsymbol{C}_{N2} + \boldsymbol{C}_{T2}(\tilde{\boldsymbol{T}})\right)\boldsymbol{P}_N &= \boldsymbol{F}_q(t), \end{aligned} \tag{37}$$

where $\boldsymbol{K}_m$ are stiffness matrices, $\hat{\boldsymbol{K}}_m$ represent thermal expansion properties, $\boldsymbol{C}_{Nm}$ contain normal directions, $\boldsymbol{C}_{Tm}$ represent sliding directions and the temperature-dependent friction model in (25) using nodal interface temperatures collected in $\tilde{\boldsymbol{T}}$, $\boldsymbol{P}_N$ contains nodal normal contact forces, $\boldsymbol{F}_a = \boldsymbol{F}_a(\dot{\theta}_1, \ddot{\theta}_1)$ represents inertia forces from (7) in (2) and, finally, $\boldsymbol{F}_q = \boldsymbol{F}_q(t)$ is the nodal forces corresponding to the pad pressure in (12).

The finite element formulation of the heat balances in (30) and (32) reads

$$\begin{aligned} \boldsymbol{C}_1\dot{\boldsymbol{T}}_1 + \left(\boldsymbol{N}(\dot{\theta}_1) + \boldsymbol{O}_1\right)\boldsymbol{T}_1 &= \boldsymbol{Q}_{C1}(\boldsymbol{T}_1, \boldsymbol{T}_2, \boldsymbol{P}_N, \dot{\theta}_1), \\ \boldsymbol{C}_2\dot{\boldsymbol{T}}_2 + \boldsymbol{O}_2\boldsymbol{T}_2 &= \boldsymbol{Q}_{C2}(\boldsymbol{T}_1, \boldsymbol{T}_2, \boldsymbol{P}_N, \dot{\theta}_1), \end{aligned} \tag{38}$$

where $\boldsymbol{C}_m$ are heat capacity matrices, $\boldsymbol{N} = \boldsymbol{N}(\dot{\theta}_1)$ is the convection matrix from the convective part in (32) and $\boldsymbol{O}_m$ represents the thermal conduction in the bodies. Furthermore,

$$\begin{aligned} \boldsymbol{Q}_{C1} &= \boldsymbol{Q}_{C1}(\boldsymbol{T}_1, \boldsymbol{T}_2, \boldsymbol{P}_N, \dot{\theta}_1) = \boldsymbol{S}_{P2}\boldsymbol{T}_2 - \boldsymbol{S}_{P1}\boldsymbol{T}_1 + \boldsymbol{Q}_{\mu 1}(\tilde{\boldsymbol{T}}, \boldsymbol{P}_N, \dot{\theta}_1), \\ \boldsymbol{Q}_{C2} &= \boldsymbol{Q}_{C2}(\boldsymbol{T}_1, \boldsymbol{T}_2, \boldsymbol{P}_N, \dot{\theta}_1) = \boldsymbol{S}_{P1}\boldsymbol{T}_1 - \boldsymbol{S}_{P2}\boldsymbol{T}_2 + \boldsymbol{Q}_{\mu 2}(\tilde{\boldsymbol{T}}, \boldsymbol{P}_N, \dot{\theta}_1), \end{aligned} \tag{39}$$

where $\boldsymbol{S}_{Pm} = \boldsymbol{S}_{Pm}(\boldsymbol{P}_N)$ represent the contact conductances in (34) and $\boldsymbol{Q}_{\mu m} = \boldsymbol{Q}_{\mu m}(\tilde{\boldsymbol{T}}, \boldsymbol{P}_N, \dot{\theta}_1)$ are frictional powers from (27). It is well known that the convection matrix $\boldsymbol{N}$ is non-symmetric, which is causing numerically instabilities. In this work, this unwanted phenomenon is stabilized by adding artificial conduction along the streamlines which for our disc are well-defined by the rotation $\dot{\theta}_1$. In such manner $(38)_1$, is modified to

$$\boldsymbol{C}_1\dot{\boldsymbol{T}}_1 + \left(\boldsymbol{N}(\dot{\theta}_1) + \boldsymbol{O}_1 + \boldsymbol{R}(\dot{\theta}_1)\right)\boldsymbol{T}_1 = \boldsymbol{Q}_{C1}(\boldsymbol{T}_1, \boldsymbol{T}_2, \boldsymbol{P}_N, \dot{\theta}_1), \tag{40}$$

where $\boldsymbol{R} = \boldsymbol{R}(\dot{\theta}_1)$ contains artificial conduction according to the streamline-upwind approach.

The finite element discretization of the contact law in (22) becomes

$$\boldsymbol{P}_N = \frac{\boldsymbol{P}_N + r\,(\boldsymbol{\delta} - \boldsymbol{\delta}_w) + |\boldsymbol{P}_N + r\,(\boldsymbol{\delta} - \boldsymbol{\delta}_w)\,|}{2}, \tag{41}$$

where

$$\boldsymbol{\delta} = \boldsymbol{C}_{N1}\boldsymbol{d}_1 + \boldsymbol{C}_{N2}\boldsymbol{d}_2 \tag{42}$$

and $\boldsymbol{\delta}_w = \boldsymbol{\delta}_w(\tilde{\boldsymbol{T}}, \boldsymbol{P}_N, \dot{\theta}_1)$ contains nodal wear gaps governed by Archard's wear law in (23).

5. Sequential approach

For given temperatures $\boldsymbol{T}^n = \{\boldsymbol{T}_1^n, \boldsymbol{T}_2^n\}$ and given angular rates $\dot{\theta}_1^n$, $\ddot{\theta}_1^n$, we find $\boldsymbol{d}^{n+1} = \{\boldsymbol{d}_1^{n+1}, \boldsymbol{d}_2^{n+1}\}$ and $\boldsymbol{P}_N^{n+1}$ by solving (37) and (41) by using a non-smooth Newton algorithm with an Armijo line-search procedure.[2] We represent these equations with

$$\boldsymbol{h}(\boldsymbol{d}^{n+1}, \boldsymbol{P}_N^{n+1}, \boldsymbol{T}^n, \dot{\theta}_1^n, \ddot{\theta}_1^n) = \boldsymbol{0}. \tag{43}$$

Equation (1) is treated in time by applying the average acceleration method. Letting $\boldsymbol{\theta}^n$, $\dot{\boldsymbol{\theta}}^n$, $\ddot{\boldsymbol{\theta}}^n$ and $\boldsymbol{\theta}^{n+1}$, $\dot{\boldsymbol{\theta}}^{n+1}$, $\ddot{\boldsymbol{\theta}}^{n+1}$ represent the approximations of $\boldsymbol{\theta}(t)$ at time t_n and t_{n+1}, respectively, the procedure for this is governed by

$$\boldsymbol{J}^{\text{eff}}\ddot{\boldsymbol{\theta}}^{n+1} = \boldsymbol{M}^{\text{eff}}(\boldsymbol{P}_N^{n+1}, \boldsymbol{\theta}^n, \dot{\boldsymbol{\theta}}^n, \ddot{\boldsymbol{\theta}}^n), \tag{44}$$

where

$$\begin{aligned} \boldsymbol{J}^{\text{eff}} &= \left[\boldsymbol{J} + \frac{\Delta t^2}{4}\boldsymbol{K}\right], \\ \boldsymbol{M}^{\text{eff}}(\boldsymbol{P}_N^{n+1}, \boldsymbol{\theta}^n, \dot{\boldsymbol{\theta}}^n, \ddot{\boldsymbol{\theta}}^n) &= \boldsymbol{M}^{n+1} - \hat{\boldsymbol{M}}^n, \end{aligned} \tag{45}$$

$\Delta t = t_{n+1} - t_n$ is the time step,

$$\hat{\boldsymbol{M}}^n = \boldsymbol{K}\boldsymbol{\theta}^n + \Delta t \boldsymbol{K}\dot{\boldsymbol{\theta}}^n + \frac{\Delta t^2}{4}\boldsymbol{K}\ddot{\theta}^n,$$
$$\dot{\boldsymbol{\theta}}^{n+1} = \dot{\boldsymbol{\theta}}^n + \frac{\Delta t}{2}(\ddot{\boldsymbol{\theta}}^n + \ddot{\boldsymbol{\theta}}^{n+1}) \tag{46}$$

and

$$\boldsymbol{\theta}^{n+1} = \boldsymbol{\theta}^n + \Delta t \dot{\theta}^n + \frac{\Delta t^2}{4}(\ddot{\boldsymbol{\theta}}^n + \ddot{\boldsymbol{\theta}}^{n+1}). \tag{47}$$

The time rates appearing in $(38)_2$ and (40) are treated by applying the trapezoidal rule, i.e. given $\boldsymbol{T}^n$, $\dot{\boldsymbol{T}}^n$ and $\dot{\boldsymbol{T}}^{n+1}$,

$$\boldsymbol{T}^{n+1} = \boldsymbol{T}^n + \Delta t \left((1-\tau)\dot{\boldsymbol{T}}^n + \tau \dot{\boldsymbol{T}}^{n+1} \right), \tag{48}$$

where we set $\tau = 2/3$ in order to obtain Galerkin's approach. (48) utilized in $(38)_2$ and (40) yields

$$\left(\boldsymbol{N} + \boldsymbol{O}_1 + \boldsymbol{R} + \frac{1}{\tau \Delta t}\boldsymbol{C}_1 \right) \boldsymbol{T}_1^{n+1} = \boldsymbol{Q}_1^{\text{old}} + \boldsymbol{Q}_{C1},$$
$$\left(\boldsymbol{O}_2 + \frac{1}{\tau \Delta t}\boldsymbol{C}_2 \right) \boldsymbol{T}_2^{n+1} = \boldsymbol{Q}_2^{\text{old}} + \boldsymbol{Q}_{C2}, \tag{49}$$

where

$$\boldsymbol{Q}_m^{\text{old}} = \frac{(1-\tau)}{\tau}\boldsymbol{C}_m \dot{\boldsymbol{T}}^n + \frac{1}{\tau \Delta t}\boldsymbol{C}_m \boldsymbol{T}_m^n. \tag{50}$$

By using (39), we can write (49) as

$$\boldsymbol{A}(\boldsymbol{P}_N, \dot{\theta}_1)\boldsymbol{T}^{n+1} = \boldsymbol{Q}(\tilde{\boldsymbol{T}}, \boldsymbol{P}_N, \theta_1), \tag{51}$$

where

$$\boldsymbol{A}(\boldsymbol{P}_N, \dot{\theta}_1) = \begin{bmatrix} \boldsymbol{N} + \boldsymbol{O}_1 + \boldsymbol{R} + \frac{1}{\tau \Delta t}\boldsymbol{C}_1 + \boldsymbol{S}_{P1} & -\boldsymbol{S}_{P2} \\ -\boldsymbol{S}_{P1} & \boldsymbol{O}_2 + \frac{1}{\tau \Delta t}\boldsymbol{C}_2 + \boldsymbol{S}_{P2} \end{bmatrix},$$
$$\boldsymbol{Q}(\tilde{\boldsymbol{T}}, \boldsymbol{P}_N, \theta_1) = \begin{Bmatrix} \boldsymbol{Q}_1^{\text{old}} + \boldsymbol{Q}_{\mu 1} \\ \boldsymbol{Q}_2^{\text{old}} + \boldsymbol{Q}_{\mu 2} \end{Bmatrix}. \tag{52}$$

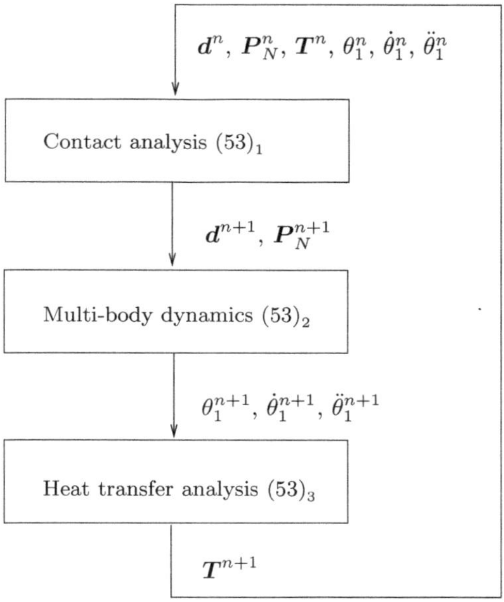

Fig. 3. Flowchart of the sequential approach.

Summarized, for a given state $\boldsymbol{d}^n$, $\boldsymbol{P}_N^n$, $\boldsymbol{T}^n$, $\boldsymbol{\theta}^n$, $\dot{\boldsymbol{\theta}}^n$, $\ddot{\boldsymbol{\theta}}^n$ at time t_n, our proposed sequential approach for finding the next state at time t_{n+1} reads

$$\begin{aligned} \boldsymbol{h}(\boldsymbol{d}^{n+1}, \boldsymbol{P}_N^{n+1}, \boldsymbol{T}^n, \dot{\theta}_1^n, \ddot{\theta}_1^n) &= \boldsymbol{0}, \\ \boldsymbol{J}^{\text{eff}} \ddot{\boldsymbol{\theta}}^{n+1} &= \boldsymbol{M}^{\text{eff}}(\boldsymbol{P}_N^{n+1}, \boldsymbol{\theta}^n, \dot{\boldsymbol{\theta}}^n, \ddot{\boldsymbol{\theta}}^n), \\ \boldsymbol{A}(\boldsymbol{P}_N^{n+1}, \dot{\theta}_1^{n+1})\boldsymbol{T}^{n+1} &= \boldsymbol{Q}(\tilde{\boldsymbol{T}}^n, \boldsymbol{P}_N^{n+1}, \theta_1^{n+1}). \end{aligned} \tag{53}$$

The sequential approach is also illustrated by the flowchart in Fig. 3.

6. Numerical implementation

The Eulerian-based thermo-flexible multi-body approach presented above in (5) and Fig. 3 is implemented in an in-house toolbox (Brake43P) using Matlab and Fortran and runs on 64-bits Windows. So far, the toolbox has been developed for a symmetric disc as shown in Fig. 4, running with constant angular velocity and constant pad pressure. This implementation has been validated in collaboration with Volvo Truck for a vented disc–pad system to a heavy duty truck for a load case representing hard braking downhill causing severe frictional heating. When the stiffer back-plate of

Fig. 4. Symmetric model of the vented disc and a heat band from simulation of hard braking downhill using the proposed sequential approach.

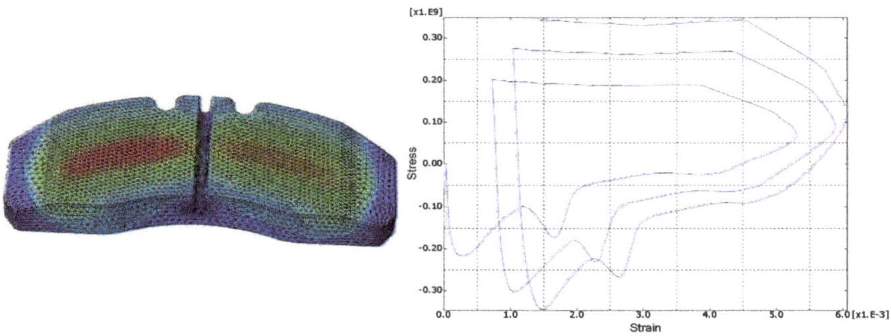

Fig. 5. Two hot spots conducting into the pad and circumferential residual stresses.

the pad was included in the model and a running-in procedure with wear was performed according to the test procedure, we obtained good agreement in temperature history of the disc between results from heat camera and simulations.[19] A typical heat band from such a simulation at the end of braking is shown in Fig. 4.[b] In the simulation, two bands appear in the beginning and then quickly merge to one band with step gradients. This is also in agreement with results from testing. The corresponding temperature field of the pad is shown in Fig. 5. At the beginning the temperature field is almost uniform over the surface and then increases rapidly in level developing to two hot spots, which then conducts into the pad.

One major advantage of the proposedf Eulerian-based approach is that the contact interface is always well-defined and a fine mesh can therefore

[b]Available at: `https://www.youtube.com/watch?v=_Sh9FoXEfKM` shows an animation of the development of such a heat band using Brake43P. Accessed on 18-03-2016.

easily be used at the contact zone without any need for adopting an adaptive mesh strategy. Figure 4 clearly shows how a finer mesh is used at the contact interface. Another major advantage is the computational time compared to the time obtained using a total Lagrangian formulation. The CPU-time for hard braking of 45s including cooling takes approximately 2 CPU-hrs on an ordinary PC. This is significantly quicker compared to a total Lagrangian formulation, which might end up in CPU-days for this problem.

The heat band developed in the disc will cause residual tensile stresses in the circumferential direction after cooling. Cycles of such stresses might cause radial cracking of the disc. We have studied these kind of residual stresses in several papers,[17,18] by exporting the temperature history obtained by the toolbox to Abaqus/Standard or LS-Dyna, and then performing J2-plasticity analysis using temperature-dependent material properties for grey cast iron.[16] A stress history from such an analysis using

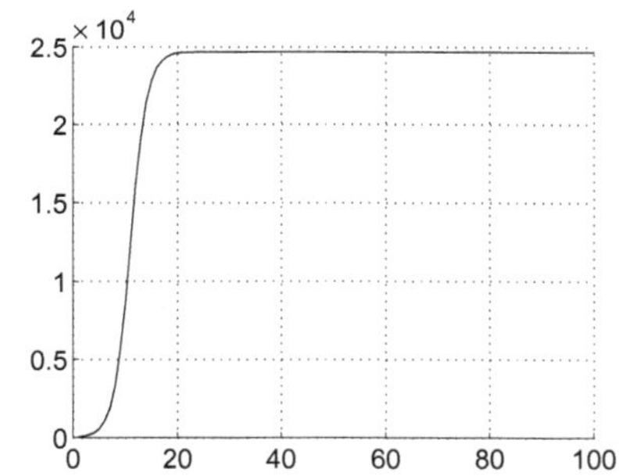

(a) Clamping force

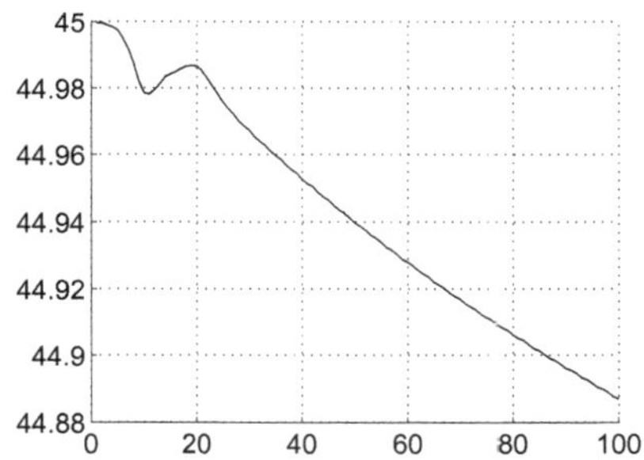

(b) Angular velccity

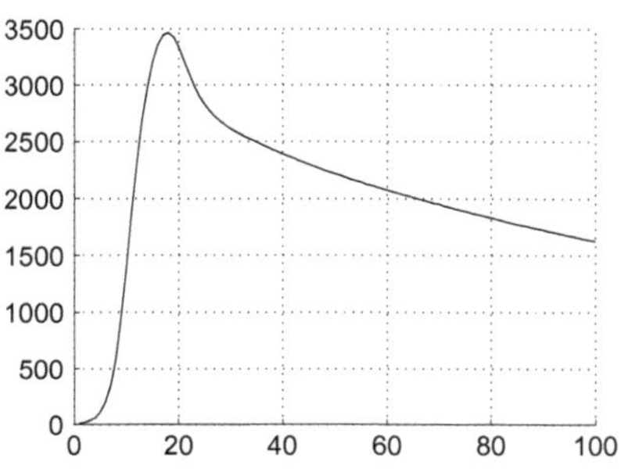

(c) Brake moment

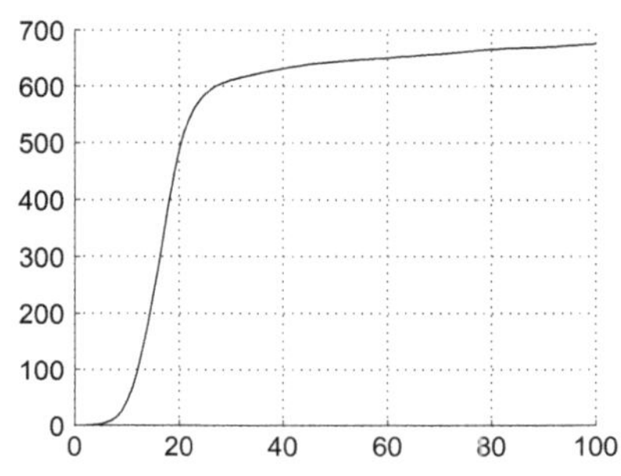

(d) Maximum temperature

Fig. 6. History of data generated by the toolbox. Notice the fading in brake torque which is a result of the friction model with fading properties.

LS-Dyna is shown in Fig. 5 for J2-plasticity with nonlinear isotropic hardening during three cycles of hard braking and cooling.

With the thermo-flexible multi-body approach developed in this work, we are no longer restricted to study braking downhill with constant angular velocity and constant pad pressure, but can instead start from an initial angular velocity with a pad pressure history according to a test procedure. In Fig. 6, we have tested the model according to the test procedure for hard braking, where J_1 is defined by the disc and J_2 represents the assembly of flywheels for this test case. Notice the fading obtained in brake torque due to the temperature-dependent friction model suggested in (25). We have also initiated virtual tests of other breaking scenarios using the proposed methodology. Preliminary results are most promising. For instance, we can study heat band development and corresponding thermal stresses for braking from high speed to zero velocity. Hopefully, we have the financial possibility to present such results in a near future.

7. Concluding remarks

A methodology for performing virtual dynamometer tests of disc brake systems is suggested and implemented. The disc and flywheels of the dynamometer are represented with a rigid multi-body model, which in turn is coupled to a thermo-mechanical model of a disc–pad system, where the disc is formulated in an Eulerian framework. The implementation of the proposed approach is most promising, and hopefully financially support makes it possible to continue this work in the near future.

Acknowledgements

I am grateful for the financial support from Vinnova (www.vinnova.se) and Volvo Truck, and for the collaboration with Per Hasselberg and Asim Rashid.

References

1. N. Strömberg, L. Johansson, and A. Klarbring, Derivation and analysis of a generalized standard model for contact, friction and wear, *International Journal of Solids Structures*, **33**, 1817–1836 (1996).
2. N. Strömberg, An augmented Lagrangian method for fretting problems, *European Journal of Mechanics, A/Solids*, **16**, 573–593 (1997).
3. P. Christensen, A. Klarbring, J. S. Pang, and N. Strömberg, Formulation and comparison of algorithms for frictional contact problems,

International Journal for Numerical Methods in Engineering, **42**, 145–173 (1998).
4. N. Strömberg, A Newton method for three-dimensional fretting problems, *International Journal of Solids Structures*, **36**, 2075–2090 (1999).
5. N. Strömberg, Finite element treatment of two-dimensional thermoelastic wear problems, *Computer Methods in Applied Mechanics and Engineering*, **177**, 441–455 (1999).
6. P. Ireman, A. Klarbring, and N. Strömberg, Finite element algorithms for thermoelastic wear problems, *European Journal of Mechanics, A/Solids*, **21**, 423–440 (2002).
7. P. Ireman, A. Klarbring, and N. Strömberg, A model of damage coupled to wear, *International Journal of Solids and Structures*, **40**, 2957–2974 (2003).
8. P. Ireman, A. Klarbring, and N. Strömberg, Gradient theory of damage coupled to frictional contact and wear, and its numerical treatment, *Computer Modeling in Engineering Science*, **52**, 125–158 (2009).
9. O. Lundvall, N. Strömberg, and A. Klarbring, A flexible multi-body approach for frictional contact in spur gears, *Journal of Sound and Vibration*, **278**, 479–499 (2004).
10. N. Strömberg, A method for structural dynamic contact problems with friction and wear, *International Journal for Numerical Methods in Engineering*, **58**, 2371–2385 (2003).
11. N. Strömberg, An implicit method for frictional contact, impact and rolling, *European Journal of Mechanics, A/Solids*, **24**, 1016–1029 (2005).
12. N. Strömberg, Contact, impact and rolling of a nonlinear elastic body on rigid foundations, *Proceedings of Multibody Dynamics, International Conference on Advances in Computational Multibody Dynamics*, June 21–24, Madrid, Spain (2005).
13. N. Strömberg, Frictional contact/impact between a hyperelastic body and moving rigid obstacles, *Proceedings of the 3rd European Conference on Computational Mechanics*, June 5–9, Lisbon, Portugal (2006).
14. N. Strömberg, An eulerian approach for simulating frictional heating in disc-pad systems, *European Journal of Mechanics, A/Solids*, **30**, 673–683 (2011).
15. A. Rashid, *Finite Element Modeling of Contact Problems*, Ph.D. thesis, Linköping Studies in Science and Technology, Dissertations, No. 1736, 2016.
16. E. Gustafsson, M. Hofwing, and N. Strömberg, Residual stresses in a stress lattice — experiments and finite element simulations, *Journal of Materials Processing Technology*, **209**, 4320–4328 (2009).
17. A. Rashid and N. Strömberg, An efficient sequential approach for simulation of thermal stresses in disc brakes, *Proceedings of the 15th Nordic Symposium on Tribology, NORDTRIB*, June 12–15, Trondheim, Norway (2012).
18. A. Rashid and N. Strömberg, Sequential simulation of thermal stresses in disc brakes for repeated braking, *Proceedings of the Institution of Mechanical Engineers, Part J:Journal of Engineering Tribology*, **227**, 919–929 (2013).
19. A. Rashid and N. Strömberg, Thermomechanical simulation of wear and hot bands in a disc brake by adopting an Eulerian approach, *Proceedings of Eurobrake 2013*, 17–19 June, Dresden, Germany (2013).

Chapter 6

Nonlinear Analysis With Functionally Graded Materials by Finite Block Method

J. Li, J. Jin, J. J. Yang, T. Huang and P. H. Wen

Based on the 1D differential matrix derived from the Lagrange series expansion, the Finite Block Method (FBM) was recently developed to solve both the elasticity and transient heat conduction problems of anisotropic and functionally graded materials. In this chapter, the formulation of the Lagrange FBM with boundary type in the strong form is presented and applied to nonlinear analysis including non-conforming contact and large deformation of Reissner plate problems with functionally graded materials subjected to either static or dynamic loads. The first-order partial differential matrices are only needed both in the governing equations and in the Neumann boundary condition. By introducing the mapping technique, a block of quadratic type is transformed from the Cartesian coordinate of global system to the normalized coordinate with 8 seeds. Time-dependent partial differential equations are analyzed in the Laplace transformed domain, and the Durbin's inversion method is applied to determine all the physical values in the time domain. Conforming and non-conforming contacts are investigated by using the iterative algorithm with full load technique. However, the load incremental technique and displacement control scheme are developed for the large deformation and post-buckling investigations of plates under pressure and compressive loads. Illustrative numerical examples are given and comparisons have been made with analytical solutions and numerical solutions using the finite element method (ABAQUS).

1. Introduction

Functional Graded Materials (FGM) are materials in which some particular physical properties have changed with dimensions. The use of FGM may become an important issue for developing advanced lightweight structures. In the quest for developing lightweight high-performing flexible structures, a concept has emerged to develop the structures with self-controlling and self-monitoring capabilities.[1,2] In FGM, the variation of the material properties can be predetermined by controlling the spatial distribution of the composition and the volume fraction of their constituents. Owing to the composite structure, the material properties are directionally dependent or anisotropic in general. Thus, the consideration of physically realistic problems such as transient heat conduction in anisotropic and non-homogeneous media is a complex task in general, even in the case of material linearity. There are many instances where composite materials are loaded in compression and whilst the buckling behaviours are not well defined. In addition, production processes may lead to imperfections in the composite panels which have a poorly understood influence on FGM/composite materials buckling stability. The instability of structural parts subjected to compressive loading may be regarded as a mode of failure due to excessive deformation and distortion of the structure. This mode of failure is termed buckling and is prevalent in plate structures for which the transverse dimension (thickness) is small compared with the overall length. Modern aircraft structures consist of thin sheets attached to slender stiffeners. Buckling of these lightweight members can occur at stresses well below the elastic limit. Predicting the resistance to buckling of compression and shear members is a critical part of the structural design processes. As closed-form solutions for the critical load can be found only for the simplest case of plates with regular shapes such as rectangular plate and circular plate, the numerical methods are required to deal with general engineering problems. Very few analytical solutions are available in the textbook which can be used as benchmarks.

To deal with such problems, numerical strategies including the Finite Difference Method (FDM), Finite Element Method (FEM) and Boundary Element Method (BEM)[3–4] have been developed to find the numerical solutions. Due to the complexity of the irregular boundary configurations and the material properties in general engineering problems, it is very difficult to obtain the analytical solutions. Very few analytical solutions are available in the literature, such as in books and publications, which can be used as benchmarks. In spite of the great success of the FEM and the BEM as effective numerical tools for the solution of boundary value problems in piezoelectric solids, there is still a growing interest in the development of new advanced numerical methods. In recent years, meshless formulations are becoming popular due to their high adaptation and low cost in preparing input and output data for numerical analysis.[5–22]

A new algorithm, the Finite Block Method (FBM), based on the point collocation method was proposed to solve the heat conduction problem in the functionally graded media and anisotropic materials by Wen[23] and Li *et al.*,[24] with high accuracy. The essential feature of the FBM is that the physical domain is divided into few blocks and the partial differential operators are derived easily in terms of the nodal values, using mapping technique for each block. This method is developed to solve elastic 2D and 3D elasticity of anisotropic and functionally graded materials in this chapter. Due to the physical domain that is assembled with several blocks (like elements in the FEM), continuities of stress and displacement are considered on the interfaces. In the normalized coordinate system, the Lagrange series approximation in normalized (mapping) domain is used with a uniform network to construct the first-order differential matrices. The first-order derivative matrices are constructed with nodes collocated on a straight line. Then the higher-orders of partial differential matrices can be obtained straightaway.[24,25] With the quadratic type of block to transfer real domain to normalized domain with 8 seeds for 2D, the partial differential operators in the practical domain (xoy) are calculated by the first-order differential matrices in the transformed normalized domain ($\xi o \eta$), where the collocation points are distributed uniformly. A set of linear equations from the equilibrium equations and the boundary conditions in terms of displacements is obtained to determine all the nodal values of displacement. In addition, the displacements at internal points can be obtained in terms of the outer boundary displacements, by considering governing equations. Then the stresses or tractions can be presented by the boundary displacement too. It means that the FBM has characteristics of the BEM, which is of distinct advantage to deal with the contact problems. The friction and frictionless contact problems with linear variation of coordinate is investigated in this chapter, under static and dynamic loads. In addition, the finite block Petrov–Galerkin method with mapping technique by the use of 2D Lagrange series was applied for nonlinear and post-buckling analysis with Reissner plate theory. Coupling nonlinear effects of in-plane stress and plate bending problems are considered as body forces in the deferential equations. With the quadratic type of block, the first-order partial differentials in practical domain are formulated. A set of nonlinear equations from the weak form of the governing equations is deduced and the load incremental algorithm is adapted.

2. Lagrange interpolation for 2D

Consider a set of 1D uniformly distributed nodes shown in Fig. 1(a), $\xi_i = -1 + 2(i-1)/(M-1), i = 1, 2, \ldots, M$, which are defined as location of node. Function $u(\xi)$ is approximated, by using Lagrange series, as

$$u(\xi) = \sum_{i=1}^{M} \prod_{\substack{m=1 \\ m \neq i}}^{M} \frac{(\xi - \xi_m)}{(\xi_i - \xi_m)} u_i = \sum_{i=1}^{M} F(\xi, \xi_i) u_i, \tag{1}$$

where M is the number of nodes along the ξ axis, u_i indicates the nodal value of node i. In the same way, for axis $\eta, \eta_j = -1 + 2(j-1)/(N-1)$, $j = 1, 2, \ldots, N$, one has

$$u(\eta) = \sum_{j=1}^{N} \prod_{\substack{n=1 \\ n \neq j}}^{N} \frac{(\eta - \eta_n)}{(\eta_j - \eta_n)} u_j = \sum_{j=1}^{N} G(\eta, \eta_j) u_j. \tag{2}$$

Therefore, the function $u(\xi, \eta)$ of two coordinates can be approximated by

$$\begin{aligned} u(\xi, \eta) &= \sum_{i=1}^{M} \sum_{j=1}^{N} \prod_{\substack{m=1 \\ m \neq i}}^{M} \frac{(\xi - \xi_m)}{(\xi_i - \xi_m)} \prod_{\substack{n=1 \\ n \neq j}}^{N} \frac{(\eta - \eta_n)}{(\eta_j - \eta_n)} u_{ij} \\ &= \sum_{i=1}^{M} \sum_{j=1}^{N} F(\xi, \xi_i) G(\eta, \eta_j) u_{ij}. \end{aligned} \tag{3}$$

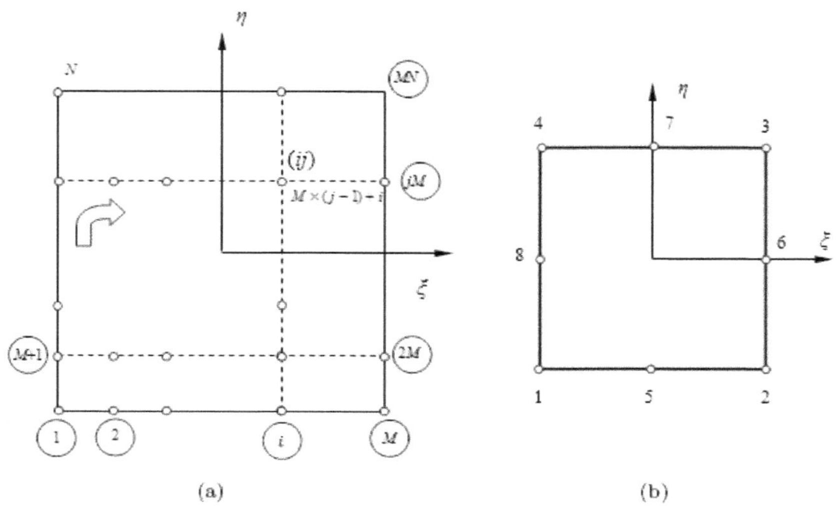

Fig. 1. 2D node distribution: (a) The number system of nodes and (b) The quadratic element and its 8 seeds.

In the global number system, the number of node (ij) is $(j-1)M+i$, as shown in Fig. 1(a). The number of total nodes is $T = M \times N$. Then, the first-order partial differential is determined easily with respect to ξ

$$\frac{\partial u}{\partial \xi} = \sum_{i=1}^{M}\sum_{j=1}^{N} \frac{\partial F(\xi,\xi_i)}{\partial \xi} G(\eta,\eta_j) u_{ij}, \tag{4}$$

where

$$\begin{aligned}\frac{\partial F(\xi,\xi_i)}{\partial \xi} &= \frac{\partial}{\partial \xi} \prod_{\substack{m=1,\\ m\neq i}}^{M} \frac{(\xi-\xi_m)}{(\xi_i-\xi_m)}\\ &= \left\langle \sum_{l=1}^{M} \prod_{k=1,\,k\neq i,\,k\neq l}^{M} (\xi-\xi_k)\right\rangle \Big/ \prod_{m=1,m\neq i}^{M} (\xi_i-\xi_m)\end{aligned} \tag{5}$$

and

$$\frac{\partial u}{\partial \eta} = \sum_{i=1}^{M}\sum_{j=1}^{N} F(\xi,\xi_i) \frac{\partial G(\eta,\eta_j)}{\partial \eta} u_{ij}, \tag{6}$$

where

$$\begin{aligned}\frac{\partial G_j(\eta,\eta_j)}{\partial \eta} &= \frac{\partial}{\partial \eta} \prod_{\substack{n=1,\\ n\neq j}}^{N} \frac{(\eta-\eta_n)}{(\eta_j-\eta_n)}\\ &= \left\langle \sum_{l=1}^{N} \prod_{k=1,\,k\neq i,\,k\neq l}^{N} (\eta-\eta_k)\right\rangle \Big/ \prod_{n=1,n\neq j}^{N} (\eta_j-\eta_n).\end{aligned} \tag{7}$$

For 2D problem, the area Ω in the Cartesian coordinate (x,y)can be mapped into a square Ω' in the normalized domain, by using the quadratic shape functions with 8 seeds, shown in Fig. 1(b). The shape functions are given by

$$N_i(\xi,\eta) = \frac{1}{4}(1+\xi_i\xi)(1+\eta_i\eta)(\xi_i\xi+\eta_i\eta-1), \qquad \text{for } i = 1,2,3,4 \tag{8a}$$

$$N_i(\xi,\eta) = \frac{1}{2}(1-\xi^2)(1+\eta_i\eta), \qquad \text{for } i = 5,7 \tag{8b}$$

$$N_i(\xi,\eta) = \frac{1}{2}(1-\eta^2)(1+\xi_i\xi), \qquad \text{for } i = 6,8. \tag{8c}$$

It guarantees that the displacements and stresses are continuous on the interface between the two blocks. Then the coordinate transform (mapping) can be written as

$$x = \sum_{k=1}^{8} N_k(\xi, \eta) x_k, \quad y = \sum_{k=1}^{8} N_k(\xi, \eta) y_k, \tag{9}$$

where (x_k, y_k) is the coordinate of seed k for the real domain. For partial differentials of function $u(x, y)$ in the Cartesian coordinate system, one has

$$\frac{\partial u}{\partial x} = \frac{1}{J}\left(\beta_{11}\frac{\partial u}{\partial \xi} + \beta_{12}\frac{\partial u}{\partial \eta}\right), \quad \frac{\partial u}{\partial y} = \frac{1}{J}\left(\beta_{21}\frac{\partial u}{\partial \xi} + \beta_{22}\frac{\partial u}{\partial \eta}\right), \tag{10}$$

where

$$\begin{aligned} \beta_{11} &= \frac{\partial y}{\partial \eta}, \quad \beta_{12} = -\frac{\partial y}{\partial \xi}, \quad \beta_{21} = -\frac{\partial x}{\partial \eta}, \\ \beta_{22} &= \frac{\partial x}{\partial \xi}, \quad J = \beta_{22}\beta_{11} - \beta_{21}\beta_{12}. \end{aligned} \tag{11}$$

Therefore, the first-order partial differentials can be obtained as

$$\begin{aligned} \frac{\partial u}{\partial x} &= \frac{1}{J}\sum_{i=1}^{M}\sum_{j=1}^{N}\left[\beta_{11}\frac{\partial F_i(\xi)}{\partial \xi}G_j(\eta) + \beta_{12}F_i(\xi)\frac{\partial G_j(\eta)}{\partial \eta}\right]u_{ij} \\ &= \sum_{l=1}^{L} K_{l,x}(\xi, \eta) u_l, \end{aligned} \tag{12a}$$

$$\begin{aligned} \frac{\partial u}{\partial y} &= \frac{1}{J}\sum_{i=1}^{M}\sum_{j=1}^{N}\left[\beta_{21}\frac{\partial F_i(\xi)}{\partial \xi}G_j(\eta) + \beta_{22}F_i(\xi)\frac{\partial G_j(\eta)}{\partial \eta}\right]u_{ij} \\ &= \sum_{l=1}^{L} K_{l,y}(\xi, \eta) u_l. \end{aligned} \tag{12b}$$

Where $ij = (j-1)M + i$. Therefore, we can also express each nodal value of partial differentials (12) in a matrix form as

$$\mathbf{u}_x = \mathbf{D}_x\mathbf{u}, \quad \mathbf{u}_y = \mathbf{D}_y\mathbf{u}, \tag{13}$$

where the vectors of nodal value

$$\begin{aligned} \mathbf{u}_x &= \left\{\frac{\partial u(x_1, y_1)}{\partial x}, \frac{\partial u(x_2, y_2)}{\partial x}, \ldots, \frac{\partial u(x_L, y_L)}{\partial x}\right\}^{\mathrm{T}}, \\ \mathbf{u}_y &= \left\{\frac{\partial u(x_1, y_1)}{\partial y}, \frac{\partial u(x_2, y_2)}{\partial y}, \ldots, \frac{\partial u(x_L, y_L)}{\partial y}\right\}^{\mathrm{T}}, \end{aligned} \tag{14}$$

in which L is the number of total collocation points ($L = M \times N$ from network, shown in Fig. 1). In addition, for higher-order partial differentials in 2D problems with respect to both coordinates ξ and η,

$$u_{xy}^{(mn)}(x_p, y_p) = \frac{\partial^{m+n} u}{\partial x^m \partial y^n}(x_p, y_p) \tag{15}$$

and we have the multi partial differentials in the matrix form, in terms of nodal values of u, as

$$\mathbf{u}_{xy}^{(mn)} = \mathbf{D}_x^m \mathbf{D}_y^n \mathbf{u}. \tag{16}$$

3. Finite block method in strong form with boundary type

Consider 2D elasticity of domain Ω with boundary Γ in functionally graded media. It is assumed that the materials are directionally dependent and all material coefficients could be dependent on the spatial coordinates in a non-homogeneous material. The equilibrium equations for 2D plane stress are

$$\begin{aligned} &\frac{\partial \sigma_x}{\partial x} + \frac{\partial \tau_{xy}}{\partial y} + b_x = 0, \\ &\frac{\partial \tau_{xy}}{\partial x} + \frac{\partial \sigma_y}{\partial y} + b_y = 0, \quad (x, y) \in \Omega \end{aligned} \tag{17}$$

and relationship between stress and strain in orthotropic and continuously non-homogeneous media are

$$\begin{aligned} \sigma_x &= Q_{11}\frac{\partial u_1}{\partial x} + Q_{12}\frac{\partial u_2}{\partial y}, \\ \sigma_y &= Q_{12}\frac{\partial u_1}{\partial x} + Q_{22}\frac{\partial u_2}{\partial y}, \\ \tau_{xy} &= Q_{66}\left(\frac{\partial u_1}{\partial y} + \frac{\partial u_2}{\partial x}\right), \end{aligned} \tag{18}$$

where

$$\begin{aligned} Q_{11} &= \frac{E_1}{1 - \nu_{12}\nu_{21}}, \quad Q_{12} = \frac{\nu_{12} E_1}{1 - \nu_{12}\nu_{21}}, \\ Q_{22} &= \frac{E_2}{1 - \nu_{12}\nu_{21}}, \quad Q_{66} = G_{12}, \end{aligned} \tag{19}$$

in which $E_1, E_2, \nu_{12}, \nu_{21}$ and G_{12} are the material coefficients, which are functions of coordinate x and y, respectively. In addition, u_k denotes the displacements and b_k are body forces. The boundary conditions can be written as

$$\begin{aligned} &u_1(x_i, y_i) = \bar{u}_1(x_i, y_i), \quad u_2(x_i, y_i) = \bar{u}_2(x_i, y_i), && (x_i, y_i) \in \Gamma_u \\ &\sigma_x n_x + \tau_{xy} n_y = \bar{t}_1(x_i, y_i), \quad \tau_{xy} n_x + \sigma_y n_y = \bar{t}_2(x_i, y_i), && (x_i, y_i) \in \Gamma_t \end{aligned} \tag{20}$$

in which all variables with bar are specified boundary values, where $\mathbf{n}(n_x, n_y)$ is the normal outward to the boundary.

Applying the mapping technique and differential matrices in (16) and substituting them into (18) and the equilibrium equation (17) results, in matrix form, as

$$\left(\mathbf{D}_x \mathbf{Q}_{11} \mathbf{D}_x + \mathbf{D}_y \mathbf{Q}_{66} \mathbf{D}_y\right) \mathbf{u}_1 + \left(\mathbf{D}_x \mathbf{Q}_{12} \mathbf{D}_y + \mathbf{D}_y \mathbf{Q}_{66} \mathbf{D}_x\right) \mathbf{u}_2 + \mathbf{b}_1 = 0, \tag{21}$$

$$\left(\mathbf{D}_y \mathbf{Q}_{12} \mathbf{D}_x + \mathbf{D}_x \mathbf{Q}_{66} \mathbf{D}_y\right) \mathbf{u}_1 + \left(\mathbf{D}_x \mathbf{Q}_{66} \mathbf{D}_x + \mathbf{D}_y \mathbf{Q}_{22} \mathbf{D}_y\right) \mathbf{u}_2 + \mathbf{b}_2 = 0, \tag{22}$$

where $\mathbf{u}_k = [u_k^{(1)}, u_k^{(2)}, \ldots, u_k^{(L)}]^T, \mathbf{b}_k = [b_k^{(1)}, b_k^{(2)}, \ldots, b_k^{(L)}]^T$ are nodal value vectors for the displacement and body force and diagonal matrix

$$\mathbf{Q}_{ij} = \mathbf{diag}\left[Q_{ij}\right] = \begin{pmatrix} Q_{ij}^{(1)} & 0 & \cdots & 0 \\ 0 & Q_{ij}^{(2)} & \cdots & 0 \\ \cdots & \cdots & \cdots & \cdots \\ 0 & 0 & \cdots & Q_{ij}^{(L)} \end{pmatrix}, \tag{23}$$

in which $Q_{ij}^{(p)}$ indicates the coefficient of material in (19) at point p. Also, Eqs. (21) and (22) can be written as

$$\mathbf{D}\mathbf{u} = -\mathbf{b}, \tag{24}$$

where $\mathbf{u} = [u_1^{(1)}, u_2^{(1)}, \ldots, u_1^{(L)}, u_2^{(L)}]^{\mathrm{T}}$, $\mathbf{b} = [b_1^{(1)}, b_2^{(1)}, \ldots, b_1^{(L)}, b_2^{(L)}]^{\mathrm{T}}$ are the nodal value vectors of the displacement and body force in the local numbering system. Consider that the collocation point $p = N \times (j-1) + i$ in the local numbering system is the collocation point p' in global numbering

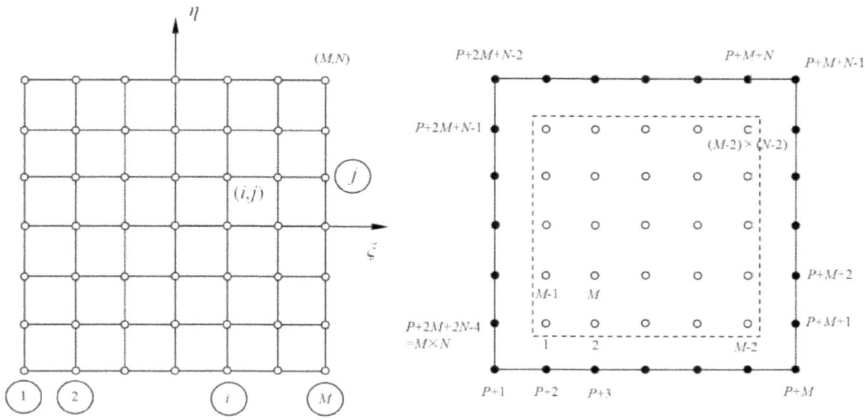

(a) (b)

Fig. 2. Transformation of numbering system in the mapping domain: (a) Local numbering system and (b) Global numbering system, $P = (M-2) \times (N-2)$.

system, therefore, in the transform matrix $\mathbf{T}$, all elements are zero except $T_{N\times(j-1)+i,p'} = 1, (i,j = 1,2,\ldots,N)$. By a simple re-arrangement of the number of nodes, displacements can be arranged, in the global system, as

$$\mathbf{u} = \mathbf{T}\mathbf{u}^*, \quad \mathbf{b} = \mathbf{T}\mathbf{b}^*, \tag{25}$$

where $\mathbf{u}^*$and $\mathbf{b}^*$are vectors of displacement and body force in global numbering system, as shown in Fig. 2(b). The governing equation in the domain becomes

$$\mathbf{T}^{-1}\mathbf{D}\mathbf{T}\mathbf{u}^* = -\mathbf{b}^*. \tag{26}$$

If the vector of displacement is arranged as

$$\mathbf{u}^* = (\mathbf{u}_I, \mathbf{u}_B)^{\mathrm{T}}, \quad \mathbf{b}^* = (\mathbf{b}_I, \mathbf{b}_B)^{\mathrm{T}}, \tag{27}$$

where $\mathbf{u}_I = [u_1^{(1)}, u_2^{(1)}, \ldots, u_1^{(P)}, u_2^{(P)}]^{\mathrm{T}}, \mathbf{u}_B = [u_1^{(P+1)}, u_2^{(P+1)}, \ldots, u_1^{(L)}, u_2^{(L)}]^{\mathrm{T}}$ and $\mathbf{b}_I = [b_1^{(1)}, b_2^{(1)}, \ldots, b_1^{(P)}, b_2^{(P)}]^{\mathrm{T}}$ in global system shown in Fig. 2(b). By considering the governing equation (24), we have

$$(\mathbf{D}_I, \mathbf{D}_B)(\mathbf{u}_I, \mathbf{u}_B)^{\mathrm{T}} = -\mathbf{b}_I, \quad (\mathbf{D}_I, \mathbf{D}_B) = \mathbf{T}^{-1}\mathbf{D}\mathbf{T}. \tag{28}$$

Then, we have

$$\mathbf{u}_I = -\mathbf{D}_I^{-1}\mathbf{b}_I - \mathbf{D}_I^{-1}\mathbf{D}_B\mathbf{u}_B. \tag{29}$$

It means that the displacements $\mathbf{u}_I$, which are unknowns at internal collocation points in the domain, can be represented in terms of the displacement $\mathbf{u}_B$ on the boundary.

In the local system, the stresses at collocation points are

$$\boldsymbol{\sigma}_x = \mathbf{S}_{xx}\mathbf{u}, \quad \boldsymbol{\sigma}_y = \mathbf{S}_{yy}\mathbf{u}, \quad \boldsymbol{\tau}_{xy} = \mathbf{S}_{xy}\mathbf{u}, \tag{30}$$

where

$$\mathbf{S}_{xx} = \begin{pmatrix} Q_{11}^{(1)}D_x^{(11)} & Q_{12}^{(1)}D_y^{(11)} & Q_{11}^{(1)}D_x^{(12)} & Q_{12}^{(1)}D_y^{(12)} & \cdots & \cdots & Q_{11}^{(1)}D_x^{(1L)} & Q_{12}^{(1)}D_y^{(1L)} \\ Q_{11}^{(2)}D_x^{(21)} & Q_{12}^{(2)}D_y^{(21)} & Q_{11}^{(2)}D_x^{(22)} & Q_{12}^{(2)}D_y^{(22)} & \cdots & \cdots & Q_{11}^{(2)}D_x^{(2L)} & Q_{12}^{(2)}D_y^{(2L)} \\ \cdots & \cdots & \cdots & \cdots & \cdots & \cdots & \cdots & \cdots \\ \cdots & \cdots & \cdots & \cdots & \cdots & \cdots & \cdots & \cdots \\ Q_{11}^{(L)}D_x^{(L1)} & Q_{12}^{(L)}D_y^{(L1)} & Q_{11}^{(L)}D_x^{(L2)} & Q_{12}^{(L)}D_y^{(L2)} & \cdots & \cdots & Q_{11}^{(L)}D_x^{(LL)} & Q_{12}^{(L)}D_y^{(LL)} \end{pmatrix}_{L\times 2L},$$

$$\mathbf{S}_{yy} = \begin{pmatrix} Q_{21}^{(1)}D_x^{(11)} & Q_{22}^{(1)}D_y^{(11)} & Q_{21}^{(1)}D_x^{(12)} & Q_{22}^{(1)}D_y^{(12)} & \cdots & \cdots & Q_{21}^{(1)}D_x^{(1L)} & Q_{22}^{(1)}D_y^{(1L)} \\ Q_{21}^{(2)}D_x^{(21)} & Q_{22}^{(2)}D_y^{(21)} & Q_{21}^{(2)}D_x^{(22)} & Q_{22}^{(2)}D_y^{(22)} & \cdots & \cdots & Q_{21}^{(2)}D_x^{(2L)} & Q_{22}^{(2)}D_y^{(2L)} \\ \cdots & \cdots & \cdots & \cdots & \cdots & \cdots & \cdots & \cdots \\ \cdots & \cdots & \cdots & \cdots & \cdots & \cdots & \cdots & \cdots \\ Q_{21}^{(L)}D_x^{(L1)} & Q_{22}^{(L)}D_y^{(L1)} & Q_{21}^{(L)}D_x^{(L2)} & Q_{22}^{(L)}D_y^{(L2)} & \cdots & \cdots & Q_{21}^{(L)}D_x^{(LL)} & Q_{22}^{(L)}D_y^{(LL)} \end{pmatrix}_{L\times 2L},$$

$$\mathbf{S}_{xy} = \begin{pmatrix} Q_{66}^{(1)} D_y^{(11)} & Q_{66}^{(1)} D_x^{(11)} & Q_{66}^{(1)} D_y^{(12)} & Q_{66}^{(1)} D_x^{(12)} & \cdots & \cdots & Q_{66}^{(1)} D_y^{(1L)} & Q_{66}^{(1)} D_x^{(1L)} \\ Q_{66}^{(2)} D_y^{(21)} & Q_{66}^{(2)} D_x^{(21)} & Q_{66}^{(2)} D_y^{(22)} & Q_{66}^{(2)} D_x^{(22)} & \cdots & \cdots & Q_{66}^{(2)} D_y^{(2L)} & Q_{66}^{(2)} D_x^{(2L)} \\ \cdots & \cdots & \cdots & \cdots & \cdots & \cdots & \cdots & \cdots \\ \cdots & \cdots & \cdots & \cdots & \cdots & \cdots & \cdots & \cdots \\ Q_{66}^{(L)} D_y^{(L1)} & Q_{66}^{(L)} D_x^{(L1)} & Q_{66}^{(L)} D_y^{(L2)} & Q_{66}^{(L)} D_x^{(L2)} & \cdots & \cdots & Q_{66}^{(L)} D_y^{(LL)} & Q_{66}^{(L)} D_x^{(LL)} \end{pmatrix}_{L\times 2L}. \tag{31}$$

Then, in the global system, one has

$$\sigma_x^* = \mathbf{T}^{-1}\mathbf{S}_{xx}\mathbf{T}\mathbf{u}^*, \quad \sigma_y^* = \mathbf{T}^{-1}\mathbf{S}_{yy}\mathbf{T}\mathbf{u}^*, \quad \tau_{xy}^* = \mathbf{T}^{-1}\mathbf{S}_{xy}\mathbf{T}\mathbf{u}^*. \tag{32}$$

Thus, the stresses either in the domain or on the boundary can be obtained in matrix form from (27) and (29), and then (32) becomes

$$\begin{aligned} \sigma_x^* &= (\sigma_{xI}, \sigma_{xB})^{\mathrm{T}} = -\bar{\mathbf{S}}_{xxI}\mathbf{D}_I^{-1}\mathbf{b}_I + (\bar{\mathbf{S}}_{xxB} - \bar{\mathbf{S}}_{xxI}\mathbf{D}_I^{-1}\mathbf{D}_B)\mathbf{u}_B \\ &= \alpha_x \mathbf{b}_I + \beta_x \mathbf{u}_B, \\ \sigma_y^* &= (\sigma_{yI}, \sigma_{yB})^{\mathrm{T}} = -\bar{\mathbf{S}}_{yyI}\mathbf{D}_I^{-1}\mathbf{b}_I + (\bar{\mathbf{S}}_{yyB} - \bar{\mathbf{S}}_{yyI}\mathbf{D}_I^{-1}\mathbf{D}_B)\mathbf{u}_B \\ &= \alpha_y \mathbf{b}_I + \beta_y \mathbf{u}_B, \\ \tau_{xy}^* &= (\tau_{xyI}, \tau_{xyB})^{\mathrm{T}} = -\bar{\mathbf{S}}_{xyI}\mathbf{D}_I^{-1}\mathbf{b}_I + (\bar{\mathbf{S}}_{xyB} - \bar{\mathbf{S}}_{xyI}\mathbf{D}_I^{-1}\mathbf{D}_B)\mathbf{u}_B \\ &= \alpha_{xy} \mathbf{b}_I + \beta_{xy} \mathbf{u}_B, \end{aligned} \tag{33}$$

in which

$$(\bar{\mathbf{S}}_{xxI}, \bar{\mathbf{S}}_{xxB}) = \mathbf{T}^{-1}\mathbf{S}_{xx}\mathbf{T}, \quad (\bar{\mathbf{S}}_{yyI}, \bar{\mathbf{S}}_{yyB}) = \mathbf{T}^{-1}\mathbf{S}_{yy}\mathbf{T}, \\ (\bar{\mathbf{S}}_{xyI}, \bar{\mathbf{S}}_{xyB}) = \mathbf{T}^{-1}\mathbf{S}_{xy}\mathbf{T}. \tag{34}$$

Boundary conditions become

$$\begin{aligned} u_k^{(i)} &= \bar{u}_k^{(i)} \qquad (x_i, y_i) \in \Gamma_u(N_u), \\ t_k^{(i)} &= \bar{t}_k^{(i)} \qquad (x_i, y_i) \in \Gamma_t(N_t). \end{aligned} \tag{35}$$

The tractions on the boundary Γ_t

$$t_1^{(i)} = \sigma_{xB}^{(i)} n_1 + \tau_{\mathrm{xy}B}^{(i)} n_2, \quad t_2^{(i)} = \tau_{xyB}^{(i)} n_1 + \sigma_{yB}^{(i)} n_2, \tag{36}$$

where $\sigma_{xB}^{(i)}, \sigma_{yB}^{(i)}, \tau_{xyB}^{(i)}$ are given in terms of the body forces and boundary displacements in (33). Obviously there are only $2 \times N_t$ linear algebraic equations from (36) above to determine the displacements $\mathbf{u}_B$ on the traction boundary Γ_t($\mathbf{u}_B$ on the displacement boundary are specified). Particularly for the displacement boundary condition only, there are no unknowns to be solved, as $\mathbf{u}_B$ are given ($N_t = 0$). The displacements and stresses can be obtained from (29) and (32) directly. The advantage of the FBM with boundary type is that the number of unknowns is reduced significantly, from $2 \times (M \times N)$ to $2 \times N_t$ (more than 1D reduction). In the general case, the equations of stresses in (32) on the boundary nodes are considered. Recall that in the BEM, the FBM with boundary type has the same characteristics, i.e. the relationship between tractions and displacements on the boundary ($\mathbf{Hu}_B = \mathbf{Gt}_B$ for BEM). However, there are no requirements of the fundamental solutions in the FBM. In the case of two and more blocks, as shown in Fig. 3, the connection conditions in the strong form should be introduced as

$$\begin{aligned} &u_1^I(x,y) = u_1^{II}(x,y), \quad u_2^I(x,y) = u_2^{II}(x,y), \\ &t_1^I(x,y) + t_1^{II}(x,y) = 0, \quad t_2^I(x,y) + t_2^{II}(x,y) = 0 \qquad (x,y) \in \Gamma_{\text{int}} \end{aligned} \tag{37}$$

For example, in the case of two blocks fully bounded, the number of total nodes on the boundary is maximum $2M_I + 2N_I + 2M_{II} + 2N_{II} - 8$ and there are the same numbers of algebraic equations from the boundary conditions for each block and connection conditions on $\Gamma_c/\Gamma_{\text{int}}$. By solving a set of linear algebraic equations in (36) and (37), the solutions of displacement

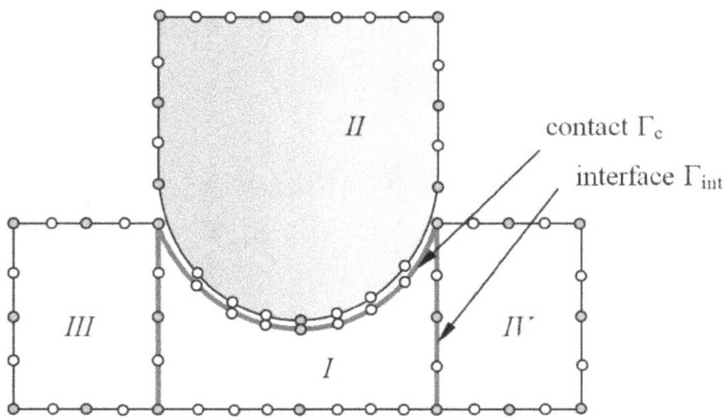

Fig. 3. Blocks for contact problem: ⊙ seeds; ○ boundary collocation points.

$\mathbf{u}_B^I$ and $\mathbf{u}_B^{II}$ for each block can be solved. For fully bounded interface, it is easy to prove that the stresses are continuous on the interface between two blocks, i.e. $\sigma_x^I = \sigma_x^{II}, \sigma_y^I = \sigma_y^{II}, \tau_{xy}^I = \tau_{xy}^{II}$. Therefore, the higher accuracy of the numerical solution is expected by using the FBM with boundary type than the normal point collocation method.

4. Finite block method in weak form (Petrov–Galerkin)

In the local Petrov–Galerkin approaches, the weak form of the governing equation over a local domain $\Omega_k$centred at point $\mathbf{x}_k$ can be written as

$$\int_{\Omega_k} \left(\frac{\partial \sigma_x}{\partial x} + \frac{\partial \tau_{xy}}{\partial y} + b_x \right) u^* d\Omega = 0,$$

$$\int_{\Omega_k} \left(\frac{\partial \tau_{xy}}{\partial x} + \frac{\partial \sigma_y}{\partial y} + b_y \right) u^* d\Omega = 0, \quad \Omega_k \in \Omega, \quad k = 1, 2, \ldots, L_{\text{in}} \tag{38}$$

where u^* is a test function. By the divergence theorem, (38) can be rewritten in a weak form as

$$\int_{\Gamma_k} (\sigma_x n_x + \tau_{xy} n_y) u^* d\Gamma - \int_{\Omega_k} \left(\sigma_x \frac{\partial u^*}{\partial x} + \tau_{xy} \frac{\partial u^*}{\partial y} - b_x u^* \right) d\Omega = 0,$$

$$\int_{\Gamma_k} (\tau_{xy} n_x + \sigma_y n_y) u^* d\Gamma - \int_{\Omega_k} \left(\tau_{xy} \frac{\partial u^*}{\partial x} + \sigma_y \frac{\partial u^*}{\partial y} - b_y u^* \right) d\Omega = 0,$$

$$\mathbf{X} \in \Omega_k \tag{39}$$

where Γ_k is the boundary of the local integral domain Ω_k. The simplest choice of the test function is a unit step function in each local domain (PGFBM1), as follows

$$u^*(\mathbf{x}) = \begin{cases} 1 & \text{at } \mathbf{x} \in \Omega_k, \\ 0 & \text{at } \mathbf{x} \notin \Omega_k. \end{cases} \tag{40}$$

Then, the governing equation in the local weak form (38) can be written as

$$\int_{\Gamma_k} (\sigma_x n_x + \tau_{xy} n_y) d\Gamma(\mathbf{x}') + \int_{\Omega_k} b_x d\Omega(\mathbf{X}') = 0,$$

$$\mathbf{X} \in \Omega_k, \quad k = 1, 2, \ldots, L_{\text{in}}$$

$$\int_{\Gamma_k} (\tau_{xy} n_x + \sigma_y n_y) d\Gamma(\mathbf{x}') + \int_{\Omega_k} b_y d\Omega(\mathbf{X}') = 0. \tag{41}$$

From (18), one has

$$\begin{aligned}\sigma_x(\mathbf{x}') &= \frac{Q_{11}}{J}\sum_{i=1}^{M}\sum_{j=1}^{N}\left(\beta_{11}\frac{\partial F_i}{\partial \xi}G_j+\beta_{12}F_i\frac{\partial G_j}{\partial \eta}\right)u^{ij}\\&\quad+\frac{Q_{12}}{J}\sum_{i=1}^{M}\sum_{j=1}^{N}\left(\beta_{21}\frac{\partial F_i}{\partial \xi}G_j+\beta_{22}F_i\frac{\partial G_j}{\partial \eta}\right)v^{ij},\\\sigma_y(\mathbf{x}') &= \frac{Q_{12}}{J}\sum_{i=1}^{M}\sum_{j=1}^{N}\left(\beta_{11}\frac{\partial F_i}{\partial \xi}G_j+\beta_{12}F_i\frac{\partial G_j}{\partial \eta}\right)u^{ij}\\&\quad+\frac{Q_{22}}{J}\sum_{i=1}^{M}\sum_{j=1}^{N}\left(\beta_{21}\frac{\partial F_i}{\partial \xi}G_j+\beta_{22}F_i\frac{\partial G_j}{\partial \eta}\right)v^{ij},\\\tau_{xy}(\mathbf{x}') &= \frac{Q_{66}}{J}\sum_{i=1}^{M}\sum_{j=1}^{N}\left(\beta_{21}\frac{\partial F_i}{\partial \xi}G_j+\beta_{22}F_i\frac{\partial G_j}{\partial \eta}\right)u^{ij}\\&\quad+\frac{Q_{66}}{J}\sum_{i=1}^{M}\sum_{j=1}^{N}\left(\beta_{11}\frac{\partial F_i}{\partial \xi}G_j+\beta_{12}F_i\frac{\partial G_j}{\partial \eta}\right)v^{ij}.\end{aligned} \tag{42}$$

In order to evaluate the local boundary integrals in (39), a local integral domain is selected as a circle of radius r centred at (ξ_k, η_k) in the normalized domain $(\xi o \eta)$, which corresponds to the point $\mathbf{X}_k$ in the real Cartesian coordinate (xoy). Therefore, the coordinate of the local integral boundary $S'(\xi_s, \eta_s)$ in the mapping domain, shown in Fig. 4 is

$$\xi_s = \xi_k + r\cos\theta, \quad \eta_s = \eta_k + r\sin\theta \tag{43}$$

and in the practical domain, the coordinate $S(x_s, y_s)$

$$x_s = \sum_{l=1}^{8} N_l(\xi_s, \eta_s)x_l, \quad y_s = \sum_{l=1}^{8} N_l(\xi_s, \eta_s)y_l. \tag{44}$$

The components of the line segment are

$$\begin{aligned}dx &= rd\theta\sum_{l=1}^{8}\left\langle -\frac{\partial N_l(\xi,\eta)}{\partial \xi}\sin\theta+\frac{\partial N_l(\xi,\eta)}{\partial \eta}\cos\theta\right\rangle x_l = f(\xi_s,\eta_s)rd\theta,\\dy &= rd\theta\sum_{l=1}^{8}\left\langle -\frac{\partial N_l(\xi,\eta)}{\partial \xi}\sin\theta+\frac{\partial N_l(\xi,\eta)}{\partial \eta}\cos\theta\right\rangle y_l = g(\xi_s,\eta_s)rd\theta.\end{aligned} \tag{45}$$

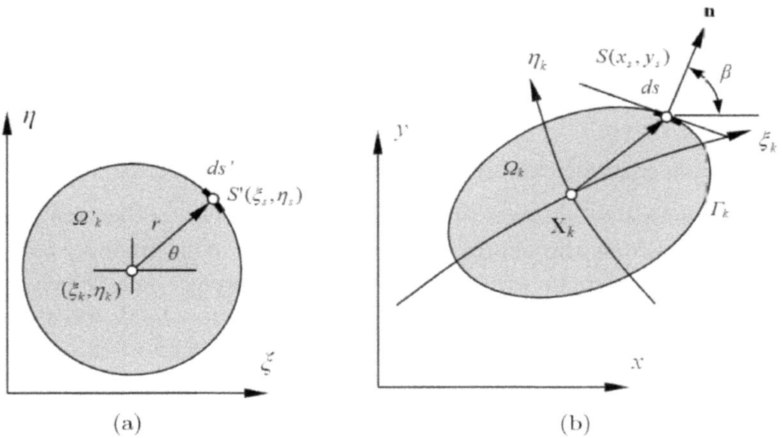

Fig. 4. Local integral domain and its boundary: (a) Circle in normalized coordinate ($\xi o \eta$) and (b) Real local integral domain (xoy).

Thus, the relationship between these two segments of length shown in Fig. 4 is

$$ds = \sqrt{dx^2 + dy^2} = \sqrt{f^2 + g^2} r d\theta = \sqrt{f^2 + g^2} ds'. \tag{46}$$

Therefore, the tangential at point $S(x_s, y_s)$ in the practical domain is obtained by

$$\alpha = \tan^{-1} \frac{dy}{dx} = \frac{f}{g} \tag{47}$$

and the components of normal outward to the boundary in Eq. (39) are

$$n_x = \cos\beta, \quad n_y = \sin\beta, \quad \beta = \alpha - \pi/2. \tag{48}$$

The weak form of governing equation (38) can be written approximately as

$$\int_0^{2\pi} r\left(\sigma_x n_x + \tau_{xy} n_y\right)\sqrt{f^2 + g^2} d\theta + b_x \Omega_k = 0,$$

$$\int_0^{2\pi} r\left(\tau_{xy} n_x + \sigma_y n_y\right)\sqrt{f^2 + g^2} d\theta + b_y \Omega_k = 0.$$

$$\mathbf{X} \in \Omega_k, \quad k = 1, 2, \ldots, L_{\text{in}}. \tag{49}$$

Then, (41) can be arranged in matrix form as

$$\mathbf{A}_1 \mathbf{u} = -\mathbf{\Omega b} \tag{50}$$

where $\mathbf{u} = \{u^1, v^1, u^2, v^2, \ldots, u^L, v^L\}^{\mathrm{T}}, \boldsymbol{\Omega} = \mathbf{diag}[\Omega_1, \Omega_1, \ldots, \Omega_{L_{\mathrm{in}}}, \Omega_{L_{\mathrm{in}}}]$ are the nodal value vectors of the displacement and matrix of the local integral domain area.

5. Contact analysis by finite block method in strong form

In the numerical simulation of contact problems, two bodies are brought into contact along the common boundary under compressive load. It is essential when solving a contact problem that relatively small elements are prescribed inside the potential contact area. For zero body force contact problem, we have tractions (along the normal and tangential directions) on a smooth boundary for each block

$$\mathbf{t}_B = \mathbf{H}\mathbf{u}_B \tag{51}$$

where

$$\begin{aligned} \mathbf{t}_B &= (t_n^{(1)}, t_t^{(1)}, \ldots, t_n^{(Q)}, t_t^{(Q)}), \\ H_{2i-1,j} &= \beta_x^{(j)} n_1^2 + \beta_y^{(j)} n_2^2 + 2\beta_{xy}^{(j)} n_1 n_2, \\ H_{2i,j} &= \left(\beta_x^{(j)} - \beta_y^{(j)}\right) n_1 n_2 + \beta_{xy}^{(j)} \left(n_1^2 - n_2^2\right), \quad Q = N_u + N_t, \end{aligned} \tag{52}$$

in the global numbering system. The contact analysis procedure by using the BEM, introduced by Man *et al.*[26,27,28] is valid for the FBM. For frictional contact, there are three modes which include Separation, Slip and Stick. In this chapter, the iterative and total loading strategy is employed for fractional contact problems. Contact problem is studied with only one total load step and performing iterations until the stick and slip area are found and all the contact conditions are satisfied. If Stick mode is assumed, then both traction equilibrium and displacement compatibility must be enforced, as prescribed by the contact constraints below:

$$t_n^I - t_n^{II} = 0, \quad t_t^I - t_t^{II} = 0, \quad u_t^I + u_t^{II} = 0, \quad u_n^I + u_n^{II} = f(x, y). \tag{53}$$

Also Stick condition $t_t^I \le \mu t_n^I$ should be satisfied, μ indicates the coefficient of friction and $f(x, y)$ is the normal gap between node-pair of contact, as shown in Fig. 5. However, if the Slip mode is assumed, then Slip mode constraints must be considered by

$$t_n^I - t_n^{II} = 0, \quad t_t^I - t_t^{II} = 0, \quad t_t^I \pm \mu t_n^I = 0, \quad u_n^I + u_n^{II} = f(x, y). \tag{54}$$

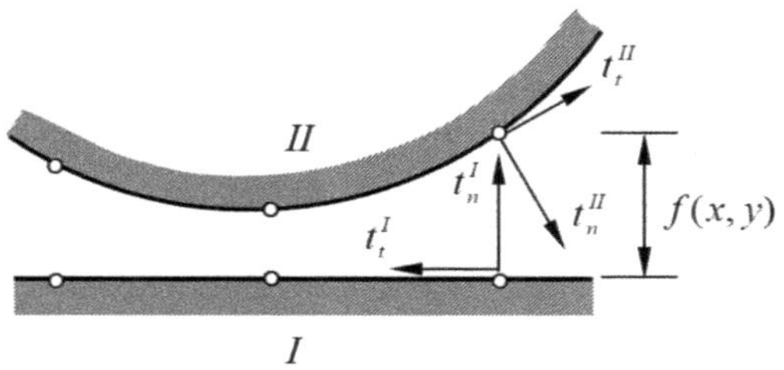

Fig. 5. Definition of the normal and initial normal gap for non-conforming contact.

If μ is set to zero, the relationships of the frictionless contacts are obtained as

$$t_n^I - t_n^{II} = 0, \quad t_t^I = t_t^{II} = 0, \quad u_n^I + u_n^{II} = f(x, y), \tag{55}$$

in this case, a frictionless problem can be solved easily without any decision being required regarding the tangential tractions for conforming contact. Therefore, for frictionless and frictional contact problems, we use two strategies as the following:

5.1. *The seven steps of the iterative procedure for non-conforming frictionless contact for total loading*

Step 1. Computations of matrices **H** for each block;
Step 2. Guess contact zone with Slip mode;
Step 3. Apply total loading;
Step 4. Solve displacements on the boundary and calculate tractions;
Step 5. Observe the separation points and check the contact tractions in the slip zone and opening displacements in the Separation zone;
Step 6. If $|t_n^I| < \varepsilon$ and $|u_n^I + u_n^{II} - f(x, y)| < \varepsilon, \varepsilon$ is the small value in the iteration, stop the iteration;
Step 7. Modify the boundary conditions:
If $t_n^I < 0$ and two surfaces overlap, $|u_n^I + u_n^{II}| > f(x, y)$, increase the size, otherwise, decrease; go to Step 3.

5.2. *The eight steps of the iterative procedure for non-conforming frictional contact for total loading*

Step 1. Computations of matrices **H** for each block;
Step 2. Guess contact zone with Stick;
Step 3. Apply total loading with the new contact zone and the new contact modes for each pair of node;

Step 4. Solve displacements on the boundary and calculate tractions;

Step 5. Re-determine contact modes for each pair in the contact zone:
If $t_n^I > 0 \rightarrow$ Separation mode;
If $t_n^I < 0$ and $|t_t^I| < \mu|t_n^I| \rightarrow$ Stick mode;
If $t_n^I < 0$ and $|t_t^I| > \mu|t_n^I| \rightarrow$ Slip mode;

Step 6. Modify the locations between the Slip/Stick modes (fix separation point):
If there is no contact mode change for each pair of node, go to Step 7; otherwise go to Step 3;

Step 7. Check the separation point:
If $|t_n^I| < \varepsilon$ and $|u_n^I + u_n^{II} - f(x, y)| < \varepsilon$, stop the iteration;

Step 8. Modify the locations of separation and location between the Slip/Stick modes:
If $t_n^I < 0$ and two surfaces overlap, $|u_n^I + u_n^{II}| > f(x, y)$, increase the size of contact zone, otherwise, decrease the size; go to Step 3.

6. Contact analysis under impact load

Finally, consider a 2D elasticity of domain Ω with boundary Γ in functionally graded media. The equilibrium equations for 2D problems are

$$\frac{\partial \sigma_x}{\partial x} + \frac{\partial \tau_{xy}}{\partial y} + b_x = \rho \frac{\partial^2 u_1}{\partial t^2} + c\frac{\partial u_1}{\partial t}, \tag{56}$$

$$\frac{\partial \tau_{xy}}{\partial x} + \frac{\partial \sigma_y}{\partial y} + b_y = \rho \frac{\partial^2 u_2}{\partial t^2} + c\frac{\partial u_2}{\partial t}, \qquad (x, y) \in \Omega \tag{57}$$

where ρ indicates the mass density and c the damping coefficient. The boundary conditions are given as

$$\begin{aligned} u_k(x_i, y_i, t) &= \bar{u}_k(x_i, y_i, t), \quad (x_i, y_i) \in \Gamma_u \\ t_k(x_i, y_i, t) &= \bar{t}_k(x_i, y_i, t), \quad (x_i, y_i) \in \Gamma_t \end{aligned} \tag{58}$$

for $t > 0$ with the initial conditions

$$u_k(x_i, y_i, 0) = u_k^0(x_i, y_i), \quad \dot{u}_k(x_i, y_i, 0) = v_k^0(x_i, y_i) \quad (x_i, y_i) \in \Omega, \tag{59}$$

where $t = 0$. Application of the Laplace transform to the equilibrium equation gives

$$\begin{aligned} &\frac{\partial \tilde{\sigma}_x}{\partial x} + \frac{\partial \tilde{\tau}_{xy}}{\partial y} + \tilde{b}_1 = \rho\left(s^2\tilde{u}_1 - su_1^0 - v_1^0\right) + cs\tilde{u}_1, \\ &\frac{\partial \tilde{\tau}_{xy}}{\partial x} + \frac{\partial \tilde{\sigma}_y}{\partial y} + \tilde{b}_2 = \rho\left(s^2\tilde{u}_2 - su_2^0 - v_2^0\right) + cs\tilde{u}_2 \quad (x, y) \in \Omega, \end{aligned} \tag{60}$$

where the Laplace transform of function $f(t)$ is defined as

$$\tilde{f}(s) = \int_0^\infty f(t)e^{-st}dt \tag{61}$$

The transformed displacements and tractions satisfy the following boundary conditions:

$$\begin{aligned} \tilde{u}_k(x_i, y_i, s) &= \tilde{\bar{u}}_k(x_i, y_i, s), \quad (x_i, y_i) \in \Gamma_u \\ \tilde{t}_k(x_i, y_i, s) &= \tilde{\bar{t}}_k(x_i, y_i, s), \quad (x_i, y_i) \in \Gamma_t \end{aligned} \tag{62}$$

Introducing the mapping technique and differential matrices in (56) and (57) give

$$\begin{aligned} &\left(\mathbf{D}_x\mathbf{Q}_{11}\mathbf{D}_x + \mathbf{D}_y\mathbf{Q}_{66}\mathbf{D}_y - s^2\boldsymbol{\rho} + s\mathbf{c}\right)\tilde{\mathbf{u}}_1 \\ &\quad + \left(\mathbf{D}_x\mathbf{Q}_{12}\mathbf{D}_y + \mathbf{D}_y\mathbf{Q}_{66}\mathbf{D}_x\right)\tilde{\mathbf{u}}_2 = -\tilde{\mathbf{b}}_1 - s\boldsymbol{\varphi}_1 - \boldsymbol{\theta}_1, \end{aligned} \tag{63}$$

$$\begin{aligned} &\left(\mathbf{D}_y\mathbf{Q}_{12}\mathbf{D}_x + \mathbf{D}_x\mathbf{Q}_{66}\mathbf{D}_y\right)\tilde{\mathbf{u}}_1 \\ &\quad + \left(\mathbf{D}_x\mathbf{Q}_{66}\mathbf{D}_x + \mathbf{D}_y\mathbf{Q}_{22}\mathbf{D}_y - s^2\boldsymbol{\rho} + s\mathbf{c}\right)\tilde{\mathbf{u}}_2 = -\tilde{\mathbf{b}}_2 - s\boldsymbol{\varphi}_2 - \boldsymbol{\theta}_2, \end{aligned} \tag{64}$$

where $\boldsymbol{\rho} = \mathbf{diag}[\rho], \mathbf{c} = \mathbf{diag}[c], \boldsymbol{\varphi}_k = \mathbf{diag}[\rho u_k^0], \boldsymbol{\theta}_k = \mathbf{diag}[\rho v_k^0]$. Also Eqs. (63) and (64) can be written as

$$\tilde{\mathbf{D}}(s)\tilde{\mathbf{u}} = -\tilde{\mathbf{b}}. \tag{65}$$

Following the same procedure for static case, the displacements at the internal nodes can be represented in terms of the displacement on the boundary nodes as

$$\tilde{\mathbf{u}}_I = -\tilde{\mathbf{D}}_I^{-1}\tilde{\mathbf{b}}_I - \tilde{\mathbf{D}}_I^{-1}\tilde{\mathbf{D}}_B\tilde{\mathbf{u}}_B, \tag{66}$$

in which $(\tilde{\mathbf{D}}_I, \tilde{\mathbf{D}}_B) = \mathbf{T}^{-1}\tilde{\mathbf{D}}(s)\mathbf{T}$. Thus, for the zero body force contact problem, we have tractions (along the normal and tangential directions) on a smooth boundary for each block in the Laplace transform domain

$$\tilde{\mathbf{t}}_B = \tilde{\mathbf{H}}(s)\tilde{\mathbf{u}}_B, \tag{67}$$

where $\tilde{\mathbf{H}}(s)$ can be determined from (52) following the same technique as for the static case in the Laplace domain. Same again, there are only $2 \times N_t$ linear algebraic equations in (54) to determine the displacements $\tilde{\mathbf{u}}_B$ on the boundary of traction Γ_t. Suppose there are $(K+1)$ samples in the transformation space $s_k, k = 0, 1, \ldots, K$ and the same number of transformed values $\tilde{f}(s_k)$ evaluated by the FBM. Then, $f(t)$ can be determined by the

Laplace inversion techniques. A simple and accurate method proposed by Durbin[29] is adopted as follows:

$$f(t)=\frac{2e^{\sigma t/T}}{T}\left[-\frac{1}{2}\tilde{f}(s_0)+\sum_{k=0}^{K}\mathrm{Re}\left\{\tilde{f}(s_k)e^{2\pi ikt/T}\right\}\right], \tag{68}$$

where $\tilde{f}(s_k)$denotes the transformed variable in the Laplace domain, the parameter of the Laplace transform is chosen as: $s_k=(\sigma+2\pi\, ik)/T$ ($i=\sqrt{-1}$). Two free normalized parameters in s_k, i.e. σ and T should be specified. The selection of the parameters T depends on the observing period in the time domain and $\sigma=5$. In the following examples, all variables are normalized for the convenience of the analysis.

7. Large deformation for FGM plates

Consider a flat plate of constant thickness h with domain Ω enclosed with boundary Γ subjected to transverse load q. On the neutral plane of the plate $z=z_0$, on which the normal bending stress is zero, the in-plane strains of 2D elasticity can be written as[30,31]

$$\varepsilon_{x0}=\varepsilon_{x0}^{(l)}+\varepsilon_{x0}^{(n)},\quad \varepsilon_{y0}=\varepsilon_{y0}^{(l)}+\varepsilon_{y0}^{(n)},\quad \gamma_{xy0}=\gamma_{xy0}^{(l)}+\gamma_{xy0}^{(n)}, \tag{69}$$

where subscripts (l) and (n) denote the linear and nonlinear contributions which are defined as

$$\varepsilon_{x0}^{(l)}=\frac{\partial u_x}{\partial x},\quad \varepsilon_{y0}^{(l)}=\frac{\partial u_y}{\partial y},\quad \gamma_{xy0}^{(l)}=\frac{\partial u_x}{\partial y}+\frac{\partial u_y}{\partial x}, \tag{70a}$$

for linearity and

$$\varepsilon_{x0}^{(n)}=\frac{1}{2}\left(\frac{\partial w}{\partial x}\right)^2,\quad \varepsilon_{y0}^{(n)}=\frac{1}{2}\left(\frac{\partial w}{\partial y}\right)^2,\quad \gamma_{xy0}^{(n)}=\frac{1}{2}\frac{\partial w}{\partial x}\frac{\partial w}{\partial y}. \tag{70b}$$

for nonlinearity, where u_x and u_y are displacements in neutral plane and w is deflection of plate. Consider an FGM plate of uniform thickness h, the displacements in space are assumed as

$$\tilde{u}_x=u_x+(z-z_0)\phi_x,\quad \tilde{u}_y=u_y+(z-z_0)\phi_y,\quad \tilde{w}=w, \tag{71}$$

where ϕ_x,ϕ_y are rotations. Therefore, the strains are given from Eq. (14) as

$$\tilde{\varepsilon}_x=\varepsilon_{x0}+(z-z_0)\frac{\partial\phi_x}{\partial x},\tilde{\varepsilon}_y=\varepsilon_{y0}+(z-z_0)\frac{\partial\phi_y}{\partial y},$$

$$\tilde{\gamma}_{xy} = \gamma_{xy0} + (z - z_0)\left(\frac{\partial \phi_x}{\partial y} + \frac{\partial \phi_y}{\partial x}\right), \tag{72}$$

$$\tilde{\gamma}_{xz} = \phi_x + \frac{\partial w}{\partial x}, \tilde{\gamma}_{yz} = \phi_y + \frac{\partial w}{\partial y},$$

For plane stress state of orthotropic materials linearity, we have the Hook's law in matrix form

$$\begin{bmatrix} \tilde{\sigma}_x \\ \tilde{\sigma}_y \\ \tilde{\tau}_{xy} \\ \tilde{\tau}_{xz} \\ \tilde{\tau}_{yz} \end{bmatrix} = \mathbf{E}(z) \begin{bmatrix} \tilde{\varepsilon}_x \\ \tilde{\varepsilon}_y \\ \tilde{\gamma}_{xy} \\ \tilde{\gamma}_{xz} \\ \tilde{\gamma}_{yz} \end{bmatrix}, \tag{73}$$

where material coefficient matrix (orthotropic media)

$$\mathbf{E}(z) = \begin{bmatrix} E_x/e & E_x\nu_{yx}/e & 0 & 0 & 0 \\ E_y\nu_{xy}/e & E_y/e & 0 & 0 & 0 \\ 0 & 0 & G_{xy} & 0 & 0 \\ 0 & 0 & 0 & G_{xz} & 0 \\ 0 & 0 & 0 & 0 & G_{yz} \end{bmatrix}, \tag{74}$$

in which $e = 1 - \nu_{xy}\nu_{yx}$, E_x and E_y are the Young's moduli referring to the axes x and y respectively. G_{xy}, G_{xz} and G_{yz} are shear moduli and ν_{xy}, ν_{yx} are Poisson's ratios. With regard to functionally graded materials, there are two different assumptions. The first assumption is that the material properties are graded along the plate thickness except the Poisson ratios as

$$P(z) = P_b + (P_t - P_b)(z/h)^n, \tag{75}$$

and second assumption is

$$P(z) = P_b \exp(z\delta/h), \quad \delta = \ln(P_t/P_b), \tag{76}$$

where P_t and P_b denote the material properties either E or G in (74) on the top and bottom surfaces respectively, and n is a parameter that indicates the material variation profile. In this chapter, Poisson's ratios of the plate are assumed constants for simplicity. Under pure bending circumstance, we have zero resultant force along x-axis as

$$\int_0^h \tilde{\sigma}_x(x,y,z)dz = 0. \tag{77}$$

It results the location of the neutral plane as

$$z_0 = \frac{(n+1)(n\beta+2)}{2(n+2)(n\beta+1)}h, \tag{78}$$

for the first assumption in (75), and

$$z_0 = \left(\frac{1}{1-\beta} + \frac{1}{\ln\beta}\right)h, \tag{79}$$

for the second assumption in (76), where $\beta = P_b/P_t$. It is clear that the neutral plane is not the middle plane ($z = h/2$) of the plate except $\beta = 1$.

A new coordinate system is established on the neutral plane as shown in Fig. 6, i.e. $\bar{z} = z - z_0$, and the internal bending moments, shear forces and in-plane forces are defined as

$$\begin{bmatrix} M_x \\ M_y \\ M_{xy} \end{bmatrix} = \int_{-z_0}^{h-z_0} \begin{bmatrix} \tilde{\sigma}_x \\ \tilde{\sigma}_y \\ \tilde{\tau}_{xy} \end{bmatrix} \bar{z}d\bar{z}, \quad \begin{bmatrix} Q_x \\ Q_y \end{bmatrix} = \kappa \int_{-z_0}^{h-z_0} \begin{bmatrix} \tilde{\tau}_{xz} \\ \tilde{\tau}_{yz} \end{bmatrix} d\bar{z},$$

$$\begin{bmatrix} N_x \\ N_y \\ N_{xy} \end{bmatrix} = \int_{-z_0}^{h-z_0} \begin{bmatrix} \tilde{\sigma}_x \\ \tilde{\sigma}_y \\ \tilde{\tau}_{xy} \end{bmatrix} d\bar{z}, \tag{80}$$

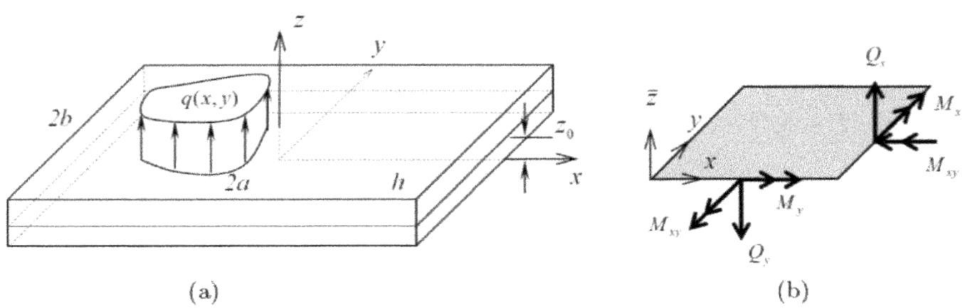

Fig. 6. (a) A rectangular plate and coordinate system and (b) sign convention of bending moments and shear forces for the plate on neutral plane.

where coefficient $\kappa = 5/6$ with Reissner plate theory. Substituting (73) into (80) yields

$$N_x = A_{11}\left[\frac{\partial u_x}{\partial x} + \frac{1}{2}\left(\frac{\partial w}{\partial x}\right)^2\right] + A_{12}\left[\frac{\partial u_y}{\partial y} + \frac{1}{2}\left(\frac{\partial w}{\partial y}\right)^2\right]$$

$$= N_x^{(l)} + N_x^{(n)}, \tag{81a}$$

$$N_y = A_{21}\left[\frac{\partial u_x}{\partial x} + \frac{1}{2}\left(\frac{\partial w}{\partial x}\right)^2\right] + A_{22}\left[\frac{\partial u_y}{\partial y} + \frac{1}{2}\left(\frac{\partial w}{\partial y}\right)^2\right]$$

$$= N_y^{(l)} + N_y^{(n)}, \tag{81b}$$

$$N_{xy} = B\left[\frac{\partial u_x}{\partial y} + \frac{\partial u_y}{\partial x} + \frac{1}{2}\left(\frac{\partial w}{\partial x}\right)\left(\frac{\partial w}{\partial y}\right)\right] = N_{xy}^{(l)} + N_{xy}^{(n)}, \tag{81c}$$

for in-place elasticity and

$$M_x = D_{11}\frac{\partial \phi_x}{\partial x} + D_{12}\frac{\partial \phi_y}{\partial y}, \quad M_y = D_{21}\frac{\partial \phi_x}{\partial x} + D_{22}\frac{\partial \phi_y}{\partial y},$$

$$M_{xy} = S\left(\frac{\partial \phi_x}{\partial y} + \frac{\partial \phi_y}{\partial x}\right), \tag{82a}$$

$$Q_x = C_1\left(\phi_x + \frac{\partial w}{\partial x}\right), \quad Q_y = C_2\left(\phi_y + \frac{\partial w}{\partial y}\right), \tag{82b}$$

for plate bending, where A, B, D, S and C are related to material coefficients and thickness of plate. As far as the first material assumption in (75) is concerned, these coefficients are given as

$$A_{11} = \frac{E_{1t}g}{e}, \quad A_{11} = \frac{E_{1t}\nu_{21}g}{e} = A_{21}, \quad A_{11} = \frac{E_{2t}g}{e}, \quad B = G_{12}g,$$

$$D_{11} = \frac{E_{1t}f}{e}, \quad D_{12} = \frac{E_{1t}\nu_{21}}{e}f = D_{21}, \quad D_{22} = \frac{E_{2t}f}{e}, \tag{83}$$

$$S = G_{12t}f, \quad C_1 = \kappa G_{13t}g, \quad C_2 = \kappa G_{23t}g,$$

where

$$f = \left(\frac{n\beta + 3}{3(n+3)}h - \frac{n\beta + 2}{2(n+2)}z_0\right)h^2, \quad g = \frac{n\beta + 1}{n+1}h. \tag{84}$$

For large deformation of plate nonlinear analysis, five equilibrium equations are given by

$$\frac{\partial N_x}{\partial x} + \frac{\partial N_{xy}}{\partial y} + b_x = 0, \quad \frac{\partial N_{xy}}{\partial x} + \frac{\partial N_y}{\partial y} + b_y = 0, \tag{85a}$$

$$\frac{\partial M_x}{\partial x} + \frac{\partial M_{xy}}{\partial y} - Q_x - m_x^{(n)} = 0, \quad \frac{\partial M_{xy}}{\partial x} + \frac{\partial M_y}{\partial y} - Q_y - m_y^{(n)} = 0, \tag{85b}$$

$$\frac{\partial}{\partial x}\left(Q_x + m_x^{(n)}\right) + \frac{\partial}{\partial y}\left(Q_y + m_y^{(n)}\right) + q = 0, \tag{85c}$$

where b_x, b_y are in-plane body forces, nonlinear terms

$$m_x^{(n)} = N_x \frac{\partial w}{\partial x} + N_{xy}\frac{\partial w}{\partial y}, \quad m_y^{(n)} = N_{xy}\frac{\partial w}{\partial x} + N_y \frac{\partial w}{\partial y}, \tag{86}$$

and q is pressure on the plate surface. In the local Petrov–Galerkin approaches, the weak form of governing equations over a local domain Ω_s^i centred at point $\mathbf{x}_i(x_i, y_i)$ shown in Fig. 3 can be written as

$$\begin{aligned}&\int_{\Omega_s^i}\left(\frac{\partial N_x}{\partial x} + \frac{\partial N_{xy}}{\partial y} + b_x\right) u^* d\Omega = 0,\\ &\int_{\Omega_s^i}\left(\frac{\partial N_{xy}}{\partial x} + \frac{\partial N_y}{\partial y} + b_y\right) u^* d\Omega = 0,\end{aligned} \tag{87a}$$

$$\begin{aligned}&\int_{\Omega_s^i}\left(\frac{\partial M_x}{\partial x} + \frac{\partial M_{xy}}{\partial y} - Q_x - m_x^{(n)}\right) u^* d\Omega = 0,\\ &\int_{\Omega_s^i}\left(\frac{\partial M_{xy}}{\partial x} + \frac{\partial M_y}{\partial y} - Q_y - m_x^{(n)}\right) u^* d\Omega = 0,\end{aligned} \tag{87b}$$

$$\int_{\Omega_s^i}\left[\frac{\partial}{\partial x}\left(Q_x + m_x^{(n)}\right) + \frac{\partial}{\partial y}\left(Q_y + m_y^{(n)}\right) + q\right] u^* d\Omega = 0, \tag{87c}$$

where u^* is a test function and larger than zero in the local integral domain. By the divergence theorem, (87) can be rewritten in a weak form as

$$\begin{aligned}&\int_{\Gamma_s^i}\left(N_x n_x + N_{xy} n_y\right) u^* d\Gamma\\ &\quad - \int_{\Omega_s^i}\left(N_x \frac{\partial u^*}{\partial x} + N_{xy}\frac{\partial u^*}{\partial y} - b_x u^*\right) d\Omega = 0,\end{aligned} \tag{88a}$$

$$\int_{\Gamma_s^i}(N_{xy}n_x + N_y n_y)\,u^* d\Gamma - \int_{\Omega_s^i}\left(N_{xy}\frac{\partial u^*}{\partial x} + N_y\frac{\partial u^*}{\partial y} - b_y u^*\right)d\Omega = 0, \quad (88b)$$

$$\int_{\Gamma_s^i}(M_x n_x + M_{xy} n_y)\,u^* d\Gamma - \int_{\Omega_s^i}\left[M_x\frac{\partial u^*}{\partial x} + M_{xy}\frac{\partial u^*}{\partial y} + \left(Q_x + m_x^{(n)}\right)u^*\right]d\Omega = 0, \quad (88c)$$

$$\int_{\Gamma_s^i}(M_{xy} n_x + M_y n_y)\,u^* d\Gamma - \int_{\Omega_s^i}\left[M_{xy}\frac{\partial u^*}{\partial x} + M_y\frac{\partial u^*}{\partial y} + \left(Q_y + m_y^{(n)}\right)u^*\right]d\Omega = 0, \quad (88d)$$

$$\int_{\Gamma_s^i}\left[\left(Q_x + m_x^{(n)}\right)n_x + \left(Q_y + m_y^{(n)}\right)n_y\right]u^* d\Gamma + \int_{\Omega_s^i}\left[\left(Q_x + m_x^{(n)}\right)\frac{\partial u^*}{\partial x} + \left(Q_y + m_y^{(n)}\right)\frac{\partial u^*}{\partial y} - q u^*\right]d\Omega = 0, \quad (88e)$$

where (n_x, n_y) are components of normal outward to the boundary Γ_s^i of the integral domain Ω_s^i. The simplest choice of the test function is selected as a unit step function in each local integral domain as follows

$$u^*(\mathbf{x}) = \begin{cases} 1 & \text{at } \mathbf{x} \in \Omega_s^i, \\ 0 & \text{at } \mathbf{x} \notin \Omega_s^i. \end{cases} \quad (89)$$

Then the governing equations in (88) become

$$\int_{\Gamma_s^i}\left(A_{11}\frac{\partial u_x}{\partial x}n_x + B\frac{\partial u_x}{\partial y}n_y + A_{12}\frac{\partial u_y}{\partial y}n_x + B\frac{\partial u_y}{\partial x}n_y\right)d\Gamma = -\int_{\Omega_s^i} b_x d\Omega - \int_{\Gamma_i}\left(N_x^{(n)}n_x + N_{xy}^{(n)}n_y\right)d\Gamma, \quad (90a)$$

$$\int_{\Gamma_s^i}\left(B\frac{\partial u_x}{\partial y}n_x + A_{21}\frac{\partial u_x}{\partial x}n_y + B\frac{\partial u_y}{\partial x}n_x + A_{22}\frac{\partial u_y}{\partial y}n_y\right)d\Gamma = -\int_{\Omega_s^i} b_y d\Omega - \int_{\Gamma_i}\left(N_{xy}^{(n)}n_x + N_y^{(n)}n_y\right)d\Gamma, \quad (90b)$$

$$\int_{\Gamma_s^i}\left(D_{11}\frac{\partial\phi_x}{\partial x}n_x+S\frac{\partial\phi_x}{\partial y}n_y+D_{12}\frac{\partial\phi_y}{\partial y}n_x+S\frac{\partial\phi_y}{\partial x}n_y\right)d\Gamma$$
$$-\int_{\Gamma_i}C_1wn_xd\Gamma-\int_{\Omega_s^i}C_1\phi_xd\Omega-\int_{\Omega_s^i}m_x^{(n)}d\Omega=0, \tag{90c}$$

$$\int_{\Gamma_s^i}\left(S\frac{\partial\phi_x}{\partial y}n_x+D_{21}\frac{\partial\phi_x}{\partial x}n_y+S\frac{\partial\phi_y}{\partial x}n_x+D_{22}\frac{\partial\phi_y}{\partial y}n_y\right)d\Gamma$$
$$-\int_{\Gamma_i}C_2wn_yd\Gamma-\int_{\Omega_s^i}C_2\phi_yd\Omega-\int_{\Omega_s^i}m_y^{(n)}d\Omega=0, \tag{90d}$$

$$\int_{\Gamma_s^i}\left(C_1\phi_xn_x+C_2\phi_yn_y+C_1\frac{\partial w}{\partial x}n_x+C_2\frac{\partial w}{\partial y}n_y\right)d\Gamma$$
$$=-\int_{\Omega_s^i}qd\Omega-\int_{\Gamma_i}\left(m_x^{(n)}n_x+m_y^{(n)}n_y\right)d\Gamma. \tag{90e}$$

For the sake of simplicity, the integral area in the mapping domain is selected as a circle of radius R centred at point (ξ_i,η_j) which corresponds to the point $\mathbf{x}_k$ in physical domain. Therefore, the coordinates of local integral boundary $S'(\xi_s,\eta_s)$ in the mapping domain shown in Fig. 4 are

$$\xi_s=\xi_i+R\cos\theta,\quad \eta_s=\eta_j+R\sin\theta, \tag{91}$$

and the coordinate $S(x_s,y_s)$

$$x_s=\sum_{l=1}^{8}N_l(\xi_s,\eta_s)x_l,\quad y_s=\sum_{l=1}^{8}N_l(\xi_s,\eta_s)y_l. \tag{92}$$

The components of the segment are

$$dx=Rd\theta\sum_{l=1}^{8}\left(-\frac{\partial N_l(\xi,\eta)}{\partial\xi}\sin\theta+\frac{\partial N_l(\xi,\eta)}{\partial\eta}\cos\theta\right)x_l=f(\xi_s,\eta_s)Rd\theta,$$
$$dy=Rd\theta\sum_{l=1}^{8}\left(-\frac{\partial N_l(\xi,\eta)}{\partial\xi}\sin\theta+\frac{\partial N_l(\xi,\eta)}{\partial\eta}\cos\theta\right)y_l=g(\xi_s,\eta_s)Rd\theta. \tag{93}$$

Thus, the segments in two space are

$$ds=\sqrt{dx^2+dy^2}=\sqrt{f^2+g^2}Rd\theta=\sqrt{f^2+g^2}ds', \tag{94}$$

and the tangential at point $S(x_s,y_s)$ in the practical domain is obtained by

$$\alpha=\tan^{-1}\frac{dy}{dx}=\frac{f}{g}, \tag{95}$$

and the components of normal outward to the boundary in (93) are

$$n_x = \cos\beta, \quad n_y = \sin\beta, \quad \beta = \alpha - \pi/2. \tag{96}$$

The weak form of governing equations (90), at domain point $\mathbf{x}(x_i, y_i) i = 1, 2, \ldots L$, can be written as

$$\int_0^{2\pi} \sum_{l=1}^{L} (A_{11} K_{l,x} n_x u_{xl} + B K_{l,y} n_y u_{xl} + A_{12} K_{l,y} n_x u_{yl} + B K_{l,x} n_y u_{yl}) J d\theta = -b_x \Omega_s^i - \int_0^{2\pi} (N_x^{(n)} n_x + N_{xy}^{(n)} n_y) J d\theta, \tag{97a}$$

$$\int_0^{2\pi} \sum_{l=1}^{L} (B K_{l,y} n_x u_{xl} + A_{21} K_{l,x} n_y u_{xl} + B K_{l,x} n_x u_{yl} + A_{22} K_{l,y} n_y u_{yl}) J d\theta = -b_y \Omega_s^i - \int_0^{2\pi} (N_{xy}^{(n)} n_x + N_y^{(n)} n_y) J d\theta, \tag{97b}$$

$$\int_0^{2\pi} \sum_{l=1}^{L} (D_{11} K_{l,x} n_x \phi_{xl} + S K_{l,y} n_y \phi_{xl} + D_{12} K_{l,y} n_x \phi_{yl} + S K_{l,x} n_y \phi_{yl}) J d\theta - \int_0^{2\pi} C_1 w n_x J d\theta - C_1 \phi_x \Omega_s^i - m_x^{(n)} \Omega_s^i = 0, \tag{97c}$$

$$\int_0^{2\pi} \sum_{l=1}^{L} (S K_{l,y} n_x \phi_{xl} + D_{21} K_{l,x} n_y \phi_{xl} + S K_{l,x} n_x \phi_{yl} + D_{22} K_{l,y} n_y \phi_{yl}) J d\theta - \int_0^{2\pi} C_2 w n_y J d\theta - C_2 \phi_y \Omega_s^i - m_y^{(n)} \Omega_s^i = 0, \tag{97d}$$

$$\int_0^{2\pi} (C_1 K_l n_x \phi_{xl} + C_2 K_l n_y \phi_{yl} + (C_1 K_{l,x} n_x + C_2 K_{l,y} n_y) w_l) J d\theta = -q \Omega_s^i - \int_0^{2\pi} (m_x^{(n)} n_x + m_y^{(n)} n_y) J d\theta, \tag{97e}$$

where Jacob coefficient $J = \sqrt{f^2 + g^2}R$, K_l, $K_{l,x}$ and $K_{l,y}$ are shape function in (12) and its partial differentials.

8. Load and displacement incremental algorithms

As all integral functions in (97) are regular and any standard Gaussian integral formula is available over above integrals, Eq. (97) can be written in the form of summation as following, for each collocation point $\mathbf{x}(x_i, y_i)(i = 1, 2, \ldots L)$, as

$$\begin{aligned}
&\sum_{l=1}^{L} u_{xl} \sum_{k=1}^{K} (A_{11k} K_{lk,x} n_{xk} + B_k K_{lk,y} n_{yk}) W_k J_k \\
&\quad + \sum_{l=1}^{L} u_{yl} \sum_{k=1}^{K} (A_{12} K_{lk,x} n_{xk} + B_k K_{lk,y} n_{yk}) W_k J_k \\
&\quad = -b_x \Omega_s^i - \sum_{l=1}^{L} N_{xl}^{(n)} \sum_{k=1}^{K} W_k J_k K_{lk} n_{xk} - \sum_{l=1}^{L} N_{xyl}^{(n)} \sum_{k=1}^{K} W_k J_k K_{lk} n_{yk},
\end{aligned} \tag{98a}$$

$$\begin{aligned}
&\sum_{l=1}^{L} u_{xl} \sum_{k=1}^{K} (B_k K_{lk,y} n_{xk} + A_{21} K_{lk,x} n_{yk}) W_k J_k \\
&\quad + \sum_{l=1}^{L} u_{yl} \sum_{k=1}^{K} (B_k K_{lk,x} n_{xk} + A_{22k} K_{lk,y} n_{yk}) W_k J_k \\
&\quad = -b_x \Omega_s^i - \sum_{l=1}^{L} N_{xl}^{(n)} \sum_{k=1}^{K} W_k J_k K_{lk} n_{xk} - \sum_{l=1}^{L} N_{xyl}^{(n)} \sum_{k=1}^{K} W_k J_k K_{lk} n_{yk},
\end{aligned} \tag{98b}$$

$$\begin{aligned}
&\sum_{l=1}^{L} \phi_{xl} \sum_{k=1}^{K} (D_{11k} K_{lk,x} n_{xk} + S_k K_{lk,y} n_{yk}) W_k J_k \\
&\quad + \sum_{l=1}^{L} \phi_{yl} \sum_{k=1}^{K} (D_{12k} K_{lk,x} n_{xk} + S_k K_{lk,y} n_{yk}) W_k J_k \\
&\quad - \sum_{l=1}^{L} w_l \sum_{k=1}^{K} C_k K_{lk,x} n_{xk} - C_i \phi_{xi} \Omega_s^i - m_x^{(n)} \Omega_s^i = 0,
\end{aligned} \tag{98c}$$

$$\sum_{l=1}^{L}\phi_{xl}\sum_{k=1}^{K}\left(S_kK_{lk,y}n_{xk}+D_{21k}K_{lk,x}n_{yk}\right)W_kJ_k$$
$$+\sum_{l=1}^{L}\phi_{yl}\sum_{k=1}^{K}\left(S_kK_{lk,x}n_{xk}+D_{22k}K_{lk,y}n_{yk}\right)W_kJ_k$$
$$-\sum_{l=1}^{L}w_l\sum_{k=1}^{K}C_kK_{lk,x}n_{xk}-C_i\phi_{xi}\Omega_s^i-m_y^{(n)}\Omega_s^i=0, \tag{98d}$$

$$\sum_{l=1}^{L}\phi_{xl}\sum_{k=1}^{K}C_{1k}K_{lk}n_{xk}W_kJ_k+\sum_{l=1}^{L}\phi_{yl}\sum_{k=1}^{K}C_{2k}K_{lk}n_{yk}W_kJ_k$$
$$+\sum_{l=1}^{L}w_l\sum_{k=1}^{K}\left(C_{1k}K_{lk,x}n_{xk}+C_{2k}K_{lk,y}n_{yk}\right)W_kJ_k$$
$$=-q\Omega_s^i-\sum_{l=1}^{L}m_{xl}^{(n)}\sum_{k=1}^{K}W_kJ_kK_{lk}n_{xk}-\sum_{l=1}^{L}m_{yl}^{(n)}\sum_{k=1}^{K}W_kJ_kK_{lk}n_{yk}, \tag{98e}$$

where $K_{lk}, K_{lk,x}$ and $K_{lk,y}$ are shape functions, and its partial differentials at Gaussian integral points (ξ_k,η_k), (n_{xk},n_{yk}) are components of normal on the local integral boundary Γ_s^i as shown in Fig. 4(b) and are given in (48), $W_k$is Gaussian integral coefficient, new Jacobi $J_k=\pi R\sqrt{f_k^2+g_k^2}$ and $\Omega_s^i$is the area of integral area centred at point $\mathbf{x}(x_i,y_i)$. Finally, the system of equations Eq. (41) for domain collocation points can be arranged in matrix form as

$$\mathbf{AX}=\mathbf{F}, \tag{99}$$

where $\mathbf{X}=\{\mathbf{X}_1,\mathbf{X}_2,\ldots,\mathbf{X}_L\}^{\mathrm{T}}$, $\mathbf{X}_i=\{u_{xi},u_{yi},\phi_{xi},\phi_{yi},w_i\}$, $\mathbf{F}=\{\mathbf{F}_1,\mathbf{F}_2,\ldots,\mathbf{F}_L\}^{\mathrm{T}}$ and $\mathbf{F}_i=\{f_{1i},f_{2i},f_{3i},f_{4i},f_{5i}\}$ are nodal value vectors of displacement and body forces, respectively. From Eq. (41), one has

$$f_{1i}=-b_x\Omega_s^i-\sum_{l=1}^{L}N_{xl}^{(n)}\sum_{k=1}^{K}W_kJ_kK_{lk}n_{xk}-\sum_{l=1}^{L}N_{xyl}^{(n)}\sum_{k=1}^{K}W_kJ_kK_{lk}n_{yk}, \tag{100a}$$

$$f_{2i}=-b_y\Omega_s^i-\sum_{l=1}^{L}N_{xl}^{(n)}\sum_{k=1}^{K}W_kJ_kK_{lk}n_{xk}$$
$$-\sum_{l=1}^{L}N_{xyl}^{(n)}\sum_{k=1}^{K}W_kJ_kK_{lk}n_{yk}, \tag{100b}$$

$$f_{3i} = -m_x^{(n)}\Omega_s^i, f_{4i} = -m_y^{(n)}\Omega_s^i, \tag{100c}$$

$$f_{5i} = -q\Omega_s^i - \sum_{l=1}^{L} m_{xl}^{(n)} \sum_{k=1}^{K} W_k J_k K_{lk} n_{xk}$$
$$- \sum_{l=1}^{L} m_{yl}^{(n)} \sum_{k=1}^{K} W_k J_k K_{lk} n_{yk}. \tag{100d}$$

For boundary nodes, we have to consider five boundary conditions of displacements/rotations $\mathbf{X}_i$ and tractions $\{t_{xi}, t_{yi}, p_{xi}, p_{yi}, Q_{ni}\}$, where

$$\begin{aligned} t_{xi} &= N_{xi}n_x + N_{xyi}n_y, t_{yi} = N_{xyi}n_x + N_{yi}n_y, \\ p_{xi} &= M_{xi}n_x + M_{xyi}n_y, p_{yi} = M_{xyi}n_x + M_{yi}n_y, \\ Q_{ni} &= Q_{xi}n_x + Q_{yi}n_y. \end{aligned} \tag{101}$$

For example, for clamped boundary conditions, we have

$$\mathbf{X}_i = \{u_{xi}, u_{yi}, \phi_{xi}, \phi_{yi}, w_i\} = \{\mathbf{0}\}. \tag{102}$$

For traction boundary conditions, the equations introduced in the system equations are

$$\mathbf{T}_i = \{t_{xi}, t_{yi}, p_{xi}, p_{yi}, Q_{ni}\} = \{\bar{t}_{xi}, \bar{t}_{yi}, \bar{p}_{xi}, \bar{p}_{yi}, \bar{Q}_{ni}\}, \tag{103}$$

where vector $\{\bar{t}_{xi}, \bar{t}_{yi}, \bar{p}_{xi}, \bar{p}_{yi}, \bar{Q}_{ni}\}$ indicates the given boundary traction/moment/shear force. However, for the nodes collocated on the interface between two blocks, we need to consider the connection of deformation and five equilibrium equations, i.e. $\mathbf{X}_l^{(I)} = \mathbf{X}_l^{(II)}$ and $\mathbf{T}_l^{(I)} = -\mathbf{T}_l^{(II)}$. Although we have the same number of governing equations in weak form in domain plus boundary conditions and the same number of displacements/rotations, we are not able to determine the same number of unknowns $\mathbf{X}_i$ due to nonlinearity. To solve the nonlinear system, the load incremental scheme is adopted and (99) is treated as a linear system by

$$\mathbf{AX}^{(m+1)} = \mathbf{F}^{(m)}, \quad (m = 1, 2, \ldots, M_L) \tag{104}$$

where m indicates the step of load increment and M_L is the number of the final step. It means that the nonlinear terms are approximated by in-plane stresses and bending deflection obtained in previous step.

The *Incremental load algorithm* for large deformation of plate bending problem can be summarized as

Step 1: Let $m = 1$, $q = q^{(m)} = \Delta q_0$ (where Δq_0 is constant increment of transverse load and q_0 is the final pressure) and nonlinear terms $(m_{xi}^{(nm)}, m_{yi}^{(nm)})$ are settled to be zero for each node;
Step 2: Solve linear algebraic systems for plate bending problem with boundary conditions, calculate displacement/rotation $(\phi_{xi}^{(m)}, \phi_{yi}^{(m)}, w_i^{(m)})$ and determine nonlinear terms for in-plane 2D problem $(N_{xi}^{(nm)}, N_{yi}^{(nm)}, N_{xyi}^{(nm)})$ with shape function;
Step 3: Solve linear algebraic systems for 2D elasticity with boundary conditions, and calculate displacements $(u_{xi}^{(m)}, u_{yi}^{(m)})$ and membrane forces $(N_{xi}^{(m)}, N_{yi}^{(m)}, N_{xyi}^{(m)})$ in the domain;
Step 4: Evaluate all nonlinear terms $(m_{xi}^{(nm)}, m_{yi}^{(nm)})$ with shape functions;
Step 5: Let $q = q^{(m+1)} = (m + 1)\Delta q_0$;
Step 6: If transverse load $q^{(m+1)} < q_0$, let $m = m + 1$ and go to Step 2;
Step 7: Print results for each step and terminate.

The *Incremental displacement control algorism* for post-buckling analysis plate bending problem can be summarized as

Step 1: Let $m = 0$, $q = \varepsilon$ (where ε is disturbance of plate), $\bar{N}_x = \Delta N_x^0$ (where ΔN_x^0 is increment of in-plane compressive load and N_x^0 is final load along x-axis and no other in-plane loads are assumed);
Step 2: Solve linear algebraic systems for plate bending problem with boundary conditions, calculate rotation and deflection $(\phi_{xi}^{(m)}, \phi_{yi}^{(m)}, w_i^{(m)})$ and determine nonlinear terms $(N_{xi}^{(nm)}, N_{yi}^{(nm)}, N_{xyi}^{(nm)})$ for in-plane 2D problem with shape function;
Step 3: Let $\bar{N}_x = \bar{N}_x^{(m)}$ and solve 2D elasticity in-plane problem and determine nonlinear terms $(m_{xi}^{(nm)}, m_{yi}^{(nm)})$;
Step 4: Let $\bar{N}_x = \bar{N}_x^{(m+1)} = (m + 1)\Delta N_x^0$. If compressive load $\bar{N}_x^{(m+1)} < N_x^0$, let $m = m + 1$ and go to Step 2;
Step 5: Print results for each step and terminate.

If we change N_x^0 to u_x^0 in the post-buckling computational procedure above, the displacement control strategy can be fulfilled straightaway. By the use of in-plane compressive increment load/displacement control techniques, the nonlinear problems are solved simply without iteration procedure. However, the numerical solutions are dependent of the amount of load increment. The accuracy and convergence comparisons are shown in the following examples.

9. Numerical examples for contact

Example 9.1 (Flat punch on an elastic foundation). Firstly, consider a 2D flat punch resting on an elastic foundation as shown in Fig. 7(a). As the symmetry, half of the punch and foundation are modelled as shown in Fig 7(b). The punch has width $2w$ and height h and the foundation has width $2W$ and height H. Uniformly distributed compressive load p_0 is applied on the top surface of the punch. To compare with analytical solution, isotropic materials are assumed with the ratio $\kappa(=E_p/E_f)$, where E_p and E_f are Young's modulus for the punch and foundation, respectively. Poisson's ratios are selected as $\nu_p = \nu_f = 0.3$. The geometry of punch $h = 1.6w$ and foundation $H = 2w$ and $W = 1.6w$. Three blocks are introduced, as shown in Fig. 7(b). For reference, the analytical solution of frictionless contact stress for a rigid punch subject to pressure load p_0 in an infinite half plane is considered.[28]

$$t_n = \frac{2p_0}{\pi\sqrt{1 - x^2/w^2}}. \tag{105}$$

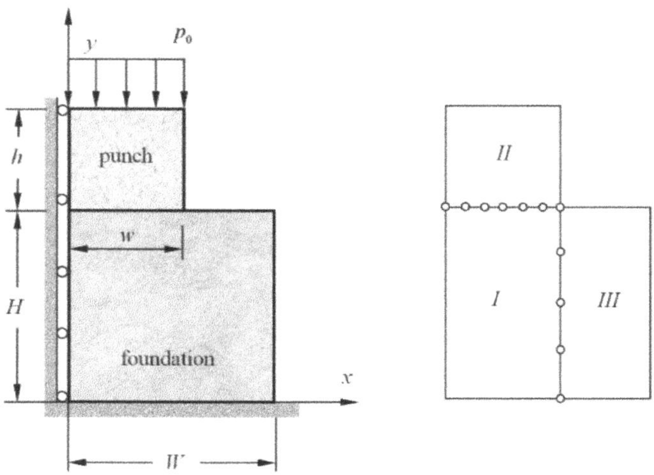

(a) (b)

Fig. 7. An elastic punch resting on a flat foundation: (a) Geometry and (b) Three blocks and uniformly distributed nodes on the contact surface (*I–II*) and interface (*I–III*).

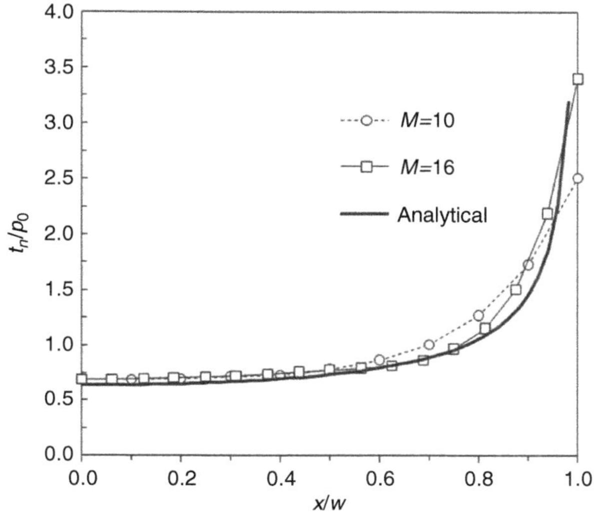

Fig. 8. Comparison with analytical solution of the normalized normal traction distribution t_n/p_0 in contact zone for different numbers of node when the ratio of $E_p/E_f = 100$.

Figure 8 shows the numerical solutions ($E_p/E_f = 100$) with analytical solutions for two different densities of node, i.e. $M = N = 10$ and 16. By observing the singular stress near the corner of punch, it is clear that for larger number of nodes, the degree of accuracy is higher as expected. Figure 9 presents the results of contact stress for isotropic materials for both punch and foundation. It can be observed that the normal traction decrease as the punch becomes stiffer $E_p/E_f = 100$ and the normal traction tends to a constant equal to the applied load p_0 as the punch is much softer $E_p/E_f = 0.01$. At the corner of the punch, the normalized tractions become much higher, as a singularity exists in the elasticity media. Considering the effect of the foundation with functionally graded materials, we assume that a punch with ratio$E_p/E_0 = 100$, the Young's modulus properties of the foundation $E_f(x) = E_0 + (E_R - E_0)x/W$ and Poisson ratios are assumed to be constant $\nu_f = \nu_p = 0.3$, here E_0 and E_R indicate the values of the Young's modulus at the centre and edge of the right-hand side of the foundation. In this case, the geometry parameters are selected as $h/w = 1$, $H/w = 3$ and $W/w = 3$. The variation of contact stress for different ratio $\kappa(= E_R/E_0)$ when $M = N = 16$ is presented in Fig. 10. The contact stress at origin decreases when ratio κ increases. In addition, the singular contact stress at the corner of the punch can be seen clearly.

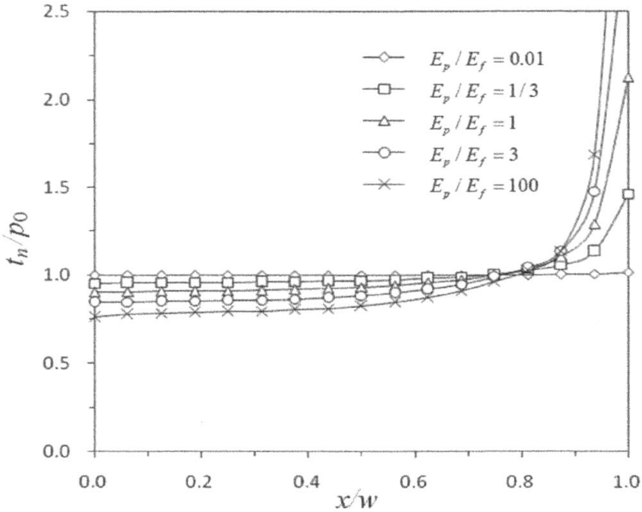

Fig. 9. Normalized normal traction distribution t_n/p_0 in contact zone with different ratio of $\lambda = E_p/E_f$.

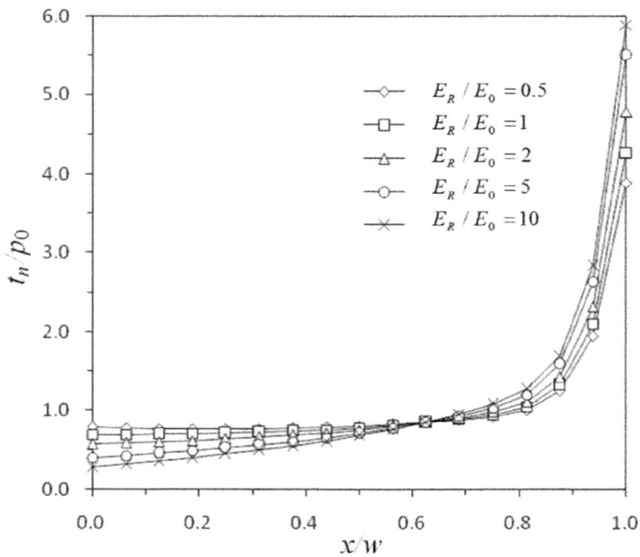

Fig. 10. Normalized normal traction distribution t_n/p_0 in contact zone with different ratios of $\kappa = E_R/E_0$.

Example 9.2 (Curved punch on an elastic foundation). In this example, a punch with curved surface (cylinder with radius R) resting on an elastic foundation, as shown in Fig. 11(a), is considered. As the symmetry, the half of the punch and foundation is modelled in Fig. 11(b) with four blocks. The gap between the punch and elastic foundation can be described approximately as

$$f(x, y) = \frac{x^2}{2R}. \tag{106}$$

The half punch of width $2w$, height h is subjected to a pressure load p_0 on the top and rests on the foundation of width $2W$ and height H. The punch and foundation are considered as isotropic materials with the ratio $\kappa(= E_p/E_f)$ and Poisson's ratios are given as $\nu_p = \nu_f = 0.3$. The dimensions of punch are selected $w/R = 2/5, h/R = 3/5$ and foundation $H/R = 2/5$ and $W/R = 3/5$. Four blocks are introduced, as shown in Fig. 11(b). The width of blocks I and II is d, which is the length of the contact region. It is a typical non-conforming contact problem with unknown contact zone $0 \leq x \leq d$. For frictionless contact, there is Slip mode only. However, for frictional contact ($\mu \neq 0$), there are two contact modes, i.e. Stick and Slip. Uniformly distributed nodes in the mapping domain and the numbers of node are selected as $M_I = M_{II} = 16, M_{III} = M_{IV} = 10, N_I = N_{III} = 16$

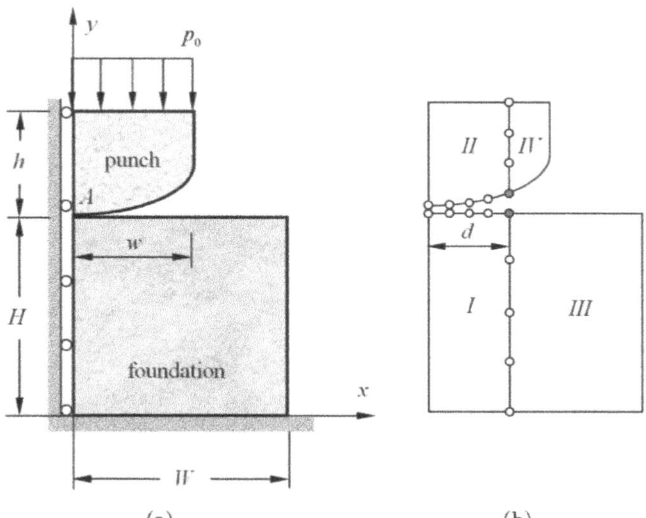

Fig. 11. Cylinder punch resting on a flat foundation: (a) Geometry and (b) Four blocks.

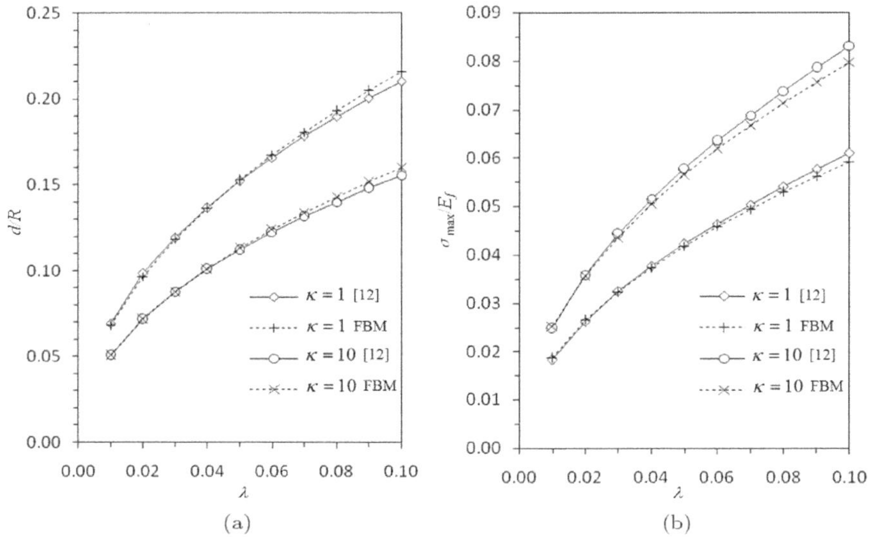

Fig. 12. Comparison with analytical solutions: (a) Normalized size of contact zone d/R and (b) Normalized maximum contact stress $\sigma_{\max}/E_f$.

and $N_{II} = N_{IV} = 12$. An analytical solution is available for an elastic cylindrical punch resting on the semi-infinite foundation[32]

$$d/R = 0.8\sqrt{2\lambda C_\kappa w/R},$$

$$\sigma_{\max}/E_f = 0.798\sqrt{\lambda w/2RC_\kappa}, \quad C_\kappa = \frac{1-\nu_p^2}{\kappa} + 1 - \nu_f^2, \tag{107}$$

$\sigma_{\max}$ represents the maximum contact stress occurring at the centre of contact zone $A(0, H)$, non-dimensional load factor $\lambda = P/E_f w$, where P is applied load on the punch, and therefore $p_0 = P/2w$. The normalized contact zone d/R and maximum stress $\sigma_{\max}/E_f$ are shown in Fig. 12 for two different ratios of $\kappa(= 1, 10)$. For small values of the applied load P, excellent agreements with analytical solution have been achieved both for the size of the contact zone and for the maximum contact stress. When $\lambda = 0.1$, the relative errors increase to 5% due to the effects from the size of foundation, boundary constrain conditions and also from the curvature approximation of the punch. For frictional contact problem, Fig. 13 shows the normal stress t_n/E_f and shear stress $t_t/\mu E_f$ in the contact zone with three different friction coefficients $\mu(= 0.005, 0.01, 0.02)$. In these cases, the dimensions of the punch and foundation are the same, as described in the frictionless case and $\kappa = 1$, $\lambda = 0.01$ respectively. With friction,

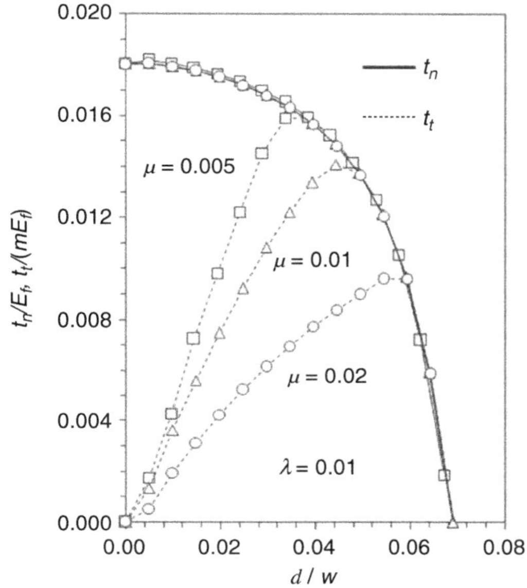

Fig. 13. Normalized contact stresses t_n/E_f and $t_t/\mu E_f$ for different friction coefficients.

the problem becomes nonlinear as both Stick and Slip conditions will occur simultaneously on different parts of the contact surface. The tangential traction components are normalized by μE_f in Fig. 13. Inside the stick zone, the tangential components are less than the normal traction, i.e. $t_t < \mu t_n$. At the point of transition from the stick mode to the slip mode, the normal and tangential components are equal, i.e. $t_t = \mu t_n$. In addition, we also observed that when friction is involved, the tangential tractions at the stick/slip partition are highly sensitive to the discretization of the stick/slip region.[28]

Example 9.3 (Impact between flat punch and foundation with velocity). Consider a flat punch with initial velocity $v_2^0 = V$ moving forward to a static elastic foundation as shown in Fig. 14. To simplify the problem, it is assumed that the punch and foundation will move together after impact, when $t > 0$, i.e. the contact mode is the tick mode after impact (fully bonded). Therefore, a nonlinear problem is simplified to a linear problem. However, the solution is correct for a short time before the contact normal stresses become large than zero (separation). The geometry is the same as that in Example 7.1. Both isotropic material and functionally graded materials are considered with the ratio $\psi(= E_R/E_0)$; the mass

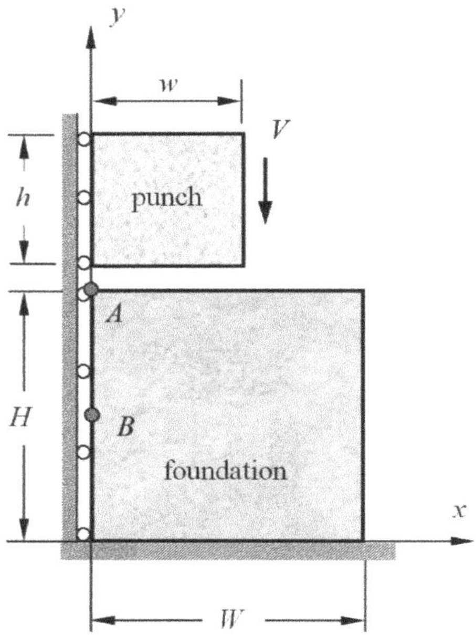

Fig. 14. Impact between the foundation and punch with speed V.

densities for both punch ρ_p and foundation $\rho_f(=\rho_0)$ are constants. The punch is assumed to be isotropic with the ratio $\kappa(=E_p/E_0=\rho_p/\rho_0)$. To verify the accuracy of the FBM with analytical solutions, we consider a 1D problem first. In this case, two blocks are used only and dimensions are selected as $w/h=H/h=1$. The Poisson's ratio is zero for the punch and foundation, and ψ is free parameter. There are $T(=12\times 12)$ uniformly distributed notes for each block. The number of boundary nodes for each block is $Q(=4N-4=44)$. If the foundation is simplified to an elastic spring with functionally graded media

$$E(y)=E_0+\frac{E_0-E_B}{H}y. \tag{108}$$

Then, the deformation on the top of the foundation (spring) caused by a tensile load P is

$$\Delta=\frac{P}{A}\int_{-H}^{0}\frac{dy}{E(y)}=\frac{PH}{A(E_0-E_B)}\ln\left(\frac{E_0}{E_B}\right), \tag{109}$$

Table 1. Comparisons of period, maximum deformation and contact stress between the FBM and the "spring" model.

ψ	$T/2$	$T^*/2$	$y_{\max}/V$	$y^*_{\max}/V$	$\sigma_{\max}/E_0V$	$\sigma^*_{\max}$
0.1	50.25	51.60	16.00	16.33	6.25	6.57
1.0	31.42	31.40	10.00	9.97	10.00	9.92
5.0	19.93	20.0	6.34	6.36	15.77	15.74

*indicates the numerical solution by FBM.

where A is the area of cross-section of the foundation. The stiffness of "spring" is obtained as

$$k = \frac{P}{\Delta} = \frac{(E_0 - E_B)A}{H\ln(E_0/E_B)}. \tag{110}$$

From the equations above, the moving of the punch obeys

$$\rho_p A h \frac{d^2y}{dt^2} + ky = 0, \quad \frac{d^2y}{dt^2} + \omega^2 y = 0,$$
$$\omega = \sqrt{(E_0 - E_A)/Hh\rho_p \ln(E_0/E_B)}. \tag{111}$$

The period of vibration is given analytically as

$$T = 2\pi/\omega = 2\pi\sqrt{Hh\rho_p \ln(E_0/E_B)/(E_0 - E_B)}, \tag{112}$$

with elasticity parameter $\psi = 0.1, 1, 5$ and $\kappa = 100$. So the velocity of angle $\omega = 0.06252, 0.1, 0.15765$, and the half-period for these three cases is listed in Table 1. Finally the movement of the punch is given, by considering the initial condition as

$$y(t) = -\frac{V}{\omega}\sin\omega t, \tag{113}$$

and the impact stress between the punch and "spring"

$$\sigma(t) = \rho_p h\omega V \sin\omega t. \tag{114}$$

The time-dependent normalized displacement and contact stress at point A are shown in Figs. 15 and 16 for different ratios ψ. For comparison, the analytical solution of "spring" mode in (64) and (65) are plotted as well. The agreement for deformation is excellent for $\psi \geq 1$ and the relative error of period in the case of $\psi = 0.1$ is around 2.5%. In addition, there are oscillations for both $\psi = 0.1$ and 1 shown in Fig. 15, for the contact stress by the FBM. It is because the elastic waves travel between the top and bottom surfaces in the punch. However, the average of contact stress

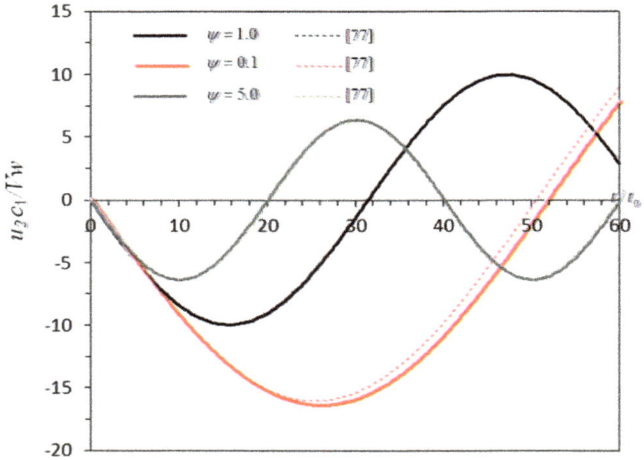

Fig. 15. Normalized displacement on the top $u_2^A c_1/Vw$ for $\kappa = 100$ and different ratios $\psi(= E_B/E_0)$.

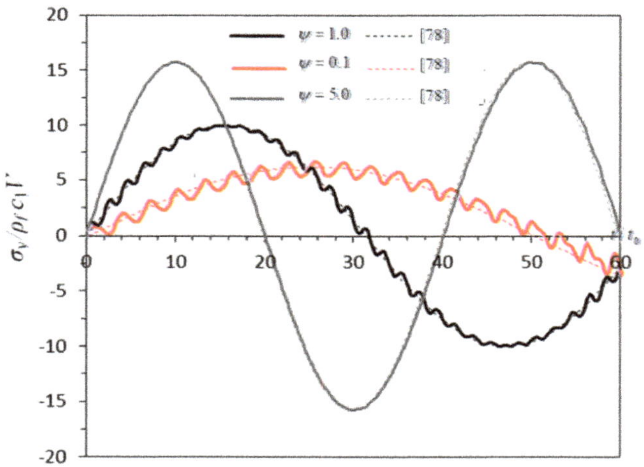

Fig. 16. Normalized stress on the top $\sigma_y^A/(\rho_f c_1 V)$ for $\kappa = 100$ and different ratios $\psi(= E_B/E_0)$.

is very close to the solution of the "spring" model. The magnitudes of the oscillation are reduced significantly for large value of factor $\psi(> 5)$. Table 1 shows the periods, maximum deformations and contact stresses between the FBM and the analytical solutions of the "spring" model.

Next, consider the geometry of punch $h/w = 2$ and foundation $H/w = 2$ and $W/w = 4$. Three blocks are introduced, as shown in Fig. 7(b). There

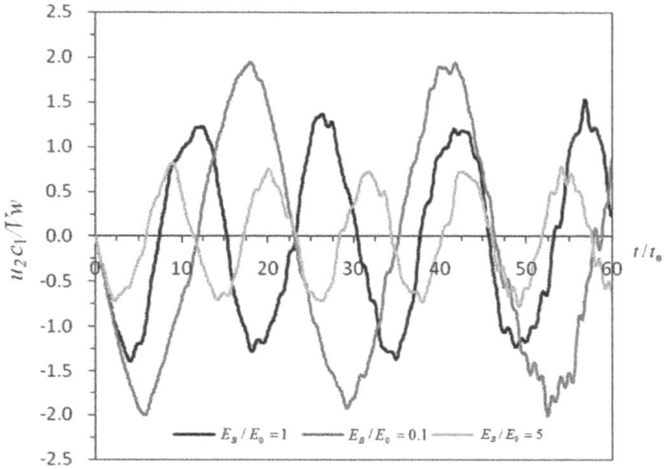

Fig. 17. Normalized displacement on the top A $u_2(A)c_1/wV$ for $\kappa = 1$ and different ratios $\psi(= E_B/E_0)$.

are $T(= 17 \times 17)$ nodes uniformly distributed for each block. The number of boundary nodes for each block is $Q(= 4N - 4 = 64)$. The punch is assumed to be an isotropic material, with constants E_p, ρ_p and ν_p. The property of the foundation is assumed to be a linear function of the y axis only, i.e. $E_f(y) = E_0 - (E_B - E_0) \times y/H$, E_0 and E_B are the Young's modulus at the top and bottom of the foundation and the Poisson ratios are constants $\nu_f(= \nu_p = 0.3)$. The variations of the normalized displacements $u_2(A)c_1/wV$ and $u_2(B)c_1/wV$ versus the normalized time t/t_0 are presented in Figs. 17–23 for different ratios $\kappa(= E_p/E_0 = \rho_p/\rho_0)$, where $t_0 = w/c_1$, c_1 is the speed of the longitudinal wave and is defined as

$$c_1 = \sqrt{\frac{(1-\nu)E_0}{(1+\nu)(1-2\nu)\rho_0}}. \tag{115}$$

In the Laplace inversion Durbin method, the free parameters are selected as $T = 20t_0$ and $\sigma = 5$. The number of samples in the Laplace domain K is larger than 25 ($K = 50$ in this chapter). It has been observed that the selection of these free parameters in the Durbin's method is very stable in elastodynamics. From these results by the FBM, we can see that for a short time, the variations of displacement are linear for both points. However, for point B, the displacement is almost zero before the fastest elasticity wave (longitudinal wave) arrived as expected, i.e. $t/t_0 < 1$.

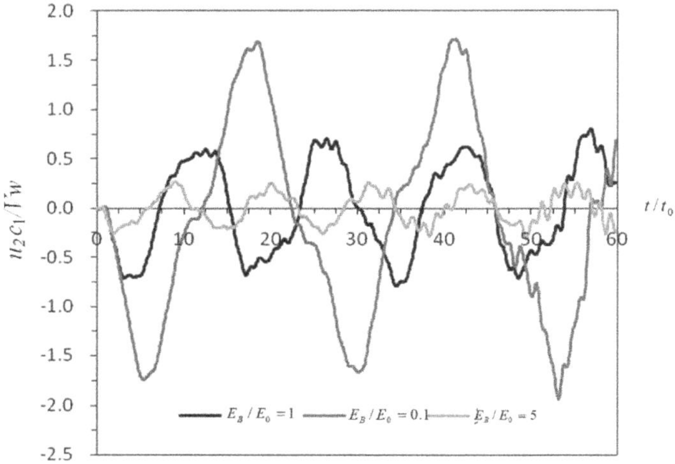

Fig. 18. Normalized displacement on the top B $u_2(B)c_1/wV$ for $\kappa = 1$ and different ratios $\psi(= E_B/E_0)$.

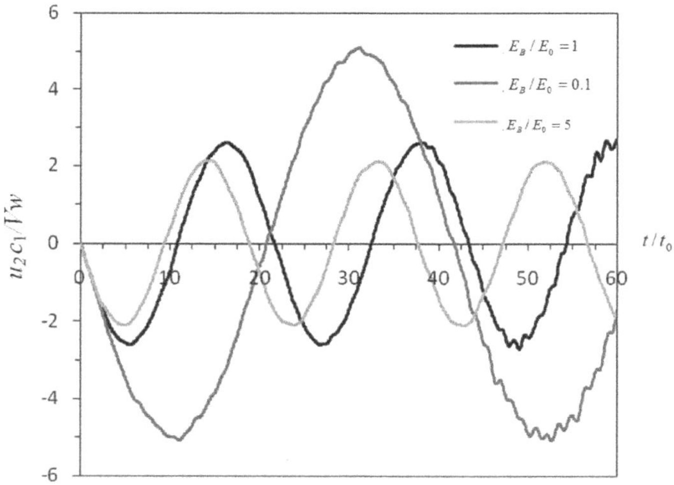

Fig. 19. Normalized displacement on the top A $u_2(A)c_1/wV$ for $\kappa = 5$ and different ratios $\psi(= E_B/E_0)$.

10. Numerical examples for large deformation of plate

In the following examples, without specific definition for free parameters, a square plate with width $2a$ under different working conditions and different media is observed. Node density $M = N = 11$ and the radius of local

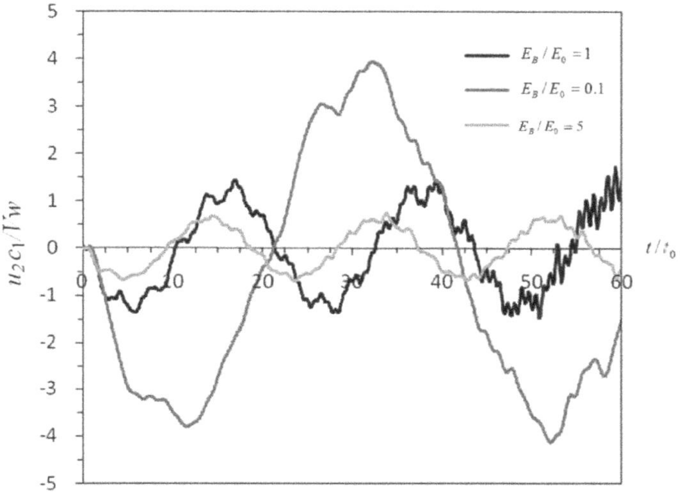

Fig. 20. Normalized displacement on the top B $u_2(B)c_1/wV$ for $\kappa = 5$ and different ratios $\psi(= E_B/E_0)$.

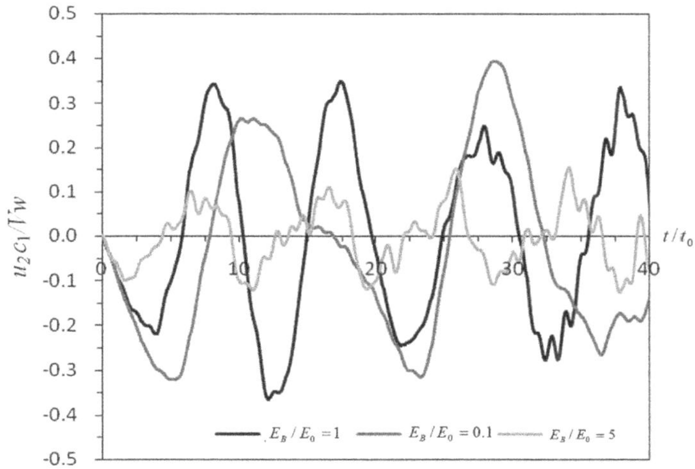

Fig. 21. Normalized displacement on the top A $u_2(B)c_1/wV$ for $\kappa = 0.1$ and different ratios $\psi(= E_B/E_0)$.

integral domain $R = a/(M-1)$ with the area of local integral domain $\Omega_s^i = \pi R^2$. The number of Gaussian integral point is chosen as $K = 180$ with $W_k = 1/K$ (constant). In addition, orthotropic media is considered only in this chapter as: $E_x = E, E_x/E_y = \varphi$, where φ is free parameter.

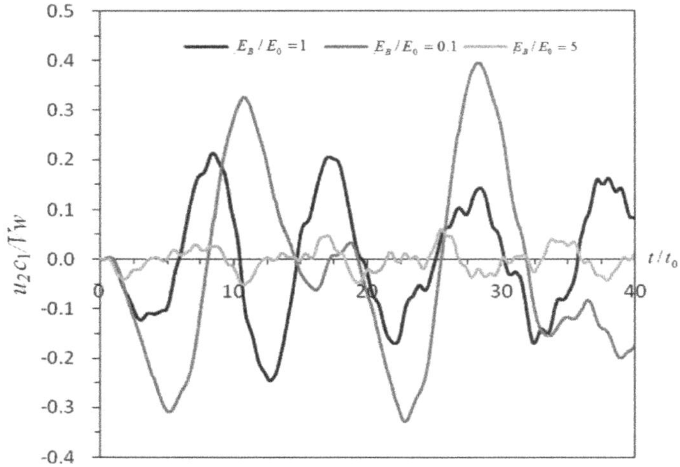

Fig. 22. Normalized displacement on the top B $u_2(B)c_1/wV$ for $\kappa = 0.1$ and different ratios $\psi(= E_B/E_0)$.

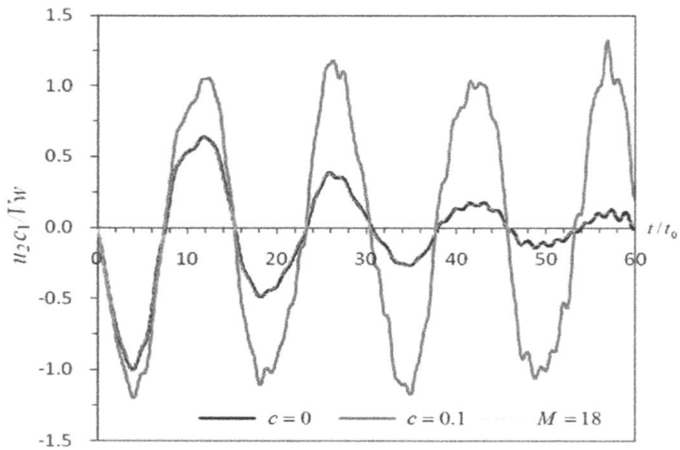

Fig. 23. Normalized displacement on the top A $u_2(B)c_1/wV$ for $\kappa = 1$, $\psi = 1$ with damping factor $c = 0.1$.

Poisson ratios are assumed as $\nu_{xy} = \nu = 0.316$ and shear modulus $G_{xy} = G_{xz} = G_{yz} = E/2(1+\nu)$ for simplicity.

Example 10.1 (Simply supported square plate under uniform pressure). Consider a simply supported square plate subjected to a uniformly distributed pressure q_0. It is assumed that the displacements and deflection are zero with free rotation. The boundary conditions along all edges are

$$u_x = u_y = w = \phi_y = M_x = 0 \quad \text{when} \quad x = \pm a,$$

$$u_x = u_y = w = \phi_x = M_y = 0 \quad \text{when} \quad y = \pm a.$$

The ratio h/a is selected as 0.1 and 0.2, respectively. Firstly, consider material as isotropic with the ratios $\varphi = 1$, $\beta = 1$. In the case of $\beta = 1$, the neutral plane is coincident with the middle plane, i.e. $z_0 = h/2$. Normalized transverse load is defined as $\bar{q} = 16q_0a^4/Eh^4$ and the incremental pressure is $\Delta\bar{q} = 0.2$. The normalized deflection $w_{\max}/h$ for different thickness are shown in Figs. 24(a) and 24(b) for two different ratios of h/a, respectively. The results given by both meshless point collocation method (Wen and Hon, 2007[31]) and FEM (ABAQUS) are presented in the figures for comparison. Obviously the divergence occurs with PCM earlier than that with FBPGM, which means that the FBPGM with Lagrange series interpolation is more stable than PCM. To observe the effect with graded material along the thickness, the parameters $\varphi = 1$ and $h/a = 0.1$ and the ratios of β and parameter n are free variables. The maximum deflections $w_{\max}/h$ at the centre of plate are plotted in Figs. 25(a) and 25(b).

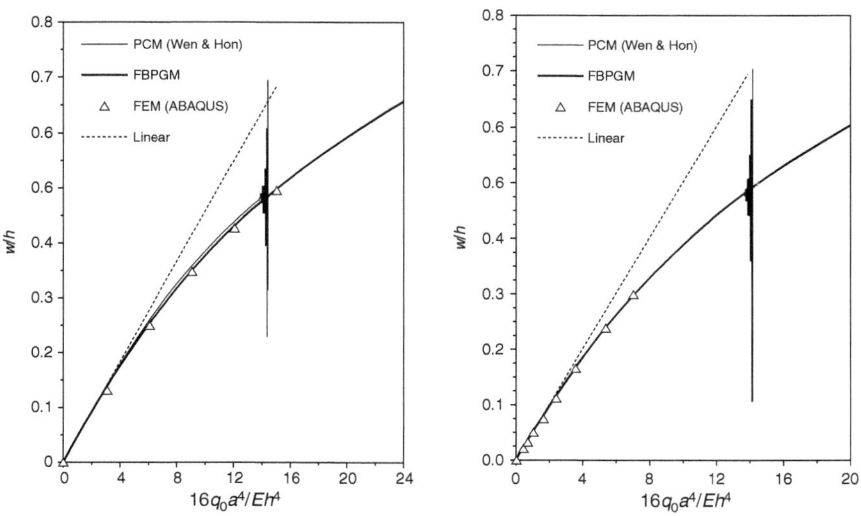

(a) $h/a = 0.1$ (b) $h/a = 0.2$

Fig. 24. Deflection of simply supported plate w/h versus normalized pressure load $16q_0a^4/Eh^4$.

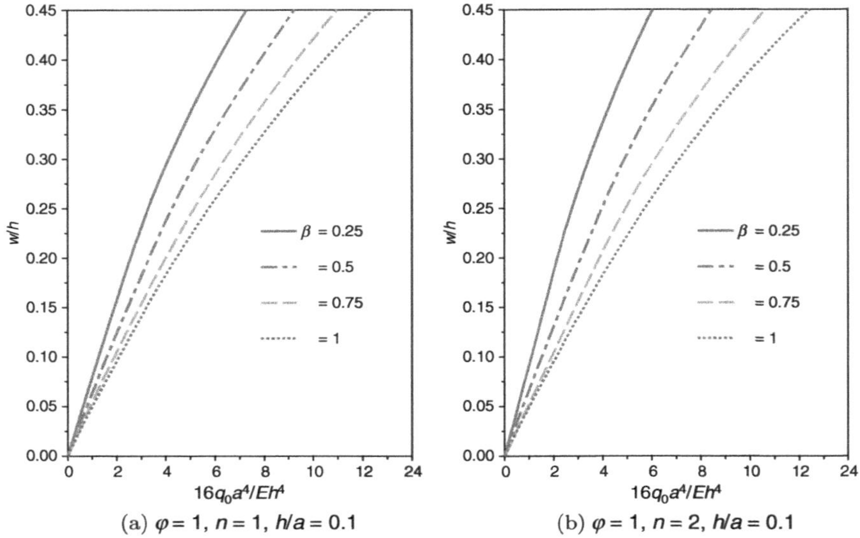

Fig. 25. Deflection of simply supported plate w/h versus normalized pressure load $16q_0a^4/Eh^4$.

Example 10.2 (Clamped square plate subjected to uniform pressure). Similar to the previous example, a clamped square plate of width $2a$ is subjected to a uniform pressure on the surface and the ratio h/a is selected as 0.1 and 0.2, respectively. Again, in order to compare with FEM, the material is chosen as isotropic with $\varphi = 1$, $\beta = 1$. The normalized transverse pressure is defined as $\bar{q} = 16q_0a^4/Eh^4$ and $\Delta\bar{q} = 0.2$. Normalized deflection $w_{\max}/h$ for different thickness are shown in Figs. 26(a) and 26(b), and the results given by PCM (Wen and Hon, 2007[31]) and FEM (ABAQUS) are plotted in the same figure for comparison. Secondly, if the thickness $h/a = 0.1$ and $\varphi = 0.5$ (orthotropic graded material), the maximum deflections $w_{\max}/h$ at the centre of plate are presented in Figs. 27(a) and 27(b) versus different ratios of β and parameter n.

To study the convergence of the method, two factors are observed. The first factor observed is the size of integral area Ω_s^l, i.e. different radius of integral domain R with Δ, $\Delta/2$ and $\Delta/10$ respectively, where $\Delta = 2a/(M-1)$. In this case, the rest of parameters are chosen as $h/a = 0.2$, $\varphi = 1$ and $\beta = 1$. The deflections at the centre of plate are shown in Fig. 28(a). It is obvious that the difference between last two cases ($\Delta/2$ and $\Delta/10$) are

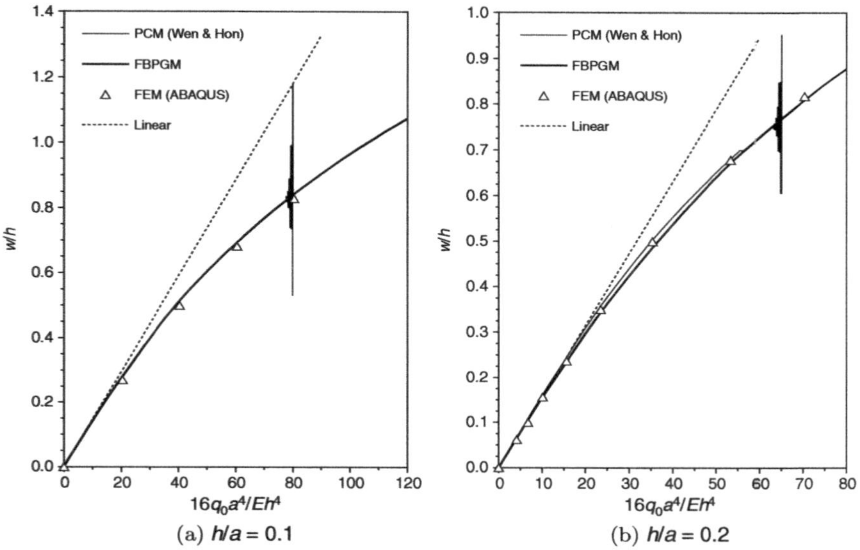

Fig. 26. Deflection of clamped plate w/h versus normalized pressure load $16q_0a^4/Eh^4$.

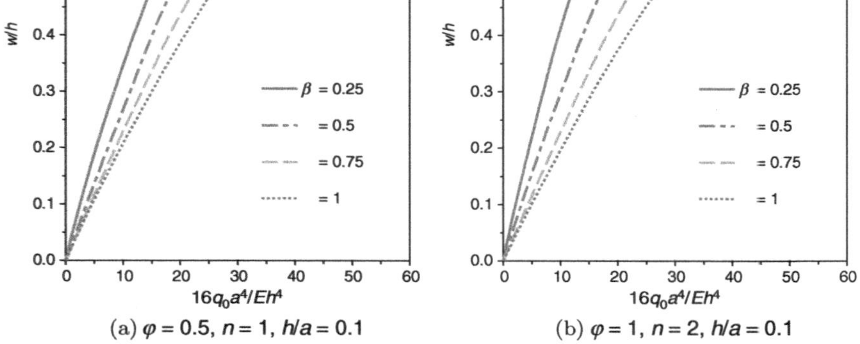

Fig. 27. Deflection of clamped plate w/h versus normalized pressure load $16q_0a^4/Eh^4$.

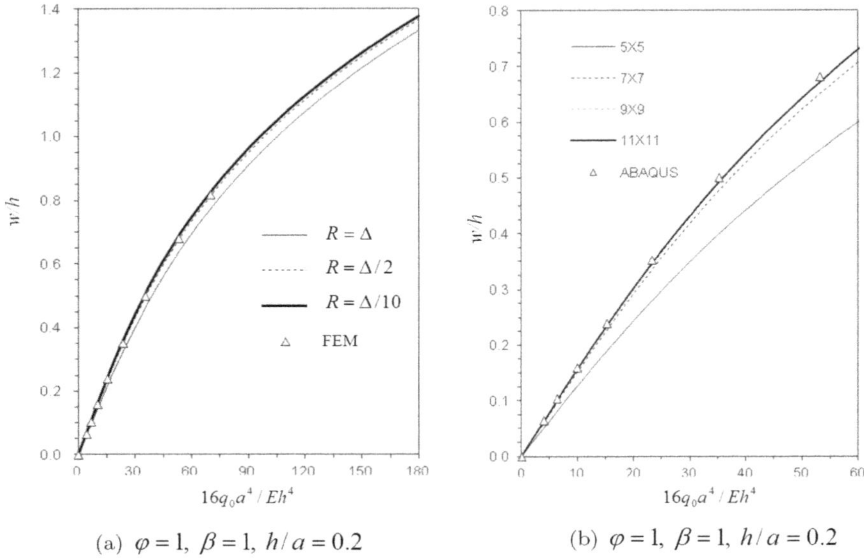

Fig. 28. Sensitivity and convergence observations (a) Size of integral domain and (b) Density of node.

too small to be distinguished. The second factor observed is the density of node, i.e. $M \times N$ are chosen as $5 \times 5, 7 \times 7, 9 \times 9$ and 11×11, respectively. The normalized maximum deflections of plate at centre point $w_{\max}/h$ are plotted in Fig. 28(b). It is clear that the results for node distribution with 9×9 and 11×11 are almost the same.

Example 10.3 (Clamped quarter of ring with homogeneous material). Consider part of ring with inner radius a and outer radius b, and abduction angle $\pi/2$ subjected to a uniform pressure q_0. Two blocks are utilized in this example with interface line AB shown in Fig. 29(a). For the sake of convenience of comparison with other method, isotropic material is considered, i.e. $\varphi = 1, \beta = 1$. The ratio h/a is selected as 0.05, a and b are chosen as 1m and 2m, the radius of integral domain $R = (b-a)/2(M-1)$, Young's modulus $E_x = E_y = 10^7$Pa and Poisson ratio $\nu = 0.3$. Normalized transverse loads are specified as $q_0 = 2500$Pa, 5000Pa, 7500Pa and 10000Pa. The increment of applied load as $\Delta q = 62.5$Pa. The normalized deflections of plate $w(r/a-1, \pi/4)/a$ along the interface AB are plotted in the Fig. 29(b). Obviously a good agreement between the FEM (ABAQUS) and FBPGM has been achieved, which also verifies the accuracy of the present method.

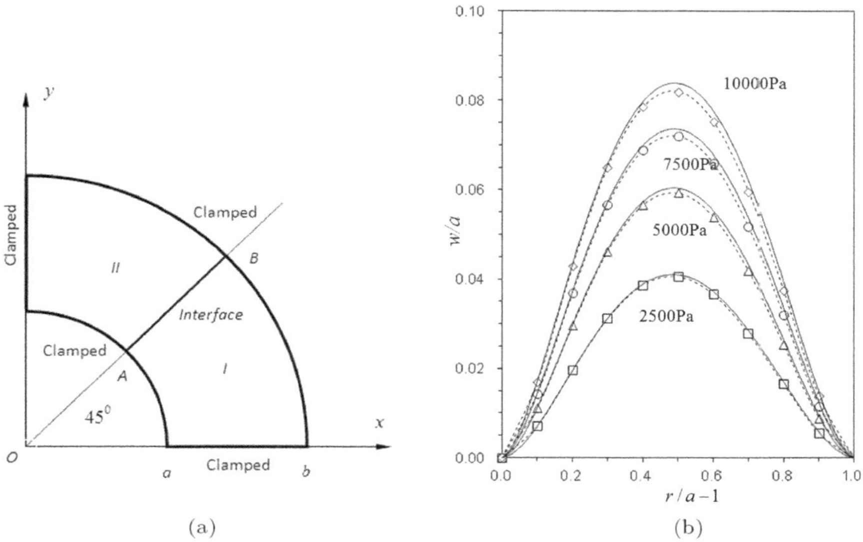

Fig. 29. Deflection along the straight line AB versus different pressures on the surface q_0: (a) Geometry and (b) Results and comparison (solid lines for FBPGM and dash lines for FEM with ABAQUS).

Example 10.4 (Post-buckling analysis for simply supported square plate). Consider a square plate of width $2a$ subjected to uniformly distributed in-plane compressive load σ_0 ($N_x = -\sigma_0 h$). The ratio $h/a = 0.1$ and material constants are chosen as $E_x = E_y = 3 \times 10^7$Pa, Poisson ratio $\nu = 0.316$, $\varphi = 1$ and $\beta = 1$. Two loading techniques, i.e. compressive load incremental load and displacement controls, are compared in this example. For the post-buckling of plate, the deflection dependents on the imperfection factor q^*. With different imperfection factors $q^* = \alpha q_0^*$, in which $q_0^* = 0.118Eh^4/(2a)^4$ and α is free parameter, Fig. 30(a) shows the variation of the maximum deflection of plate against the normalized in-plane stress $\sigma_x/\sigma_{\text{cr}}$ at the centre of plate. In this case, the elastic buckling load of the first mode $\sigma_{\text{cr}} = 27411$Pa, the increment of load $\Delta\sigma_x/\sigma_{\text{cr}} = 3.65 \times 10^{-3}$ and increment of displacement $\Delta u_x/a = 1.8274 \times 10^{-5}$. In Fig. 30(a), the results given by load control are plotted as solid line and displacement control as a dashed line. However, the displacement control is more stable. Furthermore, it is clear that the deflection of the plate tends to become unstable when the compressive load σ_0 approaches the elastic buckling load σ_{cr}. The same problem was studied by Wen and Aliabadi[33] with BEM. The results of FBPGM and BEM are presented in Fig. 30(b) when $\alpha = 1$ for comparison

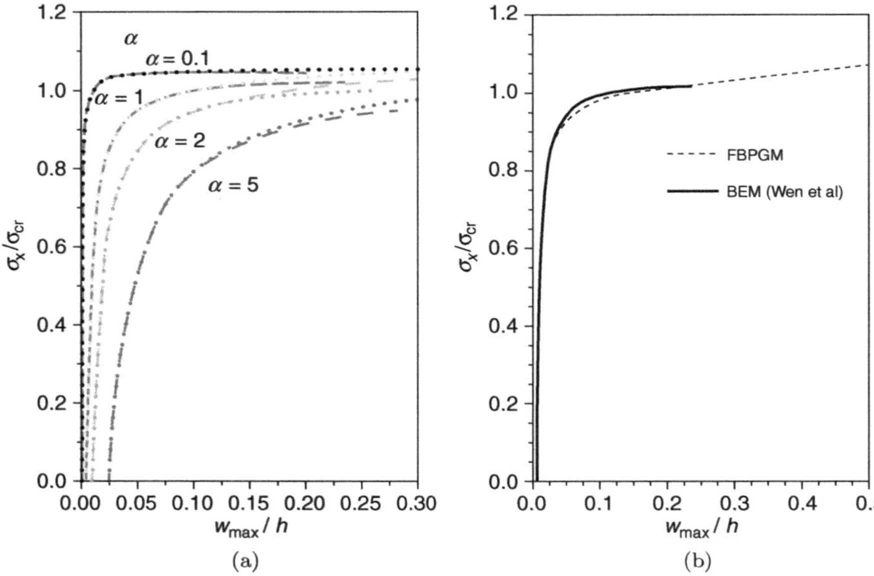

Fig. 30. Deformation of plate under axial load/displacement controls: (a) axial load (solid lines) or displacement control (dashed lines) with different imperfection parameters α and (b) comparsion with BEW[33] when $\alpha = 1$.

and the average relative error is less than 2%, where the increment load scheme is applied for both methods.

11. Summary

The FBM was developed to two dimensional nonlinear problems including frictional contact of functionally graded material and post-buckling of Reissner plate. The iterative procedure for non-conforming frictionless contact and load increment algorism were adopted to solve nonlinear equations in strong form. The post-buckling analysis was fulfilled by both load increment and displacement control algorithms (in weak form). The essential features of the FBM in this chapter can be summarized as:

(1) The physical domain is divided into few blocks with 8-node quadratic shape function;
(2) First-order partial differential matrices $\mathbf{D}_x$ and $\mathbf{D}_y$ are easily obtained by the Lagrange series interpolation;

(3) For irregular domain, the mapping technique is applied for each block and the traction equations can be used to determine the boundary displacements using the FBM with boundary type;
(4) Nonlinear problems with functionally graded materials aniso-tropic media or all kind of boundary value problems of partial differential equations with variable coefficients can be analyzed easily;
(5) Contact with static and dynamic loads and post-buckling of Reissner plate can be studied directly with load increment technique;
(6) FBM is a flexible, simple and suitable method for certain complicated nonlinear partial differential equations such as functionally graded materials and Winkler elastic foundation, etc.

References

1. S. Suresh and A. Mortensen, *Fundamentals of Functionally Graded Materials*, Institute of Materials, London (1998).
2. Y. Miyamoto, W. A. Kaysser, B. H. Rabin, A. Kawasaki, and R. G. Ford, *Functionally Graded Materials; Design, Processing and Applications*, Kluwer Academic Publishers, Dordrecht (1999).
3. J. H. Kim and G. H. Paulino, Isoparametric graded finite elements for nonhomogeneous isotropic and orthotropic materials, *Journal of Applied Mechanics ASME*, **69**, 502–514 (2002).
4. M. H. Aliabadi, *The Boundary Element, Applications in Solids and Structures*, John Wiley and Sons, Chichester, England (2002).
5. S. N. Atluri, *The Meshless Method (MLPG) for Domain & BIE Discretizations*, Tech Science Press (2004).
6. X. Ling and T. Belytschko, Thermal softening induced plastic instability in rate-dependent materials, *Journal of the Mechanics and Physics of Solids*, **57**(4), 788–802 (2009).
7. J. P. Ponthot and T. Belytschko, Arbitrary Lagrangian–Eulerian formulation for element–free Galerkin method, *Computer Methods in Applied Mechanics and Engineering*, **152**, 19–46 (1998).
8. P. Lancaster and K. Salkauskas, Surfaces generated by moving least square methods, *Mathematical Computation*, **37**, 141–158 (1981).
9. B. Nayroles, G. Touzot, and P. Villon, Generalizing the finite element method: Diffuse approximation and diffuse elements, *Computational Mechanics*, **10**, 307–318 (1991).
10. T. Belytschko, Y. Y. Lu, and L. Gu, Element-free Galerkin method, *International Journal for Numerical Methods in Engineering*, **37**, 229–256 (1994).
11. M. A. Golgerg, C. S. Chen, and S. R. Karur, Improved multiquadric approximation for partial differential equations, *Enginering Analysis with Boundary Elements*, **18**, 9–17 (1996).
12. L. Marinand and D. Lesnic, The method of fundamental solutions for nonlinear functionally graded materials, *International Journal of Solids and Structure*, **44**, 6878–6890 (2007).

13. E. J. Kanse, A scattered data approximation scheme with applications to computational fluid-dynamics I Surface approximations and partial derivative estimates, *Computer & Mathematics with Applications*, **19**(8–9), 127–145 (1991).
14. E. J. Kanse, A scattered data approximation scheme with applications to computational fluid-dynamics II Surface Solutions to parabolic, hyperbolic and elliptic partial differential equations, *Computer & Mathematics with Applications*, **19**(8–9), 147–161 (1991).
15. Y. C. Hon and X. Z. Mao, A multiquadric interpolation method for solving initial value problems, *Journal of Scientific Computing*, **12**, 51–55 (1997).
16. J. Li, A. H. D. Cheng, and C. S. Chen, A comparing of efficiency and error convergence of multiquadric collocation method and finite element method, *Engineering Analysis with Boundary Elements*, **27**, 251–257 (2003).
17. M. Fleming, Y. A. Chu, B. Moran, T. Belytschko, Y. Y. Lu, and L. Gu, Enriched element-free Galerkin methods for crack-tip fields, *International Journal for Numerical Methods in Engineering*, **40**, 1483–1504 (1997).
18. J. Sladek, V. Sladekand, and Ch. Zhang, Transient heat conduction analysis in functionally graded materials by the meshless local boundary integral equation method, *Computer Material Science*, **28**, 494–504 (2003).
19. J. Sladek, V. Sladek, and S. N. Atluri, Meshless local Petrov-Galerkin method for heat conduction problem in an anisotropic medium, *CMES — Computer Modeling in Engineering & Sciences*, **6**, 309–318 (2004).
20. V. Sladek, J. Sladek, and Ch. Zhang, Local integro-differential equations with domain elements for the numerical solution of partial differential equations with variable coefficients, *Journal of Engineering Mathematics*, **51**, 261–282 (2005).
21. J. Sladek, V. Sladek, and Ch. Zhang, Transient heat conduction analysis in functionally graded materials by the meshless local boundary integral equation method, *Computer Material Science*, **28**, 494–504 (2003).
22. J. Sladek, V. Sladek, and S. N. Atluri, Meshless local Petrov-Galerkin method for heat conduction problem in an anisotropic medium, *CMES — Computer Modeling in Engineering & Sciences*, **6**, 309–318 (2004).
23. P. H. Wen and M. H. Aliabadi. An improved meshless collocation method for elastostatic and elasto-dynamic problems, *Journal Communications in Numerical Methods in Engineering*, **24**(8), 635–651 (2008).
24. P. H. Wen, P. Cao, and T. Korakianitis, Finite block method in elasticity, *Engineering Analysis with Boundary Elements*, **46**, 116–125 (2014).
25. M. Li and P. H. Wen, Finite block method for transient heat conduction analysis in functionally graded media, *International Journal for Numerical Methods in Engineering*, **99**, 372–390 (2014).
26. K. Man, M. H. Aliabadi, and D. P. Rooke, BEM frictional contact analysis: An incremental loading technique, *Computers and Structures*, **47**, 893–905 (1993).
27. K. Man, M. H. Aliabadi, and D. P. Rooke, BEM frictional contact analysis: Modelling considerations, *Engineering Analysis with Boundary Elements*, **11**, 77–85 (1993).

28. K. Man, *Contact Mechanics using Boundary Elements*, Topics in Engineering, Computational Mechanics Publications, Southampton (1994).
29. F. Durbin, Numerical inversion of Laplace transforms: An efficient improvement to Dubner and Abate's method, *The Computer Journal*, **17**, 371–376 (1975).
30. P. H. Wen, M. H. Aliabadi, and A. Young, A boundary element method for dynamic plate bending problems, *Journal of Solids and Structure*, **37**, 5177–5188 (1999).
31. P. H. Wen and Y. C. Hon, Geometrically nonlinear analysis of Reissner-Mindlin plate by meshless computation, *CMES: Computer Modeling in Engineering & Sciences*, **21**(3), 177–191 (2007).
32. R. L. Hardy, Multiquadric equations of topography and other irregular surface, *Journal of Geophysical Research*, **76**, 1905–1915 (1971).
33. P. H. Wen and M. H. Aliabadi, Large deformation analysis of Reissner plate by boundary element method, *Computers and Structure*, **83**, 870–879 (2005).

Chapter 7

A Coupled Finite Element Material Point Method for Large Deformation Problems

Y. Lian and X. Zhang

In the dynamic analysis of mechanics, some of the large deformation problems such as contact and impact problems involve not only extreme deformations with material damage and separation but also finite and small deformations without material damage and separation. Consequently, it brings challenges to traditional finite element method in terms of element distortion and mesh entanglement due to extreme deformation and to advanced mesh-free methods in terms of low computational efficiency, especially when the size of finite/small deformation region is considerable. This chapter gives a state-of-the-art review of the coupled finite element material point (CFEMP) method, which takes both advantages of traditional finite element method (FEM) and a type of mesh-free methods, material point method (MPM), and avoids their disadvantages with the aim to overcome the challenges, and examples of its applications to the large deformation problems are also presented. The CFEMP method includes two variants, static CEMP based on a contact method and adaptive CFEMP further based on an element-to-particle conversion algorithm and alternative coupling method. In addition, MPM is also introduced in detail for completeness.

1. Introduction

In this chapter, a state-of-the-art review of the coupled finite element material point (CFEMP) method[1–4] is presented and examples of its applications to mechanics problems involving large deformation (with material damage and separation) are given. CFEMP consists of finite element method (FEM) and material point method (MPM) and is developed as a numerical method aimed at large deformation problems.

The large deformation problems such as contact, impact, explosion and slope failure problems bring many challenges for existing numerical methods in terms of accuracy, efficiency, stability and so on. These challenges are associated with the form of the governing equations used to model the problems. If the Eulerian description is used, then the challenges are specifically embodied instability associated with the advection term, interface capturing or tracking, and tracking of internal history variables. If the Lagrangian description is used, the challenges further depend on the specified numerical methods. For example, FEM is limited by the element distortion and mesh entanglement, and arbitrary Lagrangian–Eulerian (ALE) methods were developed to alleviate the bottleneck of numerical methods based on both descriptions but with new issues associated with mesh moving algorithm for 3D problems. Till the 1990s, the computational community focused on and developed advanced mesh-free methods, such as smoothed hydrodynamics particle (SPH) method,[5] reproducing kernel particle (RKPM) method,[6,7] element-free Galerkin (EFG) method,[8,9] just name a few, which shed light on these challenges. The basic idea of all the mesh-free methods is to discretize material domain by a set of particles instead of mesh and then it avoids the element distortion and mesh entanglement that occurs to mesh-based numerical methods. However, the mesh-free methods suffer from their inherent shortcomings such as high computational cost associated with shape functions, complexity to impose essential boundary conditions, integration problems if it is based on weak form, and stability issue if it is based on strong form. As a result, much effort has been devoted to couple these methods with FEM[10–13] to take advantages of both mesh-free and FEM. For the time being, the coupling algorithms include master–slave coupling, coupling via mixed interpolation, coupling via Lagrange multipliers, coupling via a bridge domain, and so on. Contributions to this topic have been reviewed in Ref. [14].

Among mesh-free methods, MPM[15] is a special one in which the material domain is discretized by a set of body-fixed (Lagrangian) particles, or material points, which are covered by a background grid. The particles carry all the state variables, and the background grid is used to update variables carried by particles and is convenient for the calculation of integration. It should be noted that MPM applies a given shape function associated with the grid cells instead of constructing them from a set of discrete particles. As a consequence, MPM shows higher efficiency over other meshless methods.[16] Moreover, it is easy to impose essential boundary conditions if the linear shape functions of FEM are used. For the time being, significant effort has been devoted to the development of MPM in terms of contact methods,[17–20] explicit/implicit time integration,[20–23] background grid types[24,25] and shape functions,[26,27] coupled with other numerical methods[1,28–30] and so on. Based on these contributions, MPM has found applications in the solution of a wide variety of problems including explosion,[21,31–33] impact and penetration,[18,19] geomechanics,[34,35] cracking expansion,[27,36,37] fluid–structure interaction problems,[1,4,38] multiphase flows,[39] incompressible flows,[30] just to name a few. In Section 2, MPM is introduced in detail including a derivation of the discrete equations, time integration, algorithms for particles' stress update, shape functions and numerical implementations.

From Section 2 it can be found that in each time step MPM is similar to FEM but with two fundamental differences. The first one is that the particles are taken as quadrature points instead of Gaussian points used in FEM; the other one is that the background grid is only temporarily embedded in and deforms with the material domain in each time step whereas the mesh is permanently embedded in the material domain in FEM. The two differences provide MPM with advantage in terms of no mesh entanglement but also cause the disadvantage in terms of expensive computation cost for the material region with finite deformation. Therefore CFEMP method was developed to take full advantages of both MPM and FEM with the aim to model extreme deformation problems where the size of finite deformation region is considerable. There are two variants, static CFEMP method and adaptive CFEMP method, which are introduced in Sections 3 and 4, respectively. In the first variant, an individual body that involves extreme deformation is initially discretized by particles solving by MPM, while the individual body only with finite deformation is initially discretized by mesh solving by FEM. During the simulation process, the discretization to each body remains the same, and the contact between two bodies with different discretizations is handled by a contact method through the background grid. It is clear that such a coupling method does not fully use FEM to the whole material region. To this end, the second variant was developed based on the first one. In the adaptive CFEMP method, bodies are allowed to be initially discretized only by mesh solving by FEM, and during the simulation process part of them can be adaptively or automatically switched to particle discretization solving by MPM if the quality of the element in terms of accuracy is not acceptable. Such a switch is fulfiled by an element-to-particle conversion algorithm and a coupling method for the particles and the rest part of the elements. As a result, the material region with finite deformation is only modelled by FEM, while the material region with extreme deformation is modelled by MPM after the element-to-particle conversion and then the adaptive CFEMP method could take full advantages of both numerical methods.

2. Material point method

In this section, the details of MPM are introduced in terms of derivation of the discrete equations, time integration, particles' stress update algorithm, shape functions, and numerical implementations.

2.1. *Governing equations for solid mechanics of interest*

For a given material domain Ω as shown in Fig. 1(a), the governing equations in updated Lagrangian formulation consist of mass conservation:

$$\rho(\boldsymbol{X},t)J(\boldsymbol{X},t) = \rho_0(\boldsymbol{X}), \tag{1}$$

momentum conservation:

$$\sigma_{ij,j} + \rho f_i = \rho \ddot{u}_i, \tag{2}$$

boundary conditions:

$$\begin{cases} (n_j\sigma_{ij})|_{\Gamma_t} = \bar{t}_i \\ u_i|_{\Gamma_v} = \bar{u}_i \end{cases} \tag{3}$$

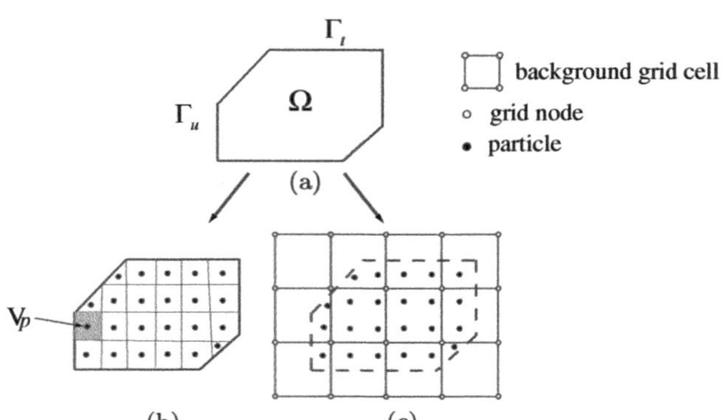

Fig. 1. (a) A material domain. (b) Material points discretization and (c) Material points covered by a uniform background grid.

and initial conditions:

$$\begin{cases} \dot{u}(\boldsymbol{X},0) = \dot{u}_0(\boldsymbol{X}) \\ u(\boldsymbol{X},0) = u_0(\boldsymbol{X}) \end{cases}, \tag{4}$$

where ρ is the current density, J is the Jacobian determinant, $\boldsymbol{X}$ is the Lagrangian coordinate of a material point. Subscripts i and j denote the component of the space with Einstein summation convention, subscript 0 signifies the initial value, the comma denotes differentiation and the superimposed dot indicates the time derivatives. σ_{ij} is the Cauchy stress, f_i is the body force per unit mass, u_i is the displacement, and n_j is the unit outward normal to the boundary. Γ_t and Γ_u stand for the prescribed traction boundary and displacement boundary of Ω, respectively, overbar and subscript 0 signify that the functions are given.

2.2. *Space discretization*

The derivation of the MPM can be cast in variation, or weak form[40] as follows. Taking the virtual displacement δu_i as test functions, the weak form of the momentum equation reads

$$\int_\Omega \rho \ddot{u}_i \delta u_i \mathrm{d}\Omega + \int_\Omega \sigma_{ij} \delta u_{i,j} \mathrm{d}\Omega - \int_\Omega \rho f_i \delta u_i \mathrm{d}\Omega - \int_{\Gamma_t} \bar{t}_i \delta u_i \mathrm{d}\Gamma = 0. \tag{5}$$

For convenience of derivation, the above Eq. (5) may be rewritten as:

$$\int_\Omega \rho \ddot{u}_i \delta u_i \mathrm{d}\Omega + \int_\Omega \rho \sigma^s_{ij} \delta u_{i,j} \mathrm{d}\Omega - \int_\Omega \rho f_i \delta u_i \mathrm{d}\Omega - \int_{\Gamma_t} \rho \bar{t}^s_i \delta u_i \mathrm{d}\Gamma = 0 \tag{6}$$

where $\sigma^s_{ij} = \sigma_{ij}/\rho$ is referred to as specific stress, and $\bar{t}^s_i = \bar{t}_i/\rho$ is specific boundary traction force.

The essence of the discretization procedure is to represent the material domain by a set of material points, or particles as shown in Fig. 1(b). The terms material point and particle will be used interchangeably throughout this chapter. Each particle carries all the information associated with the material subdomain, V_p centred on it. Therefore, the material density can be approximated as

$$\rho(x) = \sum_{p=1}^{n_p} m_p \delta(x_i - x_{ip}) \tag{7}$$

where n_p stands for the total number of particles, m_p is the mass of particle p, x_{ip} is the coordinate of particle p in ith direction, δ is Dirac delta function.

Substituting Eq. (7) into the weak form Eq. (6) yields:

$$\sum_{p=1}^{n_p} m_p \ddot{u}_{ip}\delta u_{ip} + \sum_{p=1}^{n_p} m_p \sigma_{ijp}^s \delta u_{ip,j} - \sum_{p=1}^{n_p} m_p f_{ip}\delta u_{ip} - \sum_{p=1}^{n_p} m_p \bar{t}_{ip}^s h^{-1}\delta u_{ip} = 0, \tag{8}$$

where $u_{ip} = u_i(x_p)$, $\delta u_{ip,j} = \delta u_{i,j}(x_p)$, $\sigma_{ijp}^s = \sigma_{ij}^s(x_p)$, $f_{ip} = f_i(x_p)$, $\bar{t}_{ip}^s = \bar{t}_i^s(x_p)$ and h denotes the thickness of the artificial boundary layer introduced only for the convenience of derivation.

In order to update the particle's variable instead of solving equations on particle itself, the background grid is introduced to cover all the particles as shown in Fig. 1(c), on which velocity field and momentum equations represented by particles are reproduced in each time step. To solve Eq. (8), there are two major stages.

In the first stage, let the background grid be embedded in the material domain represented by the particles. Then the material information can be mapped to the background grid, e.g. mass and momentum, to initialize the nodal value of the background grid. Next, the momentum equations are reproduced and solved on the background grid nodes. In other words, it seems like the particles are rigidly attached to the background grid and they deform with the grid. In order to introduce background grid to Eq. (8), let the displacements of particles and their derivatives be expressed as the function of the grid nodal value through given shape functions, N_I as

$$u_{ip} = \sum_{I=1}^{n_g} N_{Ip} u_{iI}, \tag{9}$$

$$u_{ip,j} = \sum_{I=1}^{n_g} N_{Ip,j} u_{iI}, \tag{10}$$

where subscript I denotes the value of node I, $N_{Ip} = N_I(x_p)$ is the value of shape function of grid node I evaluated at particle p, and n_g represents the total number of grid nodes. It should be noted that both Eqs. (9) and (10) also exist to δu_i.

Substituting Eqs. (9) and (10) into Eq. (8) and invoking the arbitrary property of δu_{iI} yields:

$$\dot{p}_{iI} = f_{iI}^{\text{int}} + f_{iI}^{\text{ext}} \quad I = 1, 2, \ldots, n_g, \tag{11}$$

which are the discrete equations to the momentum equation Eq. (2). Here p_{iI} denotes the grid nodal momentum and defined as

$$p_{iI} = m_I v_{iI}, \tag{12}$$

where $v_{iI} = \dot{u}_{iI}$ stands for nodal velocity, and a lumped mass matrix is used indicating by nodal mass m_I which is calculated by

$$m_I = \sum_{p=1}^{n_p} m_p N_{Ip}. \tag{13}$$

In Eq. (11), f_{iI}^{int} represents the nodal internal force and is calculated by

$$f_{iI}^{\text{int}} = -\sum_{p=1}^{n_p} N_{Ip,j}\sigma_{ijp}\frac{m_p}{\rho_p}, \tag{14}$$

f_{iI}^{ext} is the nodal external force and consists of two contributions as follows.

$$f_{iI}^{\text{ext}} = \sum_{p=1}^{n_p} N_{Ip}m_p f_{ip} + \sum_{p=1}^{n_p} N_{Ip}\bar{t}_{ip}h^{-1}\frac{m_p}{\rho_p}. \tag{15}$$

It is clear that the momentum equations are reproduced on the background grid where the particles serve as integration points. The first stage is finished by solving momentum equations Eq. (11), which is detailed in Section 2.3. In the second stage, the results of momentum equations, i.e. updated velocity and acceleration, are mapped back to the particles to update their positions and velocities, respectively, which are also detailed in Section 2.3.

2.3. *Time discretization*

For the problem of interest in this chapter, the explicit time integration method, e.g. central difference time integration algorithm, is used to solve the Eq. (11).

As shown in Fig. 2, the notations used here are defined as $t^{k+1} = t^k + \Delta t^{k+1/2}$, $t^{k+1/2} = t^k + \Delta t^{k+1/2}/2 = t^{k-1/2} + \Delta t^k$ and $\Delta t^k = (\Delta t^{k-1/2} + \Delta t^{k+1/2})/2$. In addition, a superscript k denotes the value of a variable at the time t^k.

Given u_{ip}^k and $v_{ip}^{k-1/2}$, from Eq. (11), the nodal momentum at time $t^{k+1/2}$ can be updated by

$$p_{iI}^{k+1/2} = p_{iI}^{k-1/2} + f_{iI}^k \Delta t^k, \tag{16}$$

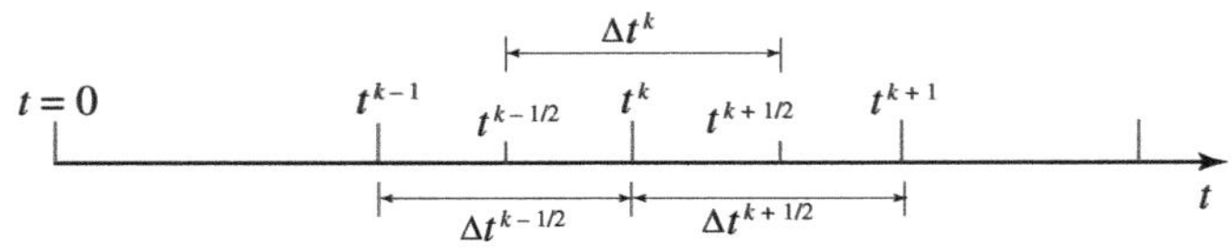

Fig. 2. Illustration of time step size used in the time integration algorithm.

where f_{iI}^k is the nodal force given in Eqs. (14) and (15), and the nodal velocity $p_{iI}^{k-1/2}$ are mapped from $v_{ip}^{k-1/2}$ via

$$p_{iI} = \sum_{p=1}^{n_p} m_p v_{ip} N_{Ip}. \tag{17}$$

The second stage is to update the variables carried by particles, e.g. position and velocity. Using the results of Eq. (16), the velocities and the positions of particles at time $t^{k+1/2}$ and t^{k+1} are updated, respectively, as follows.

$$v_{ip}^{k+1/2} = v_{ip}^{k-1/2} + \Delta t^k \sum_{I=1}^{n_g} f_{iI}^k N_{Ip}^k / m_I^k \tag{18}$$

and

$$x_{ip}^{k+1} = x_{ip}^k + \Delta t^{k+1/2} \sum_{I=1}^{n_g} p_{iI}^{k+1/2} N_{Ip}^k / m_I^k. \tag{19}$$

Central difference time integration algorithm is a type of conditionally stable algorithm, and then the time step should satisfy CFL condition. In MPM, the time step is determined by

$$\Delta t = \alpha \frac{l}{\max\limits_p (c_p + |\mathbf{v}_p|)}, \tag{20}$$

where α is the time step scale factor, l represents the characteristic length of the cell and usually takes the value of background grid cell size Δx, and c_p and $\mathbf{v}_p$ are particle wave speed and velocity, respectively.

Once the second stage is finished, the background grid is detached with the particles via setting the nodal variables of background grid as 0. This indicates that the background grid does not carry any permanent material information, and just serves as the calculations of the momentum equation and particles strain as shown in Section 2.4. Therefore, it avoids mesh entanglement because the deformation of the cell in each time step is finite due to the small time step and has the potential to model extreme deformation problems, where the traditional FEM often encounters element distortion which leads to numerical difficulties.

2.4. *Particle stress update*

Under the framework of explicit time integration outlined in Section 2.3, the stress at time $t^{k+1/2}$ may be updated by

$$\sigma_{ij}^{k+1} = \sigma_{ij}^{k} + \dot{\sigma}_{ij}^{k-1/2}\Delta t^{k+1/2}, \tag{21}$$

where $\dot{\sigma}_{ij}$ is the material time derivative of the stress. $\dot{\sigma}_{ij}$ is determined by

$$\dot{\sigma}_{ij} = \sigma_{ij}^{\nabla} + \sigma_{il}\Omega_{lj} + \sigma_{jl}\Omega_{li}, \tag{22}$$

where σ_{ij}^{∇} is the Jaumann (co-rotational) stress rate, and Ω_{ij} is the spin tensor. σ_{ij}^{∇} is determined from the strain rate $\dot{\varepsilon}_{ij}$ by a constitutive model. Usually, the stress is decomposed into the deviatoric part and the pressure, which are updated by a constitutive law and an equation of state (EOS), respectively.

In MPM, the particles serve as integration points and then the stress is updated at each particle based on the strain rate and spin tensor of the same particle. In order to take account of the influence from others particles, the particle's strain rate and spin tensor are calculated through the grid nodal velocity as follows.

The strain rate is by

$$\dot{\varepsilon}_{ijp} = \frac{1}{2}\sum_{I=1}^{8}(N_{Ip,j}v_{iI} + N_{Ip,i}v_{jI}), \tag{23}$$

and spin tensor by

$$\Omega_{ijp} = \frac{1}{2}\sum_{I=1}^{8}(N_{Ip,j}v_{iI} - N_{Ip,i}v_{jI}). \tag{24}$$

Considering the source of grid nodal velocity, there are three options to update the particle stress based on the order of stress calculation and integration of momentum equation.

(1) The first option is to update the particle stress at the beginning of the time step, where the following nodal velocity, $v_{iI}^{k-1/2}$ is used.

$$v_{iI}^{k-1/2} = p_{iI}^{k-1/2}/m_I^n = \sum_{p=1}^{n_p} m_p v_{ip}^{k-1/2} N_{Ip}^k / m_I^k. \tag{25}$$

This option is referred to as the update-stress-first (USF) algorithm.

(2) The second option is to update the particle stress at the end of the time step, where the grid nodal velocity, $v_{iI}^{k+1/2} = p_{iI}^{k+1/2}/m_I^k$ is used. Consequently, this option is referred to as the update-stress-last (USL) algorithm.

(3) The third option is to update the particle stress at the end of the time step but from the updated particle velocities interpolated to the grid, namely, the nodal velocity is obtained as

$$v_{iI}^{k+1/2} = \sum_{p=1}^{n_p} p_{ip}^{k+1/2} N_{Ip}^k / m_I^k. \tag{26}$$

Therefore, this option is referred to as the modified-update-stress-last (MUSL) algorithm.

As shown in the work by Bardenhagen[41] and Nairn,[36] USL may cause instability issues and induce more dissipation, while USF and MUSL conserve energy better.

In addition, artificial bulk viscosity q[42] is applied to treat shock waves, which is defined as:

$$q = \begin{cases} c_0 \rho l^2 (\dot{\varepsilon}_{kk})^2 - c_1 \rho l c \dot{\varepsilon}_{kk} & \text{if } \dot{\varepsilon}_{kk} < 0, \\ 0 & \text{if } \dot{\varepsilon}_{kk} \geq 0, \end{cases} \tag{27}$$

where c_0 and c_1 are dimensionless constants, c is the local sound speed, $\dot{\varepsilon}_{kk}$ is the trace of the strain rate tensor. Similar to the constitutive response, q is evaluated at the particles. Then the particle stress is augmented by

$$\sigma_{ijp}^{k+1} = \sigma_{ijp}^{k+1} + q_p \delta_{ij}, \tag{28}$$

where δ_{ij} the Kronecker delta function.

2.5. *Shape functions*

As shown in both Sections 2.2 and 2.3, the shape function plays an important role for the interpolation between background grid and particles.

A general form of shape function for MPM is defined as

$$S_{Ip} = \frac{1}{V_p} \int_{\Omega_p \cap \Omega} \Psi_p(\boldsymbol{x}) N_I(\boldsymbol{x}) d\Omega, \tag{29}$$

where $N_I(\boldsymbol{x})$ stands for a linear shape function used in FEM, and $\Psi_p(\boldsymbol{x})$ is referred to as a characteristic function of the particle proposed in generalized interpolation material point (GIMP) method by Bardenhagen *et al.*[26] It defines the space occupied by the particle itself and can be thought of as the spatially varying volume fraction of that particle. Moreover $\Psi_p(\boldsymbol{x})$ satisfies partition of unity as follows.

$$\sum_{p}^{n_p} \Psi_p(\boldsymbol{x}) = 1. \tag{30}$$

It is clear that the shape function is dependent on $\Psi_p(\boldsymbol{x})$. Here, two types of $\Psi_p(\boldsymbol{x})$ are introduced.

In the first type, $\Psi_p(\boldsymbol{x}) = \delta(\boldsymbol{x} - \boldsymbol{x}_p)V_p$ which leads to $S_{Ip} = N_{Ip}$. Therefore, the influence domain of the particle is limited in the the cell that it is located in without considering the space occupied by the material associated with the particle as shown in Fig. 3(a). For example, the shape function of 8-node hexahedral element could be used for 3D problem:

$$N_I = \frac{1}{8}(1 + \xi\xi_I)(1 + \eta\eta_I)(1 + \zeta\zeta_I) \quad I = 1, 2, \ldots, 8, \tag{31}$$

where (ξ_I, η_I, ζ_I), either -1 or $+1$, denote the natural coordinates of node I. (ξ, η, ζ), which stand for the natural coordinates of a given point, i.e. the particle. Such a shape function satisfies Kronecker delta property and owns high computational efficiency. However, due to the discontinuity of its first-order derivative, the so-called cell crossing noise exists and will cause

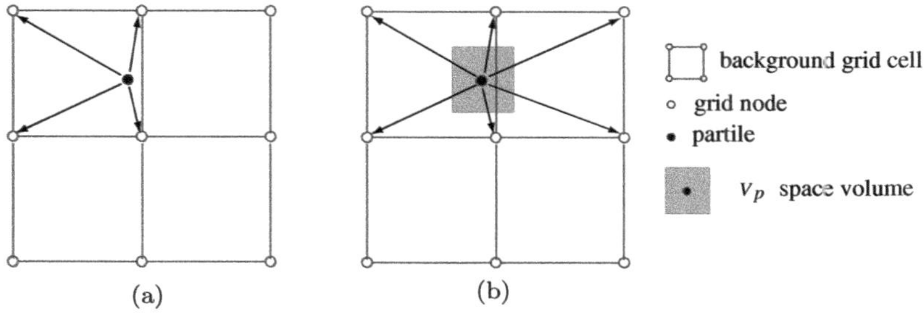

Fig. 3. Influence domain of particle: (a) Limited to the cell it's located in and (b) Beyond the cell it's located in.

stress oscillation. One of the methods to relieve the noise is to put more particles in each cell.

In the other type, the shape function is based on not only linear interpolation function but also the domain size occupied by material represented by the particle. As shown in Fig. 3(b), the influence domain of the particle is extended to all the grid cells that have an overlap with the space occupied by V_p. Take 1D case as an example, let

$$\Psi_p(x) = \begin{cases} 1 & x \in \Omega_p, \\ 0 & x \notin \Omega_p, \end{cases} \tag{32}$$

$$N_I(x) = \begin{cases} 0 & x - x_I \leq -L, \\ 1 + (x - x_I)/L & -L < x - x_I \leq 0, \\ 1 - (x - x_I)/L & 0 < x - x_I \leq L, \\ 0 & L < x - x_I, \end{cases} \tag{33}$$

where L stands for the cell size of the background grid.

Substitute Eqs. (33) and (32) to Eq. (29) to yield

$$S_{Ip} = \begin{cases} 0 & x_p - x_I \leq -L - l_p, \\ \frac{(L+l_p+(x_p-x_I)^2)}{4Ll_p} & -L - l_p < x_p - x_I \leq -L + l_p, \\ 1 + \frac{x_p - x_I}{L} & -L + l_p < x_p - x_I \leq -l_p, \\ 1 - \frac{(x_p-x_I)^2+l_p^2}{2Ll_p} & -l_p < x_p - x_I \leq l_p, \\ 1 - \frac{x_p - x_I}{L} & l_p < x_p - x_I \leq L - l_p, \\ \frac{(L+l_p(x_p-x_I)^2)}{4Ll_p} & L - l_p < x_p - x_I \leq L + l_p, \\ 0 & L + l_p < x_p - x_I, \end{cases} \tag{34}$$

where l_p represents a half length of the characteristic length of particle p. One example of such a shape function is shown in Fig. 4 based on the assumptions that there are two particles in each cell and the size of the space occupied by the particle is treated as constant.

It is straightforward to extend this to 3D case as follows.

$$S_{Ip}(x) = S_{Ip}(\xi) S_{Ip}(\eta) S_{Ip}(\zeta). \tag{35}$$

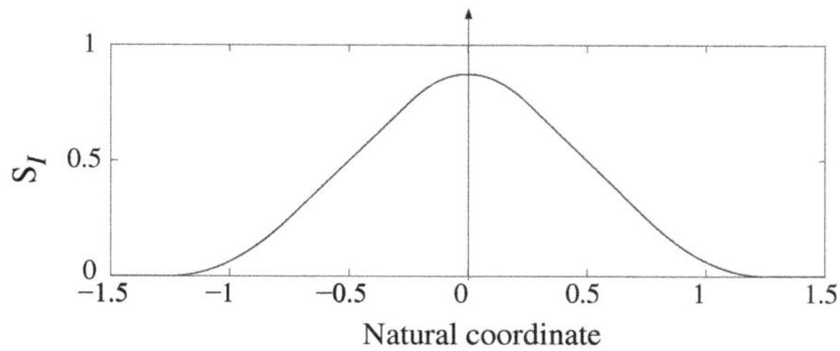

Fig. 4. A shape function in C^1.

Such a shape function in C^1 is better to alleviate the cell crossing noise but with higher computational cost. Moreover, it does not satisfy the Kronecker delta property.

2.6. *Numerical implementations*

The detailed implementations of MPM based on USF and MUSL algorithms are presented here for a time step k.

(1) Redefine the background grid and calculate grid nodal mass and momentum as follows

$$m_I^k = \sum_{p=1}^{n_p} m_p N_{Ip}^k, \tag{36}$$

$$p_{iI}^{k-1/2} = \sum_{p=1}^{n_p} m_p v_{ip}^{k-1/2} N_{Ip}^k. \tag{37}$$

(2) Impose boundary conditions.
According to the specified boundary conditions, set the corresponding component of nodal momentum of grid nodes that are located along the boundary of material as 0. Taking fixed boundary conditions as an example, let $p_{iI}^{k-1/2} = 0$.
(3) If USF algorithm is used, then update particles' stress.
(4) Calculate grid nodal internal force.
(5) Integrate grid nodal momentum equation and impose boundary conditions again.
(6) Update particles' positions and velocities.
(7) If MUSL algorithm is used, then update particles' stress.

(a) Recalculate grid nodal velocities based on the updated particles' velocities as follows.

$$v_{iI}^{k+1/2} = \sum_{p=1}^{n_p} p_{ip}^{k+1/2} N_{Ip}^k / m_I^k. \tag{38}$$

(b) Calculate particles' strain rate and spin tensor based on $v_{iI}^{k+1/2}$ as follows

$$D_{ijp}^{k+1/2} = \frac{1}{2}\sum_{I=1}^{n_g}(N_{Ip,j}^k v_{iI}^{k+1/2} + N_{Ip,i}^k v_{jI}^{k+1/2}), \tag{39}$$

$$\Omega_{ijp}^{n+1/2} = \frac{1}{2}\sum_{I=1}^{n_g}(N_{Ip,j}^k v_{iI}^{k+1/2} - N_{Ip,i}^k v_{jI}^{k+1/2}). \tag{40}$$

(c) Use specified constitutive law and EOS to update particles' deviatoric stress and pressure, respectively, based on $D_{ijp}^{k+1/2}$ and $\Omega_{ijp}^{k+1/2}$.

(8) Discard the deformed background grid. Return to step 1 for the next time step.

3. Static-coupled finite element material point method

As outlined in Section 2, MPM has the potential to model extreme deformation problems with material damage and separation, where the traditional FEM often encounters mesh distortion and element entanglement which lead to numerical difficulties, but it is with less accuracy and efficiency than FEM for problems with small deformation. Therefore, it is desirable to model the continuum body with extreme deformation by MPM and the continuum body with small deformation by FEM, respectively.

In this section, the first variant of the CFEMP method is introduced and is referred to as CFEMP for short in the following.

3.1. *Methodology*

As shown in Fig. 5(a), two continuum bodies labeled r and s, respectively, come into contact at the position indicated by γ_c. Assuming the material region Ω_s of body s will undergo extreme deformation, while the material region Ω_r of bodies r only with small deformation, bodies r and s are discretized by MPM and FEM, respectively, as shown in Fig. 5(b).

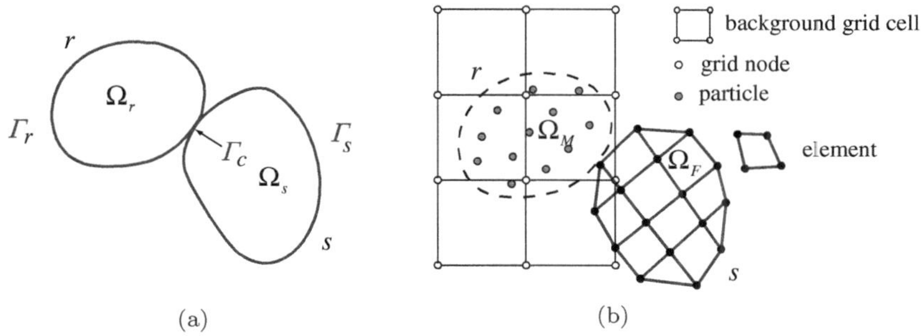

Fig. 5. (a) Two bodies coming into contact and (b) Discretization of two bodies with different methods.

In CFEMP, a contact algorithm is implemented on the background grid of MPM to handle the contact between the FEM body, Ω_{F} and MPM body, Ω_{M}. In each time step, the MPM body and FEM body are first updated independently to obtain their trial values of nodal variables, as if they were not in contact with each other. If the momentum of an MPM particle and a FEM node are projected to the same grid node, e.g. I, two bodies contact at the grid point. Then the contact force is calculated and imposed on them to prevent penetration. Therefore, the key points to implement such a coupled method are the time integration and the contact method, both of which are detailed in the following sections.

3.2. *Time integration for different discretization domain*

For both FEM and MPM domains, the central difference time integration algorithm as shown in Section 2.3 is used. In order to keep all operations synchronic in each time step, choose the minimum over all FEM and MPM domains as the global time step size.

For Ω_{M} domain: The momentum of grid node at time $t^{k+1/2}$ can be updated by

$$p_{iI}^{k+1/2} = p_{iI}^{k-1/2} + f_{iI}^{k}\Delta t^{k}, \tag{41}$$

where f_{iI}^{k} is the nodal force of grid node I at time t^{k}.

The velocities at time $t^{k+1/2}$ and the positions at t^{k+1} of particles are updated, respectively, by

$$v_{ip}^{k+1/2} = v_{ip}^{k-1/2} + \Delta t^{k} \sum_{I=1}^{n_g} f_{iI}^{k} N_{Ip}^{k}/m_I^{k} \tag{42}$$

and

$$x_{ip}^{k+1} = x_{ip}^{k} + \Delta t^{k+1/2} \sum_{I=1}^{n_g} p_{iI}^{k+1/2} N_{Ip}^{k} / m_I^k. \tag{43}$$

For Ω_F domain: The element nodes are decomposed into two types. The nodes located on the contact surface belong to the first type and are referred to as hybrid nodes. The remaining nodes belong to the second type.

The velocities of hybrid node at time $t^{k+1/2}$ are updated in a way similar to that of the particles as

$$v_{iK}^{k+1/2} = v_{iK}^{k-1/2} + \Delta t^k \sum_{I=1}^{n_g} f_{iI}^k N_{IK}^k / m_I^k + f_{iK}^{k,\Gamma} \Delta t^k / m_J, \tag{44}$$

where the subscript I denotes the grid node of background grid, f_{iI}^k is the nodal force of grid node I at time t^k, and f_{iK}^{Γ} is the hourglass resisting force if one Gaussian point is used in the element. For the second type node, the velocities at time $t^{k+1/2}$ are updated as

$$v_{iK}^{k+1/2} = v_{iK}^{k-1/2} + f_{iK}^k \Delta t^k / m_J. \tag{45}$$

The positions of all the element nodes at time t^{k+1} are updated by

$$x_{iK}^{k+1} = x_{iK}^{k} + v_{iK}^{k+1/2} \Delta t^{k+1/2}. \tag{46}$$

3.3. *Contact method*

A Lagrangian multiplier contact method is used to couple both FEM and MPM regions and implemented on the background grid. After the integration of momentum equations for both element nodes and background grid nodes, we first detect the contact points and then calculate contact forces, based on which the nodal momentum is readjusted to satisfy an impenetrability condition.

3.3.1. *Contact detection*

The contact detection can be done by checking the penetration of two velocity fields at the same point. To this end, map the mass and momentum of MPM body s and the surface element nodes of FEM body r to the same background grid. For a general case, the velocity of grid node I contributed from body b can be obtained as

$$v_{iI}^{b,k-1/2} = \frac{p_{iI}^{b,k-1/2}}{m_I^{b,k}}, \tag{47}$$

where the superscript b stands for the value mapped from body b, $m_I^{b,k}$ and $p_{iI}^{b,k-1/2}$ stand for the mass and the momentum of grid node I, respectively. It is clear that if the FEM body contacts with the MPM body in the vicinity of grid point I, then

$$(v_{iI}^{r,k-1/2} - v_{iI}^{s,k-1/2})n_{iI}^{r,k} > 0, \tag{48}$$

is satisfied. In Eq. (48), $n_{iI}^{r,k}$ represents the unit outward normal vector of body r at grid node I along the boundary.

In the FEM region, $n_{iI}^{r,k}$ may be calculated based on the outward normal vector of the element faces, while in the MPM domain, $n_{iI}^{s,k}$ may be calculated approximately by the gradient of the mass[43] as

$$n_{iI}^{s,k} = \frac{\sum_p N_{Ip,i}^k m_p}{\left|\sum_p N_{Ip,i}^k m_p\right|}. \tag{49}$$

The unit normal vectors n_{iI}^{rk} and $n_{iI}^{s,k}$ usually do not satisfy the following collinearity condition.

$$n_{iI}^{r,k} + n_{iI}^{s,k} = 0. \tag{50}$$

Therefore, they need to be modified based on several criteria to guarantee the collinearity condition. For details of these criteria, please refer to the Ref. [18].

Because in MPM the grid nodal momentum is directly calculated rather than the velocity, the contact detection condition Eq. (48) should be rewritten in terms of nodal momentum. Timing both side of Eq. (48) by $m_I^{r,k} m_I^{s,k}$ yields

$$(m_I^{s,k} p_{iI}^{r,k-1/2} - m_I^{r,k} p_{iI}^{s,k-1/2})n_{iI}^{r,k} > 0. \tag{51}$$

It should be noted that the contact grid nodes does not represent precisely the physical boundary of each body due to the mapping operation. That is to say, the contact detection condition is satisfied but the two bodies do not physically come into contact. Therefore the contact detection condition should be augmented by a physical distance detection.[19]

3.3.2. *Contact force*

For a given time step k, contact forces, $f_{iI}^{b,c,k}$ are calculated at the contact grid node I and used to prevent penetration. Correspondingly, the final nodal momentum p_{iI}^{b} of body b at the grid node I may be written as

$$p_{iI}^{b,k+1/2} = \bar{p}_{iI}^{b,k+1/2} + \Delta t^k f_{iI}^{b,c,k}, \tag{52}$$

where

$$\bar{p}_{iI}^{b,k+1/2} = p_{iI}^{b,k-1/2} + \Delta t^k f_{iI}^{b,k}, \tag{53}$$

stands for the trial value of the grid nodal momentum. $p_{iI}^{b,k-1/2}$ represents the grid nodal momentum at the beginning of each time step.

The final grid nodal momentum must satisfy the impenetrability condition

$$(m_I^{s,k} p_{iI}^{r,k+1/2} - m_I^{r,k} p_{iI}^{s,k+1/2}) n_{iI}^{r,k} = 0. \tag{54}$$

Then the normal component of contact force

$$f_I^{\text{nor},k} = f_{iI}^{r,c,k} n_{iI}^{r,k} = -f_{iI}^{s,c,k} n_{iI}^{r,k}, \tag{55}$$

can be derived by substituting Eq. (52) into the impenetrability condition as

$$f_I^{\text{nor},k} = \frac{(m_I^{s,k} \bar{p}_{iI}^{r,k+1/2} - m_I^{r,k} \bar{p}_{iI}^{s,k+1/2}) n_{iI}^{r,k}}{(m_I^{r,k} + m_I^{s,k}) \Delta t} = f_{I,1}^{\text{nor},k} + f_{I,2}^{\text{nor},k}, \tag{56}$$

where

$$f_{I,1}^{\text{nor},k} = \frac{(m_I^{s,k} p_{iI}^{r,k-1/2} - m_I^{r,k} p_{iI}^{s,k-1/2}) n_{iI}^{r,k}}{(m_I^{r,k} + m_I^{s,k}) \Delta t}, \tag{57}$$

$$f_{I,2}^{\text{nor},k} = \frac{(m_I^{s,k} f_{iI}^{r,k} - m_I^{r,k} f_{iI}^{s,k}) n_{iI}^{r,k}}{(m_I^{r,k} + m_I^{s,k})}. \tag{58}$$

Here, $f_{iI}^{s,k}$ denotes the nodal force of grid node I contributed from MPM body s, and $f_{iI}^{r,k}$ represents the contribution from FEM body r and is calculated as

$$f_{iI}^{r,k} = \sum_{J=1}^{n_h} N_{IK} f_{iK}^{r,k}, \tag{59}$$

where n_h is the total number of the hybrid nodes, and $f_{iJ}^{r,k}$ is the nodal force from the elements that connected to it.

For a case of stick contact, the tangential component of contact force

$$f_I^{\text{stick},k} = f_{iI}^{r,c,k} t_{iI}^{r,k} = -f_{iI}^{s,c,k} t_{iI}^{r,k}, \tag{60}$$

can be determined through the non-slip condition, namely

$$(m_I^{s,k} p_{iI}^{r,k+1/2} - m_I^{r,k} p_{iI}^{s,k+1/2}) t_{iI}^{r,k} = 0, \tag{61}$$

where $t_{iI}^{r,k}$ is the unit tangential vector at grid node I along the boundary, which may be obtained as

$$t_{iI}^{r,k} = \frac{(v_{iI}^{r,k-1/2} - v_{iI}^{s,k-1/2}) - (v_{jI}^{r,k-1/2} - v_{jI}^{s,k-1/2}) n_{jI}^{r,k} n_{iI}^{r,k}}{\left|(v_{iI}^{r,k-1/2} - v_{iI}^{s,k-1/2}) - (v_{jI}^{r,k-1/2} - v_{jI}^{s,k-1/2}) n_{jI}^{r,k} n_{iI}^{r,k}\right|}. \tag{62}$$

Then the tangential contact force can be derived by substituting Eq. (52) into Eq. (61) as

$$\begin{aligned} f_I^{\text{stick},k} &= \frac{(m_I^{s,k} \bar{p}_{iI}^{r,k+1/2} - m_I^{r,k} \bar{p}_{iI}^{s,k+1/2}) t_{iI}^{r,k}}{(m_I^{r,k} + m_I^{s,k}) \Delta t}, \\ &= \frac{(m_I^{s,k} p_{iI}^{r,k-1/2} - m_I^{r,k} p_{iI}^{s,k-1/2}) t_{iI}^{r,k}}{(m_I^{r,k} + m_I^{s,k}) \Delta t} \\ &\quad + \frac{(m_I^{s,k} f_{iI}^{r,k} - m_I^{r,k} f_{iI}^{s,k}) t_{iI}^{r,k}}{(m_I^{r,k} + m_I^{s,k})}. \end{aligned} \tag{63}$$

For a case of slip contact, Coulomb friction model is applied here to calculate the friction force as $\mu f_I^{\text{nor},k}$ where μ is a friction coefficient. Then the tangential contact force can be determined as

$$f_I^{\tan,k} = \min(\mu f_I^{\text{nor},k}, f_I^{\text{stick},k}). \tag{64}$$

Finally, the total contact force applied on body b at grid node I can be expressed as

$$f_{iI}^{b,c,k} = f_I^{\text{nor},k} n_{iI}^{b,k} + f_I^{\tan,k} t_{iI}^{b,k}. \tag{65}$$

3.3.3. *Mesh requirements*

As outlined in Sections 3.3.1 and 3.3.2, the contact detection and contact force calculation are based on background grid node, which bring two requirements to the mesh. One is that the background grid is required to cover the whole material domain. The other one is that the element size used to in Ω_{F} should match the cell size used in Ω_{M}, e.g., less than two times of the cell size, otherwise there will be penetration occurring. If the so-called point-to-surface contact method is used, then the second requirement can be ignored. Following this line, Chen *et al.*[44] improved the CFEMP method. However, it's better to still keep the match requirement of element size and cell size with the concern of the consistency of spatial discretization resolution level.

3.4. *Numerical implementation*

Because the new background grid is defined at the beginning of the next time step, the impenetrability condition Eq. (54) may not exist though it exists at the end of the last time step on the deformed background grid. As a consequence, the nodal velocity $v_{iI}^{b,k-1/2}$ used to update the stresses may violate the impenetrability condition Eq. (54), which may introduce a disturbance to the system,[18] if the USF algorithm is used.

In order to avoid the artificial disturbance, the grid nodal momentum $p_{iI}^{s,k-1/2}$ of contact grid node I needs to be adjusted to satisfy the impenetrability condition Eq. (54). In order to do this, the first part of the normal contact force, $f_{iI,1}^{\text{nor},k}$ in Eq. (56) is used here to get $\tilde{p}_{iI}^{s,k-1/2}$ as

$$\tilde{p}_{iI}^{s,k-1/2} = p_{iI}^{s,k-1/2} + \Delta t^k f_{iI,1}^{\text{nor},k}. \tag{66}$$

Correspondingly, the velocities of hybrid nodes in Ω_{F} are reset as

$$\tilde{v}_{iJ}^{r,k-1/2} = v_{iJ}^{r,k-1/2} - \Delta t^k \sum_{I=1}^{n_g} f_{iI,1}^{\text{nor},k} N_{IJ}^k / m_I^{r,k}. \tag{67}$$

The detailed implementation of the method is presented as follows, where USF algorithm is used in MPM.

(1) Redefine the background grid.
Loop over all the particles in MPM body s to map their mass and the momentum to grid nodes as follows

$$m_I^{s,k} = \sum_p m_p^s N_{Ip}^k, \tag{68}$$

$$p_{iI}^{s,k-1/2} = \sum_p m_p^s v_{ip}^{s,k-1/2} N_{Ip}^k. \tag{69}$$

(2) Impose the boundary conditions on background grid for MPM.
(3) Adjust nodal velocity to satisfy impenetrability condition before getting the trial result of momentum equations.

(a) Reproduce the velocity field of Ω_F on the background grid by mapping the nodal mass and momentum of element nodes along the surface of body r to the background grid.
(b) Detect the contact grid nodes and hybrid nodes based on the velocity fields. Meanwhile, adjust the momentum of contact grid nodes and hybrid nodes by Eqs. (66) and (67), respectively.
(c) Impose boundary conditions again if the contact grid node and hybrid nodes located on the surface with boundary conditions.

(4) Integrate momentum equations to get the trial solution for both Ω_M and Ω_M.

(a) Update the stress of particles and Gaussian points for MPM and FEM, respectively.
(b) Calculate the background grid nodal force for Ω_M and finite element nodal force for Ω_F.
(c) Integrate momentum equations on background grid and element nodes, respectively.

(5) Detect contact grid nodes and get their final momentum results.

(a) Reproduce the velocity field of Ω_F on the background grid by mapping the nodal mass and momentum of element nodes along the surface of body r to the background grid and detect the contact grid nodes and hybrid nodes.
(b) Loop over all contact grid nodes to calculate $f_{I,2}^{\text{nor},k}$, the second term of the normal contact forces, by Eq. (58) and the tangential contact force $f_I^{\tan,k}$ by Eq. (64).
It should be noted that $f_{I,1}^{\text{nor},k}$ is calculated and imposed at Step (3) already.
(c) Loop over all the grid points to update their momenta by

$$p_{iI}^{g,k+1/2} = \tilde{p}_{iI}^{g,k-1/2} + \Delta t^k (f_{iI}^{g,k} + f_{I,2}^{\text{nor},k} n_{iI}^{g,k} + f_I^{\tan,k} t_{iI}^{g,k}), \quad (70)$$

and apply the boundary conditions of the background grid.

(6) Update the position and velocity of particles and element nodes as normal, where Eqs. (44) and (46) are used for hybrid nodes.
(7) Discard the deformed background grid. Return to step 1 for the next time step.

3.5. *Numerical examples*

The CFEMP has been used in the application to impact/penetration problems and fluid–structure interaction problems.[1] Here, we presented a numerical example of the penetration of thick plates by projectiles at a 30° oblique angle with different impact velocities, the experiment of which was conducted by Piekutowski *et al.*[45]

In the experiment, 26.3 mm-thick, 6061-T651 aluminium plates and ogive-nose steel rods as shown in Fig. 6 were used. The cross-section area of the plate is set as 110 × 110 mm in the simulation model.

In the simulation, the projectile is discretized by an unstructured mesh consisting of hexahedral elements and modelled by an elasto-plastic constitutive law, while the target plate is discretized by particles and modelled by Johnson–Cook model for deviatoric stress and Mie–Gruneisen EOS for pressure. The material constants for the projectile and target plate are listed in Tables 1 and 2, respectively. The material failure of the target plate is taken into account by setting the deviatoric components of the stress tensor to zero when the effective plastic strain reaches the plastic strain $\varepsilon_{\text{fail}} = 1.6$. The friction between projectile and target was ignored.

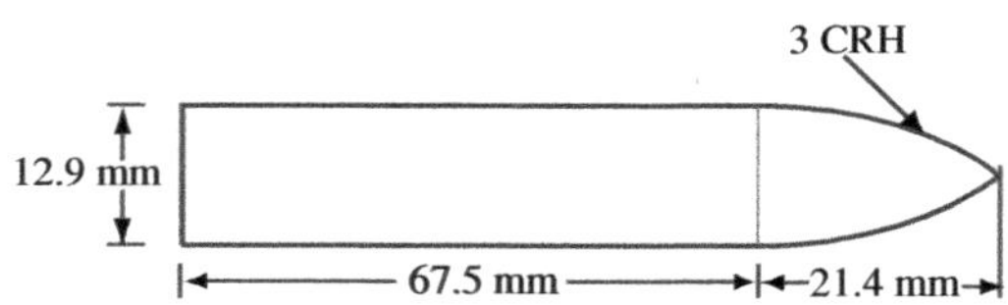

Fig. 6. Schematic of the ogive-nose projectile (from Ref. [1]).

Table 1. Material constants of projectile.

ρ(g/mm^3)	E (GPa)	μ	σ_y (GPa)	E_T (GPa)
0.00785	202	0.3	1.43	14.759

Table 2. Material constants of target plate.

ρ(g/m^3)	E (MPa)	μ	A (MPa)	B (MPa)	n	C	m
0.0027	69	0.3	262	52.1	0.41	0	0.859
c_0(mm/ms)	s	γ_0	T_{melt}(K)	T_{room}(K)			
5,350	1.34	2.0	875	293			

Table 3. The projectile's residual velocities (m/s).

v_0	Experiment	CFEMP
400	217	229
446	288	293
575	455	456
730	655	631

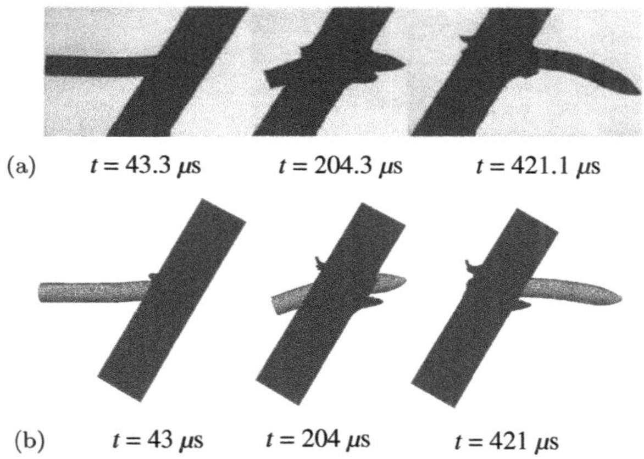

Fig. 7. The perforation process for the case with the impact velocity, v_0 = 400m/s (from Ref. [1]).

Table 3 lists the projectile's residual velocity from the experiments and the simulations by CEMP, where four cases in terms of different impact velocities were conducted. It can be found that the numerical results are in good agreement with the experimental data. Furthermore, the perforation processes for the first and third cases are displayed in Figs. 7 and 8, respectively. In each figure, the top part shows three snapshots obtained in the experiment via X-ray, while the bottom part is numerical results by CFEMP. It is clear that both experiment data and numerical results demonstrate that the global deformation modes of the projectile depend on the impact velocity through the perforation process.

4. Adaptive coupled finite element material point method

In this section, we introduce the second variant of CFEMP method, the adaptive CFEMP method.[3] Compared with the first variant, it takes full

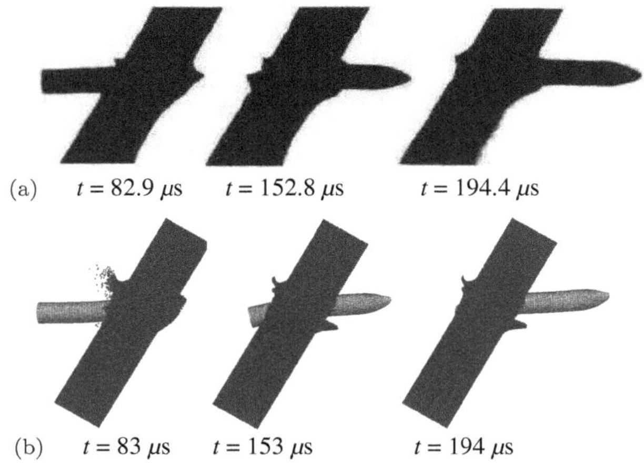

Fig. 8. The perforation process for the case with the impact velocity, $v_0 = 575$m/s (from Ref. [1]).

advantages of both FEM and MPM and turns out to be a powerful numerical method with the target to extreme deformation problems in terms of accuracy and efficiency. In the following, this variant is referred to as AFEMP method for short.

4.1. *Methodology*

In AFEMP, the whole material domain Ω shown in Fig. 9(a) is initially discretized into finite elements as shown in Fig. 9(b). During the simulation process, some elements with extreme deformation can be automatically detected and then adaptively converted into particles solving by MPM as shown in Fig. 9(c). In such a way, FEM could be used to all the material region without extreme deformation instead of only the body with finite deformation as is used in the first variant of CFEMP method. Moreover, the material region with extreme deformation modelled by MPM is activated on the fly. That is to say, AFEMP method is an adaptive combination of both FEM and MPM in an optimizer sense and then takes their advantages fully. To this end, three algorithms are needed in AFEMP method. The first one is an element–particle conversion algorithm which converts elements to particles; the second one is a coupling method that couples the FEM and MPM used for the same continuum body; the last one is the coupling method used in the first variant, which handles the contact between different continuum bodies with different or same discretization

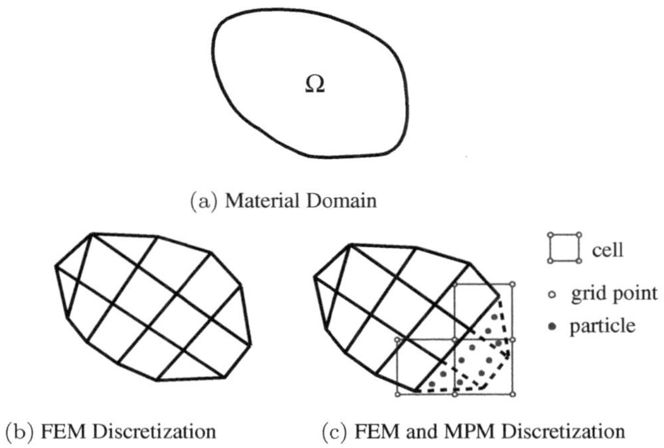

Fig. 9. A material domain with AFEMP discretization (from Ref. [3]).

schemes. Therefore, the mesh requirements mentioned in Section 3.3.3 also exist to AFEMP method.

As the third algorithm is presented in Section 3, only the first and second algorithms are elaborated in the following sections.

4.2. *Element–particle conversion algorithm*

The conversion algorithm consists of two parts. The first one is the criterion of the conversion; the other one is how to convert an element to particles.

The aim to convert the element into particle is to avoid element distortion, so any variable relevant to this could be used as the criterion. Here, two of them are introduced. One is a geometrical variable defined as the ratio of minimum value to maximum value of element lateral length (for 2D) or surface area (for 3D). This variable is a general one and then can be used for all the problems. The other one is the effective plastic strain of Gaussian point associated with the element. So this can only be used in problems involving plastic deformation. Once the value of the selected variable of element exceeded the user-defined threshold, the element is converted to particles.

Given that the conservations of mass, momentum and energy are maintained, the conversion algorithm further depends on the number of Gaussian points within the element and number of particles used to replace the element. In AFEMP, one Gaussian point integration is used and 2 particles in each coordinate direction are used to replace a hexahedral element for the

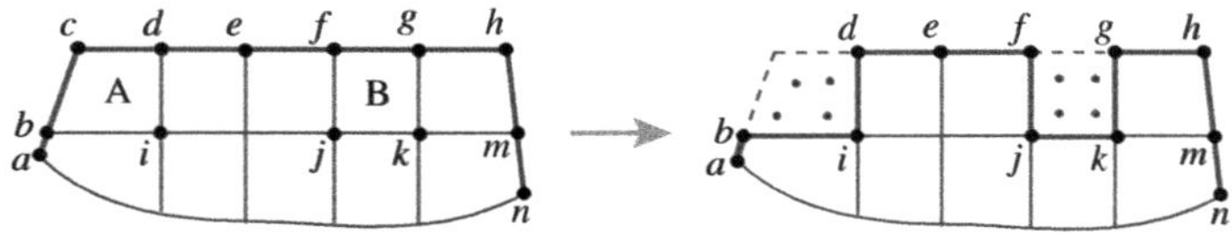

Fig. 10. Conversion of finite elements to particles.

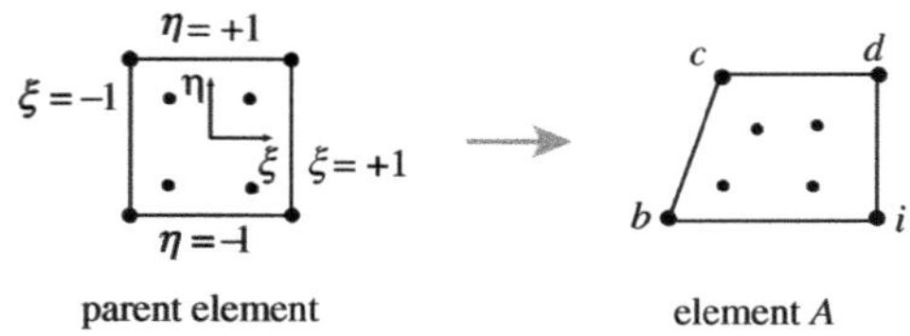

Fig. 11. Calculation of particles position.

3D problem. Besides, some additional work is needed to be done during the conversion process for the following coupling method.

For the sake of clarity to show the conversion process in detail, take a quadrilateral element as an example and it is straightforward to extend this to the hexahedral element. Figure 10 shows a mesh consisting of quadrilateral elements and a boundary defined by FE nodes $a, b, \ldots,$ and n. Let elements A and B be the elements that satisfy the criterion for conversion. Firstly, elements A and B are removed from the finite elements list and each replaced by four particles. Then, the mass, volume and internal energy of the removed elements are separated to the four particles equally following the corresponding conservation law. Correspondingly, the natural coordinates of the particles within the element are set as $(\pm 0.5, \pm 0.5)$ as shown in Fig. 11. Therefore, the positions of the particles are determined as

$$x_{ip} = \sum_{I=1}^{4} N_I(\pm 0.5, \pm 0.5) x_{iI}, \tag{71}$$

where x_{iI} is the position of the FE node I. However, the velocities of the particles are set as the velocity of the adjacent FE nodes for simplicity. As for the stress, strain and other state variables of history, all the particles take same values of the Gaussian point. Next, the FE node c which is not connected to any elements is removed from the FE nodes list, while the FE nodes b, i, d, f, j, k and g located at the interface between MPM particles and remaining finite elements are labelled transition node from whose nodal

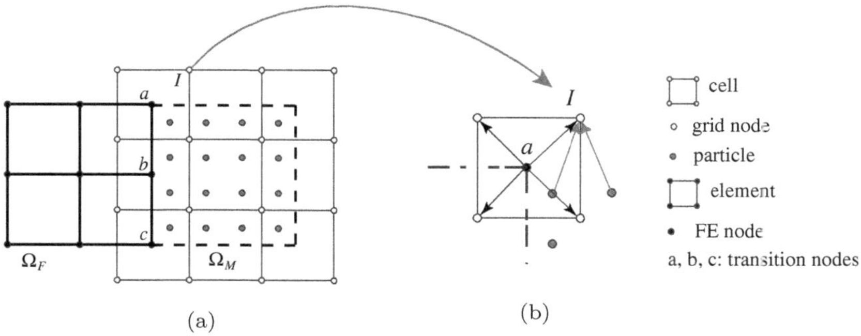

Fig. 12. Coupling between FEM and MPM: (a) Illustration of discretization and (b) Grid node I and its associates.

mass the mass contribution of the removed elements is subtracted. Finally, put the background grid to cover all the particles and the transition nodes.

Based on such a conversion, one Gauss point quadrature is replaced by four particles quadrature with the conservations of mass, momentum and energy, and the finite element mesh is replaced by the background grid.

4.3. *Coupling between finite elements and MPM particles*

The coupling between the remaining finite elements and converted particles for the same continuum body is based on the transition nodes, which are used to implement the continuity and interplay between two discretization domains. To this end, the momentum equations of transition nodes are solved on the background grid together with the particles. That is to say, the continuity and interplay are guaranteed through the single-value fields of the background grid. For the sake of clarity, the coupling method is detailed by taking a 2D problem as an example.

As shown in Fig. 12(a), a material domain is discretized by the combination of finite elements and particles, where the element nodes a, b and c are the transition nodes defined in Section 4.2. In each time step, the nodal mass, momentum and force of transition nodes are mapped to the background grid nodes along with MPM particles. As shown in Fig. 12(b), the grid nodal mass is obtained by

$$m_I = \sum_{p=1}^{n_p} N_{Ip} m_p + \sum_{t=1}^{n_t} N_{It} m_t, \tag{72}$$

where the subscript t represents the transition node, n_t is the total number of transition nodes. Similarly, the grid nodal momentum is:

$$p_{iI} = \sum_{p=1}^{n_p} N_{Ip} p_{ip} + \sum_{t=1}^{n_t} N_{It} p_{it}. \tag{73}$$

Then the velocity field reproduced on the background grid is used for the stress update of particles and the Gaussian point within the element connected to the transition nodes. To this end, the velocity of the transition nodes are reset as

$$v_{it} = \sum_{I=1}^{n_g} N_{tI} v_{iI}. \tag{74}$$

After stress update, calculate the background grid nodal force and all element nodal force as normal. An additional work is to map the nodal force of transition nodes to the background grid as an external force, which means the external nodal force of background grid is calculated by

$$f_{iI}^{\text{ext}} = \sum_{p=1}^{n_p} m_p N_{Ip} b_{ip} + \sum_{t=1}^{n_t} N_{It} f_{it}, \tag{75}$$

where f_{it} is the nodal force of transition node without the hourglas-resisting force.

After solving the momentum equations on background grid, the velocities and positions of transition nodes are updated, respectively, by

$$v_{it}^{k+1/2} = v_{it}^{k-1/2} + \Delta t^k \sum_{I=1}^{n_g} f_{iI}^k N_{pI}^k / m_I^k + f_{it}^{\Gamma,k} / m_t \Delta t^k, \tag{76}$$

$$x_{it}^{k+1} = x_{it}^k + \Delta t^{k+1/2} \left(\sum_{I=1}^{n_g} p_{iI}^{k+1/2} N_{pI}^k / m_I^k + f_{it}^{\Gamma,k} / m_t \Delta t^k \right), \tag{77}$$

where $f_{it}^{\Gamma,k}$ is the hourglass-resisting force. It should be noted that the hourglass-resisting force is not mapped to the background grid and used in Eq. (77) and then plays its role better to control hourglass modes of the element.

By treating the transition nodes as particles when updating their position and velocity, the displacement and velocity fields are consistent along the interface between the FEM domain and the MPM domain.

4.4. *Numerical implementation*

Without concern about the contact between different bodies, the detailed implementation of AFEMP method in a time step can be summarized as follows.

(1) If there are converted particles, redefine the background grid, and map the mass momentum of all MPM particles and transition nodes to the background grid by Eqs. (72) and (73), respectively.
(2) Reset the velocity of the transition nodes by the velocity field of the background grid according to Eq. (74).
(3) Update the stress of particles and element Gaussian points.
(4) Calculate the FEM nodal force by

$$f_{ie}^{k} = f_{ie}^{k,\mathrm{int}} + f_{ie}^{k,\mathrm{ext}} + f_{ie}^{k,\Gamma} \tag{78}$$

and the MPM grid nodal force by

$$f_{iI}^{k} = \sum_{p=1}^{n_p} m_p N_{Ip}^{k} b_{ip}^{k} - \sum_{p=1}^{n_p} N_{Ip,j}^{k} \sigma_{ijp} \frac{m_p}{\rho_p} + \sum_{t=1}^{n_t} N_{It}^{k} f_{it}^{k}. \tag{79}$$

(5) Integrate the momentum equation. It should be noted that the momentum equation of transition nodes is solved on the background grid rather than on themselves.
(6) Update the position and velocity particles and element nodes based on corresponding equations.
(7) Convert elements into particles, if any.
Based on the algorithm presented in Section 4.2, convert elements into particles. Meanwhile, update transition nodes list and delete the FE nodes which are not connected to any elements.
(8) Discard the deformed background grid for MPM. Return to step 1 for the next time step.

4.5. *Numerical examples*

Two applications of AFEMP method are presented: one is a numerical investigation of steel plates penetrated by long rods; the other is slope failure modelling.

4.5.1. *Projectile penetration*

In the penetration experiment conducted by Holmberg *et al.*,[46] 75-mm-long, 5-mm-diameter WHA rod projectile and 5/9-mm-thick steel plate

Table 4. Material constants for Johnson-Cook model.

	A(GPa)	B(GPa)	n	C	m	T_{ref}(K)	T_{melt}(K)
Target plate	0.75	1.15	0.49	0.014	1	293	1700
Projectile	1.05	0.177	0.12	0.0275	1	293	1723

Table 5. Material constants for Gruneisen EOS.

	c(m/s)	s	γ
Target plate	3570	1.92	1.8
Projectile	4030	1.24	1.8

were used. The oblique angle of the projectile to the target plate was 30°. Two cases are conducted here. In the first case with the target plate thickness of 5 mm, the impact velocity of the projectile is 1500 m/s; in the second case with the target thickness of 9mm, the impact velocity of the projectile is 2500 m/s.

In the simulation model, the cross-section area of the plate is set as 150 mm × 150 mm. Johnson–Cook model and Gruneisen EOS are applied to model the material of projectile and target plate, and the material constants are listed in Tables 4 and 5, respectively. The material failure is taken into account by an effective plastic strain failure model with the threshold of 0.9 for both projectile and target plate.[47]

As shown in Fig. 13(a), both rod and plate are discretized by hexahedral elements. The total number of elements for the target is 56,250 for the first case and 101,250 for the second case. For both cases, 1824 hexahedral elements are used to discretize the projectile. The effective plastic strain is taken as the conversion criterion with the threshold of 0.9. One element is converted into 8 MPM particles when its effective plastic strain exceeds 0.9 during the simulation process as shown in Fig. 13(b).

The numerical results in terms of residual length ratio and residual velocity ratio are compared with experimental data in Table 6, which shows that numerical results obtained by AFEMP method agree well with experimental data. In addition, the residual parts of the projectile obtained from AFEMP method are also in good agreement with the experimental data as shown in Fig. 14.

Besides, Table 6 also lists the results from MPM in which the cell size and particle space are set to 1 mm and 0.5 mm, respectively, as that used

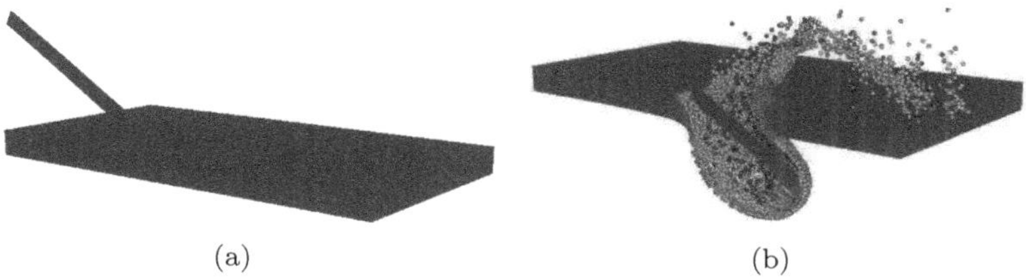

Fig. 13. (a) The initial discretization of the WHA projectile and target plate and (b) A snapshot at time of 3.25s (from Ref. [3]).

Table 6. Comparison between numerical results and experimental data.

Case	Result of	Residual length ratio	Residual velocity ratio
I	Experiment	0.85	0.97
	AFEMP	0.82	0.96
	MPM	0.85	0.96
II	Experiment	0.76	0.99
	AFEMP	0.72	0.97
	MPM	0.74	0.97

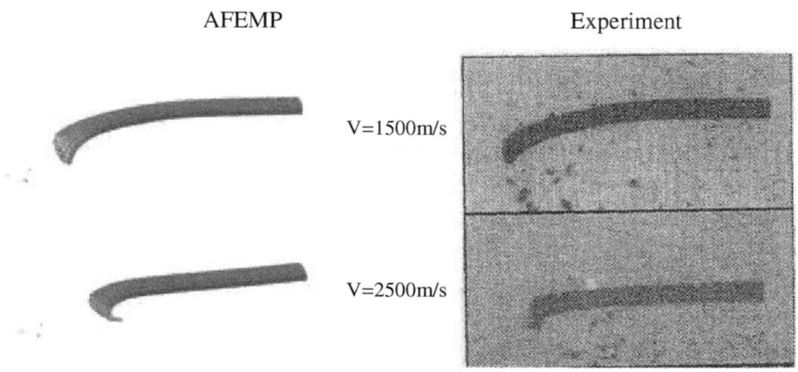

Fig. 14. Residual parts of projectiles (from Ref. [3]).

in AFEMP method. Table 7 lists the computational cost by both AFEMP method and MPM for the second case. By the comparisons shown in the two tables, AFEMP is demonstrated to be much more efficient than MPM while keeps the same level of accuracy.

Table 7. Computational cost by MPM and AFEMP.

	$\Delta t_{\max}(\mu s)$	$\Delta t_{\min}(\mu s)$	Steps	CPU (s)
MPM	6.45×10^{-2}	3.71×10^{-2}	1180	2241
AFEMP	3.67×10^{-2}	1.52×10^{-2}	2208	838

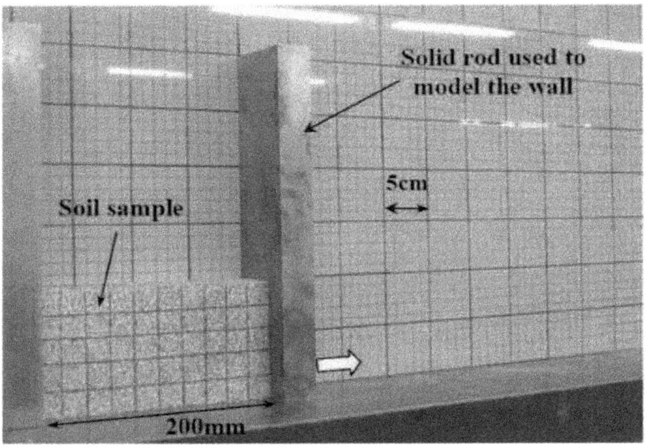

Fig. 15. Arrangement of soil collapse experiment (from Ref. [3]).

4.5.2. *Slope failure*

The final numerical example is a 2D soil collapse experiment conducted by Bui *et al.*[48] As shown in Fig. 15, many small aluminium bars of diameters 1.0mm and 1.5mm were used to mimic a 2D slope failure problem. By quickly removing the right wall horizontally to one side, the aluminium bars flowed down to the right side due to the gravity and reached the final state as shown in Fig. 16.

In this example, Drucker–Prager law is used as the soil model with material constants listed in Table 8.[48] In order to study this problem by 3D code base of AFEMP method, the plane strain assumption is applied. Around 3200 hexahedral elements are used to discretize the material domain with the element size of 2.5 mm. Correspondingly, the cell size of the background grid is set as 2.5 mm also. The effective plastic strain is used as the conversion criterion with the threshold of 1.2.

Figure 16 shows the final shape of the aluminium bars after the experiment and the soil from AFEMP method, and they are in a good

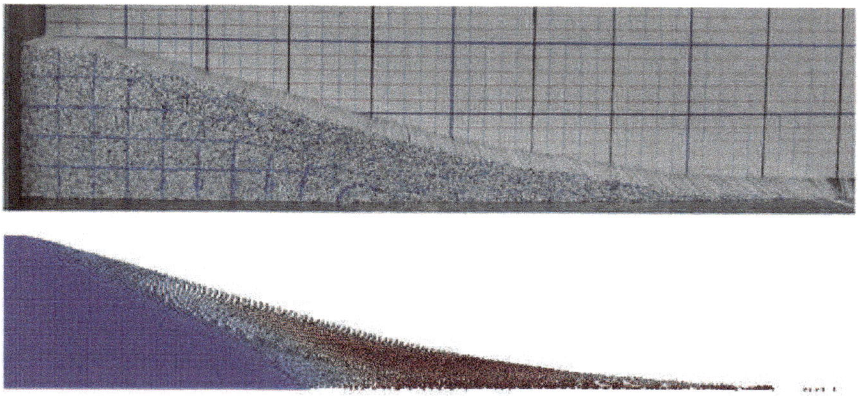

Fig. 16. The top sub-figure is the final configuration of the aluminum bars of the experiment; and the bottom sub-figure is the result of final shape of the solid with AFEMP (from Ref. [3]).

Table 8. Material constants for Drucker–Prager model.

ρ(g/mm^{-3})	K(MPa)	c	ϕ	ψ
2.65$\times$10^{-3}	0.7	0	19.8	0

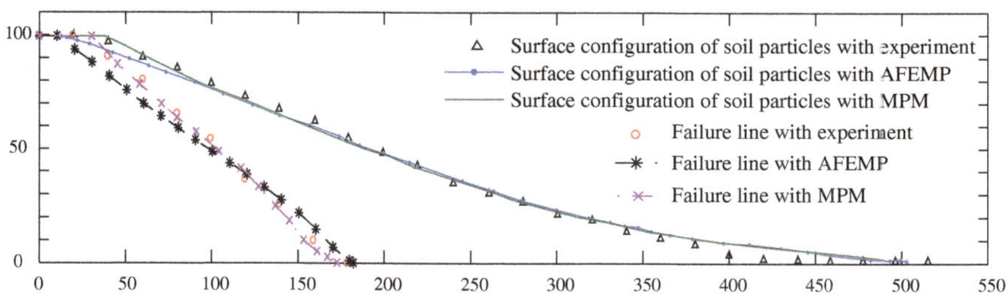

Fig. 17. Final surface configurations and failure lines (from Ref. [3]).

agreement. Moreover, the surface configurations and failure lines obtained by AFEMP method are quantitatively compared with experimental results in Fig. 17, where the results from MPM with the cell size of 2.5mm and particle space of 1.25mm are presented for comparison. It is shown that numerical results from both AFEMP method and MPM are very close to each other and agree well with experimental data. Furthermore, the computational cost required by AFEMP method and MPM are listed in

Table 9. Computational cost by MPM and AFEMP.

	$\Delta t_{\max}(\mu s)$	$\Delta t_{\min}(\mu s)$	Steps	CPU (s)
MPM	48.41	29.94	34765	1791
AFEMP	48.41	22.85	47402	1123

Table 9, which also demonstrates that AFEMP method is more efficient than MPM.

5. Conclusions

In this chapter, we introduced the CFEMP as well as MPM. CFEMP is proposed to take full advantages of FEM and MPM, and then both finite element discretization and material point discretization are used.

The CFEMP includes three algorithms aside from FEM and MPM. They are the contact algorithm, element–particle conversion algorithm and coupling algorithm between finite elements and material point, both of which belong to the same continuum body, respectively. The contact method, a type of point-to-point method, is implemented on MPM background grid to handle the contact between the continuum bodies with same discretizations or different discretizations. The element-to-particle conversion algorithm is used to make a switch to MPM from FEM for the material region with extreme deformation. The last one is used to couple FEM and MPM for the same continuum body. They are introduced in detail through two variants of CFEMP, static CFEMP method and adaptive CFEMP method. As a result, CFEMP allows one to use MPM to a continuum body that will be with extreme deformation and FEM for a continuum body with only finite deformation without material damage, or use FEM to the whole material domain at initial time where material region with extreme deformation in the subsequent simulation process will be solved by MPM. In addition, it should be noted that the finite element size and the background grid cell size should match to each other due to the point-to-point contact algorithm and the consistency of spatial discretization resolutions of both FEM and MPM. Three applications of CFEMP including the penetration of thick plate by a projectile and slope failure problems are presented to show the accuracy and efficiency of CFEMP.

Based on the methodology and algorithms presented in this chapter, it could be found that CFEMP could serve as a framework to adaptively couple FEM and MPM to propose a numerical method with higher accuracy and efficiency to the general problem of solid mechanics. Especially,

CFEMP is a powerful numerical method for transient problems involving contact, impact and penetration.

References

1. Y. P. Lian, X. Zhang, and Y. Liu, Coupling of finite element method with material point method by local multi-mesh contact method, *Computer Methods in Applied Mechanics and Engineering*, **200**, 3482–3494 (2011).
2. Y. P. Lian, X. Zhang, and Y. Liu, Coupling between finite element and material point method for problems with extreme deformation, *Theoretical & Applied Mechanics Letters*, **2**, 021003 (2012).
3. Y. P. Lian, X. Zhang, and Y. Liu, An adaptive finite element material point method and its application in extreme deformation problems, *Computer Methods in Applied Mechanics and Engineering*, **241–244**(1), 275–285 (2012).
4. Y. P. Lian, Y. Liu, and X. Zhang, Coupling of membrane element with material point method for fluid-membrane interaction problems, *International Journal of Mechanics Materials in Design*, **10**, 199–211 (2014).
5. J. J. Monaghan, An introduction to SPH, *Computer Physics Communications*, **48**, 89–96 (1988).
6. W. K. Liu and Y. J. Chen, Wavelet and multiple scale reproducing kernel methods, *International Journal for Numerical Methods in Fluids*, **21**(10), 901–931 (1995).
7. W. K. Liu, S. Jun, and Y. F. Zhang, Reproducing kernel particle methods, *International Journal for Numerical Methods in Engineering*, **20**, 1081–1106 (1995).
8. T. Belytschko, Y. Y. Lu, and L. Gu, Element free Galerkin methods, *International Journal for Numerical Methods in Engineering*, **37**, 229–256 (1994).
9. X. F. Pan and H. Yuan, Computational algorithm and application of element-free Galerkin methods for nonlocal damage models, *Engineering Fracture Mechanics*, **77**, 2640–2653 (2010).
10. G. R. Johnson, Linking of Lagrangian particle methods to standard finite element methods for high velocity impact computations, *Nuclear Engineering and Design*, **150**, 265–274 (1994).
11. T. Belytschko, D. Organ, and Y. Krongauz, A coupled finite element-element-free Galerkin method, *Computational Mechanics*, **17**, 186–195 (1995).
12. T. D. Vuyst, R. Vignjevic, and J. C. Campbell, Coupling between meshless and finite element methods, *International Journal of Impact Engineering*, **31**, 1054–1064 (2005).
13. W. K. Liu, R. A. Uras, and Y. Chen, Enrichment of the finite element method with reproducing kernel particle method, *Journal of Applied Mechanics*, **135**, 143–166 (1997).
14. T. Rabczuk, S. P. Xiao, and M. Sauer, Coupling of mesh-free methods with finite elements: Basic concepts and test results, *Communications in Numerical Methods in Engineering*, **22**, 1031–1065 (2006).

15. D. Sulsky, Z. Chen, and H. L. Schreyer, A particle method for history-dependent materials, *Computer Methods in Applied Mechanics and Engineering*, **118**(1–2), 179–196 (1994).
16. S. Ma, X. Zhang, and X. M. Qiu, Comparison study of MPM and SPH in modeling hypervelocity impact problems, *International Journal of Impact Engineering*, **36**, 272–282 (2009).
17. S. G. Bardenhagen, J. E. Guilkey, K. M. Roessig, J. U. Brackbill, W. M. Witzel, and J. C. Foster, An improved contact algorithm for the material point method and application to stress propagation in granular material, *Computer Modeling in Engineering and Sciences*, **2**(4), 509–522 (2001).
18. P. Huang, X. Zhang, S. Ma, and X. Huang, Contact algorithms for the material point method in impact and penetration simulation, *International Journal for Numerical Methods in Engineering*, **85**(4), 498–517 (2011).
19. Z. Ma, X. Zhang, and P. Huang, An object-oriented MPM framework for simulation of large deformation and contact of numerous grains, *CMES-Computer Modeling in Engineering & Sciences*, **55** (1), 61–87 (2010).
20. Z. Chen, X. Zhang, X. Qiu, and Y. Liu, A frictional contact algorithm for implicit material point method, *Computer Methods in Applied Mechanics and Engineering*, **321**, 124–144 (2017).
21. Y. P. Lian, X. Zhang, X. Zhou, S. Ma, and Y. L. Zhao, Numerical simulation of explosively driven metal by material point method, *International Journal of Impact Engineering*, **38**, 237–245 (2011).
22. B. Wang, P. Vardon, M. Hicks, and Z. Chen, Development of an implicit material point method for geotechnical applications, *Computers and Geotechnics*, **71**, 159–167 (2016).
23. T. Charlton, W. Coombs, and C. Augarde, igimp: An implicit generalised interpolation material point method for large deformations, *Computers & Structures*, **190**, 108–125 (2017).
24. Y. P. Lian, X. Zhang, F. Zhang, and X. X. Cui, Tied interface grid material point method for problems with localized extreme deformation, *International Journal of Impact Engineering*, **70**, 50–61 (2014).
25. Y. Lian, P. Yang, X. Zhang, Y. Liu, and P. Huang, A mesh-grading material point method and its parallelization for problems with localized extreme deformation, *Computer Methods in Applied Mechanics and Engineering*, **289**, 291–315 (215).
26. S. G. Bardenhagen and E. M. Kober, The generalized interpolation material point method, *Computer Modeling in Engineering Sciences*, **5**(6), 477–495 (2004).
27. Y. Liang, B. Tamas, X. Zhang, and Y. Liu, Material point method with enriched shape function for crack problems, *Computer Methods in Applied Mechanics and Engineering*, **322**, 541–562 (2017).
28. X. Zhang, K. Y. Sze, and S. Ma, An explicit material point finite element method for hyper velocity impact, *International Journal for Numerical Methods in Engineering*, **66**, 689–706 (2006).
29. X. X. Cui, X. Zhang, K. Y. Sze, and X. Zhou, An alternating finite difference material point method for numerical simulation of high explosive

explosion problems, *CMES — Computer Modeling in Engineering & Sciences*, **92**, 507–538 (2013).
30. F. Zhang, X. Zhang, K. Sze, Y. Lian, and Y. Liu, Incompressible material point method for free surface flow, *Journal of Computational Physics*, **330**, 92–110 (2017).
31. W. Q. Hu and Z. Chen, Model-based simulation of the synergistic effects of blast and fragmentation on a concrete wall using the mpm, *International Journal of Impact Engineering*, **32**(12), 2066–2096 (2006).
32. Y. X. Wang, Z. Chen, and M. Sun, Simulation of explosion and shock involving multiple materials based on the material point method, *Explosion and Shock Waves*, **28**(2), 154–160 (2008).
33. Y. X. Wang, H. G. Beom, M. Sun, and S. Lin, Numerical simulation of explosive welding using the material point method, *International Journal of Impact Engineering*, **38**, 51–60 (2011).
34. H. W. Zhang, K. P. Wang, and Z. Chen, Material point method for dynamic analysis of saturated porous media under external contact/impact of solid bodies, *Computer Methods in Applied Mechanics and Engineering*, **198**, 1456–1472 (2009).
35. S. Andersen and L. Andersen, Modelling of landslides with the material-point method, *Computational Geosciences*, **14**, 137–147 (2010).
36. J. A. Nairn, Material point method calculations with explicit cracks, *Computer Modeling in Engineering and Sciences*, **4**, 649–663 (2003).
37. P. F. Yang, Y. Gan, X. Zhang, Z. Chen, W. Qi, and P. Liu, Improved decohesion modeling with the material point method for simulating crack evolution, *International Journal of Fracture*, **186**, 177–184 (2014).
38. A. R. York II, D. Sulsky, and H. L. Schreyer, The material point method for simulation of thin membranes, *International Journal for Numerical Methods in Engineering*, **44**, 1429–1456 (1999).
39. D. Z. Zhang, Q. Zou, W. B. VanderHeyden, and X. Ma, Material point method applied to multiphase flows, *Journal of Computational Physics*, **227**, 3159–3173 (2008).
40. D. Sulsky, Z. Chen, and H. L. Schreyer, A particle method for history-dependent materials, *Computer Methods in Applied Mechanics and Engineering*, **118**(1–2), 179–196 (1994).
41. S. G. Bardenhagen, Energy conservation error in the material point method for solid mechanics, *Journal of Computational Physics*, **180**, 383–403 (2002).
42. J. O. Hallquist, *LS-DYNA Theoretical Manual*. Livermore Software Technology Corporation (1998).
43. S. G. Bardenhagen, J. U. Brackbill, and D. Sulsky, The material-point method for granular materials, *Computer Methods in Applied Mechanics and Engineering*, **187**(3–4), 529–541 (2000).
44. X. Q. Z. P. Chen, X. Zhang, and Y. Lian, Improved coupling of finite element method with material point method based on a particle-to-surface contact algorithm, *Computer Methods in Applied Mechanics and Engineering*, **293**, 1–19 (2015).

45. A. J. Piekutowski, M. J. Forrestal, K. L. Poormon, and T. L. Warren, Perforation of aluminum plates with ogive-nose steel rods at normal and oblique impacts, *International Journal of Impact Engineering*, **18**, 877–887 (1996).
46. L. Holmberg, P. Lundberg, and L. Westerling. An experimental investigation of WHA long rods penetrating oblique steel plates, In *14th International Symposium on Ballistics*, Quebec City, Canada (1993).
47. Y. K. Park and E. P. Fahrenthold, A kernel free particle-finite element method for hypervelocity impact simulation, *International Journal for Numerical Methods in Engineering*, **63**, 737–759 (2005).
48. H. H. Bui, R. Fukagawa, K. Sako, and S. Ohno, Lagrangian meshfree particles method (SPH) for large deformation and failure flows of geomaterials using elastic-plastic soil constitutive model, *International Journal for Numerical and Analytical Methods*, **32**, 1537–1570 (2008).

Chapter 8

Fracture and Contact in the Material Point Method: New Approaches and Applications*

M. A. Homel and E. B. Herbold

A recently developed approach for using the material point method to model complex phenomena involving fracture, contact and large deformation is presented. Motivations, applications and implementation considerations are discussed, along with the benefits and limitations of the method. With this approach, we demonstrate that it is possible to model complex behaviours such as fragmentation and comminution while avoiding the numerical artifacts and limitations of other approaches.

1. Motivation

In dynamic applications such as an impact into soil or ceramic armour the response characteristics may be dominated by the mechanisms of fracture, comminution and the post-failure response of a granular material. For example, Fig. 1 shows comminuted material along shear cracks in a silicon carbide sample subjected to dynamic loading.[1] Photographic images suggest that intact glass samples typically fracture after passage of an initial pressure pulse and before projectiles pass through.[2–6] Therefore, the deceleration of a projectile will depend on its interaction with an almost fully dense collection of broken glass fragments. This interaction depends on the level of confinement as these fragments interact only through compressive and shear loading.

*This work was performed under the auspices of the U.S. Department of Energy by Lawrence Livermore National Laboratory under Contract DE-AC52-07NA27344. LLNL-BOOK-725109-DRAFT.

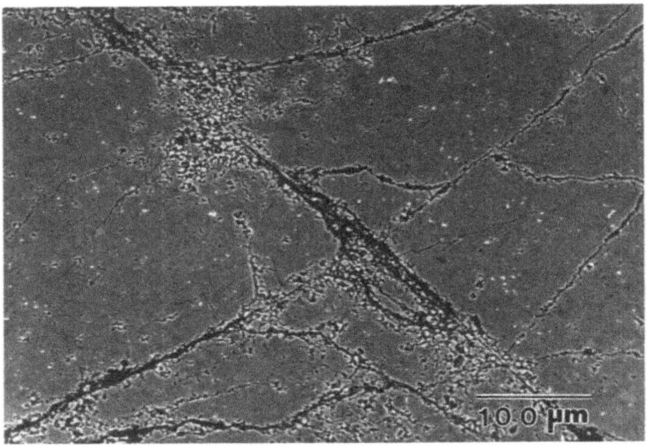

Fig. 1. Comminution visible along the lengths of cracks formed after explosive loading of silicon carbide. Image used with permission from Ref. [1].

Continuum constitutive models of granular compaction are generally inadequate to predict the material response in complex loading in large pressure ranges involving compaction, shear bulking, stress transients, softening and thermal effects; some of which are present in recent instrumented experiments.[7] For example, there is a class of large-deformation constitutive models for brittle granular materials that derive the response as a function of a scalar "breakage" or damage parameter, which describes the transition of the particle size distribution from an initial state to a fully compacted material.[8] There is a lack of understanding about the evolution of a particle size distribution throughout a dynamic loading event. Thus, having a capability for predictive grain-scale simulation of comminution makes it possible to fill a critical knowledge gap in our understanding of the mechanics of granular systems.

To simulate comminution, it is necessary to model fracture of material and subsequent frictional contact between material fragments. This is complicated by the need to account for physics across multiple length scales as the dimensions of a representative material element may be many orders of magnitude greater than that of the finest particles produced. Inevitably, there are computational limitations to the range of particle sizes that can be resolved, and it is therefore necessary to use a continuum approximation to describe the material response of a collection of the finest particles. As these particles flow and/or compress, the continuum deformation can be quite large, introducing an additional challenge to numerical simulation.

2. Limitations of other methods

A variety of computational methods to simulate fracture and large deformation have been developed for applications in solid mechanics, including Lagrangian finite element methods (FEM), generalized/extended finite element method (GFEM/XFEM), Eulerian and arbitrary Lagrangian–Eulerian (ALE) methods and mesh-free particle methods, but each has its limitations with respect to either accuracy or robustness when accounting for large-deformation and fracture with subsequent frictional contact in complex 3D geometries.

Lagrangian FEM can accommodate fracture through the introduction of cohesive zones,[9] element deletion,[10] or element-splitting algorithms,[11,12] but such approaches may have significant mesh dependencies or may not be able to handle large displacements post-failure, which is common in comminuting materials. Improved cohesive zone methods have been shown to reduce mesh dependencies, but may not reproduce complex crack interaction phenomena.[13] Figure 2 shows a Lagrangian finite element simulation of comminution in a brittle granular material, where element splitting with cohesive zones and common-plane contact are used to simulate fracture and frictional effects, respectively. This method provides an accurate model of the initial material failure and accommodates large sliding, but robustness may become an issue for severely distorted elements.

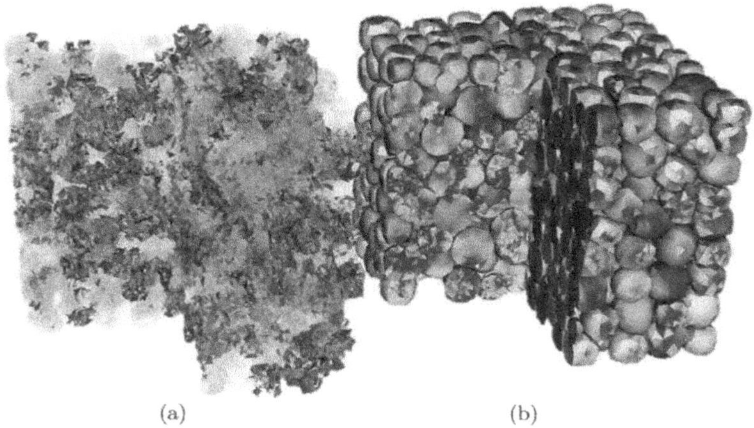

Fig. 2. Lagrangian finite element simulation of comminution of a brittle granular material, using element splitting a common plane contact. (a) Contact common planes indicate where element splitting has been activated and (b) Contours show von Mises stress in the grains.

The generalized/extended finite element method (GFEM/XFEM) reduces or eliminates mesh dependencies,[14] but typically does not allow for large-displacement frictional contact at fracture surfaces: though this capability is not explicitly prohibited by the method. Additionally, the computational geometry required to accommodate arbitrary branching and coalescence of multiple fracture networks in a 3D Lagrangian framework has proven to be a persistent challenge in the mechanics community[15] and the computational cost becomes prohibitive as the number of cracks increase.[16] More importantly, all of these Lagrangian methods may fail under large deformation due to mesh entanglement, an important consideration when modelling processes such as dynamic fragmentation and comminution.

Eulerian and arbitrary Lagrangian–Eulerian (ALE) methods are ideal for simulating large deformations, but may perform poorly with continuum damage models requiring advection of internal state variables. This is an important consideration for anisotropic material models or those that derive the mechanical response from a statistical distribution of flaws or cracks in the material, both of which require special treatment for advection in a multi-material framework. Figure 3 illustrates two examples of such complications due to advection in Eulerian simulations of brittle failure. In the first example, spatially heterogeneous material properties (e.g. strength or flaw density) may be maintained with special reference to the initial configuration to mitigate mesh dependence in continuum models of brittle failure.[17] Figure 3(a) shows an initial heterogeneous distribution of material strength, which is degraded due to advection errors in an ALE simulation of plate impact.[17] Additional complications arise when describing fracture and frictional contact in a Eulerian framework.[18]

Figure 3(b) shows numerical "healing" of damaged material in an Eulerian simulation (fourth-order Godunov scheme from GEODYN) of a 5 mm square block of SiO_2 moving at 1 km/s through a vacuum. The grid spacing is 0.125 mm and a crack is initialized in the centre of the block by specifying completely damaged cells along part of a column. The diffusion of the initially localized damage naturally occurs as the block moves 40 mm to the right, which effectively "heals" the SiO_2.

Mesh-free particle methods are ideally suited to treat fracture or other discontinuities due to the absence of advection errors and the freedom to naturally suppress a tensile interaction between discrete particles. For example, smooth particle hydrodynamics (SPH) has been shown to perform reasonably well in penetration simulations, though extensions of the original SPH method (e.g. reproducing kernel particle methods (RKPMs)) are

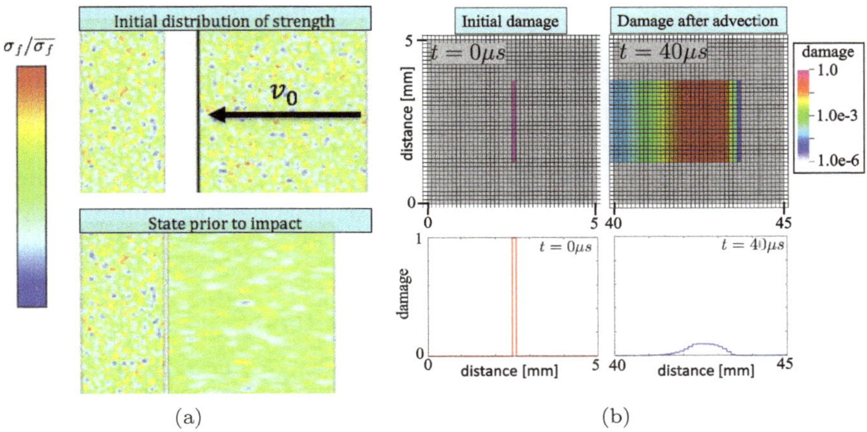

Fig. 3. Illustration of advection errors in Eulerian simulations of brittle failure showing (a) an initial heterogeneous distribution of material strength is degraded due to advection errors in an ALE simulation of plate impact,[17] and (b) numerical "healing" of damaged material in an Eulerian (fourth-order Godunov scheme from GEODYN) simulation of a 5mm square block of SiO_2 moving at 1 km/s to the right through a vacuum. The grid spacing is 0.125mm and a crack is initialized in the centre of the block by specifying completely damaged cells along part of a column. The diffusion of the initially localized damage naturally occurs as the block moves 40mm to the right, which effectively "heals" the SiO_2.

needed to alleviate inconsistencies such as the tensile instability and poor conservation of angular momentum.[19,20] Another example is the particle-cracking method, which has been shown to be predictive of experimental results in expanding thick-walled cylinder tests[21] and in concrete pullout tests.[22] However, meshless methods can be computationally expensive in large deformation problems, and in some cases may exhibit non-physical numerical fracture. A relatively new meshless approach is peridynamics, which has been shown to perform reasonably well in fracture under tensile loading,[23] but currently has a limited capability to predict the post-failure compressive and frictional response of a material.

An alternative to mesh-free particle methods is the material point method (MPM),[24] which tracks Lagrangian particles on an Eulerian grid, and which has been previously shown to be accurate for solid mechanics problems involving high rates,[25,26] large-deformations,[27] frictional contact,[28] and complex constitutive models.[29] The MPM has previously been extended to simulate fracture using either an explicit description of cracks[30,31] or cohesive laws,[32] but such methods are challenging to implement for complex intersecting fracture networks in 3D.

A new method has been developed to simulate fracture and frictional self-contact in the MPM, without requiring an explicit description of the crack geometry or an *a priori* definition of contacting bodies.[33] A conceptual description of this particular method including the approach, application, limitations and areas of future development are provided in the remainder of this chapter. The present discussion is tailored to a reader with an understanding of common numerical approaches such as the FEM, but who may be less familiar with the details of the MPM.

3. Background on the material point method

The material point method is a mixed Eulerian–Lagrangian approach in which the material state is tracked at Lagrangian particles and the equations of motion are solved on an Eulerian background grid.[24] The MPM algorithm can be conceptualized as four key steps, as shown in Fig. 4. At the start of a time step, the problem state is defined entirely by the particle data (mass, volume, position, velocity, deformation gradient, stress and any constitutive model state variables). (i) At the beginning of the time step, the mass, momentum and forces are mapped from the particles to the

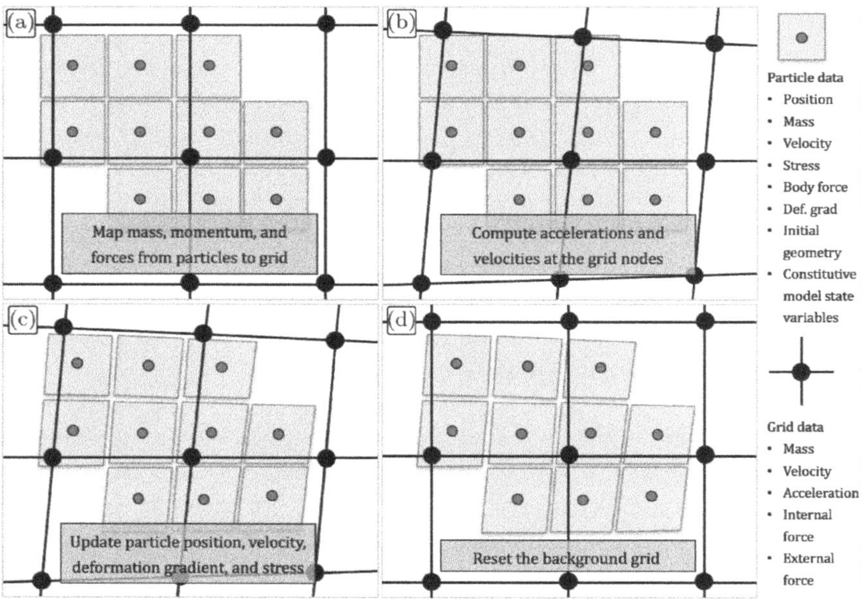

Fig. 4. Conceptual diagram of the key steps in the material point method (MPM).

nodes of a background computational mesh (the "background grid"). (ii) The nodal accelerations and velocities at the grid are updated. (iii) These values are then used to update the particle positions, velocities, deformation gradient and stresses and (iv) Conceptually, we can think of resetting the background grid at the end of each time step, though algorithmically it is not necessary to compute the deformed grid state. Thus, over time, the particles move and deform relative to this fixed computation mesh.

The standard weak formulation of the conservation of momentum applies on the background grid, leading to the following equations for the nodal mass and internal force.

$$m_i = \int_\Omega S_i^\star(\boldsymbol{x})\rho dV, \tag{1}$$

$$f_i^{\text{int}} = -\int_\Omega \boldsymbol{\sigma} \cdot \nabla S_i^\star(\boldsymbol{x}) dV. \tag{2}$$

$S_i^\star(\boldsymbol{x})$ is the finite element shape function for the ith grid node. In the FEM, the integrals are evaluated as a sum of integrals over element domains, which are evaluated numerically (e.g. through integration at Gauss points within each element). In the MPM, integral equations are evaluated as sums of integrals over particle domains.

A complete description of the MPM including a connection to the traditional finite element "weak" form and a generalization of various MPM approaches is given by Sadeghirad *et al.*[34] Key steps are summarized below to provide a framework for discussing the modifications needed to achieve the desired fracture and contact response. We describe the method in the context of an explicit time integration, though implicit formulations have been demonstrated.[35,36]

At the start of the time step, the mass and momentum are mapped from the particles to a background grid. The mass (m_i) and velocity ($\boldsymbol{v}_i$) for the ith node of the background grid are:

$$m_i = \sum_p \overline{S_{ip}^\star} m_p, \tag{3}$$

$$\boldsymbol{v}_i = \frac{1}{m_i} \sum_p \overline{S_{ip}^\star} m_p \boldsymbol{v}_p, \tag{4}$$

where m_p and $\boldsymbol{v}_p$ are the mass and velocity of the pth particle, and $\overline{S_{ip}^\star}$ is the average of the ith grid shape function $S_i^\star$ over the pth particle domain. In the original formulation of the MPM,[24] this average was computed with single-point integration, but this "Dirac delta" formulation led to spurious cell-crossing errors. The generalized interpolation material point (GIMP) method[37] improves accuracy by representing the particle domain as a finite domain, but does not fully eliminate cell crossing errors. The convective particle domain interpolation (CPDI) method[38] improves accuracy of the force calculations and particle-grid mapping over earlier GIMP methods by representing the particle integration domain as a deforming parallelepiped, for which the shape function averages can be readily computed. In our work, we use the CPDI formulation along with a domain scaling modification[39] that allows for efficient parallelization, while controlling non-physical numerical fracture. The background grid is a structured, Cartesian mesh, with trilinear shape functions, which simplifies evaluation of $S_i^\star$.

The internal and external forces (neglecting surface tractions) at the ith grid node are:

$$\boldsymbol{f}_i^{\text{ext}} = \sum_p \overline{S_{ip}^\star} \boldsymbol{b}_p m_p, \tag{5}$$

$$\boldsymbol{f}_i^{\text{int}} = -\sum_p \overline{\nabla S_{ip}^\star} \cdot \boldsymbol{\sigma}_p V_p, \tag{6}$$

where $\boldsymbol{b}_p$ is the body force acting on the pth particle, V_p is the particle volume, $\boldsymbol{\sigma}_p$ is the Cauchy stress at the particle and $\overline{\nabla S_{ip}^\star}$ is the average of the gradient of the ith grid shape function over the pth particle domain. The accelerations are computed on the background grid,

$$\boldsymbol{a}_i = \frac{\boldsymbol{f}_i^{\text{int}} + \boldsymbol{f}_i^{\text{ext}}}{m_i}, \tag{7}$$

from which the updated nodal velocities are computed,

$$\bar{\boldsymbol{v}}_i^{n+1} = \boldsymbol{v}_i^n + \boldsymbol{a}_i \Delta t. \tag{8}$$

In Eq. (8) the updated velocity, $\bar{\boldsymbol{v}}_i^{n+1}$, neglects the effect of contact forces. As described in the following section, contact forces modify the velocity field to compute a corrected $\boldsymbol{v}_i^{n+1}$ from $\bar{\boldsymbol{v}}_i^{n+1}$. The updated grid velocity field is used to compute the updated velocity gradient at the particle,

$$\boldsymbol{L}_p^{n+1} = \Sigma_i \boldsymbol{v}_i^{n+1} \otimes \overline{\nabla S_{ip}^\star}. \tag{9}$$

The updated particle velocity gradient is used to update the particle deformation gradient,

$$\boldsymbol{F}_p^{n+1} = \boldsymbol{F}_p^n + \left(\boldsymbol{L}_p^{n+1} \cdot \boldsymbol{F}_p^n\right) \Delta t. \tag{10}$$

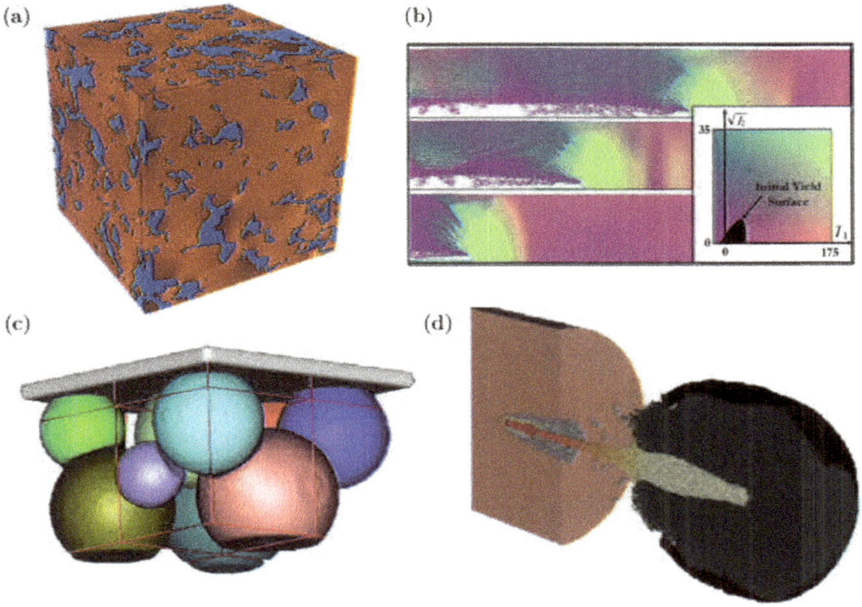

Fig. 5. Illustration of the four key advantages of MPM. (a) MPM model of fluid saturated sandstone created from a CT scan.[42] (b) MPM simulation of penetration into a target material with a complex constitutive model where avoidance of advection errors is essential.[26] (c) Large-deformation contact frictional contact[43] and (d) A 3D shaped charge formation/penetration problem involving extreme deformation, contact, multiple materials and phase changes.[26]

The constitutive model is then called with the updated deformation gradient and velocity gradient to compute the updated particle stress $\boldsymbol{\sigma}_p^{n+1}$ and internal state variables. This is the "update stress last" (USL) formulation, which has previously been shown to be most accurate for dynamic simulation with explicit time-stepping schemes.[40] The end-of-step nodal velocities and accelerations are used to update the position and velocity of the particles,

$$\boldsymbol{v}_p^{n+1} = \boldsymbol{v}_p^n + \Sigma_i \overline{S_{ip}^\star} \boldsymbol{a}_i \Delta t, \tag{11}$$

$$\boldsymbol{x}_p^{n+1} = \boldsymbol{x}_p^n + \Sigma_i \overline{S_{ip}^\star} \boldsymbol{v}_i^{n+1} \Delta t. \tag{12}$$

There are four frequently cited benefits of the material point method,[41] as illustrated in Fig. 5.

(1) Model generation is relatively simple and eliminates the need for mesh generation. For example, it is straightforward to accommodate voxelized phase topologies directly from X-ray computed microtomography (XCMT) data (*cf.* Ref. 42).
(2) The material state history is stored at the particles, so that advection errors are eliminated, allowing for the use of complex constitutive models.
(3) The background grid is reset at each time step, so the method is robust for large deformation.
(4) Frictional contact between material regions is efficient and easily implemented as described in the following section.

An example of the robust capabilities of the MPM is its use in simulating the penetration of a shaped charge jet into saturated rock, which involves extreme deformation of both the jet and target material, contact between the jet and penetration channel and which requires a complex constitutive model to describe the target response.[26,42] However, while these are challenging problems, they are relatively simple with respect to the topology of the failure and contact surfaces. For more complicated or evolving topologies, new approaches are needed, as described below. In discussing advanced in the material point method, it is important to remember the four key advantages to the MPM, to ensure that new approaches don't introduce requirements or limitations that compromise these advantages.

3.1. *Traditional approach for contact in the material point method*

The MPM algorithm outlined in Eqs. (3)–(12) computes the response for material in a single velocity field. Within this velocity field, the method naturally produces a no-slip contact condition that prevents interpenetration between materials. This behaviour is similar to the response of a Eulerian hydrocode, which also assumes that each computational cell contains a single velocity gradient when multiple materials are present.

If frictional contact or separation is an important consideration between two or more regions or material groups, then the material in each contact group is assigned to its own velocity field. The solution within a contact-group velocity field is identical to that for a single-field system, except when material from multiple velocity fields maps to the same node. In this case, a contact force $\boldsymbol{f}_i^{\text{ct}}$ is computed at the shared node, which modifies the velocity in Eq. (8). This is illustrated in Fig. 6.

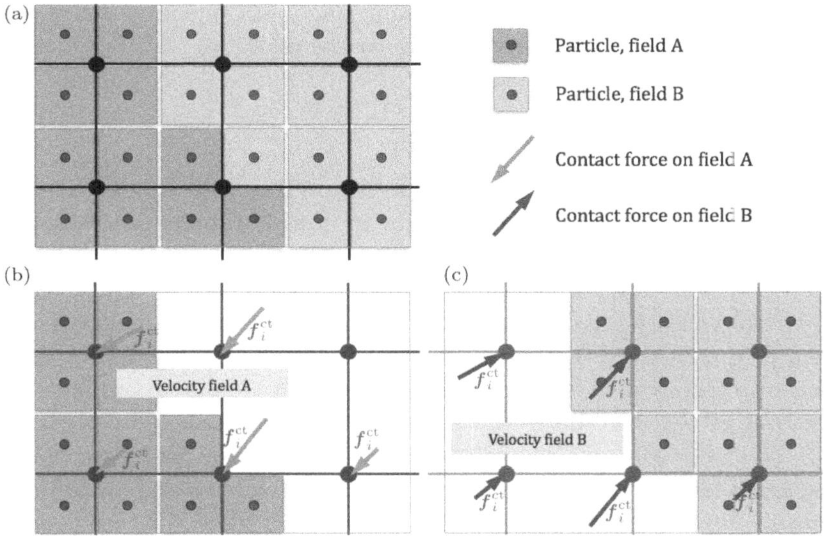

Fig. 6. Diagram of the traditional approach for contact with prescribed material contact groups: (a) Contact between two material groups, (b) particles in first velocity field with nodal contact force from second field and (c) particles in second velocity field with nodal contact force from first field.

The nodal contact force is computed to (i) prevent interpenetration, (ii) allow frictional sliding if the slip condition is met and (iii) allow separation of the contact pair. For a given node at which two or more velocity fields have nonzero mass, there is a pairwise nodal interaction forces between each field. In our work, we dynamically repartition material into separate velocity fields at each node to define fracture surfaces or to accommodate frictional self contact. Once the material has been appropriately partitioned, the subsequent contact algorithm for the step is identical to that for contact between predetermined material pairs as described by Bardenhagen *et al.*[44] The method was later modified[25] to improve accuracy by rigorous enforcement of an inter-penetrability condition through the step.

To compute the contact forces at a grid node, we first compute the outward surface normal for the material in each velocity field ($\varsigma = 1, 2$) at that node based on the normalized local density gradient in the field.

$$\boldsymbol{n}_{\varsigma i} = \frac{\sum_p \overline{\boldsymbol{\nabla} S^*_{ip}} m_{\varsigma p}}{\| \sum_p \overline{\boldsymbol{\nabla} S^*_{ip}} m_{\varsigma p} \|}, \tag{13}$$

where $m_{\varsigma p}$ is the mass of the pth particle in the ςth velocity field.

The total of the mass and momentum from both velocity fields are used to compute the centre-of-mass velocity at the node.

$$\boldsymbol{v}_i^{cm} = \frac{\Sigma_{\varsigma=1}^{2} m_{\varsigma i} \bar{\boldsymbol{v}}_{\varsigma i}}{\Sigma_{\varsigma=1}^{2} m_{\varsigma i}}, \tag{14}$$

where $\bar{\boldsymbol{v}}_{\varsigma i}$ indicates the velocity for the ςth velocity field before accounting for contact forces. This lets us define the inter-penetrability criterion, based on the surface-normal component of the velocity, for each field relative to the centre of mass. If interpenetration will occur, i.e. if $(\boldsymbol{v}_i^{cm} - \bar{\boldsymbol{v}}_{\varsigma i}) \cdot \boldsymbol{n}_{\varsigma i} < 0$, a normal contact force is computed that restores the normal velocity of each field relative to the centre of mass to zero. For frictional contact, the tangential component of the contact force may be computed consistent with the contact law, but this is still an open area of research.

The frictional contact capability in the MPM can be readily extended to accommodate an arbitrary number of contact groups by adding additional velocity fields and computing the pairwise contact forces for each combination of the material groups. For N velocity fields, there are $N(N-1)$ possible contact pairs, so the computational cost and memory requirements for the contact algorithm become more significant as N becomes large. This cost can be mitigated by only computing interactions for fields that are active at a single node. A more important concern is that while pairwise force calculations ensure conservation of linear momentum, they do not strictly prevent interpentrability for $N > 2$ fields.

When using the multifield method to model frictional contact in granular materials (*cf.* Ref. 28) it is necessary to initially partition the grains into separate velocity fields so that neighbouring grains occupy separate fields. The four-colour theorem states that any 2D domain can be partitioned to avoid edge-to-edge contact of matching fields with no more than four fields.[45] In 3D, there is no analogous theorem limiting the number of required fields to ensure frictional contact between all entities. Figure 7 illustrates initial material colouring in 2D and 3D to allow for frictional contact.

For granular systems, fractured bodies or porous materials, even if the geometry can be initialized such that neighbouring bodies are in separate velocity fields at contact surfaces, subsequent deformation and granular flow can change the contact points. This situation arises in a number of engineering applications in addition to the comminution and granular flow problem that is the focus of the discussion. For example, additive manufacturing has been used to create porous structures with unusual mechanical properties,

(a)

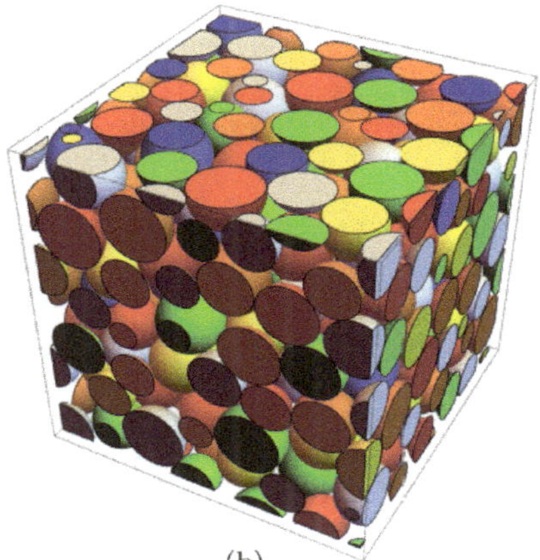

(b)

Fig. 7. Examples of *a priori* material colouring in 2D (a) and 3D (b).

often resulting from large displacement and/or contact of thin members within the structure. Figure 8 shows a traditional multifield MPM simulation of dynamic crushing of a 2D hex foam, illustrating the frictional contact capabilities and self-contact limitations of the traditional MPM method. While this example could be modelled using large-deformation Lagrangian FEM (e.g. Ref. 46), more extreme dynamic events (e.g. ballistic impact into a foam) would be very challenging with existing methods.

Another example is simulation of the de-wetting of paper pulp, where water is expelled through a "press-felt" material comprising a complex entanglement of fibres, as shown in Fig. 9. In this example, the complex topology makes it very difficult to partition the fibres into contact fields to prevent a non-physical no-slip contact at fibre–fibre interfaces.

In contrast, if a surface detection contact approach is used (such as the new approach described in Section 5), granular flow can be simulated from a single material group (divided into two velocity fields) with surface particle flags computed from the kernel sum. Using either the surface-flagged particles or the damage field, the field-gradient partitioning method lets us dynamically repartition material into contact pairs for initial surfaces or those created by fracture processes. This allows for more realistic representation of self-contact in porous materials, as well as a representation of frictional dissipation in comminution and post-fracture granular flow.

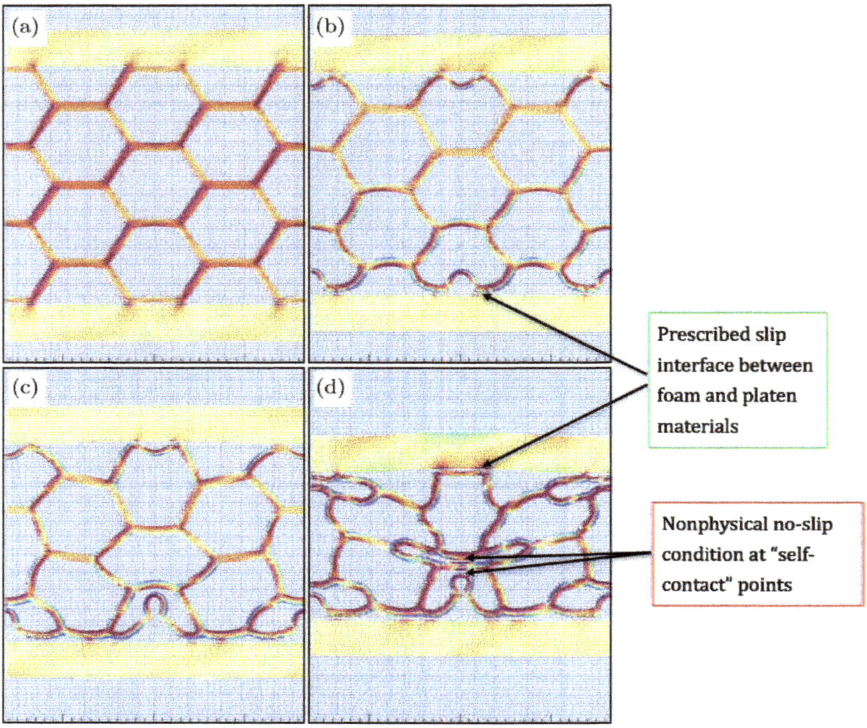

Fig. 8. Traditional multifield MPM simulation of dynamic compression of a 2D hex foam between two stiff platens. The foam and platen materials are in separate velocity fields allowing for frictional contact at the platen surface, but self-contact at internal surfaces of the foam produces a non-physical no-slip condition.

3.2. *Traditional approach for fracture in the material point method*

There are an abundance of continuum constitutive models of varying complexity that have been developed to model brittle failure in a broad spectrum of materials.[47] Such continuum damage/softening models are readily implemented in 3D, naturally accommodate branching and coalescence of failure surfaces, and can produce realistic patterns of brittle failure when combined with spatially heterogeneous and size-dependent strength.[48] However, when implemented in an MPM framework, these continuum models may not perform as desired.

The need to make "close observations in the material failure zone when softening occurs" was noted by Sulsky *et al.*,[24] when the MPM was first

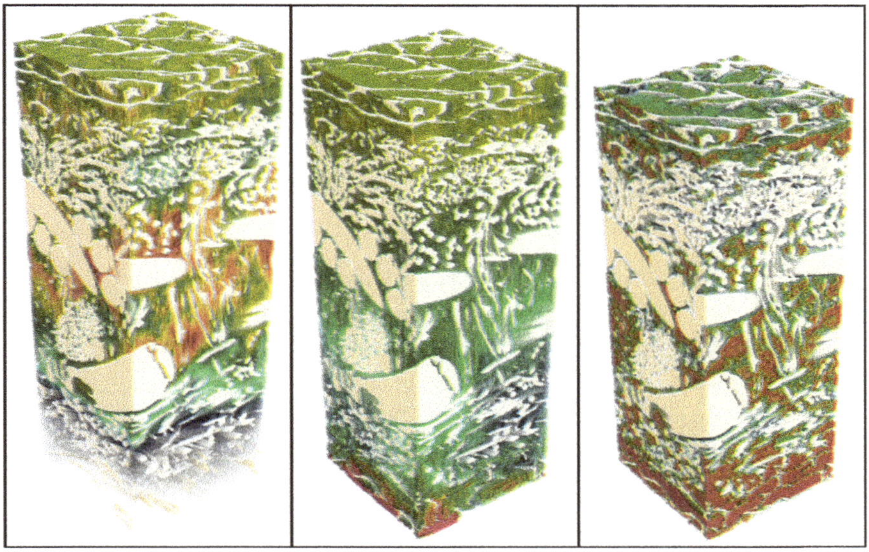

Fig. 9. Traditional single-field MPM simulation of press-felt de-wetting showing contours of the fluid velocity magnitude during pressing. The felt fibres (white) are in a single velocity field, creating a non-physical no-slip contact.

introduced. The potential for poor performance is because multiple material points within a single grid cell share a common velocity field, but do not necessarily have common material properties or loading history; spurious stresses may occur when both failed and intact material share a grid cell.

The problematic configuration is illustrated in Fig. 10, which shows a tensile bar with a single row of failed particles. Physically, we would expect the bar to separate at the failure plane when subjected to a tensile load. However, for a traditional single-field MPM solution, this will not occur. The damaged particles share a background grid cell with undamaged particles, which means that within a time-step both the damaged and undamaged particles are subjected to the same velocity gradient and corresponding increment in tensile strain. This strain produces no stress in the failed particle, but it will produce a stress in the intact material, corresponding to a force that holds the bar in place. The intact particles that share a grid cell with the failed particles experience a larger than normal strain, and the damage in these cells will evolve, causing the damaged region to spread across multiple grid cells.

This velocity-field-smearing effect can produce unexpected and undesirable behaviour as described.[39] A pathological example is shown in

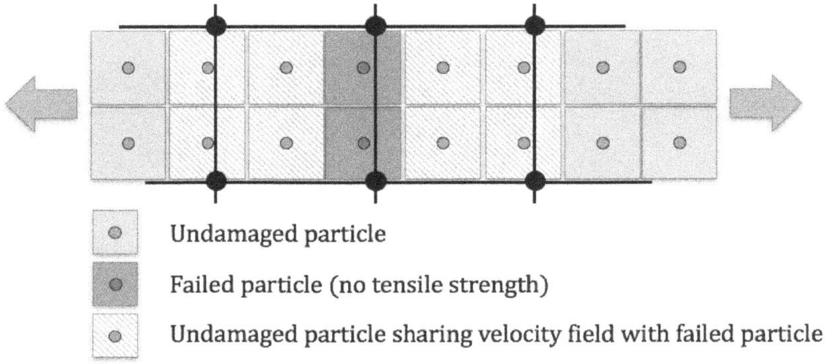

Fig. 10. Schematic of the velocity-field smearing error, where a failed particle shares a grid cell (and velocity field) with an intact particle, leading to a non-physical progression of damage.

Fig. 11(a), where a tensile bar with a single row of failed particles is subjected to a tensile load. If there is no initial motion, the damage spreads across a few rows of particles as the material separates. If the particle integration domains are frozen in the failed material, this will result in a separation of the material at the failure surface. However, the work required to create a failure surface will be greater than that required to fail a single row of material and will generally be inconsistent with the assumptions used to derive the failure model (e.g. some characteristic length). Even more troubling, is the case shown in Fig. 11(b), where the same bar is initialized with a uniform velocity. As the failed particles enter new grid cells, the damage propagates to the next layer of particles, resulting in a "smearing" of the damage field.

A tensile member containing a single layer of failed (or even deleted) particles can still support a tensile load, when the particle width is less than the grid cell spacing (as is generally the case). As a result, simulating fracture via particle deletion (akin to the element deletion practice in FEM[10]) can introduce significant error in addition to the loss of mass and momentum in the system.

An accurate treatment of fracture requires a representation of strong discontinuities in the velocity field. Enrichment of the MPM velocity field has been explored by introducing additional degrees of freedom at the particles,[34] as a way to describe interfaces between regions with different material properties. This type of approach has not yet been extended to model propagating fractures.

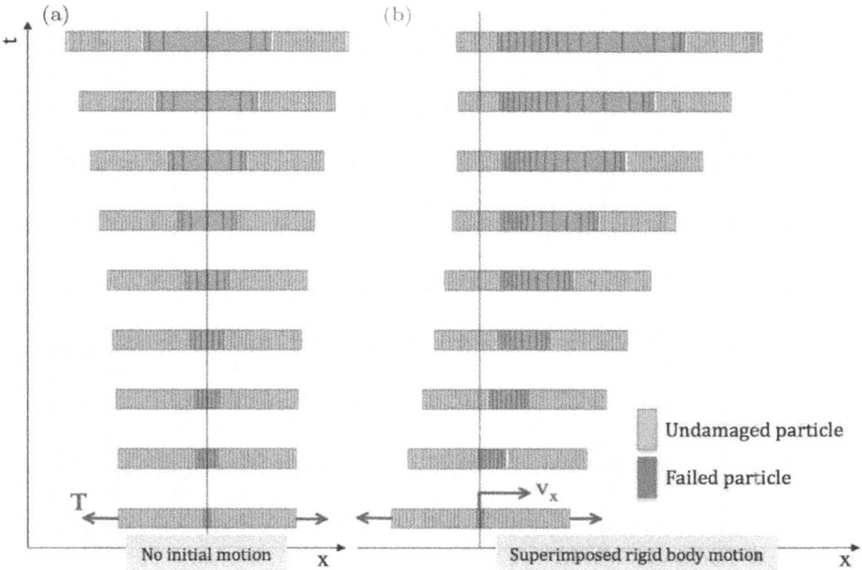

Fig. 11. Example of the velocity field smearing error in a tensile bar that is initially stationary (a) and a bar with a superimposed rigid body motion (b). Adapted from Ref. [39].

Fracture has been modelled in the MPM by Nairn *et al.*,[30,31] through introduction of explicit cracks that are used to partition the particles into two contacting velocity fields at each node, depending on the relative locations of the particle and crack to the node. This is illustrated in Fig. 12. Once the particles have been partitioned at a fracture node, the solution for the step is identical to that for frictional contact described in the previous section, except that the crack geometry can be used to exactly define the surface normal, rather than relying on a density-gradient calculation.

With explicit cracks, a variety of criteria can be used to nucleate or evolve the fracture surfaces; the explicit fracture approach has been shown to work well for single cracks in 2D or 3D,[31] but it becomes difficult to implement for complex 3D networks of branching and coalescing fractures. Additionally, while it is possible to update the crack position and orientation through time, this does not provide a framework for subsequent generalized frictional contact of fracture surfaces that have separated from their original opposing face to contact another body. Other approaches involving cohesive zones between regions[49] suffer similar limitations.

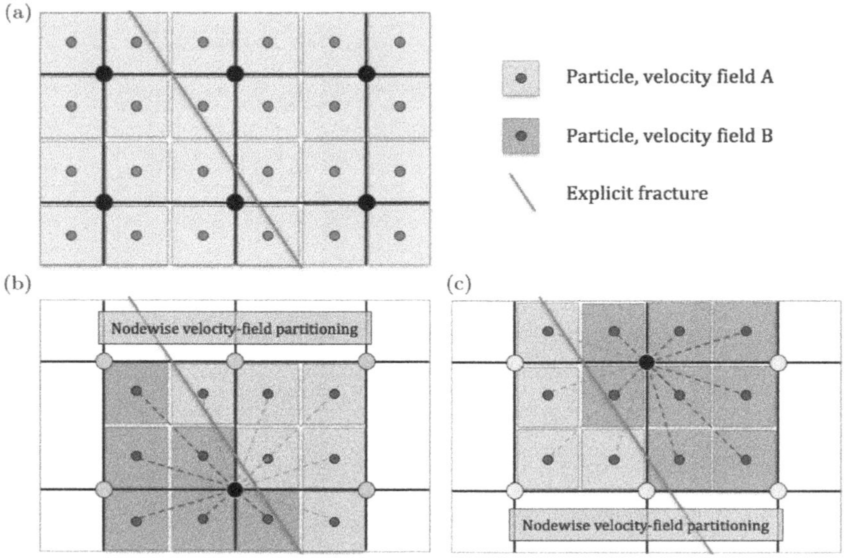

Fig. 12. Partitioning of material into two fields around a fracture depends on whether the crack lies between a particle and a node.[30] (a) Explicitly defined fracture in a plane of material. (b) Particle partitioning at top centre node and (c) Particle partitioning at bottom centre node. The field to which a given particle maps may vary from node to node.

This has motivated the development of the new damage-field gradient (DFG)-partitioning method,[33] which provides a means for dynamic repartitioning of material into contact groups based on the current distribution of damage in the material. This approach mitigates the errors of a single velocity field, allowing the use of continuum damage models without the complexity of explicitly tracking fractures. Frictional effects can be readily enforced for failed surfaces during subsequent contact, regardless of whether these surfaces were originally generated from the same fracture.

4. Field gradient partitioning

Field-gradient partitioning is a recently developed method for achieving fracture and fragmentation in the MPM, by using the gradients of a scalar damage field to partition material into contact pairs along fracture surfaces. The DFG-partitioning method is inherently 3D and naturally supports branching and coalescence of fracture networks. Partitioning material based on the gradient of an arbitrary scalar field (e.g. the density) enables frictional "self-contact", meaning sliding contact between surfaces which were originally part of the same velocity field (e.g. pore surfaces, or evolving fracture surfaces). The field-gradient partitioning method can be implemented efficiently within an existing MPM framework, it is amenable to efficient parallelization and, in most cases, it requires only a moderate increase in computational expense.

4.1. *Construction of the scalar damage field*

We will first describe the field-gradient partitioning approach as it relates to a continuum damage field, and then demonstrate how the concept is extended to the "self-contact" problem. Consider the 2D notched tensile bar shown in Fig. 13(a), where the green particles are undamaged, and the red particles are fully damaged, with no tensile strength. When subjected to a tensile load, the correct physical response is for the bar to separate in the region of failed particles, creating a new fracture surface connecting the two notches. We observe that if we were to construct a vector field where at each particle the vector pointed in the direction of increasing damage, those vectors would be normal to the desired fracture plane. This is the conceptual basis for the DFG-partitioning method.

Applying this concept, the particles along a fracture are partitioned into contact pairs based on the gradient of a continuous damage field, constructed from the some scalar damage measure ($\mathcal{D}_p$) in the particle. Generally, the term "damage" refers to a loss of stiffness, whereas "softening" refers to a loss of strength, but here we shall use the term more generally as a measure of material failure, so that a "fully damaged region" supports no tensile stress. We construct a smooth scalar damage field by summing particle-centred kernel functions,

$$\mathcal{D}(\boldsymbol{x}) = \sum_{p=1}^{n_p} \mathcal{D}_p \omega(||\boldsymbol{x} - \boldsymbol{x}_p||/r_p), \tag{15}$$

where ω is a kernel function with support radius r_p, and $\boldsymbol{x}_p$ is the position of the particle centre. For this work, a simple spherically symmetric cubic kernel is defined based on a normalized distance $\bar{r} = ||\boldsymbol{x} - \boldsymbol{x}_p||/r_p$,

$$\omega(\bar{r}) = \begin{cases} 1 - 3\bar{r}^2 + 2\bar{r}^3 : 0 \leq \bar{r} \leq 1, \\ 0 \qquad\qquad\quad\;\; : \bar{r} > 1. \end{cases} \tag{16}$$

The neighbour radius is the diagonal length of a background grid cell, $r_p = \sqrt{(\Delta x)^2 + (\Delta y)^2 + (\Delta z)^2}$ in 3D and $r_p = \sqrt{(\Delta x)^2 + (\Delta y)^2}$ for 2D plane-strain. As shown in Fig. 13(b), the simple kernel field has significant

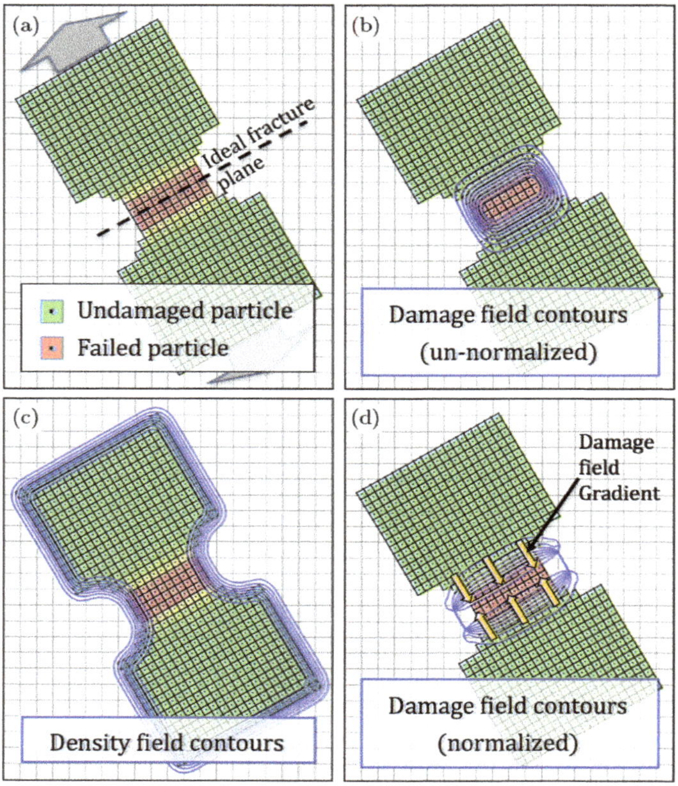

Fig. 13. Damage field for a tensile bar. (a) Particle colours indicate damage. (b) Overlaid contours show the *unnormalized* kernel damage field $\mathcal{D}(\boldsymbol{x})$. (c) Surface detection with kernel sum $\mathcal{S}(\boldsymbol{x})$ and (d) *Normalized* kernel damage field, $\overline{\mathcal{D}}(\boldsymbol{x})$. Normalization removes the edge effects so that the DFG over the domain of the solid material describes the orientation of the failure surface.

edge effects, which would complicate the analysis of damage and fracture. To eliminate edge effects, we divide by the kernel sum,

$$\mathcal{S}(\boldsymbol{x}) = \sum_{p=1}^{n_p} \omega(||\boldsymbol{x} - \boldsymbol{x}_p||/r_p), \tag{17}$$

which is equivalent to a normalized density gradient if the particles all have a uniform density. The contours of the kernel sum correspond to the outer contours of the material geometry, as shown in Fig. 13(c). Thus the kernel sum gradient can be used to identify surface particles, which is useful when initializing a model for frictional self contact, as described in Section 5.

The kernel sum is used to normalize the damage field, $\overline{\mathcal{D}}(\boldsymbol{x}) = \mathcal{D}(\boldsymbol{x})/\mathcal{S}(\boldsymbol{x})$, so that it accurately captures only the effects of the damage geometry, as shown in Fig. 13(d). The normalized damage field is undefined outside of the support domain of the kernel functions where $\mathcal{S}(\boldsymbol{x}) = 0$, but the damage field and its gradient are only evaluated at particle centres, where the field is well defined.

This type of kernel field construction is similar to that used in SPH,[50] but here we are only using the direction of the damage field gradient to locate particles relative to fracture surfaces, and the kernel field is not used in the solution of the momentum equation. As such, we do not require that the kernel functions have partition of unity, and do not require the more complex corrections used in the (RKPM).[51] The need to calculate the kernel field (and maintain the requisite neighbour list) introduces an additional cost, but we have found that this can be managed through efficient coding in a parallel framework and doesn't present a barrier to adoption of the method.[33]

4.2. *Field-gradient material partitioning*

Once the scalar damage field is defined, we then partition the particles into contact pairs by the location of the particles relative to the fracture surface. While it is possible to use topology simplification (*cf.* Ref. 52) to explicitly identify the fracture surfaces and intersection points from critical points (minima, maxima and saddles), doing so would introduce unnecessary cost and complexity.

Intuitively, it is appealing to partition the material into regions, and then compute the contact between those regions. For example, one might assign each particle to a velocity field by defining an arbitrary vector $\boldsymbol{n}^\star$, and then computing for each particle $\mathbf{sgn}[(\boldsymbol{\nabla}\bar{\mathcal{D}}_p) \cdot \boldsymbol{n}^\star]$. Zero values indicate an undamaged region, but when both positive- and negative-valued particles map to the same node, this indicates a contact surface. In some cases, this approach will produce blocks of material on either side of a crack, but the method fails when the damaged region is parallel to $\boldsymbol{n}^\star$, as shown in Fig. 14(a).

Instead, we take advantage of the background grid used in the MPM algorithm, and compare the direction of the DFG for each of the particles mapping to a given node. If opposing gradients exist, the node is potentially at a contact surface and must allow for the possibility of frictional sliding or separation.

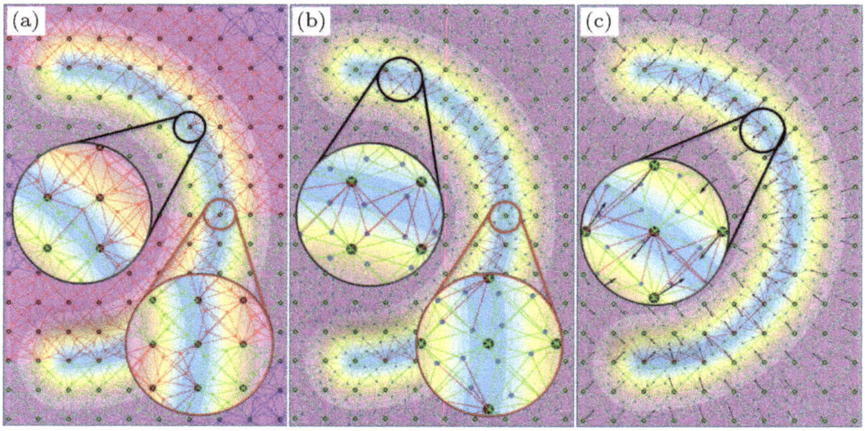

Fig. 14. Comparison of approaches for damage field partitioning around a curved crack highlighting acceptable (black) and erroneous (red) nodal partitioning. Particles are blue dots, grid nodes are black circles. The colour of the line segment joining each particle and node indicates the $\varsigma = 1, 2$ partitioning of the particle at that node. (a) Contact groups based on orientation of the particle gradient relative to an arbitrary vector $\{\boldsymbol{n}^\star\} = \{1, 0\}$ produces unacceptable results for fracture planes parallel to $\{\boldsymbol{n}^\star\}$. (b) Node-wise $\text{sgn}(\nabla\bar{\mathcal{D}}_p \cdot \nabla\bar{\mathcal{D}}_g)$ partitioning gives a desirable results except at the crack tip and when $\nabla\bar{\mathcal{D}}_g = \mathbf{0}$. (c) Node-wise $\text{sgn}(\nabla\bar{\mathcal{D}}_p \cdot \nabla\bar{\mathcal{D}}_g)$ partitioning using a nonlocal $\nabla\bar{\mathcal{D}}_g$ (black arrows) gives the desirable result.

Because the contact forces in the MPM are computed node-by-node, and it is not necessary that the ordering of the contact partitioning (i.e. whether the $\varsigma = 1$ field is above or below the crack) be consistent along the fracture surface, it is only necessary that the orientation of the surface normal for each field at the node be consistent with the direction of the normal to the fracture or contact surface. Thus node-wise partitioning can be applied as was done for explicit cracks (Fig. 12).

Evaluating the gradient of the damage field at a grid node, $\bar{\mathcal{D}}_g = \bar{\mathcal{D}}(\boldsymbol{x}_g)$, we can partition the particles as they are mapped to each node based on the sign of the dot product of the damage field gradient at the particle and at the node, $\mathbf{sgn}[(\boldsymbol{\nabla}\bar{\mathcal{D}}_p)\cdot(\boldsymbol{\nabla}\bar{\mathcal{D}}_g)]$. In many cases, this gives the desired partitioning, but this simple method fails when the grid node lies at the maximum of the scalar field and $\boldsymbol{\nabla}\bar{\mathcal{D}}_g = \mathbf{0}$, as shown in Fig. 14(b). This pathological case can occur quite frequently at the border of a fully damaged region.

As we view the contours of a damage field (Figs. 14–15), we can readily identify the desired orientation of a fracture at a grid node, but this is because our eye captures the trends over a region, not simply the gradient at a point. To capture this non-local effect, and since the partitioning method

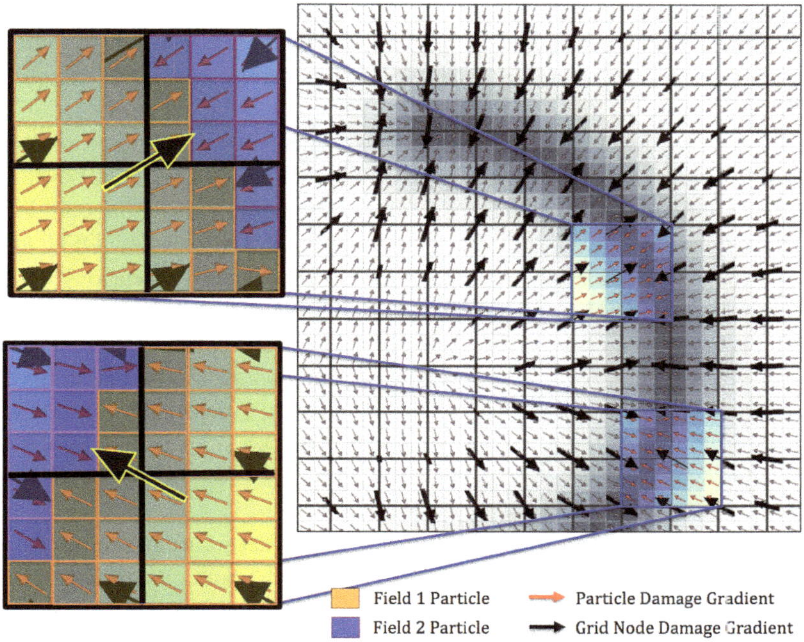

Fig. 15. Node-wise DFG-partitioning around a curved crack using a non-local gradient measure at the grid node. Inset images show the particle partitioning around a single grid node. The direction of the non-local damage gradient at the grid node ($\boldsymbol{\nabla}\bar{\mathcal{D}}_i^{\star}$) is shown as a highlighted black arrow. The surrounding particles are coloured based on their partitioning at the node, as determined from the relative orientation of $\boldsymbol{\nabla}\bar{\mathcal{D}}_i^{\star}$ to $\boldsymbol{\nabla}\bar{\mathcal{D}}_p$.

depends only on the direction of $\boldsymbol{\nabla}\bar{\mathcal{D}}_g^{\star}$ and not its magnitude, we formally define a non-local gradient at the grid node $\boldsymbol{\nabla}\bar{\mathcal{D}}_g^{\star}$ as a multiple of the unit vector ($\boldsymbol{n}_i^{\star}$) that maximizes the integral of $|\boldsymbol{n}_i^{\star} \cdot \nabla\mathcal{D}(\boldsymbol{x})|$ over the neighbourhood surrounding the grid node. In general there will be at least two solutions for $\boldsymbol{n}_i^{\star}$, and either can be used in the solution as long as a consistent definition is used for all particles mapping to the node. This approach eliminates the errors that occurred in the previous two examples, as shown in Fig. 14(c).

Finding the optimal value of $\boldsymbol{n}_i^{\star}$ is computationally expensive, but we have found that for determining the partitioning orientation it is sufficient to define the nonlocal DFG to be the gradient of the damage field evaluated at the point $\boldsymbol{x}^{\max}$ within some radius (r_g) for which the gradient has the greatest magnitude: $\boldsymbol{\nabla}\bar{\mathcal{D}}_i^{\star} = \boldsymbol{\nabla}\bar{\mathcal{D}}(\boldsymbol{x}_i^{\max})$, where

$$\boldsymbol{x}_i^{\max} = \arg\max_{x \in R}[\nabla\mathcal{D}(x)] \ni \|\boldsymbol{x} - \boldsymbol{x}_i\| < r_g. \tag{18}$$

This formulation was chosen because it allows for an efficient approximate evaluation. For a given node, $\nabla\bar{\mathcal{D}}_g^\star$ is set equal to the value of $\nabla\bar{\mathcal{D}}_p^\star$ for the particle (mapping to that node) that has the largest DFG magnitude, $\|\nabla\bar{\mathcal{D}}_p^\star\|$.

A detailed view of the partitioning of a curved-crack damage field using the approximate non-local grid-node DFG is shown in Fig. 15. In our simulations, we defined the neighbour radius to be equal to the diagonal length of the background grid cell. The red arrows show the scaled DFG at the particles (there are three particles-per-grid-cell in each direction) and the black arrows show the nodal DFG ($\nabla\bar{\mathcal{D}}_g^\star$), which is approximately normal to the crack surface along the entire length of the crack, as desired. While it is necessary that $\nabla\bar{\mathcal{D}}_g^\star$ be approximately normal to the crack surface, the direction can be reversed without affecting the algorithm. At the crack tip, the direction of $\nabla\bar{\mathcal{D}}_g^\star$ is no longer normal to the semicircular crack, but here the separability condition (described in the next section) would not be met, so the orientation of $\nabla\bar{\mathcal{D}}_g^\star$ is irrelevant.

Once the DFG has been defined at the particles and grid nodes, it can be used to partition particle contributions at each grid node as shown in the inset frames of Fig. 15. This is done within the loops over particles in which data are mapped between the particles and grid. For a given particle–node pair, the dot product of the DFG at the particle and grid node is computed, the sign of which is used to determine the field to/from which the particle will map.

It is important to note that while the damage field gradient vector is approximately normal to the failure surface, it is used only to define the topology by partitioning the material into contact groups and it does not define the surface normal. Once the material is partitioned, the normal for subsequent contact interactions is defined by Eq. (13). As such, we expect the accuracy of the surface normals to be comparable to that of the surface representation in generalized MPM contact, improving with the resolution of the discretization relative to the curvature of the surface.

This approach for material partitioning assumes that a local contact plane can be defined at each node to describe the contact interactions of two material regions (i.e. the fracture faces) at that node. For crack intersection and branching, there can be three or more regions of material meeting at a single point. However, a crack in a smeared-damage model has a finite width, since fracture and contact are resolved only to the resolution of the grid cell. As a result, crack intersections generally manifest as grid cells containing fully damaged particles and the progression and branching of cracks

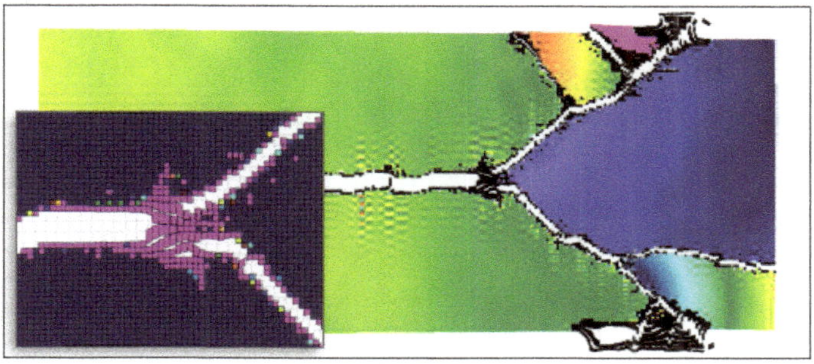

Fig. 16. Crack branching with the damage-field partitioning (DFP) method showing contours of velocity with an inset detail showing particle damage, with fully damaged particles in purple.

occur naturally. Figure 16 shows an example of crack branching in a notched plate subject to distributed tensile surface on the upper and lower surfaces.

We see that generally the fracture separates, leaving a row of damaged particles on each failed surfaces, as desired. This ensures that if separated failed surfaces contact later in the simulation, DFG-partitioning will activate and ensure frictional contact. The branch point in the fracture manifests as a mass of failed particles 1–2 grid cells across. The size of this branch region will vary with the alignment of the particles and grid, but since it generally accounts for a small fraction of the failed particles in a simulation it is not a significant source of error or mesh dependence.

In general, we find that the DFG-partitioning method has significantly less mesh dependence in the fracture response relative to the single-field method, as a result of eliminating the velocity-field smearing error. This is demonstrated in Fig. 17, which shows the results of a notched impact simulated using both single-field and DFG-partitioning methods at three mesh resolutions. For this simulation, we use a constitutive model for the notched bar that derives a softening response from the fracture energy G_f and a characteristic length. As such, we can compute the error in the simulation from an energy balance over the fracture event. There is some inherent error in energy conservation (even in an elastic collision) due to an inability to capture high-frequency motion at low grid resolutions; this potential error was computed and is shown with the width of the green band. For the single-field solution, the energy dissipation in fracture decreases as the mesh is refined but there is significant error at low resolutions, whereas for the DFG-partitioning method the solution is more consistent and lies within the expected range at all resolutions.

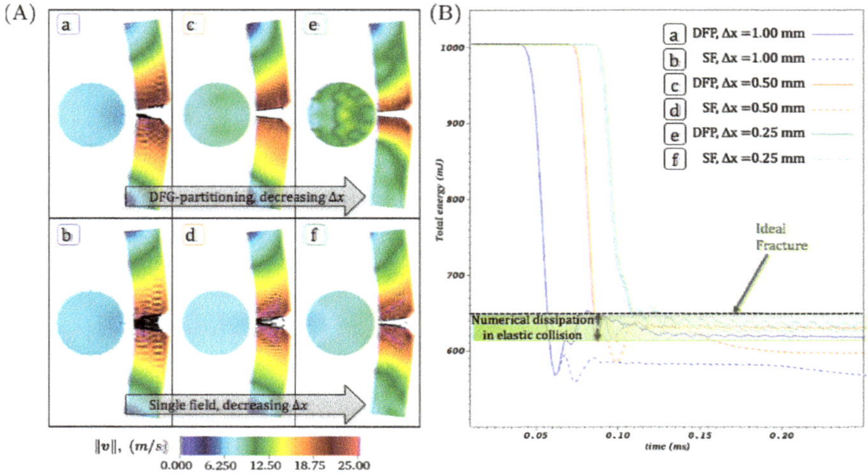

Fig. 17. A notched-impact fracture problem showing the effect of grid resolution on the velocity field (A) and total energy conservation and (B) for single-field (SF) and DFP solutions.

Details of this verification problem are given in Ref. [33] along with results for a Kalthoff fracture simulation, which demonstrates that the DFG-partitioning method produces the correct crack speed in dynamic fracture simulations.

4.3. *Separability condition*

The DFG-partitioning is used to assign the material at each side of a fracture into separate velocity fields at each node. The existence of damage does not necessarily imply that the material should be allowed to separate, since the particles may not have failed $\mathcal{D}_p < 1$, or there may not be sufficient failed particles to traverse the region surrounding the node. A *separability condition* is defined to determine whether the damage state at the node should be treated as a crack (i.e. two fields in contact), or whether the material state should locally be described by a single velocity field. For the topology to be consistent with the constitutive model predictions, the material should be allowed to separate only once a "surface" of failure particles has formed. To estimate if this has occurred, we first evaluate the average damage ($\overline{\mathcal{D}}_{\varsigma_i}$) for each velocity field at a given node from the mass-weighted average of the particle damages for the particles mapping to that node. We also compute the maximum value ($\bar{\mathcal{D}}_i^{\max}$) of damage for the particles mapping to the node in each velocity field.

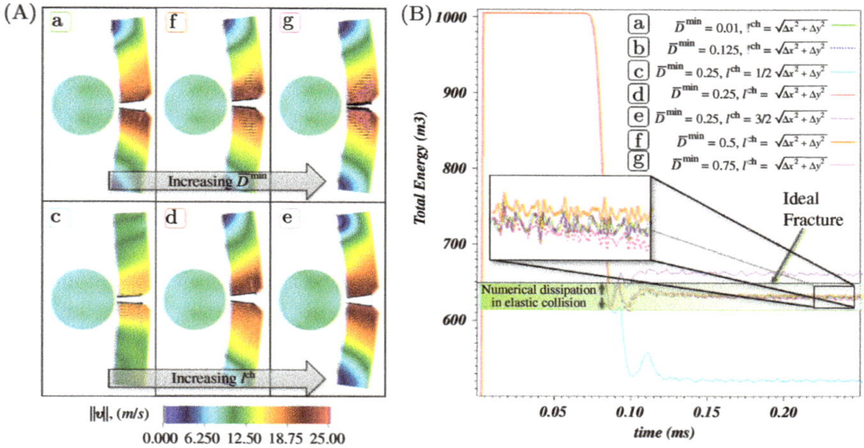

Fig. 18. Energy dissipation in a notched-impact fracture problem using DFP solutions with various values of the constitutive model characteristic length (ℓ^{ch}) and the partitioning method separability minimum damage parameter ($\bar{\mathcal{D}}^{max}$) showing (A) contours of the magnitude of the velocity field, and (B) total of the kinetic and strain energy in the system, showing the fracture energy release rate converges to the theoretical value if ℓ^{ch} is consistent with the damage law, for all values of $\bar{\mathcal{D}}^{max}$.

For the examples presented here, separability at a node requires that (i) at least one particle mapping to the node is fully damaged, i.e. $\bar{\mathcal{D}}_i^{max} = 1$, and (ii) the minimum value of the average damage exceeds a specified separability threshold $\min(\overline{\mathcal{D}}_{1i}, \overline{\mathcal{D}}_{2i}) > \bar{\mathcal{D}}^{min}$, for $0 < \bar{\mathcal{D}}^{min} < 1$. This definition of the separability criterion is somewhat arbitrary, but we have found that the fracture results are fairly insensitive to this parameter, producing a consistent energy dissipation in the formation of a fracture surface for a broad range of the separability min-damage parameter.

An example of this is shown in Fig. 18, which shows the energy dissipation in a dynamic notched-bar fracture problem for various values of the separability criterion min-damage parameter, as well as the constitutive model characteristic length parameter. The characteristic length parameter has a physical meaning, and should affect the energy dissipation per unit area of fracture surface, and indeed we see this dependency. The separability criterion min-damage is a numerical solution parameter, and we see that it has little effect on the energy dissipation in the solution, as desired.

We do observe that there is some noise in the velocity field surrounding the fracture and that this noise is made worse as the separability min-damage parameter is increased. In this work, we have considered only scalar damage models; however, it may be possible to extend the

DFG-partitioning method to accommodate models with anisotropic tensor damage. The DFG method identifies potential separation/sliding surfaces based on regions of damaged material. If an anisotropic damage model is used, the results will only be physically reasonable if the constitutive model failure plane aligns with the separation surfaces. For certain anisotropic modes of failure, this may not always be the case. To account for anisotropy in a tensor damage model, a separability criterion could be developed that accounts for the alignment of the failure planes in the material and the damage field gradient. This is a potential area for future development that would extend the applicability of this method to a broader class of materials.

5. Self-contact

One promising application for the material point method is mesoscale simulation, where complex microstructure models can be constructed directly from the voxelized data produced by CT scans. MPM is particularly well suited for modelling porous and granular materials, where localized large deformation, self-contact and frictional effects may significantly affect the bulk material response. Another area of emerging interest relates to low-density metamaterials, produced with additive manufacturing and having 3D multiscale lattice structures. In some cases these nano-truss structures are able to withstand extreme deformation, which induces frictional self-contact under normal operation.[53] Simulating this behaviour is a challenging test for traditional numerical methods.

We have already seen that by defining a kernel damage field, the material on either side of a crack can be partitioned into contact pairs. With a slight modification, we can also achieve frictional self contact with field-gradient partitioning. Using the magnitude of the gradient of the kernel sum field ($\mathcal{S}(\boldsymbol{x})$ from Eq. (17)) evaluated at a particle, we can detect surface particles in an initial porous or granular structure. We define, for each particle, a surface flag s_p that is set to 1 for particles at the surface and 0 for interior particles. Such kernel-based surface detection algorithms are common in mesh-free methods, (*cf.* Ref. 54). If the grains are initially in contact (but not bonded) it would be necessary to "paint in" the surface flag at initialization. With this approach, surface particles are readily identified.

With the surface flag so-defined, the damage field gradient can be constructed with Eq. (15), using $\max(s_p, \mathcal{D}_p)$ in place of $\mathcal{D}_p$. In this way, frictional contact and separation will occur between surface particles, even if those particles are undamaged. By using the kernel fields to detect surface particles, the thickness of the surface-flagged region is consistent with the

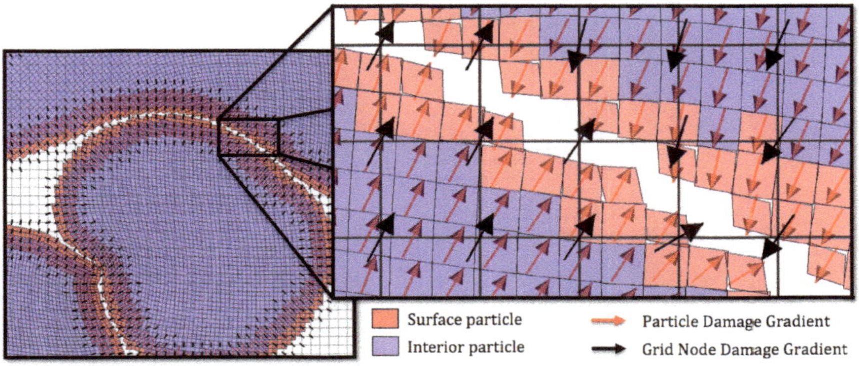

Fig. 19. Illustration of the DFG partitioning method using prescribed surface flags rather than accumulated damage to produce frictional contact at a material interface.

kernel radius in the DFG construction, which seems to improve the surface partitioning in large-deformation frictional contact. An example of this is shown in Fig. 19, which shows the direction of the DFGs for the particles and grid nodes along a contact surface between two initially separate bodies. This image is generated from the self-contact test case presented in Fig. 20.

6. Comminution and fragmentation

The verification problems give confidence in the ability of the DFG-partitioning method to accurately model dynamic fracture and large-deformation frictional contact, but the true value of the method is its unique ability to model challenging problems for which other methods would fail. Our motivation in developing this method is the need to model grain-scale fracture and friction in the comminution of granular materials. In these grain-scale simulations, a representative volume element will typically comprise hundreds or thousands of individual grains, and computational resources will generally limit the resolution available to model a single grain. As such, it is essential that the numerical method provide reasonable results for the comminution response even at relatively coarse mesh refinement, and that the results are consistent as the mesh is refined.

To test this capability, we simulate the dynamic compaction of a brittle disk between steel platens, where a uniform distributed load is applied to the platens sufficient to fracture the disk and produce a continuous reduction in particle size during granular flow. The disk is modelled in 2D

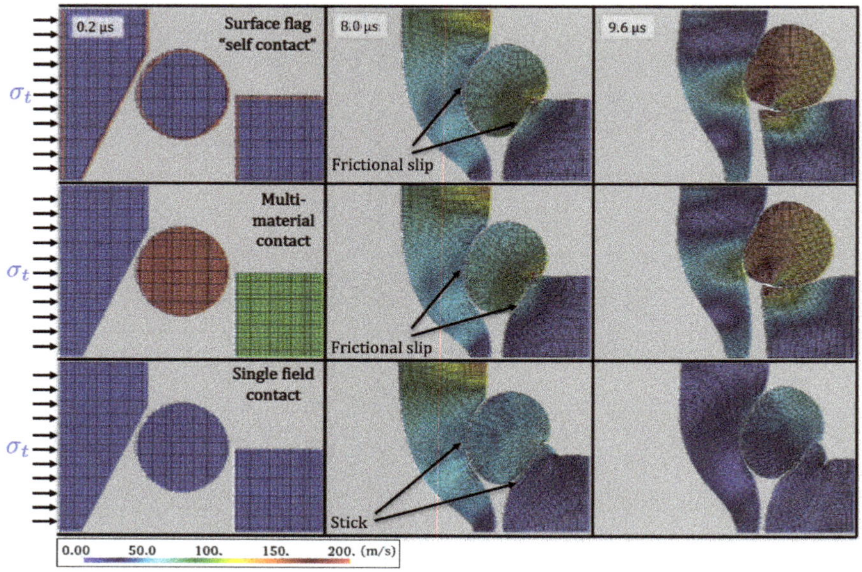

Fig. 20. Verification that the "self-contact" approach produces the same response as a prescribed multifield contact algorithm for a problem involving frictional contact with large displacement and large deformation.

plane strain, with two particles per grid cell in each direction. We simulate the compaction using DFG-partitioning and compare this to a traditional single-field solution at three grid resolutions. To achieve consistent fragmentation results in a brittle material, it is necessary to introduce spatial variability and scale effects in the strength of the material, which is done by defining a Weibull distribution of tensile strength. With this approach, the fracture pattern will differ at each grid resolution, but characteristics such as the fragment size distribution and average fragment velocity (both measures of the energy dissipation in fracture) should be consistent. Figure 21 shows the velocity field and damage in the disk at three times for each simulation method.

From Fig. 17 we see that the DFG-partitioning method produces a more consistent energy dissipation in fracture formation relative to the single-field solution. In Fig. 21, we see how that difference manifests in a simulation of comminution. The single-field solution artificially restricts fracture formation, because a fracture cannot propagate as easily from a single defect (small value of initial particle strength) or point of damage. As a result, the single-field solution fractures later (note the time for each

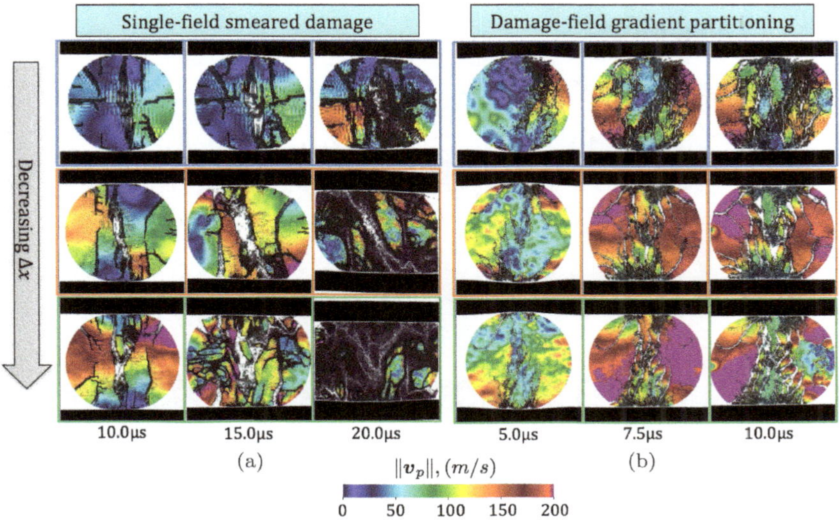

Fig. 21. Dynamic fragmentation of a brittle disk simulated using traditional single-field MPM (a) and with DFP (b). The time scales are different because the single-field solution artificially restricts fracture propagation at low resolution while the DFG-partitioning scheme produces consistent fragmentation at all resolutions.

snapshot is different for the single-field and DFP-partitioning solutions) and fracture progresses more quickly as the mesh is refined. The velocity-field smearing effect is also apparent in the single-field solution, with the width of the failure regions growing over time, even causing damage to spread over the entire domain for the fine-resolution result (where the velocities are the highest).

To quantify the results, we use an image segmentation analysis of the undamaged material to identify connected components, from which we compute a particle size distribution and an average velocity for each simulation. To facilitate the comparison, we analyze the single-field result at $t = 15.0\mu$s (before the velocity field smearing error dominates the results) and analyze the DFG-partitioning solution at $t = 10.0$ μs. Figure 22 shows the segmented fragments, particle size distribution, and average velocity for each simulation result. Because the times are different, we don't expect the fragment size distribution or velocity to be the same between the two methods, but are instead evaluating the consistency of the results at different resolutions for each method.

The results clearly show that the single-field solution has a clear trend of fragment size decreasing and fragment velocity increasing as the mesh is

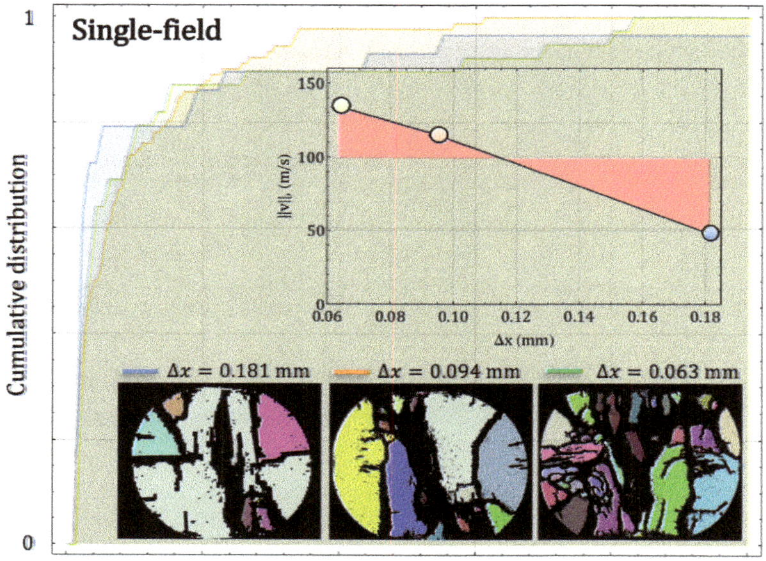

(a)

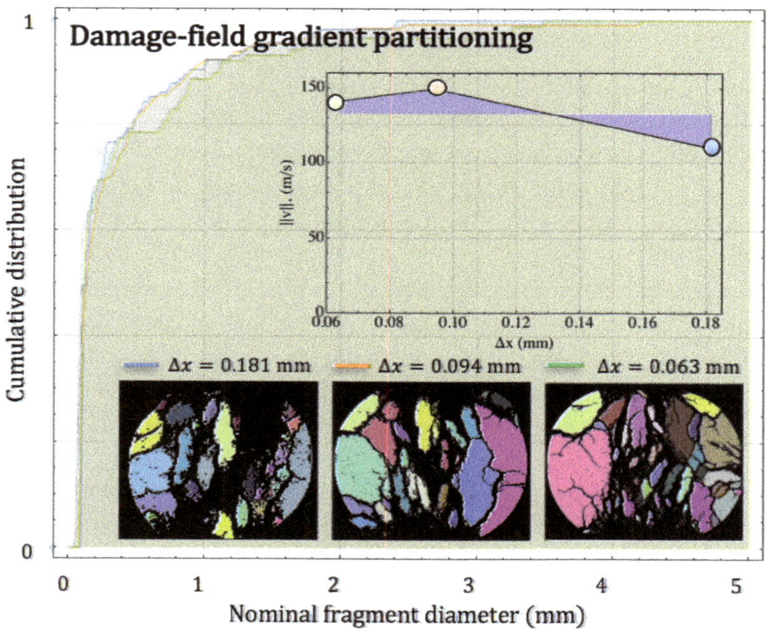

(b)

Fig. 22. Particle size distributions with average velocity and fragment geometry inset for a dynamic disk fragmentation simulation using traditional single-field MPM (a) and DFG-partitioning (b). The DFG-partitioning method gives more uniform fragment sizes and velocities as the grid is refined.

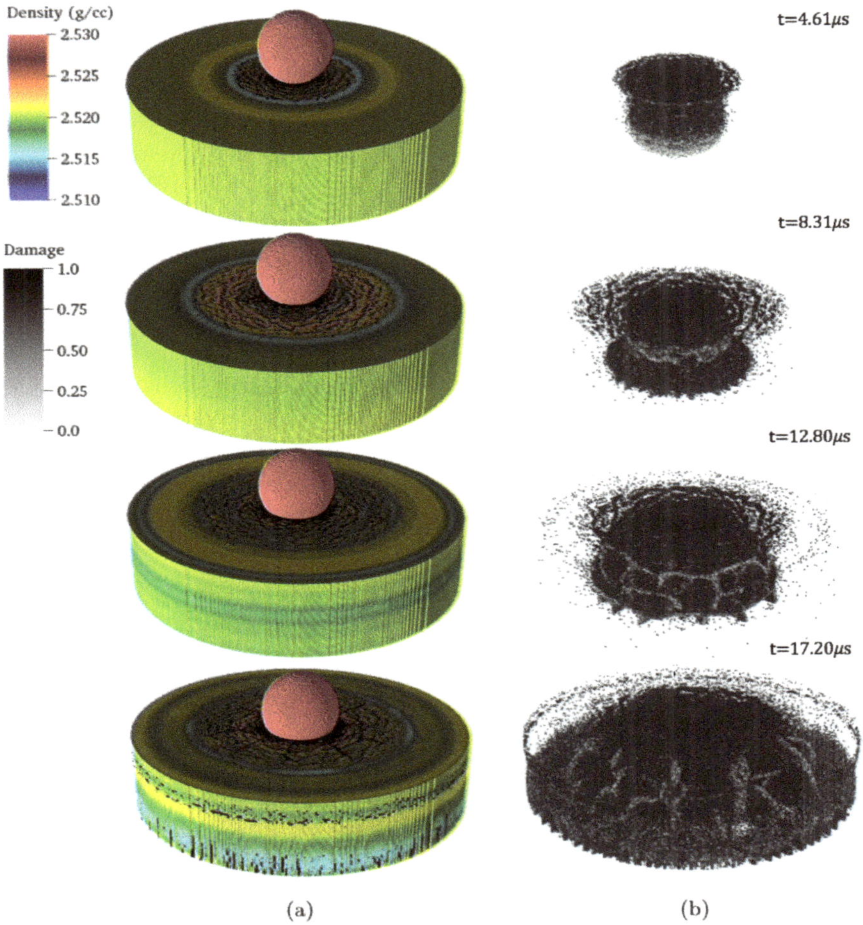

Fig. 23. Simulation of fragmentation of a brittle disk after impact by an steel sphere illustrating a 3D implementation of the DFG-partitioning method. (a) Contours show the density variation in the target material with damage overlaid in black and (b) The evolution of damage material is shown with intact material hidden.

refined. In contrast, the DFP method produces relatively consistent results at each resolution. To our knowledge, the DFG-partitioning method is the only approach that has demonstrated this capability to produce mesh-independent results of comminution in brittle materials.

While we have focused on the applications of simulating comminution for mesoscale modelling, the DFG-partitioning method also has applications in modelling other types of fracture and fragmentation problems. An example of this is shown in Fig. 23, which shows a 3D simulation using the DFG-partitioning method to model fragmentation of a thick glass disk after impact from a steel sphere. The disk has a diameter of 100 mm and a thickness of 20 mm, and the 20 mm diameter sphere impacts the surface at 300 $\mathrm{m\,s^{-1}}$. The sphere is modelled as a linear-elastic perfectly-plastic material, and disk is modelled as linear elastic with a Weibull-distributed tensile strength and a time-to-failure model, similar to the 2D fragmentation.[33] The example is included primarily to illustrate that the method works naturally in 3D without modification, but also serves to illustrate the broader class of applications for the approach.

The simulation results show the progression of failure with crazing around the impact surface and formation of a ball of damaged material under the point of impact. The damaged region grows to form a conical crack extending to the back surface, while radial cracks form at the impact surface. As failure progresses, the radial cracks grow and merge with the conical damage region, eventually leading to fragmentation of the target disk.

7. Summary and Conclusions

The DFP method we have described introduces a new capability to the computational mechanics community to simulate challenging problems involving fracture and contact without the requirements for simplistic topographies that have limited other approaches. This method has been developed to be consistent with the motivation for using the material point method, avoiding the introduction of explicit descriptions of fracture or contact surfaces that might compromise the robustness of the MPM for modelling extreme material deformation. In this way, the DFG-partitioning method addresses a long-standing limitation of the MPM when used with materials that experience softening or damage, expanding the range of applications for which the MPM is well suited.

References

1. C. J. Shih, V. F. Nesterenko, and M. A. Meyers, High-strain-rate deformation and communition of silicon carbide, *Journal of Applied Physics*, **83**(9), 4660–4671 (1998).
2. N. Bourne, On the impact and penetration of soda-lime glass, *International Journal of Impact Engineering*, **32**(1–4), 65–79 (2005).
3. N. Bourne, The shock response of float-glass laminates, *Journal of Applied Physics*, **98**(6), 063515 (2005).

4. N. Bourne, L. Forde, J. Millett, and J. Field, Impact and penetration of a borosilicate glass, *Journal De Physique IV*, **7**, 1–6 (1997).
5. N. Bourne, J. Millett, and Z. Rosenberg, Failure in a shocked high-density glass, *Journal of Applied Physics*, **80**(8), 4328 (1996).
6. J. Millett, Effect of internal strain on the propagation of failure in shock loaded soda-lime glass, *Journal of Applied Physics*, **95**(9), 4681 (2004).
7. M. Iskander, S. Bless, and M. Omidvar, *Rapid Penetration into Granular Media: Visualizing the Fundamental Physics of Rapid Earth Penetration*, 1st edn., Elsevier, Amsterdam (2015).
8. M. Rubin and I. Einav, A large deformation breakage model of granular materials including porosity and inelastic distortional deformation rate, *International Journal of Engineering Science*, **49**(10), 1151–1169 (Oct, 2011).
9. G. T. Camacho and M. Ortiz, Computational modelling of impact damage in brittle materials, *International Journal of Solids and Structures*. **33**(2), 2899–2938 (1996).
10. J.-H. Song, H. Wang, and T. Belytschko, A comparative study on finite element methods for dynamic fracture, *Computational Mechanics*, **42**(2), 239–250 (2008).
11. T. Belytschko, H. Chen, J. Xu, and G. Zi, Dynamic crack propagation based on loss of hyperbolicity and a new discontinuous enrichment, *International Journal for Numerical Methods in Engineering*, **58**(12), 1873–1905 (2003).
12. C. Annavarapu, R. R. Settgast, E. Vitali, and J. P. Morris, ScienceDirect A local crack-tracking strategy to model three-dimensional crack propagation with embedded methods, *Computer Methods in Applied Mechanics and Engineering*. **311**, 815–837 (2016).
13. R. de Borst, J. J. C. Remmers, and A. Needleman, Mesh-independent discrete numerical representations of cohesive-zone models, *Engineering Fracture Mechanics*, **73**(2), 160–177 (2006).
14. T. Belytschko, R. Gracie, and G. Ventura, A review of extended/generalized finite element methods for material modeling, *Modelling and Simulation in Materials Science and Engineering*, **17**(4), 43001 (2009).
15. S. Loehnert, D. S. Mueller-Hoeppe, and P. Wriggers, 3D corrected XFEM approach and extension to finite deformation theory, *International Journal for Numerical Methods in Engineering*, **86**(4–5), 431–452 (2011).
16. K. T. Ramesh, J. D. Hogan, J. Kimberley, and A. Stickle, A review of mechanisms and models for dynamic failure, strength, and fragmentation, *Planetary and Space Science*, **107**, 10–23 (2015).
17. E. Vitali and D. J. Benson, Modeling localized failure with arbitrary Lagrangian Eulerian methods, *Computational Mechanics*, **49**(2), 197–212 (2012).
18. E. Vitali and D. J. Benson, Contact with friction in multi-material arbitrary Lagrangian-Eulerian formulations using X-FEM, *International Journal for Numerical Methods in Engineering*, **76**(6), 893–921 (2008).
19. M. B. Liu and G. R. Liu, Smoothed particle hydrodynamics (SPH): An overview and recent developments, *Archives of Computational Methods in Engineering*, **17**(1), 25–76 (2010).

20. N. Frontiere, C. D. Raskin, and J. M. Owen, CRKSPH A conservative reproducing kernel smoothed particle hydrodynamics scheme, *Journal of Computational Physics*, **332**, 160–209 (2017).
21. T. Rabczuk, G. Zi, S. Bordas, and H. Nguyen-Xuan, A simple and robust three-dimensional cracking-particle method without enrichment, *Computer Methods in Applied Mechanics and Engineering*, **199**(37), 2437–2455 (2010).
22. T. Rabczuk and T. Belytschko, Application of particle methods to static fracture of reinforced concrete structures, *International Journal of Fracture*, **137**(1–4), 19–49 (2006).
23. S. A. Silling, Dynamic fracture modeling with a meshfree peridynamic code, In K. J. Bathe (ed.), *Computational Fluid and Solid Mechanics 2003*. Elsevier Science Ltd, Oxford, pp. 641–644 (2003).
24. D. Sulsky, Z. Chen, and H. L. Schreyer, A particle method for history-dependent materials, *Computer Methods in Applied Mechanics and Engineering*, **118**(1), 179–196 (1994).
25. P. Huang, X. Zhang, S. Ma, and X. Huang, Contact algorithms for the material point method in impact and penetration simulation, *International Journal for Numerical Methods in Engineering*, **85**(4), 498–517 (2011).
26. M. A. Homel, J. Guilkey, and R. M. Brannon, Continuum effective-stress approach for high-rate plastic deformation of fluid-saturated geomaterials with application to shaped-charge jet penetration, *Acta Mechanica*. **10**, 1007 (2015).
27. K. Kamojjala, R. Brannon, A. Sadeghirad, and J. Guilkey, Verification tests in solid mechanics, *Engineering with Computers*, **31**(2), 193–213 (2013).
28. S. G. Bardenhagen, J. E. Guilkey, K. M. Roessig, J. U. Brackbill, and W. M. Witzel, An improved contact algorithm for the material point method and application to stress propagation in granular material, *Computer Modelling in Engineering & Sciences*, **2**(4), 509–522 (2001).
29. A. L. Tonge, J. Kimberley, and K. T. Ramesh, A consistent scaling framework for simulating high rate brittle failure problems, *Procedia Engineering*, **58**, 692–701 (2013).
30. J. A. Nairn, Material point method calculations with explicit cracks, *Computer Modeling in Engineering and Sciences*, **4**, 649–663 (2003).
31. Y. J. Guo and J. A. Nairn, Three-dimensional dynamic fracture analysis using the material point method, *Computer Modelling in Engineering and Sciences*, **16**, 141–155 (2006).
32. N. P. Daphalapurkar, H. Lu, D. Coker, and R. Komanduri, Simulation of dynamics crack growth using the generalized interpolation material point (GIMP) method, *International Journal of Fracture*, **143**, 79–102 (2007).
33. M. A. Homel and E. B. Herbold, Field-gradient partitioning for fracture and frictional contact in the material point method, *International Journal for Numerical Methods in Engineering*, **109**(7), 1013–1044 (2017).
34. A. Sadeghirad, R. M. Brannon, and J. E. Guilkey, Second-order convected particle domain interpolation (CPDI2) with enrichment for weak

discontinuities at material interfaces, *International Journal for Numerical Methods in Engineering*, **95**(11), 928–952 (2013).
35. B. Wang, P. J. Vardon, M. A. Hicks, and Z. Chen, Computers and Geotechnics Development of an implicit material point method for geotechnical applications, *Computers and Geotechnics*, **71**, 159–167 (2016).
36. I. Iaconeta, A. Larese, R. Rossi, and E. Onate, An implicit material point method applied to granular flows, *Procedia Engineering*, **175**, 226–232 (2017).
37. S. G. Bardenhagen and E. M. Kober, The generalized interpolation material point method, *CMES — Computer Modeling in Engineering and Sciences*. **5**, 477–495 (2004).
38. A. Sadeghirad, R. M. Brannon, and J. Burghardt, A convected particle domain interpolation technique to extend applicability of the material point method for problems involving massive deformations, *International Journal for Numerical Methods in Engineering*, **86**(12), 1435–1456 (2011).
39. M. A. Homel, R. M. Brannon, and J. Guilkey, Controlling the onset of numerical fracture in parallelised implementations of the material point method (MPM) with convective particle domain interpolation (CPDI) domain scaling, *International Journal for Numerical Methods in Engineering* (2015).
40. P. C. Wallstedt and J. E. Guilkey, An evaluation of explicit time integration schemes for use with the generalized interpolation material point method, *Journal of Computational Physics*, **227**, 9628–9642 (2008).
41. D. Sulsky, S. J. Zhou, and H. L. Schreyer, Application of a particle-in-cell method to solid mechanics, *Computer Physics Communications*, **87**, 236–252 (1995).
42. M. A. Homel, J. E. Guilkey, and R. M. Brannon, Mesoscale validation of simplifying assumptions for modeling the plastic deformation of fluid-saturated porous material, *Journal of Dynamic Behavior of Materials*, **3**(1), 23–44 (2017).
43. J. Guilkey, T. Harman, J. Luitjens, J. Schmidt, J. Thornock, J. de St. Germain, S. Shankar, J. Peterson, and C. Brownlee, Uintah User Guide Version 1.1, *SCI Institute, University of Utah*, **UUSCI-2009** (2009).
44. S. G. Bardenhagen, J. U. Brackbill, and D. Sulsky, The material-point method for granular materials, *Computer Methods in Applied Mechanics and Engineering*, **187**(3), 529–541 (2000).
45. K. Appel and W. Haken, Every planar map is four colorable, in *Mathematical Solitaires & Games*, p. 145 (1980).
46. D. J. Benson and J. O. Hallquist, A single surface contact algorithm for the post-buckling analysis of shell structures, *Computer Methods in Applied Mechanics and Engineering*, **78**(2), 141–163 (1990).
47. J.-L. Chaboche, Continuum damage mechanics: Part IIDamage growth, crack initiation, and crack growth, *Journal of Applied Mechanics*, **55**(1), 65–72 (1988).
48. O. E. Strack, R. B. Leavy, and R. M. Brannon, Aleatory uncertainty and scale effects in computational damage models for failure and fragmentation, *International Journal for Numerical Methods in Engineering*, **102**(3–4), 468–495 (2014).

49. N. P. Daphalapurkar, H. Lu, D. Coker, and R. Komanduri, Simulation of dynamic crack growth using the generalized interpolation material point (GIMP) method, *International Journal of Fracture*, **143**(1), 79–102 (2007).
50. J. J. Monaghan, Smoothed particle hydrodynamics, *Reports on Progress in Physics*, **68**(8), 1703 (2005).
51. W. K. Liu, S. Jun, Y. F. Zhang, and Others, Reproducing kernel particle methods, *International Journal for Numerical Methods in Fluids*, **20**(8–9), 1081–1106 (1995).
52. A. Gyulassy and V. Natarajan. Topology-based simplification for feature extraction from 3D scalar fields, *Visualization, 2005, VIS 05, IEEE*, 535–542, (2005).
53. L. R. Meza, S. Das, and J. R. Greer, Strong, lightweight, and recoverable three-dimensional ceramic nanolattices, *Science*, **345**(6202), 1322–1326 (2014).
54. S. Marrone, A. Colagrossi, D. Le Touzé, and G. Graziani, Fast free-surface detection and level-set function definition in SPH solvers, *Journal of Computational Physics*, **229**(10), 3652–3663 (2010).

Chapter 9

A Cohesive-frictional Grain-boundary Technique for Microstructural Analysis of Polycrystalline Materials

I. Benedetti

The development of a 3D microstructural model for the analysis of degradation and failure in polycrystalline materials is reviewed in the present chapter. The material is explicitly modelled at the grain level, using integral equations in conjunction with a phenomenological crystal plasticity framework for the bulk grains, and with cohesive-frictional laws to represent inter-granular micro-cracking processes. The method allows to capture the initiation, development and coalescence of damage or plasticity at the aggregate scale. The formulation's key feature is the representation of the mechanical problem in terms of inter-granular variables only, which allows to reduce the computational cost of the analysis. In the present chapter, the building blocks of the technique are described and emphasis is given to the description of the adopted cohesive-frictional laws, which are the fundamental ingredient for the study of inter-granular degradation processes. The potential of the framework is illustrated through micro-cracking and crystal plasticity simulations. The effect of inter-granular friction on the macroscopic material properties is highlighted through micro-cracking simulations under compressive loading.

1. Introduction

The design of materials for advanced structural applications requires deep understanding of the material response to different operational and environmental conditions as well as the knowledge of their degradation and failure mechanisms. Phenomenological constitutive models may not always be able to predict complex materials behaviours, especially when nucleation and evolution of damage are involved. It is widely recognized that such phenomena and, in general, the macroscopic properties of a material depend on the *features of* and *interactions between* the elementary micro-constituents, which make the microstructure inherently heterogeneous. With current advancements in materials science and technology, there is interest in understanding and characterizing the link between material microstructures and macro-properties, i.e. the *structure–property relation*,[1–5] which also attracts remarkable technological interest due to the possibility of designing high-performance materials with enhanced macro-properties, e.g. fracture toughness, through manipulations of the microstructure.[6]

The characterization of the micro–macro link, under the assumption of *separation of scales*, is generally known as *material homogenization* and it can be performed using analytical, numerical or computational techniques.[7] *Computational homogenization*,[8] in particular, allows to extract *on-the-fly* the material properties associated to a generic point in a macro-continuum processing the information obtained from the analysis of micro-scale Representative Volume Elements (μRVEs), subjected to general loading and possibly undergoing internal evolution. The field is enormously benefiting from recent developments in and cross-fertilization between *microstructural materials characterization* and *High Performance Computing* (HPC), which are synergically contributing to deciphering complex materials micro-mechanisms in several application fields.[9,10] In this context, 3D material characterization, based e.g. on state-of-the-art microscopy, nano-/micro-indentation etc., provides nano-/micro-details that can be included in high-fidelity computational RVE models, requiring intense use of computing resources for deeper understanding of materials behaviour.[11–13]

In the present chapter, the development of a 3D grain-scale framework for polycrystalline materials undergoing damage and failure is reviewed. Polycrystalline materials (metals, alloys, ceramics) are widely used in engineering and there is relevant interest in understanding, modelling and predicting their degradation and failure processes.[14–16] Their microstructure, at the grain scale, is characterized by grains morphology, size distribution, anisotropy and crystallographic orientation, by the presence of flaws and porosity, and by physical and chemical properties of the inter-granular interfaces,[17] which have direct influence on the initiation and evolution of damage. In the developed framework, the representation of the interaction between the grains at the inter-granular interfaces plays a crucial role and the cohesive-frictional parameters have a direct effect on the material properties at the macro-scale.

Polycrystalline materials have been studied using both experimental[11–13,18–23] and computational techniques.[4,5,24] Until recently, the development of truly 3D computational models has been hindered by excessive computational requirements. Nowadays, however, the wider affordability of parallel HPC has impressed a remarkable acceleration to *computational micro-mechanics*, thus favouring the development of more refined models.[10,25–27] There are several scientific and technological reasons for the interest in truly 3D polycrystalline models.[16,28–31] 3D models allow better understanding of inherently 3D complex microstructural phenomena: influence of the geometry on the micro-cracking evolution; competition between different failure modes, e.g. inter- and trans-granular brittle cracking; ductile-to-brittle transition; grain-to-grain propagation of cleavage fracture;[32] plastic deformation of the individual crystals. Computational investigations complement experimental campaigns that, in the 3D case, may be particularly complex and expensive. Additionally, an increasing number of applications, in different fields, require the manufacturing of micro-components whose overall dimensions are comparable with the grain size, thus requiring accurate analysis at the grain scale level.

In this chapter, the development of a 3D boundary element framework for the analysis of degradation and failure in polycrystalline microstructures is reviewed. The polycrystalline morphology is represented through Voronoi tessellations. The microscopic degradation processes are explicitly modelled through cohesive-frictional contact laws within the context of a 3D grain-boundary formulation. The micro-scale model is based on a formulation recently developed by Sfantos and Aliabadi[33] for 2D problems and by Benedetti and Aliabadi for 3D cases.[34–37] Recent developments for micro-scale modelling have been addressed to computational enhancements[38] and to the inclusion of more complex single-grain constitutive models.[39–41] The developed framework may be employed for investigating the complex failure processes at the grain scale and, through computational homogenization, it may be used as the micro-scale block of multiscale approaches.[42,43] Due to their considerable computational requirements, fully 3D multiscale analyses are indeed quite rare in the literature and only a few methods have been developed for polycrystalline materials.[42,44–47]

The outline of the chapter is as follows. The micro-scale grain-boundary formulation is discussed in Section 2, where the integral equations for the bulk grains, the crystal plasticity constitutive equations and the cohesive-frictional interface equations are recalled. The numerical discretization and system solutions are discussed in Section 3. The results of some numerical simulations are presented in Section 4, before the Conclusions.

2. Microstructural formulation

The model for polycrystalline μRVEs has been developed by Sfantos and Aliabadi[33] for 2D problems, and by Benedetti and Aliabadi[34–37] for the 3D case. Some computational enhancements for solution speed-up have been presented by Gulizzi *et al.*,[38] while a formulation for including crystal plasticity has been presented by Benedetti *et al.*[39,40] Some specific aspects, such as grain vacancies or grain boundary cavitation, have been considered in Refs. [41, 48]. The interested readers are referred to these works for an in-depth description of the models.

2.1. *Generation of artificial morphologies*

The artificial polycrystalline microstructure is represented through 3D Voronoi tessellations,[49–51] which can be generated by using open source software packages such as `Neper`[51] (`http://neper.sourceforge.net/`) or `Voro++`[52] (`http://math.lbl.gov/voro++/`). Each crystal in the tessellation is then given an arbitrary crystallographic orientation in space and it is modelled as a generally anisotropic elastic–plastic domain. This representation captures the main morphological features of real polycrystalline aggregates and offers some advantages for the computational treatment, due to the fact that each crystal is represented as a convex polygon with flat convex polygonal faces. Some pathological aspects that may be induced by the mathematically rigorous process of tessellation can be treated by morphological regularization,[51] which gives definite computational benefits.[38] An example of Voronoi tessellation is illustrated in Fig. 1, the positions and weights of the generating seeds[38] are shown in Fig. 1(a) and the resulting tessellation is reported in Fig. 1(b).

2.2. *Governing integral equations*

The structure of the fundamental integral equations is given here. Such equations give a specific representation of the displacement fields within the individual grains. For solving the problem at hand, they must be complemented by: (a) a consistent constitutive framework accounting for the presence of plastic slip, see Section 2.3; (b) a mechanical coupling between grains at the intergranular interfaces, see Section 2.4. The

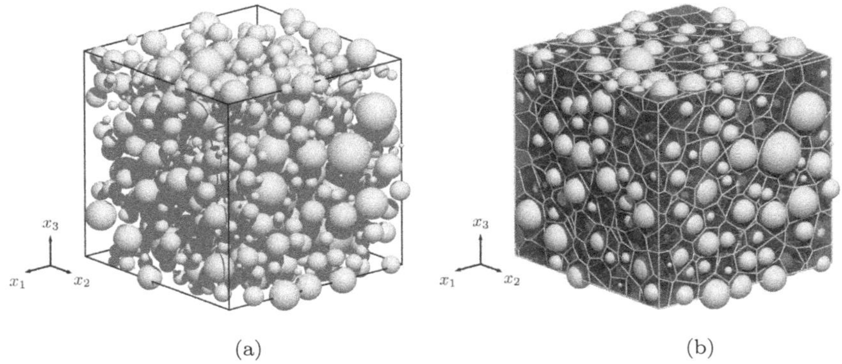

Fig. 1. 1000-grain morphology with ASTM grain size $G = 12$: (a) Seeds and weights distribution within the cubic box and (b) Resulting polycrystalline aggregate.

governing equations are presented in rate form, which allows to maintain a unified notation for both rate-dependent and rate-independent problems, due to the fact that also rate-independent problems can be generally associated with a time-like parameter that plays the role of an ordering parameter, such as the load factor λ.

Within the context of the *small strains theory*, the *total* strain rates $\dot{\varepsilon}_{ij}$ can be expressed as the sum of *elastic* and *plastic* contributions

$$\dot{\varepsilon}_{ij} = \frac{1}{2}\left(\dot{u}_{i,j} + \dot{u}_{j,i}\right) = \dot{\varepsilon}_{ij}^{e} + \dot{\varepsilon}_{ij}^{p}. \tag{1}$$

Where $\dot{u}_i$ denotes the displacements components. In the present boundary element formulation, the presence of plastic strains is addressed by adopting an *initial strain* approach,[53,54] which appears particularly suitable to the form of the crystal plasticity constitutive equations.

The first set of equations to be used in the model is that of the *displacement integral equations*. Let us consider a generic grain with volume V bounded by the grain boundary $S = \partial V$. When initial *plastic strains* $\dot{\varepsilon}_{jk}^{p}$ are present in $V_p \subset V$, the displacement integral equations can be written, in rate form, as

$$\begin{aligned} c_{ij}(\mathbf{x})\dot{u}_j(\mathbf{x}) &+ \fint_S T_{ij}(\mathbf{x},\mathbf{y})\,\dot{u}_j(\mathbf{y})dS(\mathbf{y}) \\ &= \int_S U_{ij}(\mathbf{x},\mathbf{y})\,\dot{t}_j(\mathbf{y})dS(\mathbf{y}) \\ &\quad + \int_{V_p} \Sigma_{ijk}(\mathbf{x},\mathbf{Y})\,\dot{\varepsilon}_{jk}^{p}(\mathbf{Y})dV(\mathbf{Y}) \end{aligned} \tag{2}$$

where $\mathbf{x}, \mathbf{y} \in S$ denote the *collocation* and *integration boundary* points respectively, $\mathbf{Y} \in V$ is a *volume* integration point, $\dot{u}_j$ and $\dot{t}_j$ are components of the boundary displacements and tractions, U_{ij}, T_{ij} and Σ_{ijk} are components of the displacement, traction and stress fundamental solutions respectively,[39,55] $\fint$ denotes the Cauchy principal value integral due to the presence of the strongly singular kernels T_{ij} and c_{ij} are free terms stemming from the integration procedure over the domain boundary S, which are $c_{ij} = \delta_{ij}/2$ for regular boundary points; the volume integral is performed over the interior region $V_p \subset V$ undergoing plastic deformation, i.e. where $\dot{\varepsilon}^p_{jk} \neq 0$.

If plastic slip is present within the aggregate, the displacement integral equations must be complemented by a representation of the internal stress tensor field, which triggers the plastic slip through the resolved shear. The *integral representation for the stresses* $\dot{\sigma}_{ij}$ at the generic interior point $\mathbf{X} \in V$ can be obtained applying the strain-displacement and the constitutive relationships to the displacements boundary integral equations collocated at the interior point $\mathbf{X} \in V$, which produces

$$\dot{\sigma}_{ij}(\mathbf{X}) + \int_S T^{\sigma}_{ijk}(\mathbf{X}, \mathbf{y})\, \dot{u}_k(\mathbf{y}) dS(\mathbf{y}) = \int_S U^{\sigma}_{ijk}(\mathbf{X}, \mathbf{y})\, \dot{t}_k(\mathbf{y}) dS(\mathbf{y})$$
$$+ \fint_{V_p} \Sigma^{\sigma}_{ijkl}(\mathbf{X}, \mathbf{Y})\, \dot{\varepsilon}^p_{kl}(\mathbf{Y}) dV(\mathbf{Y}) + f^{\sigma}_{ijkl}\, \dot{\varepsilon}^p_{kl}(\mathbf{X}), \tag{3}$$

where U^{σ}_{ijk}, T^{σ}_{ijk} and Σ^{σ}_{ijkl} are obtained from the kernels appearing in Eq. (2) using the strain-displacement and the constitutive equations, f^{σ}_{ijkl} are the so-called *free terms*[53,54] and the Cauchy principal value integral $\fint$ is due to the presence of the strongly singular kernels Σ^{σ}_{ijkl} in the volume integral. The expressions of the kernels and free terms as well as the numerical treatment of the singular integrals are given in Ref. [39].

In Eqs. (2)–(3) the displacement and traction components are expressed for each grain with respect to a *global* reference system. An important simplification, especially in the expression of the inter-granular and boundary conditions, is attained by expressing displacements and tractions with respect to reference systems defined *locally* on each grain face. Each grain is represented as a Voronoi cell bounded by convex polygonal *flat* faces: each face carries then its own reference system and Eqs. (2)–(3) can be rewritten in terms of grain face *local* components of displacements $\dot{\tilde{u}}_i$ and tractions $\dot{\tilde{t}}_i$, where the tilde indicates the local component. For further details, see Refs. [34, 35].

2.3. *Crystal plasticity phenomenological modelling*

The boundary integral formulation used for the individual grains must be complemented by a crystal plasticity framework to model the material evolution. Crystal plasticity can be generally modelled through *phenomenological* or *physically based* approaches.[24] Moreover, the crystallographic slip within single crystals can be idealized as either *rate-dependent* or *rate-independent.* In the present work, a phenomenological rate-dependent description is adopted. In the present section, a general introductory functional description of crystal slip is briefly recalled for the sake of completeness. The interested reader is referred to Ref. [39] for the details of the formulation and the specific crystal plasticity incremental-iterative algorithm.

The slip over a specific slip plane α identified by the unit normal vector $\mathbf{n}^\alpha$, and along a specific direction identified by the unit vector $\mathbf{s}^\alpha$, is *activated* by the *Schmid resolved shear stress*

$$\tau^\alpha = s_i^\alpha \sigma_{ij} n_j^\alpha, \tag{4}$$

which triggers the corresponding *slip* or *shear rate*

$$\dot{\gamma}^\alpha = \Psi^\alpha \left(\frac{\tau^\alpha}{\tau_c^\alpha} \right), \tag{5}$$

where Ψ^α expresses a specific *flow rule* and $\tau_c^\alpha = \Phi(\boldsymbol{\gamma}, \dot{\boldsymbol{\gamma}})$ is a certain *critical* shear stress, which plays the role of a *state variable* for the slip system α. τ_c^α generally depends on the *accumulated slips* $\gamma^\beta = \int_t |\dot{\gamma}^\beta| dt$ and on the slip rates $\dot{\gamma}^\beta$ on *all* the N_s slip systems, that have been collected into the vectors $\boldsymbol{\gamma} = (\gamma^1, \cdots, \gamma^{N_s})$ and $\dot{\boldsymbol{\gamma}} = (\dot{\gamma}^1, \cdots, \dot{\gamma}^{N_s})$ to indicate the functional dependence of τ_c^α. To avoid possible confusion, the sum of the accumulated slips γ^α over all the N_s slip systems, which is defined as $\gamma \equiv \sum_\alpha \gamma^\alpha$, will be referred to as the *overall cumulative slip.*

As plastic slip accumulates, τ_c^α evolves as

$$\dot{\tau}_c^\alpha = \Xi^\alpha (\boldsymbol{\gamma}, \dot{\boldsymbol{\gamma}}), \tag{6}$$

where Ξ^α expresses some specific *hardening law.* Eqs. (4–6) hold $\forall \alpha = 1, \ldots, N_s$.

Once the slip rates (or slip increments, in rate-independent formulations) are known, the plastic strains rate (or increment) at a point in the crystal can be expressed as

$$\dot{\varepsilon}_{ij}^{p} = \frac{1}{2} \sum_{\alpha=1}^{N_s} \dot{\gamma}^{\alpha} \left(n_i^{\alpha} \, s_j^{\alpha} + s_i^{\alpha} \, n_j^{\alpha} \right). \tag{7}$$

The developed formulation can be coupled with different types of flow and hardening rules.[39]

2.4. *Inter-granular modelling*

In a polycrystalline aggregate each grain interacts with the adjacent grains through the inter-granular interfaces. To represent the continuity of the polycrystalline microstructure in the model, the crystal plasticity framework developed for the bulk grains must be complemented with suitable *inter-granular equations* expressing the relationships between displacements and traction fields at the interface between adjacent grains. Denoting with $\mathscr{I}^{ab}$ the interface between two generic grains $\mathscr{G}^a$ and $\mathscr{G}^b$, the interface conditions can be generally expressed, $\forall \boldsymbol{\eta} \in \mathscr{I}^{ab}$, as follows

$$\begin{aligned} &\Psi_i^{ab} \left[\dot{\tilde{u}}_j^{a}(\boldsymbol{\eta}), \dot{\tilde{u}}_j^{b}(\boldsymbol{\eta}), \dot{\tilde{t}}_j^{a}(\boldsymbol{\eta}), \dot{\tilde{t}}_j^{b}(\boldsymbol{\eta}) \right] = 0, \\ &\Phi_i^{ab} \left[\dot{\tilde{t}}_j^{a}(\boldsymbol{\eta}), \dot{\tilde{t}}_j^{b}(\boldsymbol{\eta}) \right] = 0, \end{aligned} \qquad i,j = 1,2,3, \tag{8}$$

where Φ_i^{ab} represent traction equilibrium equations and Ψ_i^{ab} represent either continuity, cohesive or frictional contact laws, depending on the state of the interface at the generic interface point $\boldsymbol{\eta}$;[35] note that $\boldsymbol{\eta}$ identifies a generic reference location on the interface where *two* material points belonging to *two* different grains meet: such couples of points are referred to as *interface pairs*. The interface between two grains, at a generic point, can be initially pristine or damaged,[56] and it undergoes an evolution during the progressive loading.[33,35,38,41] In the general case, it is assumed that interface pairs belong to one of the following sets: (a) *Intact or pristine set* $\mathscr{IR}$, where interface tractions have never exceeded the damage onset threshold and no separation is allowed between the grain surfaces; (b) *Damaged or cohesive set* $\mathscr{DR}$, where interface tractions have overcome the damage threshold and depend now on the accumulated damage and displacement jump, according to a certain irreversible cohesive law and (c) *Failed or cracked set* $\mathscr{FR}$, where the interface pairs have completed the damaging process and have developed into a free crack; their status is now governed by the laws of *frictional contact mechanics*. Each interface pair is associated with *six* equations. For pairs in the pristine state, $p \in \mathscr{IR}$, there are three *continuity equations*, expressing identity of displacement components, and three *equilibrium equations*, expressing identity of traction components. During the loading process and interface evolution, the equilibrium equations never change and must always be fulfilled. On the other hand, the continuity equations are replaced by the *cohesive traction–separation laws* for pairs $p \in \mathscr{DR}$ and by *frictional contact equations* for pairs $p \in \mathscr{FR}$. In the next sections, such equations are discussed for the different cases.

2.4.1. *Interface continuity equations*

For an interface pair $p \in \mathscr{IR}$, the compatibility and equilibrium equations hold

$$\begin{aligned} &\dot{\tilde{u}}_i^a + \dot{\tilde{u}}_i^b = -\delta\dot{\tilde{u}}_i^{ab} = 0, \\ &\dot{\tilde{t}}_i^a - \dot{\tilde{t}}_i^b = 0. \end{aligned} \qquad i = 1,2,3 \qquad \text{on} \quad \mathscr{I}_{ab} \equiv \mathscr{G}_a \cap \mathscr{G}_b. \tag{9}$$

The local traction components $\dot{\tilde{t}}_i^a = \dot{\tilde{t}}_i^b = \dot{\tilde{t}}_i$ at the interface point $P \in \mathscr{G}_a$ can be used to express the representative tangential and normal tractions for the pair. The local normal and tangential interface tractions are defined by

$$\dot{\tilde{\mathbf{t}}}_n = (\dot{\tilde{\mathbf{t}}} \cdot \tilde{\mathbf{n}})\tilde{\mathbf{n}} = \begin{bmatrix} 0 \\ 0 \\ \dot{\tilde{t}}_3 \end{bmatrix} \quad \text{and} \quad \dot{\tilde{\mathbf{t}}}_s = \dot{\tilde{\mathbf{t}}} - \dot{\tilde{\mathbf{t}}}_n = \begin{bmatrix} \dot{\tilde{t}}_1 \\ \dot{\tilde{t}}_2 \\ 0 \end{bmatrix}, \tag{10}$$

and their moduli are given by $t_n = \dot{\tilde{t}}_3$ and $t_s = +\sqrt{\dot{\tilde{t}}_1^2 + \dot{\tilde{t}}_2^2}$. In the above definition, unlike t_s, t_n carries a sign that has an important role in the evolution of intergranular damage: when $t_n > 0$, the interface experiences tensile tractions that contribute to spreading damage; when $t_n < 0$, the interface is subjected to compressive loads that do not contribute to the evolution of damage. Such normal and tangential interface tractions are used to define the *effective traction*

$$t_e = \left[\langle t_n \rangle^2 + \left(\frac{\beta}{\alpha} t_s \right)^2 \right]^{\frac{1}{2}}, \tag{11}$$

where $\langle \cdot \rangle$ denote the Macauley brackets, defined by $\langle x \rangle = (x + |x|)/2$, and α and β are coefficients weighing the relative contribution of opening and sliding modes in the damage process. An interface pair $p \in \mathscr{IR}$ remains in the pristine status until $t_e < T_{\max}$, where $T_{\max}$ defines a threshold value for damage initiation.

2.4.2. *Cohesive traction–separation laws for damaged interfaces*

When the effective traction exceeds the *cohesive strength* $T_{\max}$, the pair p enters the damaged region

$$t_e \geq T_{\max} \Rightarrow \mathscr{IR} \rightarrow p \rightarrow \mathscr{DR}, \tag{12}$$

and an *irreversible extrinsic cohesive law* $\dot{\tilde{t}}_i = f_i(\mathscr{H}_d, \delta\dot{\tilde{u}}_j)$, linking interface tractions and displacement jumps, is introduced; $\mathscr{H}_d$ symbolizes the dependence on the history of the decohesion process, thus accounting for the *irreversibility* of the failure process.

To follow the damage evolution, a non-dimensional *effective opening displacement*

$$d = \left[\left\langle \frac{\delta\dot{u}_n}{\delta u_n^{cr}} \right\rangle^2 + \beta^2 \left(\frac{\delta\dot{u}_s}{\delta u_s^{cr}}\right)^2\right]^{\frac{1}{2}}, \tag{13}$$

is introduced,[57] where $\delta\dot{u}_n$ is the opening displacement jump, $\delta\dot{u}_s = (\delta\dot{\tilde{u}}_1^2 + \delta\dot{\tilde{u}}_2^2)^{1/2}$ and δu_n^{cr} and δu_s^{cr} are the critical the normal and sliding displacement jumps, for which interface failure occurs in the case of pure Mode I or Mode II loading, respectively. In the mixed-mode case, the parameter β assigns different weight to the two modes. The effective opening displacement can assume values $d \in [0, 1]$, where $d = 1$ implies the failure of the considered pair.

The cohesive laws can be derived from the introduction of a scalar potential, as discussed in Ref. [35], leading to

$$\begin{cases} \dot{\tilde{t}}_i = \alpha T_{\max} \dfrac{1-d^*}{d^*} \dfrac{\delta\dot{\tilde{u}}_i}{\delta u_s^{cr}} & i = 1, 2 \\[2ex] \dot{\tilde{t}}_3 = T_{\max} \dfrac{1-d^*}{d^*} \dfrac{\delta\dot{\tilde{u}}_3}{\delta u_n^{cr}} \end{cases} \quad \text{or} \quad \begin{cases} \dot{t}_s = \alpha T_{\max} \dfrac{1-d^*}{d^*} \dfrac{\delta\dot{u}_s}{\delta u_s^{cr}}, \\[2ex] \dot{t}_n = T_{\max} \dfrac{1-d^*}{d^*} \dfrac{\delta\dot{u}_n}{\delta u_n^{cr}}, \end{cases} \tag{14}$$

where

$$d^* = \max_{\mathscr{H}_d}\{d\} \in [0, 1], \tag{15}$$

and the relationship

$$\alpha = \beta^2 \frac{\delta u_n^{cr}}{\delta u_s^{cr}}, \tag{16}$$

must be fulfilled to guarantee the existence of the cohesive potential. However, if a *potential* traction–separation law is employed, it is not possible to distinguish energetically between pure mode I and pure mode II, as the fracture energy is independent from the fracture path; on the other hand, as pointed out by some authors,[58,59] it is reasonable to assume that the work of fracture may be path dependent, as it may depend on microstructural details that make mode II inherently different from mode I, at least at a macroscopic level. The fracture energy *path-dependency* can be achieved by simply discarding Eq. (16) and assigning suitable *independent* values to the constants α and β, which correspond to relaxing the requirement of existence of a well-defined cohesive potential. In other words, the traction–separation laws can be directly assigned through Eqs. (14) and not derived from a given potential. In the present work, the constants α and β are given independently from each other, but in such a way to ensure the desired G_{II}/G_I ratio between mode II and mode I fracture energies.

Equations (14) are valid whether the considered pair is in loading, unloading or reloading. Moreover, they express an *isotropic* cohesive law, as the cohesive tangential tractions depend on the tangential displacement jumps according to the *same* law in any direction on the contact plane. It is worth noting that the normal cohesive traction component $\dot{t}_n$ is given by Eqs. (14) only when a normal opening displacement $\delta \dot{u}_n$ is present. The condition $\delta \dot{u}_n \geq 0$, that expresses the physical impossibility of grain interpenetration, must always be verified. In the case of compressive loading, the equation $\delta \dot{u}_n = 0$ must be enforced, instead of Eq. (14); the pair moves into the pure sliding loading mode and the normal traction component t_n is provided by the solution of the microstructural problem.

A graphical representation of the employed traction–separation laws is given in Fig. 2.

2.4.3. *Interface failure: Frictional contact analysis*

Once $d^* = 1$, the considered pair p leaves the damaged set and enters the cracked set:

$$d^* = 1 \Rightarrow \mathscr{DR} \rightarrow p \rightarrow \mathscr{FR}. \tag{17}$$

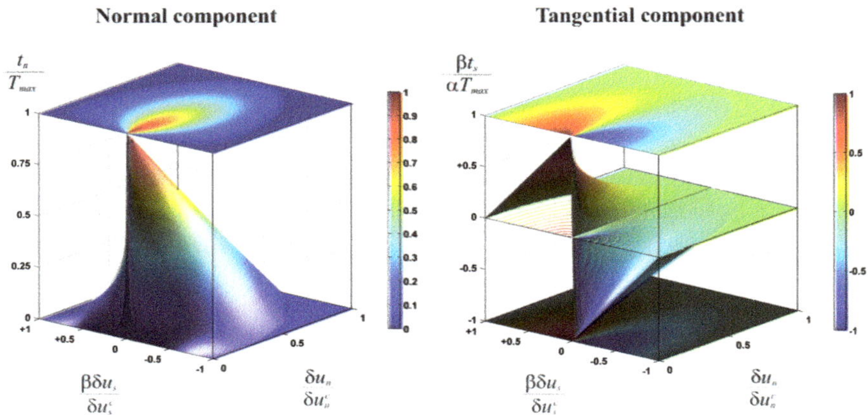

Fig. 2. Graphical representation of the normal and tangential cohesive traction-separation laws, $t_n(\delta u_n, \delta u_s)$ and $t_s(\delta u_n, \delta u_s)$.

Upon interface failure, the laws of frictional contact mechanics govern the evolution of the interface pair[60–62]: *Coulomb frictional contact* is implemented in this work. Once a micro-crack has started, the two crack surfaces can be separated or in contact.

In the case of *separation*, the condition $\delta\dot{u}_n > 0$ must always be satisfied and the following equations hold for the traction components

$$\dot{\tilde{t}}_i = 0 \quad i = 1, 2, 3 \qquad or \qquad \dot{t}_n = 0, \quad \dot{t}_s = 0. \tag{18}$$

On the other hand, if the crack surfaces are in contact, they can *stick* or *slide*. In the case of contact, the condition $\dot{t}_n \leq 0$ must always be satisfied, while the equation $\delta\dot{u}_n = 0$ is directly enforced in the system of equations, either in case of stick or slip. If the surfaces are in contact and the condition $|t_s| < \mu|t_n|$ holds, the pair is said to be in *stick* status and the following equations hold:

$$\begin{cases} \Delta\delta\dot{\tilde{u}}_i = 0 \quad i = 1, 2, \\ \delta\dot{\tilde{u}}_3 = 0, \end{cases} \quad \text{or equivalently} \quad \begin{cases} \Delta\delta\dot{u}_s = 0, \\ \delta\dot{u}_n = 0, \end{cases} \tag{19}$$

where the symbol Δ denotes the increment of a quantity in a load increment: no variation of the displacement jump is allowed if the pair is in stick condition.

If the stick condition $|t_s| < \mu|t_n|$ cannot be satisfied, then the relationship $|t_s| = \mu|t_n|$ must be enforced and the pair is said to be in *slip* status. In this case, the relative displacements $\delta\dot{\tilde{u}}_1$ and $\delta\dot{\tilde{u}}_2$ can evolve and the following equations hold:

$$\begin{cases} \dot{\tilde{t}}_1 + \mu\,\dot{\tilde{t}}_3 \cos(\theta) = 0, \\ \dot{\tilde{t}}_2 + \mu\,\dot{\tilde{t}}_3 \sin(\theta) = 0, \\ \delta\dot{\tilde{u}}_3 = 0, \end{cases} \tag{20}$$

where θ denotes the *slip angle*, defined in the plane $\{\tilde{x}_1\tilde{x}_2\}$ by $\tan(\theta) = \delta\dot{\tilde{u}}_2/\delta\dot{\tilde{u}}_1$: this angle is not *a priori* known and its value must be computed iteratively in the solution process. From the definition of slip angle and from Eqs. (20), it is apparent that the two vectors $\delta\dot{\tilde{\mathbf{u}}}_s$ and $\dot{\tilde{\mathbf{t}}}_s$ should have the same direction in the contact plane. However, since the equation $\tan(\theta) = \delta\dot{\tilde{u}}_2/\delta\dot{\tilde{u}}_1$ is not directly implemented, this requirement must be assessed as the numerical solution is available. Let β_s be the angle between $\delta\dot{\tilde{\mathbf{u}}}_s$ and $\dot{\tilde{\mathbf{t}}}_s$ on the contact plane. The slip condition is violated if $|\beta_s| \geq \pi/2$, i.e. $\dot{\tilde{\mathbf{t}}}_s \cdot \delta\dot{\tilde{\mathbf{u}}}_s \leq 0$: if this happens, the pair cannot be in slip and it is put in stick status. On the other hand, if $|\beta_s| < \pi/2$ but $|\beta_s| > 0$, then the slip angle must be suitably adjusted.

2.4.4. *Grains interface under compressive-sliding load: Cohesive-frictional laws*

In the sketched framework, an aspect deserves some attention. When the interface between two grains fails, the cohesive laws no longer apply and the frictional contact analysis is used instead. The sudden transition from cohesive to frictional behaviour is acceptable when a normal traction component $t_n > 0$ exists, but it may give raise to physical inconsistency and numerical issues if the interface is subjected to compressive-sliding loading.[59]

The issue may be better understood considering the example reported in Ref. [35]. Consider two grains subjected, in displacement control, to a compressive load $\lambda\Delta v$ and to a sliding load $\lambda\Delta u$, where λ denotes the load factor. The interface is initially undamaged and it is able to carry elastically a load until a damaging process is started; at this point, the impenetrability condition is directly enforced, while the tangential components of tractions are given by the corresponding terms in Eqs. (14). As the interface damage grows, the load-carrying capability of the interface is progressively reduced: if the model implements a switch between cohesive and friction laws, then the tangential cohesive tractions must go to zero at $d^* = 1$ before the laws of contact mechanics can be applied. After interface failure, contact mechanics apply and the contact algorithm correctly predicts a *stick* condition immediately after failure, to avoid a sudden discontinuity of the tangential traction during the loading process, followed, eventually, by the *slip* of the interface, for which $t_s = \mu|t_n|$.

However, the prediction that the interface loses its load-bearing capability before retrieving some of it through the friction μ appears slightly inconsistent. Moreover, the presence of the cusp at the transition between cohesive and contact state, and the possibility of a traction jump if the load factor is not smoothly updated, is likely to spark numerical instabilities, especially when multi-body frictional contact analysis is considered. To avoid such issues, in the present model the presence of friction is assumed also during the decohesion process, if it happens in compression: this corresponds to assuming a smooth transition between cohesive and contact state.

The gradual transition to the frictional state is achieved by introducing some frictional terms depending on the damage level in the traction–separation laws. In this case, the tangential components of traction are expressed by means of the following *cohesive-frictional laws*

$$\begin{cases} \dot{\tilde{t}}_1 = \alpha T_{\max} \dfrac{1-d^*}{d^*} \dfrac{\delta \dot{\tilde{u}}_1}{\delta u_s^{cr}} - \chi(d^*) \mu \, \dot{\tilde{t}}_3 \cos(\theta), \\[2ex] \dot{\tilde{t}}_2 = \alpha T_{\max} \dfrac{1-d^*}{d^*} \dfrac{\delta \dot{\tilde{u}}_2}{\delta u_s^{cr}} - \chi(d^*) \mu \, \dot{\tilde{t}}_3 \sin(\theta), \end{cases} \tag{21}$$

where θ is the slip angle and $\chi(d^*)$ is a suitable function modelling the soft transition from zero to full friction; in this work $\chi(d)$ is phenomenologically assumed to be

$$\chi(d) = \frac{1}{2} \left\{ 1 + \tan h \left[\frac{d - \bar{d}}{d(1 + \xi - d)} \right] \right\}, \tag{22}$$

where $0 < \bar{d} < 1$ and $0 < \xi \ll 1$ are parameters that enable one to modify, to some extent, the shape of the curve $\chi(d)$. Further details about the implemented cohesive-frictional model are given in Ref. [35].

3. Numerical discretization and solution

The numerical model for the polycrystalline problem is set up and then solved: (a) subdividing the surface and volume of each grain in the tessellation into a suitable set of non-overlapping boundary and volume elements and expressing the displacement and traction fields by means of suitable shape functions and nodal values, according with the procedures described in Refs. [38, 39]. (b) discretizing the displacement and stress integral equations, Eqs. (2–3), for each grain, on the basis of the previous elemental subdivision; (c) writing the discrete version of the interface equations, Eq. (8) and (d) using an incremental-iterative algorithm for capturing the evolution of both the elastic–plastic fields within the grains and the damaging processes at the grain interfaces.

Each grain is a convex polyhedron bounded by convex polygonal faces that is subdivided into non-overlapping triangles/quadrangles, according to the strategy described in Ref. [38]. For crystal plasticity modelling, volume meshes for the grains interior must be used: in the present work, such volume meshes for Voronoi-type microstructures have been generated using Hypermesh (`http://www.altairhyperworks.com`), which subdivides the volume into non-overlapping tetrahedra. Further details about the adopted discretization strategies are given in Refs. [38, 39].

Following Refs. [53, 63], the elastic–plastic grain-boundary problem can be expressed in terms of *accumulated* values of displacements $\mathbf{u}$, tractions $\mathbf{t}$ and plastic strains $\boldsymbol{\varepsilon}^p$. For each grain g, the collocation and discretization of the displacement integral equations, and the subsequent enforcement of the boundary conditions, produce the system

$$\mathbf{A}_g\,\mathbf{q}_g = \boldsymbol{\phi}_g(\lambda) + \mathbf{D}_g\,\boldsymbol{\varepsilon}_g^p \qquad \forall\, g = 1, \ldots, N_g, \tag{23}$$

where N_g is the total number of grains. In Eq. (23), the unknowns components of grain boundary displacements and tractions, including those corresponding to *inter-granular* points, are collected in the vector $\mathbf{q}_g$, while the components of $\boldsymbol{\phi}_g(\lambda)$ stem from linear combinations of known displacements and tractions, which in general depend on a load factor λ that plays the role of an ordering parameter in the progressive loading process; the matrix $\mathbf{D}_g$ stems from the volume integration over the region undergoing plastic flow and $\boldsymbol{\varepsilon}_g^p$ contains the components of the plastic strains within the volume discretization cells, to be determined through an iterative algorithm.[39]

The polycrystalline model is obtained writing Eq. (23) for each grain and enforcing the suitable inter-granular conditions, which leads to a system of the form:

$$\begin{bmatrix} \mathbf{A} \\ \mathbf{I}(\mathcal{D},\mathcal{H}) \end{bmatrix} \mathbf{q} = \begin{bmatrix} \boldsymbol{\phi}(\lambda) \\ \mathbf{0} \end{bmatrix} + \begin{bmatrix} \mathbf{D} \\ \mathbf{0} \end{bmatrix} \boldsymbol{\varepsilon}^p, \tag{24}$$

where: $\mathbf{A} = \text{diag}\{\mathbf{A}_1, \mathbf{A}_2, \cdots, \mathbf{A}_{N_g}\}$; $\mathbf{I}(\mathcal{D},\mathcal{H})$ is a matrix implementing the interface conditions, depending on the evolution of irreversible inter-granular damage $\mathcal{D}$ and on the loading history $\mathcal{H}$ in general; $\mathbf{q}^T = [\mathbf{q}_1^T, \ldots, \mathbf{q}_{N_g}^T]$; $\boldsymbol{\phi}(\lambda)^T = [\boldsymbol{\phi}_1^T(\lambda), \ldots, \boldsymbol{\phi}_{N_g}^T(\lambda)]$; $\mathbf{D} = \mathrm{diag}\{\mathbf{D}_1, \mathbf{D}_2, \ldots, \mathbf{D}_{N_g}\}$; $\boldsymbol{\varepsilon}^{pT} = [\boldsymbol{\varepsilon}_1^{pT}, \ldots, \boldsymbol{\varepsilon}_{N_g}^{pT}]$.

In the solution algorithm, system (24) is used in conjunction with the discretized version of Eq. (3) that, in terms of accumulated values, can be written for each grain as

$$\boldsymbol{\sigma} = \mathbf{G}_g^\sigma\, \mathbf{t}_g(\lambda) - \mathbf{H}_g^\sigma\, \mathbf{u}_g(\lambda) + \mathbf{D}_g^\sigma \boldsymbol{\varepsilon}^p \qquad \forall\, g = 1, \ldots, N_g, \tag{25}$$

which is used to evaluate the stress at each test point within each grain, to assess the possibility of crystal plastic flow. It is worth noting that, with Eq. (25) used after the solution of Eq. (24), in a *post-processing* stage within each iteration, it can be used for each grain separately, as $\mathbf{u}_g$ and $\mathbf{t}_g$ are known.

3.1. *Incremental-iterative solution of the μBVP*

The analysis of polycrystalline aggregates with crystal-plasticity and inter-granular damage evolution poses a remarkable computational task. The problem is tackled employing the incremental/iterative procedures developed in Refs. [33, 35, 39] to which the interested readers are referred for a detailed algorithmic description.

In general, the system given in Eq. (24) is used to update the accumulated values of displacements and tractions for all the aggregate and Eq. (25) used to compute the stress components at each interior test point within each grain, which then trigger, through the resolved shear stresses, the plastic flow. The algorithm given in Ref. [35] is used to update the inter-granular fields, including the damage status, while the crystal-plasticity grain boundary algorithm[39] is used to update the fields within the grains. Convergence is reached for the aggregate when the convergence criteria are satisfied for each interior test point within each grain and for each inter-granular pair.

From the purely numerical/computational point of view, System (24) is solved using the procedure in described in Refs. [35, 38]. Because the coefficients system is highly sparse, `PARDISO` (`http://www.pardiso-project.org/`)[64–66] is used as solver. In terms of computational effectiveness, further research could be addressed at the development of specialized high efficiency solvers, based for example on the use of the hierarchical format in conjunction with iterative solution strategies.[67,68]

4. Computational tests

Three numerical applications are discussed, to illustrate the scope of the described framework. The first test illustrates the effect of inter-granular friction on the mechanical behaviour of polycrystalline aggregates. The second and third examples deal with micro-cracking under tensile loading and crystal plasticity in polycrystalline aggregates.

4.1. *Micro-cracking under compressive loading: The effect of friction*

The micro-cracking of a pseudo-3D specimen subjected to unidirectional compressive load is investigated: the test gives insights about the effect of friction in polycrystalline materials.

A 20 grain SiC morphology is considered, Fig. 3, with dimensions $2W \times 2H \times 2T$ with $H/W = 3$, $4WH = N_g \bar{A}_{gr}$ and $2T = \sqrt{\bar{A}_{gr}}$, where $\bar{A}_{gr} = 126\,\mu\text{m}^2$ is the average area associated with ASTM $G = 10$ grains. The morphology is subjected, in displacement control, to unidirectional compressive load along the height H. The cohesive-frictional properties are given in Table 1. The values of α and β are adjusted so to give greater weight to mode II failure, so that in pure mode II the damaging process is activated when $t_s = T_{\max}/2$, see Eq. (11). The effect of friction on the micro-cracking patterns is investigated by considering the values $\mu = 0.0$, 0.1 and 0.2 for the friction coefficient.

Figure 4 shows the averaged macroscopic stress–strain curves for the tested values of friction. It is evident how the friction influences the load-bearing capability of the specimen, especially after micro-cracks have formed.

It is possible to observe that: (a) friction influences, to some extent, the evolution of micro-cracks, hindering the sliding between formed crack surfaces and (b) once micro-cracks are formed, the friction affects noticeably the macroscopic post-critical behaviour of the aggregate. Such observations hold their validity also for microstructures comprised of more grains. Further analyses on microstructures subjected to compression can be found in Ref. [35].

4.2. *Micro-cracking under tensile loading*

The micro-cracking of a 1000-grain cubic polycrystalline aggregate with ASTM grain size $G = 12$ is considered. The aggregate, with the edges

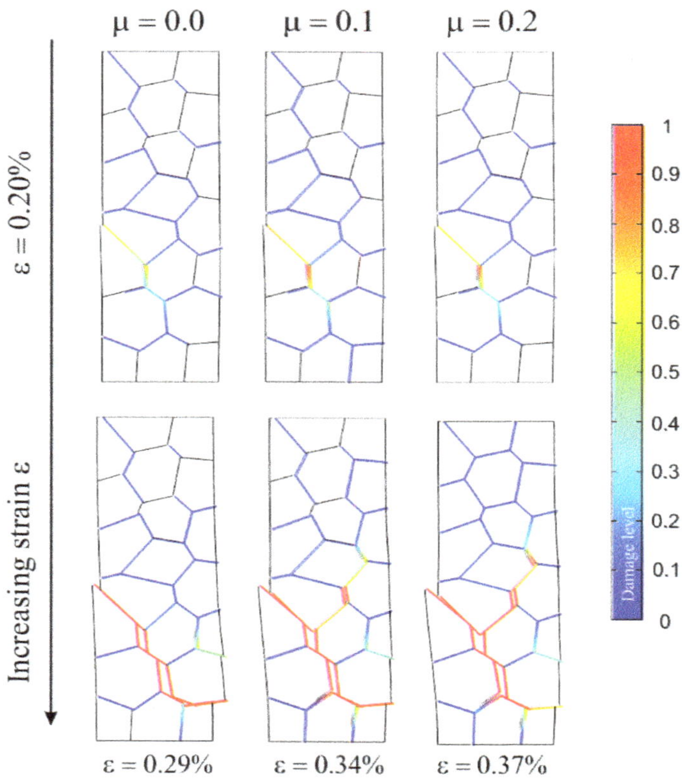

Fig. 3. Micro-cracking patterns for different values of the friction coefficient μ for the 20-grain SiC microstructure, ASTM $G = 10$, subjected to compressive load. The crack patterns are shown for the selected values of nominal applied strain $\varepsilon = 0.2\%$ and at the last simulated strain for each value of friction.

Table 1. Set of cohesive-frictional parameters adopted for the 20-grain pseudo-3D morphology subjected to compressive load.

$T_{\max}$ (MPa)	α	β	G_{II}/G_I	μ	$\bar{d}_{Eq.(22)}$	ξ
500	1/4	1/2	1	{0, 0.1, 0.2}	0.8	0

oriented along the coordinate directions, is subject to prescribed progressive uniaxial strain along the x_3 direction, as schematized in Fig. 5(a), where λ represents the load factor. The top and bottom faces of the cube are subjected to $\mathbf{u}\cdot\mathbf{n} = \lambda/2$, whereas the lateral faces are subjected to $\mathbf{u}\cdot\mathbf{n} = 0$,

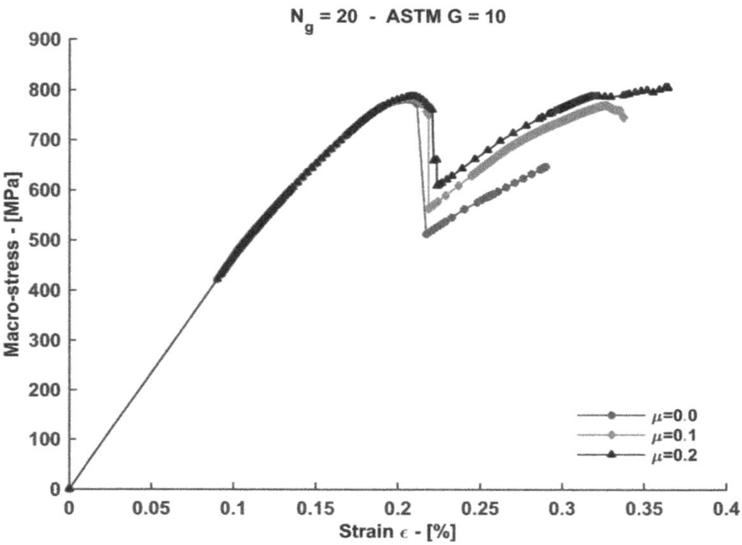

Fig. 4. Volume-averaged stress–strain curves for different values of the friction coefficient μ for the analysed 20-grain SiC microstructure subjected to compressive loading.

with $\mathbf{n}$ being the face outward unit normal. Along the remaining directions, the boundaries of the aggregate are traction free.

For this test, a SiC aggregate is considered. The non-zero elastic constants of the grains in their local reference system are $c_{1111} = c_{2222} =$ 502GPa, $c_{1122} = 95$GPa, $c_{1133} = c_{2233} = 96$GPa, $c_{3333} = 565$GPa, $c_{3232} = c_{3131} = 169$GPa and $c_{1212} = 203.5$GPa, whereas the cohesive parameters are: interface strength $T_{\max} = 500$MPa, fracture toughness $K_{IC} = 3$MPa $\sqrt{m}$, ratio between fracture toughnesses of mode II and mode I $G_{II}/G_I = 1$ and weighing parameters $\alpha = 1$ and $\beta = \sqrt{2}$. The average size of the mesh is chosen so that the characteristic element length satisfies the inequality $l_e < 10 \cdot L_{cz}$, where the cohesive zone length is estimated as $L_{cz} \approx (\pi/2)(K_{IC}/T_{\max})^2$.

Figure 5(b) shows the volume stress averages Σ_{11}, Σ_{22} and Σ_{33} as a function of the load factor λ. Figure 6 shows three snapshots of the aggregate micro-cracking at three different loading steps. In particular, the micro-cracking steps at $\lambda = 0.077$, $\lambda = 0.079$ and $\lambda = 0.086$ are reported in Figures 6(a–f), respectively. Figures 6(a), 6(c) and 6(d) show the cracked aggregate, whereas Figs. 6(b), 6(d) and 6(e) show the pattern of the micro-cracks and the corresponding level of damage.

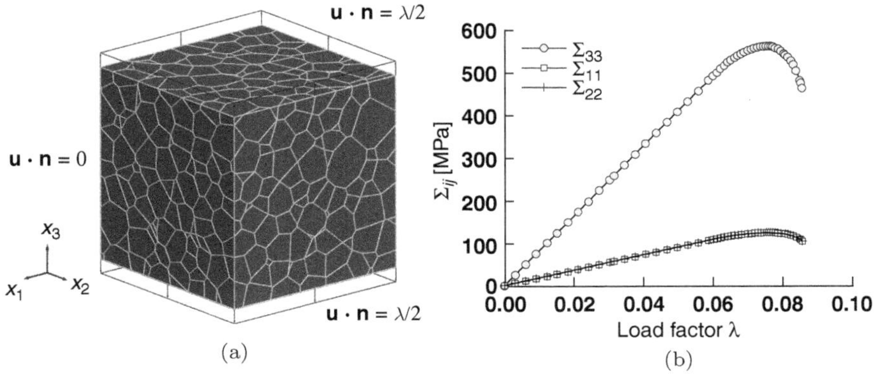

Fig. 5. (a) Outline of the boundary conditions applied to the polycrystalline aggregate; λ denotes the load factor and (b) Volume stress averages Σ_{11}, Σ_{22} and Σ_{33} as a function of the load factor λ.

4.3. *Polycrystalline plasticity simulation*

A tensile test is performed on two 100-grain *Cu* aggregates, to evaluate the plastic response of representative Cu microstructures. The two aggregates labelled as (I) and (II), shown in Figs. 7(a) and 7(d), are subjected to uniaxial tension, with uniform tractions acting over the top and bottom faces, and traction-free lateral faces. To avoid rigid motion, the rigid body modes of one grain of the aggregate are removed using the technique given in Ref. [69]. The surface meshes of the aggregates are shown in Figs. 7(b) and 7(e), while Figs. 7(c) and 7(f), show the volume meshes. The meshes are built ensuring an average number of 400 volume elements per grain, which represents a good tradeoff between solution accuracy and number of dofs.[39] Figure 8(a) reports the macro stress–strain curves for the two aggregates. The inset shows a close view of the macro response in the range 0−1.5%, with the transition from linear to plastic slip behaviour. By comparing the response of the two aggregates, it can be inferred that 100-grain aggregates can satisfactorily represent polycrystalline *Cu* undergoing small strains crystal plasticity. The solid line in Fig. 8(a) represents the linear response of the microstructure whose averaged stiffness is calculated using the Hashin–Shtrikman bounds,[70] whereas the dashed line reports the response of the aggregate using the Berveiller and Zaoui (BZ) model.[71] The BZ model matches well with the present model in capturing the elastic–plastic transition. On the other hand, the BZ model provides values of stress higher than the present model, but this appears consistent with Barbe *et al.*,[72] who reported an overestimate with respect to FEM

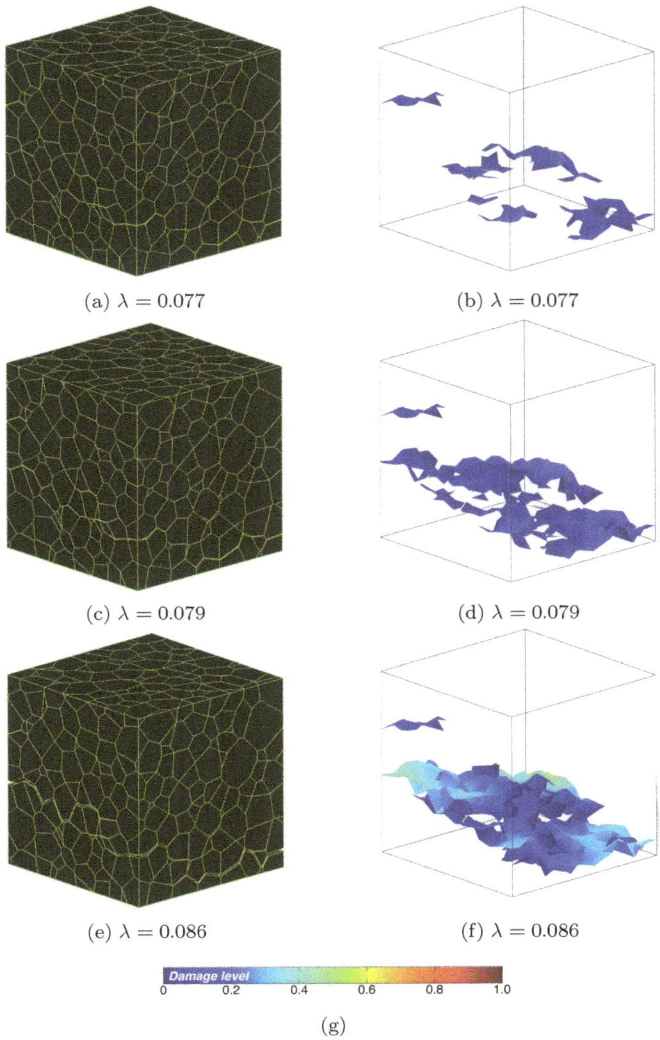

Fig. 6. Micro-cracking snapshots at the loading steps $\lambda = 0.077$ (a–b). $\lambda = 0.079$ (c–d). $\lambda = 0.089$ (e–f). (a, c, e) cracked aggregate. (b, d, f) micro-cracks pattern and (g) Colourmap of damage.

as well. Figure 8(b) shows the averaged stress–strain curve for *each* of the 100 grains of the aggregates, obtained by volume averages over *each* grain. Eventually, Figs. 9(a) and 10(a) report the contour plots of the von Mises stress and cumulative slip fields, respectively, over the external boundary

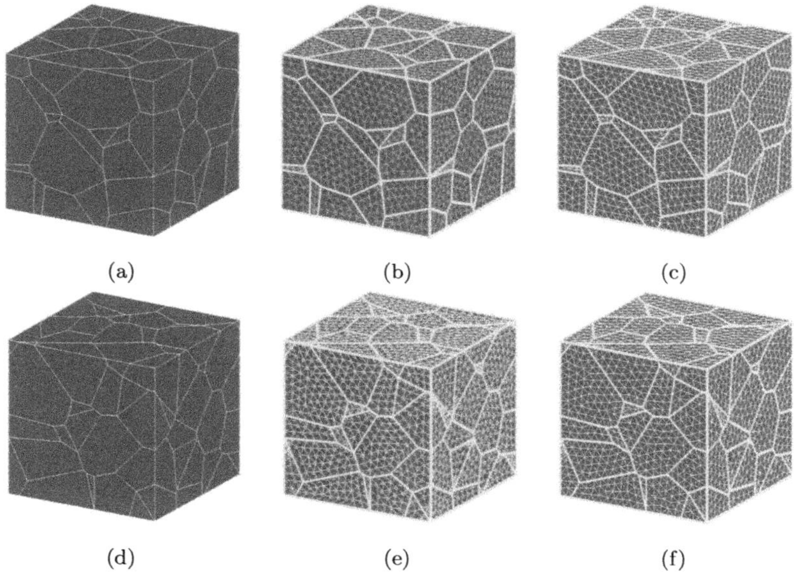

Fig. 7. (a, d) 100-grain polycrystalline microstructure (I) and (II), respectively; corresponding surface (b, e) and volume (c, f) meshes.

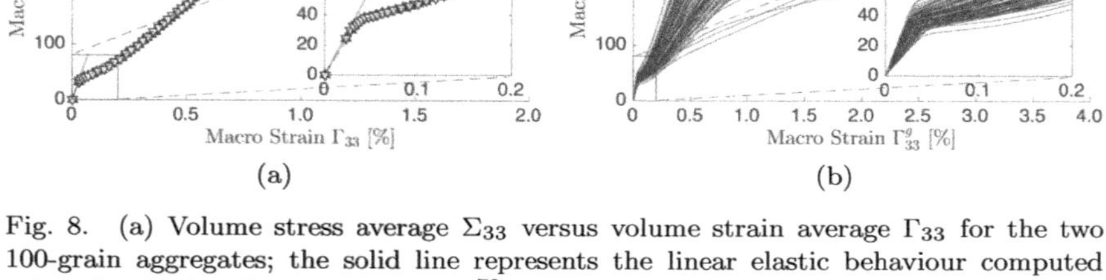

Fig. 8. (a) Volume stress average Σ_{33} versus volume strain average Γ_{33} for the two 100-grain aggregates; the solid line represents the linear elastic behaviour computed using the Hashin–Shtrikman bounds[70] and (b) Averaged stress–strain behaviour for *each* constituent grain of the aggregate (I).

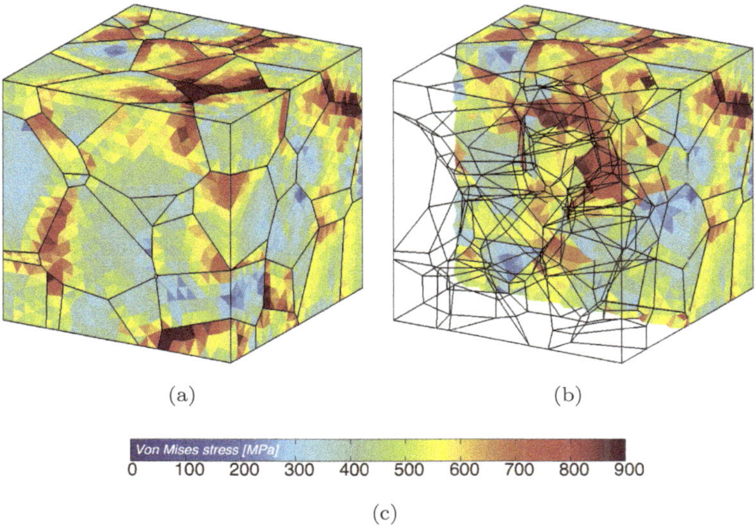

Fig. 9. (a) von Mises stress contour plots for the 100-grain polycrystalline copper aggregate (I) at the last computed step ($\Gamma_{33} = 1.5\%$). (b) Section of the aggregate and (c) Colourmap (to interpret the colour scale, the reader is referred to the web version of the article).

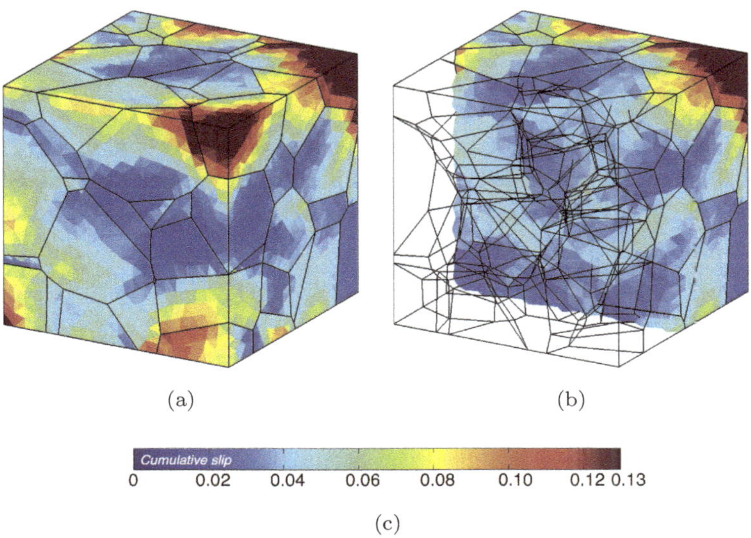

Fig. 10. (a) Cumulative slip contour plots for the 100-grain polycrystalline copper aggregate (I) at the last computed step ($\Gamma_{33} = 1.5\%$). (b) Section of the aggregate and (c) Colourmap (to interpret the colourscale, the reader is referred to the web version of the article).

of the aggregate (I), whereas Figs. 9(b) and 10(b) report the same contour plots for a section of the aggregate, giving an insight of the volume fields distribution.

5. Conclusions

The development of a computational methodology for the microstructural analysis of polycrystalline materials has been reviewed and discussed in the present chapter. The building blocks of the formulation have been discussed and the strategy for numerical integration has been described. The framework, based on a grain-scale cohesive-frictional crystal plastic formulation, allows to capture the nucleation, evolution and coalescence of micro-damage, in the form of inter-granular cracking or crystalline slip. The potential of the technique has been illustrated through micro-cracking simulations and a crystal plasticity computation. The macroscopic effect of the cohesive-frictional interfaces has been highlighted through some compressive tests. Further research could be aimed at including more sophisticated microstructural behaviours into the framework, e.g. trans-granular cracking or stress corrosion cracking, and at reducing the computational cost of the analysis.

Acknowledgements

The author would like to thank Dr. Vincenzo Gulizzi for the presented tensile micro-cracking simulations. The author is grateful to Prof. M.H. Ferri Aliabadi at Imperial College London for his support and suggestions and for the fruitful collaboration that led to part of the work presented in the chapter.

References

1. Z. Hashin, Analysis of composite materials — A survey, *ASME Journal of Applied Mechanics*, **50**, 481–505 (1983).
2. T. Mura, Micromechanics of Defects in Solids, 2nd revised edn., *Mechanics of Elastic and Inelastic Solids*, Kluwer Academic Publishers, Dordrecht, The Netherlands (1987).
3. S. Nemat-Nasser and M. Hori, *Micromechanics: Overall Properties of Heterogeneous Materials*, 2nd edn., North-Holland, Elsevier, Amsterdam, The Netherlands (1999).
4. A. Needleman, Computational mechanics at the mesoscale, *Acta Materialia*, **48**(1), 105–124 (2000).

5. T. I. Zohdi and P. Wriggers, *An Introduction to Computational Micromechanics*, Vol. 20, *Lecture Notes in Applied and Computational Mechanics*, Springer, Berlin (2005).
6. T. Watanabe and S. Tsurekawa, The control of brittleness and development of desirable mechanical properties in polycrystalline systems by grain boundary engineering, *Acta Materialia*, **47**(1516), 4171–4185 (1999).
7. V. P. Nguyen, M. Stroeven, and L. J. Sluys, Multiscale continuous and discontinuous modeling of heterogeneous materials: A review on recent developments, *Journal of Multiscale Modelling*, **03**(04), 229–270 (2011).
8. M. G. D. Geers, V. G. Kouznetsova, and W. A. M. Brekelmans, Multi-scale computational homogenization: Trends and challenges, *Journal of Computational and Applied Mathematics*, **234**(7), 2175–2182 (2010).
9. F. Barthelat, H. Tang, P. D. Zavattieri, C.-M. Li, and H. D. Espinosa, On the mechanics of mother-of-pearl: A key feature in the material hierarchical structure, *Journal of the Mechanics and Physics of Solids*, **55**(2), 306–337 (2007).
10. I. Simonovski and L. Cizelj, Towards modeling intergranular stress corrosion cracks on grain size scales, *Nuclear Engineering and Design*, **246**(0), 107–114 (2012).
11. K. M. Döbrich, C. Rau, and C. E. Krill III, Quantitative characterization of the three-dimensional microstructure of polycrystalline Al-Sn using X-ray microtomography, *Metallurgical and Materials Transactions A*, **35A**, 1953–1961 (2004).
12. M. Groeber, S. Ghosh, M. D. Uchic, and D. M. Dimiduk, A framework for automated analysis and simulation of 3D polycrystalline microstructures. Part 1: Statistical characterization, *Acta Materialia*, **56**, 1257–1273 (2008).
13. W. Ludwig, A. King, P. Reischig, M. Herbig, E. M. Lauridsen, S. Schmidt, H. Proudhon, S. Forest, P. Cloetens, S. R. du Roscoat, J. Y. Buffire, T. J. Marrow, and H. F. Poulsen, New opportunities for 3D materials science of polycrystalline materials at the micrometre lengthscale by combined use of X-ray diffraction and X-ray imaging, *Materials Science and Engineering: A*, **524**(12), 69–76 (2009).
14. H. D. Espinosa and P. D. Zavattieri, A grain level model for the study of failure initiation and evolution in polycrystalline brittle materials. Part I: Theory and numerical implementation, *Mechanics of Materials*, **35**, 333–364 (2003).
15. N. Sukumar, D. J. Srolovitz, T. J. Baker, and J. H. Prévost, Brittle fracture in polycrystalline microstructures with the extended finite element method, *International Journal for Numerical Methods in Engineering*, **56**, 2015–2037 (2003).
16. I. Benedetti and F. Barbe, Modelling polycrystalline materials: An overview of three-dimensional grain-scale mechanical models, *Journal of Multiscale Modelling*, **05**(01), 1350002 (2013).
17. B. L. Adams and T. Olson, The mesostructure — properties linkage in polycrystals, *Progress in Materials Science*, **43**, 1–88 (1998).

18. C. Zhang, A. Suzuki, T. Ishimaru, and M. Enomoto, Characterization of Three-dimensional grain structure in polycrystalline iron by serial sectioning, *Metallurgical and Materials Transactions A*, **35A**, 1927–1933 (2004).
19. M. A. Groeber, B. K. Haley, M. D. Uchic, D. M. Dimiduk, and S. Gosh, 3D reconstruction and characterization of polycrystalline microstructures using a FIB-SEM system, *Materials Characterization*, **57**, 259–273 (2006).
20. A. King, G. Johnson, D. Engelberg, W. Ludwig, and J. Marrow, Observations of intergranular stress corrosion cracking in a grain-mapped polycrystal, *Science*, **321**(5887), 382–385 (2008).
21. S. Rahimi, D. L. Engelberg, J. A. Duff, and T. J. Marrow, *In situ* observation of intergranular crack nucleation in a grain boundary controlled austenitic stainless steel, *Journal of Microscopy*, **233**(3), 423–431 (2009).
22. M. Herbig, A. King, P. Reischig, H. Proudhon, E. M. Lauridsen, J. Marrow, J.-Y. Buffire, and W. Ludwig, 3-d growth of a short fatigue crack within a polycrystalline microstructure studied using combined diffraction and phase-contrast X-ray tomography, *Acta Materialia*, **59**(2), 590–601 (2011).
23. A. King, W. Ludwig, M. Herbig, J.-Y. Buffire, A. A. Khan, N. Stevens, and T. J. Marrow, Three-dimensional in situ observations of short fatigue crack growth in magnesium, *Acta Materialia*, **59**(17), 6761–6771 (2011).
24. F. Roters, P. Eisenlohr, L. Hantcherli, D. Tjahjanto, T. Bieler, and D. Raabe, Overview of constitutive laws, kinematics, homogenization and multiscale methods in crystal plasticity finite-element modeling: Theory, experiments, applications, *Acta Materialia*, **58**(4), 1152–1211 (2010).
25. I. Simonovski and L. Cizelj, Automatic parallel generation of finite element meshes for complex spatial structures, *Computational Materials Science*, **50**(5), 1606–1618 (2011).
26. I. Simonovski and L. Cizelj, Computational multiscale modeling of intergranular cracking, *Journal of Nuclear Materials*, **414**(2), 243–250 (2011).
27. J. A. R. Bomidi, N. Weinzapfel, and F. Sadeghi, Three-dimensional modelling of intergranular fatigue failure of fine grain polycrystalline metallic mems devices, *Fatigue & Fracture of Engineering Materials & Structures*, **35**(11), 1007–1021 (2012).
28. A. G. Crocker, P. E. J. Flewitt, and G. E. Smith, Computational modelling of fracture in polycrystalline materials, *International Materials Reviews*, **50**(2), 99–124 (2005).
29. G. Smith, A. Crocker, G. Hughes, and P. Flewitt, Three-dimensional modelling of fracture in polycrystals, In E. E. Gdoutos (ed.), *Fracture of Nano and Engineering Materials and Structures*, Springer, Netherlands (2006), pp. 823–824.
30. T. J. Marrow, L. Babout, A. P. Jivkov, P. Wood, D. Engelberg, N. Stevens, P. J. Withers, and R. C. Newman, Three dimensional observations and modelling of intergranular stress corrosion cracking in austenitic stainless steel, *Journal of Nuclear Materials*, **352**(13), 62–74 (2006).
31. G. M. Hughes, G. E. Smith, P. Flewitt, and A. G. Crocker, The brittle fracture of polycrystalline zinc, *Proceedings of the Royal Society A: Mathematical, Physical and Engineering Science*, **463**(2085), 2129–2151 (2007).

32. D. Catoor and K. S. Kumar, Crackgrain boundary interactions in zinc bicrystals, *Philosophical Magazine*, **91**(16), 2154–2185 (2011).
33. G. K. Sfantos and M. H. Aliabadi, A boundary cohesive grain element formulation for modelling intergranular microfracture in polycrystalline brittle materials, *International Journal for Numerical Methods in Engineering*, **69**, 1590–1626 (2007).
34. I. Benedetti and M. H. Aliabadi, A three-dimensional grain boundary formulation for microstructural modelling of polycrystalline materials. *Computational Materials Science*, **67**, 249–260 (2013).
35. I. Benedetti and M. Aliabadi, A three-dimensional cohesive-frictional grain-boundary micromechanical model for intergranular degradation and failure in polycrystalline materials, *Computer Methods in Applied Mechanics and Engineering*, **265**(0), 36–62 (2013).
36. I. Benedetti and M. Aliabadi. A grain boundary formulation for the analysis of three-dimensional polycrystalline microstructures, In *Advances in Fracture and Damage Mechanics XI*, Vol. 525, *Key Engineering Materials*, Trans Tech Publications, Switzerland, pp. 1–4 (2013).
37. I. Benedetti and M. Aliabadi. Computational modelling of brittle failure in polycrystalline materials using cohesive-frictional grain-boundary elements, In *Advances in Fracture and Damage Mechanics XII*, Vol. 577, *Key Engineering Materials*, Trans Tech Publications, Switzerland, pp. 233–236 (2014).
38. V. Gulizzi, A. Milazzo, and I. Benedetti, An enhanced grain-boundary framework for computational homogenization and micro-cracking simulations of polycrystalline materials, *Computational Mechanics*, **56**(4), 631–651 (2015).
39. I. Benedetti, V. Gulizzi, and V. Mallardo, A grain boundary formulation for crystal plasticity, *International Journal of Plasticity*, **83**, 202–224 (2016).
40. I. Benedetti and V. Gulizzi. A novel numerical formulation for crystal plasticity. In *Advances in Fracture and Damage Mechanics XV*, Vol. 713, *Key Engineering Materials*, Trans Tech Publications, Switzerland, pp. 317–320(2016).
41. V. Gulizzi, A. Milazzo, and I. Benedetti. A micro-mechanical model for grain-boundary cavitation in polycrystalline materials. In *Advances in Fracture and Damage Mechanics XIV*, Vol. 665, *Key Engineering Materials*, Trans Tech Publications, Switzerland, pp. 65–68 (2016).
42. I. Benedetti and M. Aliabadi, Multiscale modelling of polycrystalline materials: A boundary element approach to material degradation and fracture, *Computer Methods in Applied Mechanics and Engineering*, **289**(0), 429–453 (2015).
43. I. Benedetti and M. Aliabadi. A multiscale approach to polycrystalline materials damage and failure, In *Advances in Fracture and Damage Mechanics XIII*, Vol. 627, *Key Engineering Materials*, Trans Tech Publications, Switzerland, pp. 33–36 (2015).
44. T.-S. Han and P. R. Dawson, A two-scale deformation model for polycrystalline solids using a strongly-coupled finite element methodology, *Computer Methods in Applied Mechanics and Engineering*, **196**(1316), 2029–2043 (2007).

45. E. Nakamachi, N. Tam, and H. Morimoto, Multi-scale finite element analyses of sheet metals by using SEM-EBSD measured crystallographic RVE models, *International Journal of Plasticity*, **23**(3), 450–489 (2007).
46. H. Kuramae, Y. Ikeya, H. Sakamoto, H. Morimoto, and E. Nakamachi, Multi-scale parallel finite element analyses of LDH sheet formability tests based on crystallographic homogenization method, *International Journal of Mechanical Sciences*, **52**(2), 183–197 (2010).
47. J. Rimoli and M. Ortiz, A three-dimensional multiscale model of intergranular hydrogen-assisted cracking, *Philosophical Magazine*, **90**(21), 2939–2963 (2010).
48. F. Trentacoste, I. Benedetti, and M. Aliabadi, Computational homogenization of polycrystalline materials with pores: A three-dimensional grain boundary formulation, *Journal of Multiscale Modelling*, **04**(03), 1250012 (2012).
49. G. Voronoï, Nouvelles applications des paramètres continus à la théorie des formes quadratiques. deuxième mémoire. recherches sur les parallélloèdres primitifs., *Journal für die reine und angewandte Mathematik*, **134**, 198–287 (1908).
50. F. Fritzen, T. Böhlke, and E. Schnack, Periodic three-dimensional mesh generation for crystalline aggregates based on Voronoi tessellations, *Computational Mechanics*, **43**, 701–713 (2009).
51. R. Quey, P. R. Dawson, and F. Barbe, Large scale 3D random polycrystals for the finite element method: Generation, meshing and remeshing, *Computer Methods in Applied Mechanics and Engineering*, **200**, 1729–1745 (2011).
52. C. H. Rycroft, Voro++: A three-dimensional Voronoi cell library in C++, *Chaos*, **19**, 041111 (2009).
53. C. A. Brebbia, J. C. F. Telles, and L. C. Wrobel, *Boundary Element Techniques*. Springer, Berlin-Heidelberg, (1984). DOI 10.1007/978-3-642-48860-3.
54. P. K. Banerjee, *The Boundary Element Methods in Engineering*. McGraw-Hill Book Company, Europe, Maidenhead (UK), pp. 177–188 (1994).
55. V. Gulizzi, A. Milazzo, and I. Benedetti, Fundamental solutions for general anisotropic multi-field materials based on spherical harmonics expansions, *International Journal of Solids and Structures*, **100–101**, 169–186 (2016).
56. V. Gulizzi and I. Benedetti, Micro-cracking of brittle polycrystalline materials with initial damage, *European Journal of Computational Mechanics*, **25**(1–2), 38–53 (2016).
57. M. Ortiz and A. Pandolfi, Finite-deformation irreversible cohesive elements for three-dimensional crack-propagation analysis, *International Journal for Numerical Methods in Engineering*, **44**, 1267–1282 (1999).
58. M. J. van den Bosch, P. J. G. Schreurs, and M. G. D. Geers, An improved description of the exponential xu and needleman cohesive zone law for mixed-mode decohesion, *Engineering Fracture Mechanics*, **73**(9), 1220–1234 (2006).
59. L. Snozzi and J.-F. Molinari, A cohesive element model for mixed mode loading with frictional contact capability, *International Journal for Numerical Methods in Engineering*, pp. n/a–n/a (2012).
60. K. W. Man, M. H. Aliabadi, and D. P. Rooke, Bem frictional contact analysis: Load incremental technique, *Computers & Structures*, **47**(6), 893–905 (1993).

61. K. W. Man, M. H. Aliabadi, and D. P. Rooke, Bem frictional contact analysis: Modelling considerations, *Engineering Analysis with Boundary Elements*, **11**(1), 77–85 (1993).
62. J. A. Garrido, A. Foces, and F. París, An incremental procedure for three-dimensional contact problems with friction, *Computers & Structures*, **50**(2), 201–215 (1994).
63. J. Telles and C. Brebbia, The boundary element method in plasticity, *Applied Mathematical Modelling*, **5**(4), 275–281 (1981).
64. A. Kuzmin, M. Luisier, and O. Schenk. Fast methods for computing selected elements of the greens function in massively parallel nanoelectronic device simulations. In *Euro-Par 2013 Parallel Processing*, Vol. 8097, *Lecture Notes in Computer Science*, F. Wolf, B. Mohr, and D. Mey (eds.), Springer, Berlin, Heidelberg, pp. 533–544 (2013).
65. O. Schenk, M. Bollhöfer, and R. A. Römer, On large-scale diagonalization techniques for the anderson model of localization, *Society ofr Industrial and Applied Mathematics Review*, **50**(1), 91–112 (2008).
66. O. Schenk, A. Wchter, and M. Hagemann, Matching-based preprocessing algorithms to the solution of saddle-point problems in large-scale nonconvex interior-point optimization, *Computational Optimization and Applications*, **36**(2–3), 321–341 (2007).
67. I. Benedetti, M. H. Aliabadi, and G. Davì, A fast 3D dual boundary element method based on hierarchical matrices, *International Journal of Solids & Structures*, **45**, 2355–2376 (2008).
68. I. Benedetti, A. Milazzo, and M. H. Aliabadi, A fast dual boundary element method for 3D anisotropic crack problems, *International Journal for Numerical Methods in Engineering*, **80**, 1356–1378 (2009).
69. E. Lutz, W. Ye, and S. Mukherjee, Elimination of rigid body modes from discretized boundary integral equations, *International Journal of Solids and Structures*, **35**(33), 4427–4436 (1998).
70. Z. Hashin and S. Shtrikman, A variational approach to the theory of the elastic behaviour of polycrystals, *Journal of the Mechanics and Physics of Solids*, **10**(4), 343–352 (1962).
71. M. Berveiller and A. Zaoui, An extension of the self-consistent scheme to plastically-flowing polycrystals, *Journal of the Mechanics and Physics of Solids*, **26**(5), 325–344 (1978).
72. F. Barbe, L. Decker, D. Jeulin, and G. Cailletaud, Intergranular and intragranular behavior of polycrystalline aggregates. Part 1: F.E. model, *International Journal of Plasticity*, **17**, 513–536 (2001).

Chapter 10

Piezoelectric and Magneto-electro-elastic Frictional Contact Modelling

L. Rodríguez-Tembleque, F. C. Buroni, A. Sáez
and M. H. Ferri Aliabadi

This chapter deals with the frictional contact problem for coupled piezoelectric and magneto-electro-elastic materials under indentation conditions. Due to the relevant technological applications, this topic of research has been treated in some analytical works. However, analytical solutions lack the generality of numerical methodologies, being restricted typically to simple geometries, loading conditions, idealized contact conditions and mostly taking into account transversely isotropic material symmetry with the symmetry axis normal to the contact surface. In this chapter, the numerical procedure proposed by the authors[1,2] is presented. This numerical scheme makes it possible to model the 3D frictional contact of anisotropic piezoelectric or magneto-electro-elastic materials in the presence of both electric and magnetic fields. The methodology uses the boundary element method with explicit evaluation of the fundamental solutions in order to compute the magneto-electro-elastic influence coefficients. The contact model is based on an augmented Lagrangian formulation and it uses an iterative Uzawa scheme of resolution. Conducting, semi-conducting and insulated electric and/or magnetic indentation conditions, as well as orthotropic frictional contact conditions are considered (so anisotropy is present both in the bulk and on the surface). After the validation by comparison with benchmark analytical solutions, additional exploration examples are presented and discussed in detail. Results reveal the influence of the conductivity on the indentation force and contact pressure distributions. The influence of friction in electric and magnetic potential responses has been also proved to be very significant. Moreover, tangential loads exhibit an important influence both on the maximum values of the electric and magnetic potentials as well as on their distributions.

1. Introduction

Piezoelectric (PE) or Magneto-Electro-Elastic (MEE) materials are widely used in technological applications for multiple purposes, such as sensors, actuators, filters, oscillators, phase shifters, memory devices or general smart structures, because of the coupling effects between the mechanical, electric and magnetic fields. Due to these wide and important technological applications, PE and MEE coupling has been the focus of intensive research in last years.[3–6] On the other hand, the study of the contact problem is necessary in order to address problems like positioning of micro- and nano-mechanisms and various functional devices, as well as in experimental testing and characterization of this kind of materials. Therefore, in this topic of research, it is crucial to have efficient theoretical and computational tools for modelling such coupled multifield contact problem in order to predict and understand the underlying physics in the interaction of electro-magnetic and mechanical processes.

In this context, several analytical works devoted to model the contact problem in coupled PE and MEE materials have recently appeared. Different theoretical approaches have addressed contact problems of PE materials in the presence of electric fields. Ding *et al.*,[7–9] or Giannakopoulos and Suresh[10] proposed analytical solution schemes for axisymmetric frictionless contact of PE solids under the three typical indenters (i.e. flat, conical and spherical). Later, Chen[11] studied the problem of a rigid smooth punch bonded to a transversely isotropic PE half-space, and Li and Wang[12] solved the Hertzian contact of anisotropic PE bodies. The frictionless contact in a layered PE half-space was first studied by Ramirez.[13,14] Wang *et al.*[15,16] solved the circular indentation problem on a PE layer, and Wang *et al.*[17,18] and Wu *et al.*[19] studied the indentation response of PE films. The PE indentation and surface interaction accompanied by frictional contact were considered by Makagon *et al.*,[20] Migorski *et al.*,[21] Zhou and Lee[22] and more recently by Ma *et al.*[23]

One of the first analytical works which studied the contact problem for MEE materials was by Pengfei *et al.*,[24] who obtained the Boussinesq and Cerruti solutions and further applied them to the frictional Hertz problem in transversely isotropic materials. Analytical solutions for the half-space indentation by rigid flat-ended, conical and spherical punches in transversely isotropic MEE materials have been presented and discussed in detail by Chen *et al.*[25] Indenters may be in any combination of conducting and insulating for both electric and magnetic fields. Both works are based on Fabrikant's method of potential theory for elastic materials.[26] Moving rigid punch solutions in 2D have been considered by Zhou *et al.*, both for frictionless[27,28] and frictional contact.[29–33] Ke and coworkers[34–36] presented comprehensive studies on the contact mechanics of the homogeneous or functionally graded magneto-electro-elastic materials (FGMEEMs). Most of the MEE analytical works are extensions of previous efforts related to the modelling of the contact problem for materials with electro-mechanical or magneto-mechanical coupling. For instance, some of those mentioned above are based on the mentioned works by Ding *et al.*,[7–9] or the work by Giannakopoulos and Suresh.[10] For transversely isotropic piezomagnetic (PM) materials, the frictionless axisymmetric indentation by flat rigid punch has been studied by Giannakopoulos and Parmaklis,[37] and the 2D indentation by a sliding rigid punch with flat or cylindrical profile has been presented by Zhou and Lee.[38]

However, all those analytical formulations present some limitations. They are restricted to simple geometries (e.g. 2D or axisymmetric indenters on a half-space), loading conditions, idealized contact conditions and taking into account almost uniquely transversely isotropic material symmetry with the material symmetry axis being normal to the contact surface. These limitations can be overtaken with advanced numerical formulations. Mathematical models in variational form for coupled electro-elastic contact problems and some numerical schemes based in the Finite Element Method (FEM) have been proposed by Migorski *et al.*[39] and Hüeber *et al.*[40] Quasistatic 2D contact problems between solid-foundation and a PE material and an electro-visco-elastic material have been studied, for instance, by Sofonea *et al.*,[41–44] under frictionless conditions; and by Sofonea *et al.*,[45] incorporating isotropic frictional contact conditions. Michopoulos *et al.*,[46–48] have studied the multiscale contact problem of magneto-electro-elastoplastic materials in the context of FEM, and such models have been validated against actual experiments. These works incorporate information about the roughness of the surfaces at the microscale and thermal fields are taken into account.

This chapter presents a Boundary Element Method (BEM) formulation to study 3D frictional contact of PE and MEE bodies. This methodology was initially developed to study 3D frictional contact of PE bodies in the presence of electric fields by Rodríguez-Tembleque *et al.*[1] and later extended to model the 3D frictional contact of anisotropic coupled MEE materials in presence of both electric and magnetic fields by the authors.[2] In both works, an orthotropic frictional law is assumed, so anisotropy is present both in the bulk and in the surface. Moreover, the BEM is considered to compute the elastic multifield coefficients because of its numerical suitability for solving the contact problem in finite, semi-finite or infinite domains. The ability of the BEM to accurately represent steep solution gradients is well known, and the reduction in the degrees of freedom (i.e. only considers the boundary degrees of freedom) becomes particularly attractive to handle the concerning problem, since for multifield materials the number of degrees of freedom per node notably increases. However, contrary to the FEM, conventional BEM formulations result in fully populated non-symmetric system of equations.

The chapter is organized as follows. In Section 2, the basic governing equations are presented as well as the nonlinear mechanical and magneto-electrical contact conditions. There is no explicit mention to the PE equations since they can be obtained as a particular case of the MEE equations by removing the magnetic field. The literature on BEM formulations is quite extensive, so in Section 3 we briefly present the basic ideas of the BEM with emphasis in the explicit evaluation procedure for the fundamental solutions. MEE contact discrete equations are presented in Section 4. Then, the solution scheme is sketched in Section 5. The methodology is validated by comparison with benchmark analytical solutions in Section 6, where additional exploration examples are presented and discussed in detail. Finally, the chapter is closed with a summary and some concluding remarks.

2. Coupled magneto-electro-elastic contact problem

2.1. *Governing equations*

The indentation problem formulation of a 3D MEE body $\Omega \subset \mathbb{R}^3$ with a piecewise smooth boundary $\partial\Omega$, in a Cartesian coordinate system (x_i) $(i = 1, 2, 3)$ is going to be described in this section (see Fig. 1). Assuming static loading conditions and the absence of volume forces and free electric charges, the mechanical, electrical and magnetic equilibrium equations can be written in terms of the stress $\boldsymbol{\sigma}$, the electric displacement $\boldsymbol{D}$ and the magnetic induction $\boldsymbol{B}$, as:

$$\sigma_{ij,j} = 0 \quad \text{in } \Omega, \tag{1}$$

$$D_{i,i} = 0 \quad \text{in } \Omega, \tag{2}$$

$$B_{i,i} = 0 \quad \text{in } \Omega, \tag{3}$$

where repeated dummy indices indicate summation.

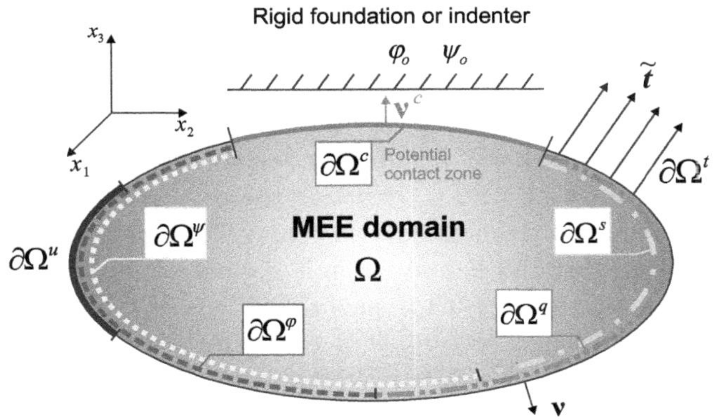

Fig. 1. The physical setting. Three partitions of the boundary $\partial\Omega$ are considered to define the mechanical, the electrical and the magnetic boundary conditions.

Small deformations are considered, so the infinitesimal strain tensor $\boldsymbol{\gamma}$, the electric field $\boldsymbol{E}$ and the magnetic field $\boldsymbol{H}$ are obtained, respectively, from derivatives of the displacements $\boldsymbol{u}$, the electric potential $\boldsymbol{\varphi}$ and the magnetic potential $\boldsymbol{\psi}$ as

$$\gamma_{ij} = (u_{i,j} + u_{j,i})/2 \quad \text{in } \Omega, \tag{4}$$

$$E_i = -\varphi_{,i} \quad \text{in } \Omega, \tag{5}$$

$$H_i = -\psi_{,i} \quad \text{in } \Omega. \tag{6}$$

In linear MEE materials, the elastic, electric and magnetic fields are coupled through the constitutive law[49]

$$\sigma_{ij} = c_{ijkl}\gamma_{kl} - e_{lij}E_l - q_{lij}H_l \quad \text{in } \Omega, \tag{7}$$

$$D_i = e_{ikl}\gamma_{kl} + \epsilon_{il}E_l + \lambda_{il}H_l \quad \text{in } \Omega, \tag{8}$$

$$B_i = q_{ikl}\gamma_{kl} + \lambda_{il}E_l + \mu_{il}H_l \quad \text{in } \Omega, \tag{9}$$

where $\boldsymbol{c}$, $\boldsymbol{\epsilon}$ and $\boldsymbol{\mu}$ denote the components of the elastic stiffness tensor, the dielectric permittivity tensor and the magnetic permeability tensor, respectively; $\boldsymbol{e}$, $\boldsymbol{q}$ and $\boldsymbol{\lambda}$ are the PE, PM and ME coupling coefficients, respectively. These tensors satisfy the following symmetries: $c_{ijkl} = c_{jikl} = c_{ijlk} = c_{klij}$, $e_{kij} = e_{kji}$, $q_{kij} = q_{kji}$, $\epsilon_{kl} = \epsilon_{lk}$, $\lambda_{kl} = \lambda_{lk}$, $\mu_{kl} = \mu_{lk}$. Moreover, the elastic constants, dielectric permittivity, magnetic permeability tensors are positive definite, and the PE, PM and ME coupling tensor are positive semi-definite.

Three partitions of the boundary $\partial\Omega$ are considered to define the mechanical, the electrical and the magnetic boundary conditions (see Fig. 1). The first one divides $\partial\Omega$ into three disjoint parts such that $\partial\Omega = \partial\Omega^u \cup \partial\Omega^t \cup \partial\Omega^c$. Here, $\partial\Omega^u$ denotes the boundary on which displacements $\tilde{u}_i$ are prescribed; $\partial\Omega^t$ denotes the boundary on which the tractions $\tilde{t}_i = \sigma_{ij}\nu_j$ are imposed, with ν_j denoting the components of the outward unit normal vector to the boundary $\partial\Omega^t$; and $\partial\Omega^c$ represents the potential contact surface, which have outward unit normal vector with components ν_i^c. The second partition of $\partial\Omega$ is such that $\partial\Omega = \partial\Omega^\varphi \cup \partial\Omega^q \cup \partial\Omega^c$, with the electrical potential $\tilde{\varphi}$ prescribed on $\partial\Omega^\varphi$, and the electrical charge flux $\tilde{q} = D_i\nu_i$ known on $\partial\Omega^q$, where ν_i is the outward unit normal vector on this part of the boundary. The last partition of $\partial\Omega$ is with regard to the magnetic boundary conditions, so $\partial\Omega = \partial\Omega^\psi \cup \partial\Omega^s \cup \partial\Omega^c$ is satisfied. In this case, magnetic potential $\tilde{\psi}$ is prescribed on $\partial\Omega^\psi$ and the normal magnetic flux $\tilde{s} = B_i\nu_i$ on the boundary $\partial\Omega^s$, with outward unit normal components ν_i.

For a well-posed problem, either Dirichlet or Neumann boundary conditions must be prescribed at each boundary point outside the contact zone $\partial\Omega^c$. In the contact zone, under small displacement assumption, a common unit normal vector ν_i^c can be considered in $\partial\Omega^c$. So, the nonlinear boundary contact conditions are:

$$\begin{aligned} \sigma_{ij}\nu_j^c &= p_i & &\text{on } \partial\Omega^c, \\ D_i\nu_i^c &= -\kappa_e(\varphi - \varphi_o) & &\text{on } \partial\Omega^c, \\ B_i\nu_i^c &= -\kappa_m(\psi - \psi_o) & &\text{on } \partial\Omega^c, \end{aligned} \tag{10}$$

where p_i are the contact traction: $p_\nu = \mathbf{p}\cdot\boldsymbol{\nu}^c$ is the normal contact pressure and $\mathbf{p}_\tau = \mathbf{p} - p_\nu\boldsymbol{\nu}^c$ the tangential contact traction; κ_e and κ_m are the electrical and magnetic conductivity coefficients, respectively; and φ_o and ψ_o denote, respectively, the electric and magnetic potentials of the indenter.

As it was mentioned in Section 1, the discrete PE contact formulation can be obtained as a particular case of the MEE equations (1–10) by removing the magnetic field.

2.2. *Normal contact conditions*

The quasistatic unilateral contact law involves Signorini's contact conditions in $\partial\Omega^c$:

$$g_\nu \geq 0, \quad p_\nu \leq 0, \quad g_\nu\, p_\nu = 0, \tag{11}$$

where $g_\nu = (g_o - u_\nu)$ is the normal gap and g_o is the initial gap between the bodies and $u_\nu = \mathbf{u} \cdot \boldsymbol{\nu}^c$. These normal contact constraints presented in (11) can be rewritten as:

$$p_\nu - \mathbb{P}_{\mathbb{R}_-}(p_\nu^*) = 0, \tag{12}$$

where $\mathbb{P}_{\mathbb{R}_-}(\bullet)$ is the normal projection function ($\mathbb{P}_{\mathbb{R}_-}(\bullet) = \min(0, \bullet)$) and $p_\nu^* = p_\nu + r_\nu g_\nu$ is the augmented normal traction. The parameters r_ν is the normal dimensional penalization parameter ($r_\nu \in \mathbb{R}^+$).

The electrical and magnetic contact boundary conditions (10) are controlled by the conductivity coefficients. These coefficients can be defined as: $\kappa_e = \kappa_e(p_\nu)$ and $\kappa_m = \kappa_m(p_\nu)$, with

$$\kappa_e(p_\nu) = \begin{cases} 0 & \text{if } p_\nu = 0, \\ \kappa_e^* & \text{if } p_\nu < 0, \end{cases} \tag{13}$$

and

$$\kappa_m(p_\nu) = \begin{cases} 0 & \text{if } p_\nu = 0, \\ \kappa_m^* & \text{if } p_\nu < 0. \end{cases} \tag{14}$$

The conductivity coefficients make it possible to describe perfect electrical and magnetic contact conditions similar to the Signorini's contact conditions. In the expressions (13) and (14), κ_e^* and κ_m^* are the conductivity parameter similar to Rodríguez-Tembleque *et al.*[1] So, according to (13) and (14), the electric and magnetic contact conditions in (10) show that when there is no contact (i.e. $p_\nu = 0$) on $\partial\Omega_c$, the normal electric and magnetic fluxes vanish (i.e. $q = D_i\nu_i^c = 0$ and $s = B_i\nu_i^c = 0$). On the other hand, when there is contact (i.e. $p_\nu < 0$), electric and magnetic charges appear.

2.3. *Tangential contact conditions*

In general, an orthotropic frictional constitutive law is considered for anisotropic MEE materials: $||\mathbf{p}_\tau||_\mu \leq |p_\nu|$ (see Fig. 2). According to Rodríguez-Tembleque *et al.* (2012, 2013), this Coulomb friction restriction can be summarized as:

$$\begin{aligned} ||\mathbf{p}_\tau||_\mu < |p_\nu| &\Rightarrow \dot{\mathbf{g}}_\tau = \mathbf{0} && \text{on } \partial\Omega_c, \\ ||\mathbf{p}_\tau||_\mu = |p_\nu| &\Rightarrow \mathbf{p}_\tau = -|p_\nu|\mathbb{M}^2\dot{\mathbf{g}}_\tau/||\dot{\mathbf{g}}_\tau||_\mu^* && \text{on } \partial\Omega_c. \end{aligned} \tag{15}$$

In the expressions above, the tangential slip can be assumed as $\dot{\mathbf{g}}_\tau \approx \Delta\mathbf{g}_\tau = \mathbf{g}_{o,\tau} - \mathbf{u}_\tau$ (i.e. $\mathbf{g}_{o,\tau}$ is the tangential translation and $\mathbf{u}_\tau = \mathbf{u} - u_\nu\boldsymbol{\nu}_c$) for

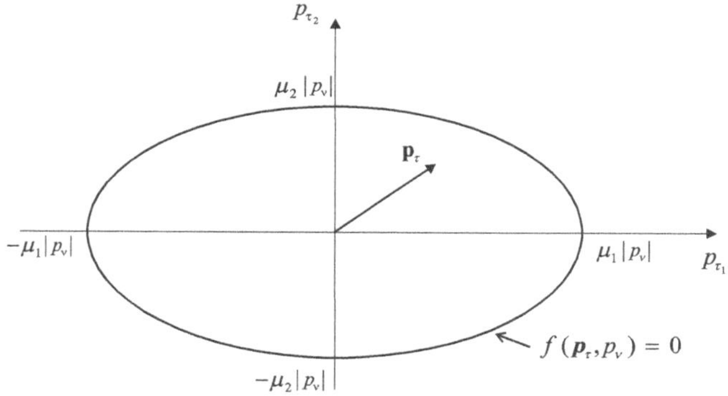

Fig. 2. Elliptic friction law: $f(\mathbf{p}_\mathcal{T}, p_\nu) = ||\mathbf{p}_\mathcal{T}||_\mu - |p_\nu| = 0$.

quasistatic contact problems, $|| \bullet ||_\mu$ denotes the elliptic norm

$$||\mathbf{p}_\mathcal{T}||_\mu = \sqrt{(p_{e_1}/\mu_1)^2 + (p_{e_2}/\mu_2)^2} \tag{16}$$

and the norm $|| \bullet ||_\mu^*$ is dual of $|| \bullet ||_\mu$

$$||\dot{\mathbf{g}}_\mathcal{T}||_\mu^* = \sqrt{(\mu_1 \dot{g}_{e_1})^2 + (\mu_2 \dot{g}_{e_2})^2}. \tag{17}$$

Finally, $\mathbb{M}$ is a diagonal matrix (i.e. $\mathbb{M} = \mathbf{diag}(\mu_1, \mu_2)$) where μ_1 and μ_2 the principal friction coefficients in the tangential directions $\{e_1, e_2\}$. In this work, $\{e_1, e_2\}$ are coincident with the x_1- and x_2-axis, respectively.

The frictional contact constraints (15) can also be formulated using contact operators as:

$$\mathbf{p}_\mathcal{T} - \mathbb{P}_{\mathbb{E}_\rho}(\mathbf{p}_\mathcal{T}^*) = 0, \tag{18}$$

where $\mathbf{p}_\mathcal{T}^* = \mathbf{p}_\mathcal{T} - r_\mathcal{T}\mathbb{M}^2\dot{\mathbf{g}}_\mathcal{T}$ ($r_\mathcal{T} \in \mathbb{R}^+$) is the augmented tangential traction and $\mathbb{P}_{\mathbb{E}_\rho}(\bullet) : \mathbb{R}^2 \longrightarrow \mathbb{R}^2$ is the tangential projection function defined in Ref. [51] as

$$\mathbb{P}_{\mathbb{E}_\rho}(\mathbf{p}_\mathcal{T}^*) = \begin{cases} \mathbf{p}_\mathcal{T}^* & \text{if } ||\mathbf{p}_\mathcal{T}^*||_\mu < \rho, \\ \rho\ \mathbf{p}_\mathcal{T}^*/||\mathbf{p}_\mathcal{T}^*||_\mu & \text{if } ||\mathbf{p}_\mathcal{T}^*||_\mu \geq \rho, \end{cases} \tag{19}$$

with $\rho = |\mathbb{P}_{\mathbb{R}_-}(p_\nu^*)|$.

3. Explicit boundary element formulation

The boundary integral equation for a 3D MEE body Ω, in absence of body forces, can be written, according to Barnett–Lothe representation,[52] as

$$c_{JK}(\mathbf{x}')u_J(\mathbf{x}') + \fint_{\partial\Omega} \check{T}_{JK}(\mathbf{x} \mid \mathbf{x}')u_J(\mathbf{x})dS(\mathbf{x})$$

$$= \int_{\partial\Omega} \check{U}_{JK}(\mathbf{x} \mid \mathbf{x}')t_J(\mathbf{x})dS(\mathbf{x}). \tag{20}$$

In Eq. (20), $\mathbf{x}' \in \partial\Omega$ is the source point and $\mathbf{x}$ is a boundary point, u_J is the extended displacement vector (see Barnett and Lothe representation in Appendix A); t_J is the extended tractions vector; c_{JK} depends on the local geometry of the boundary $\partial\Omega$ at the collocation point $\mathbf{x}'$ and is equal to $\frac{1}{2}\delta_{JK}$ for a smooth boundary at $\mathbf{x}'$ (with δ_{JK} being the extended Kronecker delta); $\check{U}_{JK}$ and $\check{T}_{JK}$ are the extended displacement fundamental solution and the extended traction fundamental solution at a boundary point $\mathbf{x}$ due to a unit extended source applied at point $\mathbf{x}'$, respectively.

The generalized equilibrium Eqs. (1–3) neglect both volume and Lorentz forces in the model, so that volume integrals do not appear in (20). However, Lorentz forces should be incorporated in the presence of high-intensity magnetic fields and moving electric charges. Since this is out of the scope of the present chapter, the interested reader may find in Ref. [53] a more involved discussion on how to deal with the body-forces arising out from Lorentz interactions in the context of BEM based electro-mechanical design.

In this work, the scheme for the evaluation of the extended fundamental solution proposed by Buroni and Sáez[54] has been implemented, which possesses the remarkable characteristics that it is exact, explicit and valid for mathematical degenerate and non-degenerate materials in the Stroh formalism context. Appendix B presents a summary of these fundamental solutions for completeness. The strongly singular integral on the left-hand side in (20) is evaluated in the Cauchy principal value sense, whereas the weakly singular integral on the right-hand side is evaluated as an improper integral.

The boundary $\partial\Omega$ can be discretized with elements and a collocation procedure on boundary integral equation (20), after the boundary conditions are considered, leading to a system of equations that, once solved, represents an approximate solution of the corresponding MEE problem.

The integral equation (20) can be written as follows:

$$
\begin{aligned}
c_{JK}(\mathbf{x}')u_J(\mathbf{x}') &+ \sum_{e=1}^{N_e}\left\{ ⨍_{\partial\Omega_e} \check{T}_{JK}(\mathbf{x}' \mid \mathbf{x})u_J(\mathbf{x})dS(\mathbf{x})\right\} \\
&= \sum_{e=1}^{N_e}\left\{ \int_{\partial\Omega_e} \check{U}_{JK}(\mathbf{x}' \mid \mathbf{x})t_J(\mathbf{x})dS(\mathbf{x})\right\}, \qquad (21)
\end{aligned}
$$

where the boundary is discretized into N_e elements of surface $\partial\Omega_e$ and the integrals over the boundary $\partial\Omega$ are replaced by the sum of the integrals over the surface of each element.

The functions u_J and t_J are approximated over each 4-node quadrilateral element $\partial\Omega_e$ using linear shape functions, as a function of the nodal values:

$$
u_J = \sum_{i=1}^{4} N_i(\xi, \eta)u_J^i, \qquad (22)
$$

$$
t_J = \sum_{i=1}^{4} N_i(\xi, \eta)t_J^i, \qquad (23)
$$

where u_J^i and t_J^i are the nodal extended displacements and tractions, respectively.

A collocation procedure on boundary integral equation (21) leads to the following system of equations:

$$
\mathbf{Hu} = \mathbf{Gt}, \qquad (24)
$$

where vectors $\mathbf{u}$ and $\mathbf{t}$ contain the values of all nodal extended displacements and tractions, respectively (i.e. $\mathbf{u}$ contains the nodal displacements and electric and magnetic potentials, and $\mathbf{t}$ contains the nodal tractions, electric charges and magnetic fluxes). Boundary conditions can be imposed rearranging the columns in $\mathbf{H}$ and $\mathbf{G}$, and passing all the unknowns to vector $\mathbf{x}$ on the left-hand side, resulting in the final system of algebraic equations $\mathbf{Ax} = \mathbf{F}$:

$$
\left[\mathbf{A}_{\mathbf{x}_e}\ \mathbf{A}_{u_c}\ \mathbf{A}_{\varphi_c}\ \mathbf{A}_{\psi_c}\ \mathbf{A}_{p_c}\ \mathbf{A}_{q_c}\ \mathbf{A}_{s_c}\right]\begin{Bmatrix} \mathbf{x}_e \\ \mathbf{u}_c \\ \boldsymbol{\varphi}_c \\ \boldsymbol{\psi}_c \\ \mathbf{p}_c \\ \mathbf{q}_c \\ \mathbf{s}_c \end{Bmatrix} = \mathbf{F}. \qquad (25)
$$

In expression (25), $\mathbf{x}_e$ collects the nodal external unknowns (i.e. the nodal unknowns which are outside the contact zone); $\mathbf{u}_c$, $\boldsymbol{\varphi}_c$ and $\boldsymbol{\psi}_c$ collect the nodal contact displacements, and electric and magnetic potentials, respectively; $\mathbf{p}_c$ contains the normal and tangential nodal contact tractions (i.e., $\mathbf{p}_\nu$ and $\mathbf{p}_\tau$); and, $\mathbf{q}_c$ and $\mathbf{s}_c$ contains the nodal electric charges and magnetic fluxes, respectively. Matrices $\mathbf{A}_{\mathrm{x}_e}$, $\mathbf{A}_{u_c}$, $\mathbf{A}_{\varphi_c}$, $\mathbf{A}_{\psi_c}$, $\mathbf{A}_{p_c}$, $\mathbf{A}_{q_c}$ and $\mathbf{A}_{s_c}$ contain the free terms and corresponding integrals in equation (21) and vector $\mathbf{F}$ contains the terms corresponding to the prescribed boundary conditions.

4. Discrete MEE contact formulation

According to the magneto-electric contact condition (10), (13) and (14), the electric charge and the magnetic flux on every contact node i can be expressed in terms of the electric and magnetic potential as:

$$(\mathbf{q}_c)_i = -\kappa_e((\mathbf{p}_\nu)_i)((\boldsymbol{\varphi}_c)_i - (\boldsymbol{\varphi}_o)_i), \tag{26}$$

$$(\mathbf{s}_c)_i = -\kappa_m((\mathbf{p}_\nu)_i)((\boldsymbol{\psi}_c)_i - (\boldsymbol{\psi}_o)_i). \tag{27}$$

So Eq. (25) can be rewritten as

$$\left[\mathbf{A}_{\mathrm{x}_e}\ \mathbf{A}_{u_c}\ \tilde{\mathbf{A}}_{\varphi_c}\ \tilde{\mathbf{A}}_{\psi_c}\ \mathbf{A}_{p_c}\right] \begin{Bmatrix} \mathbf{x}_e \\ \mathbf{u}_c \\ \boldsymbol{\varphi}_c \\ \boldsymbol{\psi}_c \\ \mathbf{p}_c \end{Bmatrix} = \tilde{\mathbf{F}}, \tag{28}$$

with

$$\tilde{\mathbf{A}}_{\varphi_c} = \mathbf{A}_{\varphi_c} - \boldsymbol{\kappa}_e(\mathbf{p}_\nu)\mathbf{A}_{q_c}, \tag{29}$$

$$\tilde{\mathbf{A}}_{\psi_c} = \mathbf{A}_{\psi_c} - \boldsymbol{\kappa}_m(\mathbf{p}_\nu)\mathbf{A}_{s_c}, \tag{30}$$

$$\tilde{\mathbf{F}} = \mathbf{F} - \boldsymbol{\kappa}_e(\mathbf{p}_\nu)\mathbf{A}_{q_c}\boldsymbol{\varphi}_o - \boldsymbol{\kappa}_m(\mathbf{p}_\nu)\mathbf{A}_{s_c}\boldsymbol{\psi}_o, \tag{31}$$

and $\boldsymbol{\kappa}_e(\mathbf{p}_\nu)$ and $\boldsymbol{\kappa}_m(\mathbf{p}_\nu)$ diagonal matrices, i.e.

$$\boldsymbol{\kappa}_e(\mathbf{p}_\nu) = \mathbf{diag}\left[\kappa_e((\mathbf{p}_\nu)_1), \ldots, \kappa_e((\mathbf{p}_\nu)_i), \ldots, \kappa_e((\mathbf{p}_\nu)_{N_c})\right], \tag{32}$$

$$\boldsymbol{\kappa}_m(\mathbf{p}_\nu) = \mathbf{diag}\left[\kappa_m((\mathbf{p}_\nu)_1), \ldots, \kappa_m((\mathbf{p}_\nu)_i), \ldots, \kappa_m((\mathbf{p}_\nu)_{N_c})\right]. \tag{33}$$

Finally, the mechanical contact restrictions (12) and (18) are defined on every contact node i as:

$$(\mathbf{p}_\nu)_i - \mathbb{P}_{\mathbb{R}_-}((\mathbf{p}_\nu)_i + r_\nu(\mathbf{g}_\nu)_i) = 0, \tag{34}$$

$$(\mathbf{p}_\tau)_i - \mathbb{P}_{\mathbb{E}_\rho}((\mathbf{p}_\tau)_i - r_\tau \mathbb{M}^2(\dot{\mathbf{g}}_\tau)_i) = \mathbf{0}, \tag{35}$$

where $\mathbf{p}_\nu$ and $\mathbf{p}_\tau$ contain the normal and tangential contact tractions of every contact node i and $\mathbf{g}_\nu$ and $\mathbf{g}_\tau$ contain the normal gap and the tangential slip vectors, respectively.

5. Solution scheme

The quasistatic MEE contact system (28–35) is solved using Uzawa's method.[55,56] To compute the variables at the load step (k), $\mathbf{z}^{(k)} = (\mathbf{x}_e^{(k)}, \mathbf{u}_c^{(k)}, \boldsymbol{\varphi}_c^{(k)}, \boldsymbol{\psi}_c^{(k)}, \mathbf{p}_c^{(k)})$, we iterate using $(n = 0, 1, 2, ...)$ index, initializing $\mathbf{z}^{(0)} = \mathbf{z}^{(k-1)}$ (i.e. $\mathbf{p}_c^{(0)} = \mathbf{p}_c^{(k-1)}$).

First, we solve the following system:

$$\left[\mathbf{A}_{\mathrm{x}_e}\ \mathbf{A}_{u_c}\ \tilde{\mathbf{A}}_{\varphi_c}\ \tilde{\mathbf{A}}_{\psi_c}\right] \begin{Bmatrix} \mathbf{x}_e \\ \mathbf{u}_c \\ \boldsymbol{\varphi}_c \\ \boldsymbol{\psi}_c \end{Bmatrix}^{(n+1)} = -\mathbf{A}_{p_c}\ \mathbf{p}_c^{(n)} + \tilde{\mathbf{F}}, \tag{36}$$

with $\tilde{\mathbf{A}}_{\varphi_c} = \mathbf{A}_{\varphi_c} - \boldsymbol{\kappa}_e(\mathbf{p}_\nu^{(n)})\mathbf{A}_{q_c}$, $\tilde{\mathbf{A}}_{\psi_c} = \mathbf{A}_{\psi_c} - \boldsymbol{\kappa}_m(\mathbf{p}_\nu^{(n)})\mathbf{A}_{s_c}$ and $\tilde{\mathbf{F}} = \mathbf{F} - \boldsymbol{\kappa}_e(\mathbf{p}_\nu^{(n)})\mathbf{A}_{q_c}\boldsymbol{\varphi}_o - \boldsymbol{\kappa}_m(\mathbf{p}_\nu^{(n)})\mathbf{A}_{s_c}\boldsymbol{\psi}_o$.

After solving the system of equations (36), contact tractions for every contact node i are updated:

$$(\mathbf{p}_\nu^{(n+1)})_i = \mathbb{P}_{\mathbb{R}_-}((\mathbf{p}_\nu^{(n)})_i + r_\nu(\mathbf{g}_\nu^{(n+1)})_i), \tag{37}$$

$$(\mathbf{p}_\tau^{(n+1)})_i = \mathbb{P}_{\mathbb{E}_\rho}((\mathbf{p}_\tau^{(n)})_i - r_\tau\ \mathbb{M}^2(\Delta\mathbf{g}_\tau^{(n+1)})_i), \tag{38}$$

where $(\mathbf{g}_\nu^{(n+1)})_i = (\mathbf{g}_{o,\nu}^{(k)})_i - (\mathbf{u}_\nu^{(n)})_i$, $(\Delta\mathbf{g}_\tau^{(n+1)})_i = (\mathbf{g}_{o,\tau}^{(k)})_i - (\mathbf{u}_\tau^{(n)})_i$ and $\rho = |(\mathbf{p}_\nu^{(n+1)})_i|$.

Finally, if $\Psi(\mathbf{z}^{(n+1)}) \leq \varepsilon$, with $\Psi(\mathbf{z}^{(n+1)}) = \max\{\|\mathbf{u}_c^{(n+1)} - \mathbf{u}_c^{(n)}\|, \|\boldsymbol{\varphi}_c^{(n+1)} - \boldsymbol{\varphi}_c^{(n)}\|, \|\boldsymbol{\psi}_c^{(n+1)} - \boldsymbol{\psi}_c^{(n)}\|, \|\mathbf{p}_c^{(n+1)} - \mathbf{p}_c^{(n)}\|\}$, the solution for the instant (k) is reached, i.e. $\mathbf{z}^{(k)} = \mathbf{z}^{(n+1)}$. Otherwise, return to (36) evaluating $\mathbf{p}_c^{(n)} = \mathbf{p}_c^{(n+1)}$.

6. Numerical results

The proposed multifield frictional contact formulation has a great potential to study the indentation response of PE and MEE materials. So, its capabilities and some numerical results are shown in this section.

First, the methodology is applied to study a transversely isotropic PE film under several indentation conditions. After the PE contact formulation is validated by comparison with the analytical results presented by Wang *et al.*,[17] the influence of the friction and the conductivity parameter are studied varying thicknesses t from a PE half-space (HS) configuration, where the contact radius a is much smaller than the thickness of the film (i.e. $a/t \to 0$), to a thin film (TF) configuration (i.e. $a/t \to \infty$).

Secondly, the MEE contact formulation is validated by comparison with the analytical results presented by Chen *et al.*,[25] where the indentation response of transversely isotropic MEE half-space is considered.

Finally, a fully orthotropic MEE solid indentation is solved under frictional and non-isolated MEE indentation conditions. In this case, the influence of the tangential load on the electric and magnetic response is studied.

6.1. *PE film: Frictionless indentation response*

In order to validate the formulation, this section studies the indentation responses of a finitely thick PE film, whose dimensions are $2L_1 \times 2L_1 \times t$ (see Fig. 3(a)), in the absence of friction. In the numerical example, a rigid spherical indenter of radius $R = 100 \times 10^{-3}$ m is subjected to a normal indentation $g_o = 5 \times 10^{-5}$ m and the PE block is assumed to be ideally bonded at the base ($x_3 = -t$). A transversely isotropic PE material (namely, $BaTiO_3$) with the symmetry axis coinciding with x_3-direction is considered, and its properties are presented in Table 1.[17] The block is discretized by 1024 linear quadrilateral boundary elements, using 16×16 elements on the $L_o \times L_o$ potential contact zone ($L_o = 5 \times 10^{-3}$ m), as Fig. 3(b) shows. In the case of PE film configurations, a more local and refined mesh is considered (see Fig. 3(c)).

Figure 4 shows a comparison between the proposed boundary element solution and the analytical solution presented by Wang *et al.*,[17] for PE HS and PE TF. Figure 4(a) shows normal contact pressure distributions on $\partial\Omega_c$ and Fig. 4(b) presents the electric potentials for $t = 80 \times 10^{-3}$ m and $t = 0.5 \times 10^{-3}$ m (i.e. $a/t = 0.031$ and $a/t = 7.5$, respectively). Magnitudes

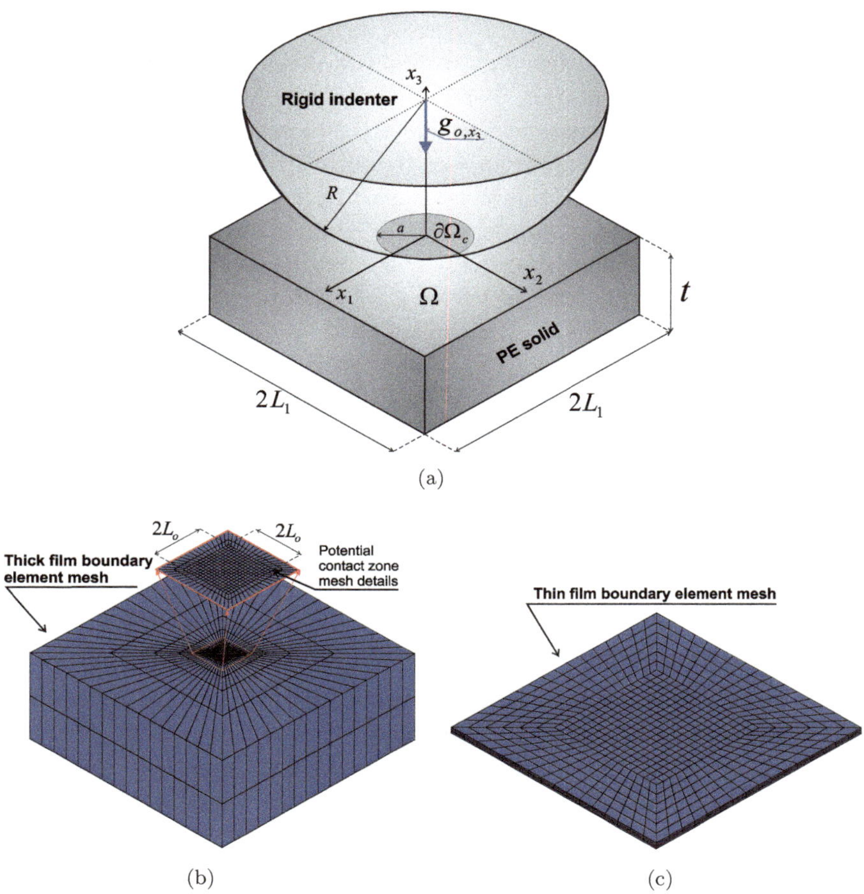

Fig. 3. (a) The physical setting: rigid indenter over a PE block. (b) Boundary element mesh details for the PE HS configuration and (c) Boundary element mesh for the PE TF configuration.

in Fig. 4 are normalized by the maximum normal contact pressure (p_o), the maximum electric potential (φ_o) and the contact radius (a_o) presented in[17] for an HS configuration. An excellent agreement between analytical and numerical solutions can be observed in both cases HS and TF. The examples have been solved considering $r_\nu = 5 \times 10^3$ for the augmented Lagrangian, and $\varepsilon = 10^{-6}$ as termination limit.

Indentation responses of a finitely thick PE film (i.e. indentation force responses and electric potential) are shown in Fig. 5. It may be seen that, from Fig. 5, the transition occurs at around $a/t \simeq 1$: Half-space solutions

Table 1. Material properties of the PE ceramic $BaTiO_3$.

Elastic coefficients (GPa)	
c_{1111}	150.00
c_{1122}	66.00
c_{1133}	66.00
c_{3333}	146.00
c_{2323}	44.00
PE coefficients (C/m^2)	
e_{113}	11.40
e_{333}	17.50
e_{311}	−4.350
Dielectric constants (10^{-9} F/m)	
ϵ_{11}	9.868
ϵ_{33}	11.151

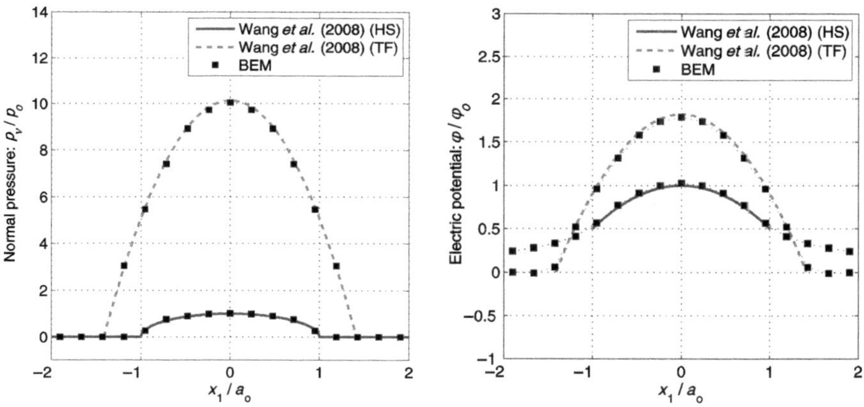

(a) (b)

Fig. 4. Comparison between the boundary element solution and the analytical solution[17] for: normal pressure (a) and electric potential (b), in PE HS and TF configurations. Magnitudes are normalized by the maximum normal contact pressure (p_o), the maximum electric potential (φ_o) and the contact radius (a_o) presented in Ref. [17] for an HS configuration. (These figures were reprinted from Rodríguez-Tembleque *et al.*[57] with permission from Elsevier.)

are good approximation of the indentation responses of a finitely thick PE film when $a/t < 0.1$, while solutions for an infinitely TF are more appropriate when $a/t \simeq 10$. These conclusions were presented in Ref. [17] for an axisymmetric frictionless insulating indentation of finitely thick PE films.

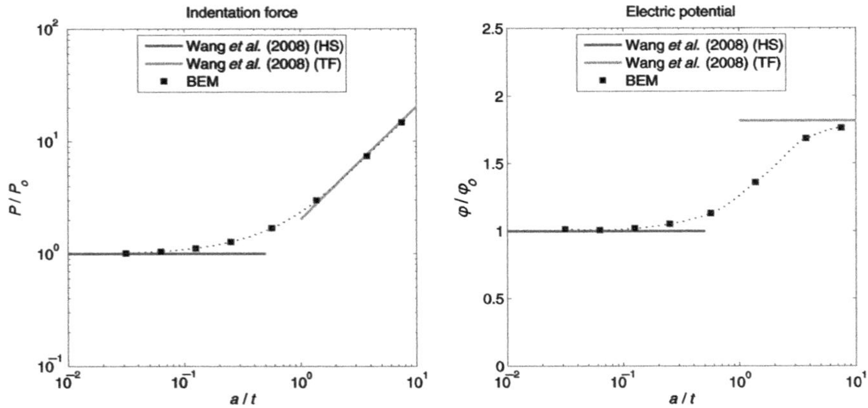

(a) (b)

Fig. 5. Normalized indentation response as function of contact radius for thick PE film subjected to frictionless insulating punch: (a) Indentation force and (b) Electric potential. Results are normalized by the maximum normal indentation load (P_o) and the maximum electric potential (φ_o) for an HS configuration. (These figures were reprinted from Rodríguez-Tembleque *et al.*[57] with permission from Elsevier.)

6.2. *PE film: Conducting indenter*

Previous studies considered frictionless insulating indentation conditions. Now we study the indentation of a finitely thick PE film by a conducting rigid spherical indenter. The influence of the conductivity parameter κ^* (13) in the indentation force and electric potential are presented in Fig. 6 for a prescribed indenter electric potential: $\varphi_o = 0$. These results are normalized relative to the insulating HS indentation.

We observe in Fig. 6(a) that indentation forces are not highly affected by the conductivity κ^* in both PE HS and TF configurations. Nevertheless, the values of the electric potential at the contact region are very affected and they tend to the indenter prescribed electric potential φ_o when the values of κ^* are higher, in finitely thick PE films configurations. In fact, when κ^* increases, the influence of a/t is neglected and PE HS and TF configurations present the same values for the electric potential.

6.3. *PE film: Frictional indentation response*

This section considers indentation responses of finitely thick PE films under frictional contact conditions. An isotropic friction law is assumed (i.e. $\mu_1 = \mu_2 = \mu$), since $BaTiO_3$ has transversely isotropic mechanical and electrical

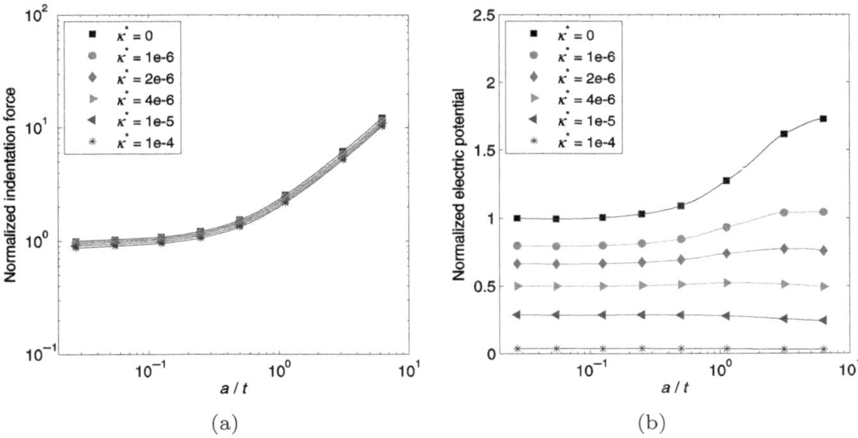

Fig. 6. Normalized indentation response as function of contact radius for thick PE film subjected to frictionless conducting punch: (a) Indentation force. (b) Electric potential and (c) Electric charges. Magnitudes are normalized by the normal contact load, the maximum electric potential computed for an isolated HS indentation and the maximum electric charge computed for a conducting HS indentation, respectively. (These figures were reprinted from Rodríguez-Tembleque *et al.*[57] with permission from Elsevier.)

material properties, and the symmetry axis is coincident with x_3-direction. The insulating indenter is subjected to a normal indentation $g_o = 5 \times 10^{-5}$m, with $r_\nu = 5 \times 10^3$, $r_\tau = 2 \times 10^5$ and $\varepsilon = 10^{-6}$.

Figures 7(a) and 7(b) present the influence of friction in the indentation force and in the electric potential of finitely thick PE film, respectively. Figure 7(a) shows the indentation force normalized relative to the frictionless HS indentation force, for different values of $\mu = \{0, 0.1, 0.2, 0.3, 0.5, 0.8\}$, and Fig. 7(b) shows the maximum electric potential normalized relative to the frictionless HS configuration. In Fig. 7(a), we can observe how indentation forces are almost not affected by the friction coefficient μ, in both PE HS and TF configurations.

However, Fig. 7(b) reveals that the influence of friction in the electric potential response is quite significant in PE HS and finitely thick PE film (i.e. $a/t \leq 1$), exhibiting an increment of 25% in the maximum electric potential for $\mu = 0.8$. Nevertheless, for infinitely TF configurations (i.e. $a/t \simeq 10$) this influence is null, as 7(b) shows. This increment is a consequence of the stress state increment inside the bulk due to the frictional tangential contact tractions presence. For an infinitely TF configurations, the tangential tractions are neglected compared to the high normal tractions that normal indentation generates.

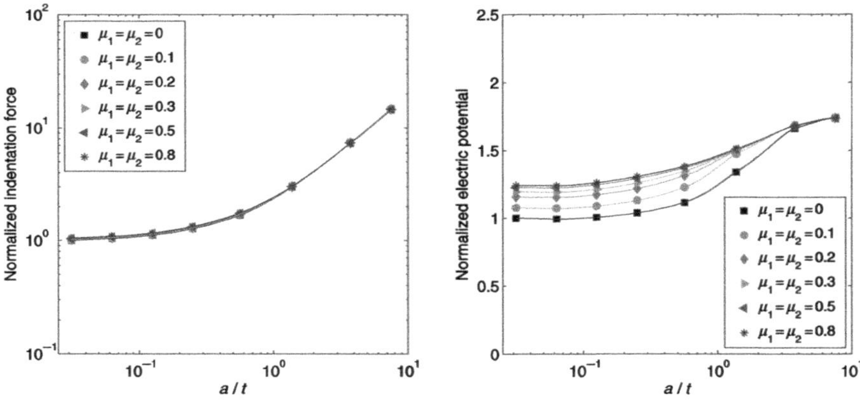

(a) (b)

Fig. 7. Normalized indentation response as function of contact radius for thick PE film subjected to frictional punch: (a) Indentation force and (b) Electric potential. Magnitudes are normalized by the normal contact load and the maximum electric potential computed for an isolated frictionless HS indentation. (These figures were reprinted from Rodríguez-Tembleque *et al.*[57] with permission from Elsevier.)

6.4. *Indentation response of a transversely isotropic MEE solid*

The spherical indentation response of a transversely isotropic MEE block, whose dimensions are $2L_1 \times 2L_1 \times L_2$ (with $L_1 = 50 \times 10^{-3}$ m and $L_2 = 120 \times 10^{-3}$ m), is solved in this section (see Fig. 8(a)). Figure 8(b) shows the boundary element mesh. The block is discretized by 1088 linear quadrilateral boundary elements, using 16×16 elements on the $L_o \times L_o$ potential contact zone ($L_o = 5 \times 10^{-3}$ m). The material considered in this example was presented by Chen *et al.*[25] It is an MEE composite material made of PE ($BaTiO_3$) and magnetostrictive ($CoFe_2O_4$) phases, whose material properties were estimated using the rule of mixture according to the volume fraction: $\chi = 0.5$. This MEE material has a symmetry axis coinciding with x_3-direction and presents no coupling between the electric and magnetic field (see Table 2).

The rigid sphere of radius $R = 100 \times 10^{-3}$m is subjected to a normal indentation of $g_{\nu,o} = 5 \times 10^{-5}$m, $g_{T,o} = 0$), and the MEE block is assumed to be ideally bonded at the base ($x_3 = -L_2$). Different frictionless indentation conditions are considered, e.g.: electrically and magnetically insulating indentation (EMII), electrically conducting and magnetically insulating indentation (ECMII), electrically insulating and magnetically conducting

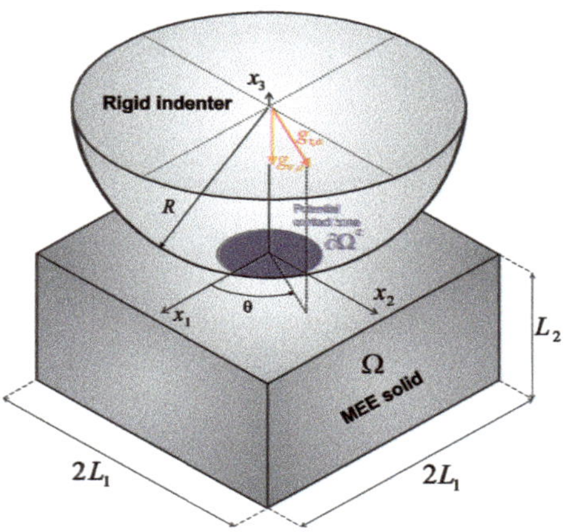

(a)

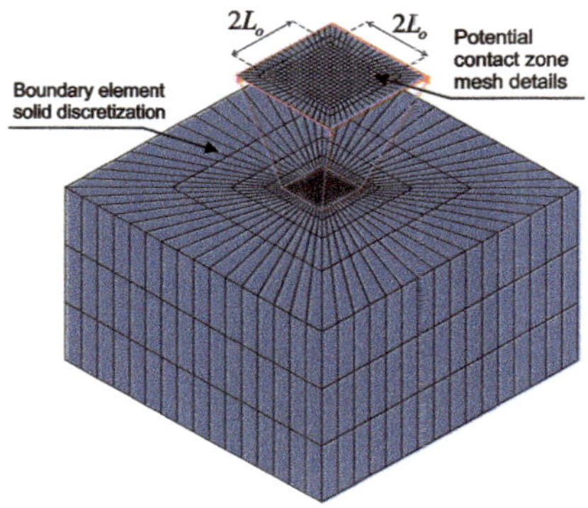

(b)

Fig. 8. (a) The physical setting: rigid indenter over a MEE domain and (b) Boundary element mesh and potential contact zone mesh details.

Table 2. Transversely isotropic MEE properties presented in Ref. [25] for ($BaTiO_3$–$CoFe_2O_4$).

Elastic coefficients (GPa)	
c_{1111}	226.00
c_{1122}	125.00
c_{1133}	124.25
c_{3333}	216.00
c_{2323}	44.15
PE coefficients (C/m^2)	
e_{113}	5.80
e_{333}	9.30
e_{311}	−2.20
PM coefficients (N/Am)	
q_{113}	275.00
q_{333}	350.00
q_{311}	290.15
Dielectric constants (10^{-9} F/m)	
ϵ_{11}	5.64
ϵ_{33}	6.35
Magnetic constants (10^{-6} Ns^2/C^2)	
μ_{11}	297.50
μ_{33}	83.50

This table was reprinted from Rodríguez-Tembleque *et al.*[2] with permission from Elsevier.

Table 3. Analytical contact relations[25] for an MEE HS indented by a spherical indenter.

Indentation condition	Resultant force (P)	Contact radius (a)
EMII	$4a^3/(3\pi\xi_{11}R)$	$\sqrt{g_{\nu,o}R}$
ECMII	$4\xi_{22}a^3/(3\pi\eta_{33}R)$	$\sqrt{g_{\nu,o}R - \xi_{12}\varphi_o R/\xi_{22}}$
EIMCI	$4\xi_{33}a^3/(3\pi\eta_{22}R)$	$\sqrt{g_{\nu,o}R - \xi_{13}\psi_o R/\xi_{33}}$
EMCI	$4a(\eta_{11}g_{\nu,o} + \eta_{21}\varphi_o + \eta_{31}\psi_o)/(3\pi\eta)$	$\sqrt{g_{\nu,o}R + (\eta_{21}\varphi_o + \eta_{31}\psi_o)R/\eta_{11}}$

Table 4. Basic indentation parameters presented in Ref. [25] for the MEE composite material.

$[\xi_{ij}]$	$\begin{bmatrix} 2.08613 & 1.77480 & 1.93900 \\ & -2.35928 & 2.85507 \\ \text{sym.} & & -1.00418 \end{bmatrix}$
$\eta = \lvert\xi_{ij}\rvert$	5.14794
$[\eta_{ij}]$	$\begin{bmatrix} 2.08613 & 1.77480 & 1.93900 \\ & -2.35928 & 2.85507 \\ \text{sym.} & & -1.00418 \end{bmatrix}$

This table was reprinted from Rodríguez-Tembleque *et al.*[2] with permission from Elsevier.

indentation (EIMCI) or electrically and magnetically conducting indentation (EMCI). Chen *et al.*[25] present the analytical solutions for a transversely isotropic MEE under all these spherical indentation conditions. For instance, the resultant indentation forces are presented in Table 3, where a is the contact radius, ξ_{ij} and η_{ij} are the MEE composite material parameters listed in Table 4 and ϕ_o and ψ_o are the indenter-prescribed electric and magnetic potential, respectively. Similar expressions can be found in Ref. [25] for the total electric charge (Q), the total magnetic charge (M), the electric potential distribution or the magnetic potential distribution.

The resultant indentation force and the total electric charge for an ECMII (i.e. $\kappa_e^* \neq 0$ and $\kappa_m^* = 0$) are shown in Fig. 9(a). We can appreciate how, when the conductivity parameter κ_e^* increases ($\kappa_e^* \to \infty$), the computed resultant indentation force and the total electric charge values converge to the analytic values P_{ECMII} and Q_{ECMII}, presented in Ref. [25]. In those cases, the prescribed electric potential was $\varphi_o = 0$. Moreover, contact variables distributions (i.e. (Fig. 9(b)) contact pressure, (Fig. 9(c)) electric

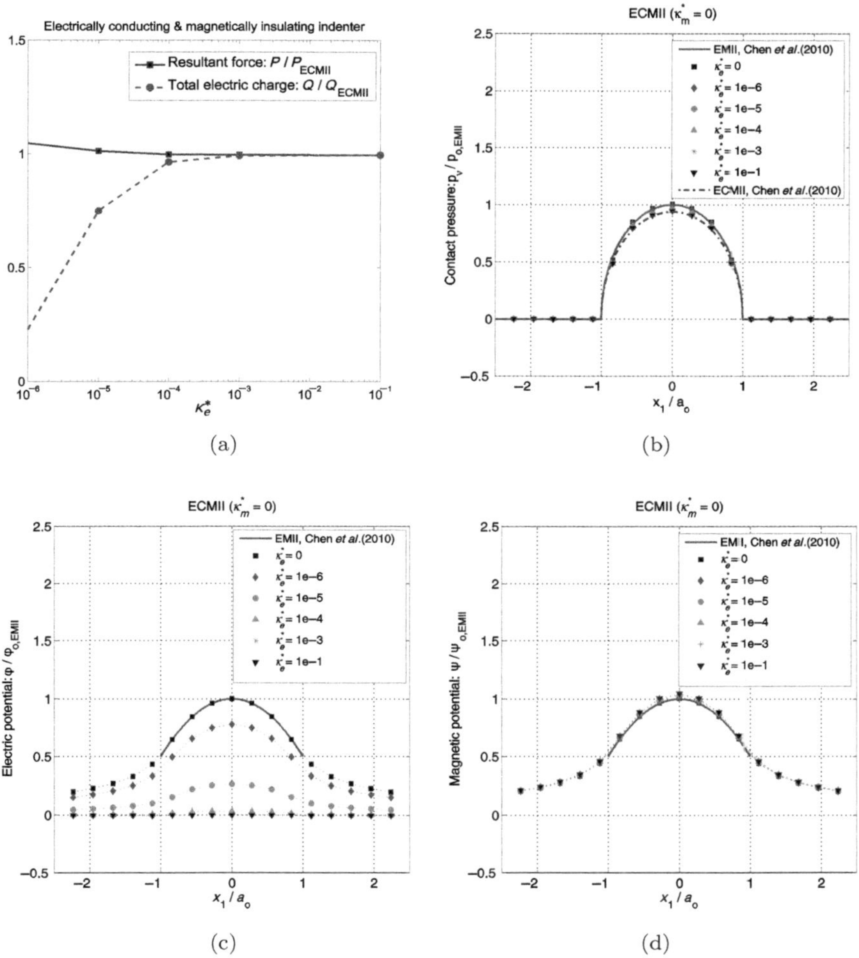

Fig. 9. (a) Influence of the electrical conductivity parameter in ECMII (i.e. $\kappa_e^* \neq 0$ and $\kappa_m^* = 0$). Normalized indentation response distributions as function of conductivity parameters (i.e. κ_e^* and κ_m^*) are presented for: (b) Dimensionless contact pressure ($p_\nu/p_{o,\text{EMII}}$). (c) Dimensionless electric potential ($\varphi/\varphi_{o,\text{EMII}}$) and (d) dimensionless magnetic potential ($\psi/\psi_{o,\text{EMII}}$). Moreover, normalized indentation responses for ECMII and EIMCI are showed. In all these figures, we compare the numerical results to the analytical solutions available in the literature for the limit cases, i.e. EMII, ECMII and EIMCI. (These figures were reprinted from Rodríguez-Tembleque *et al.*[2] with permission from Elsevier.)

potential and (Fig. 9(d)) magnetic potential) are presented as functions of conductivity parameters. For validation purposes, the analytical solutions available in the scientific literature[25] are also plotted. An excellent agreement between analytic and numerical solutions can be observed. For $\kappa_e^* = 0$ and $\kappa_m^* = 0$, perfectly insulating indentation responses (i.e. $p_{o,\text{EMII}}$, $\varphi_{o,\text{EMII}}$ and $\psi_{o,\text{EMII}}$) are obtained by the boundary elements formulation. We observe in Fig. 9(c) that, for higher values of κ_e^* ($\kappa_e^* \to \infty$, $\kappa_m^* = 0$), the value of the electric potential at the contact region tends to the prescribed electric potential $\varphi_o = 0$. However, the magnetic potential remains unaltered (see Fig. 9(d)). This is due to fact that the MEE composite material (Table 2) considered for validation in[25] presents no coupling between the electric and magnetic fields. The examples have been solved using the proposed algorithm, considering $r_\nu = 5 \times 10^3$ for the augmented Lagrangian, and $\varepsilon = 10^{-6}$ as a termination limit.

The same behaviour can be observed in Fig. 10 for the total magnetic charge, in an EIMCI (i.e. $\kappa_e^* = 0$ and $\kappa_m^* \neq 0$). Figure 10(a) shows that the computed resultant indentation force and the total magnetic charge values converge to the analytic values P_{EIMCI} and M_{EIMCI} when the conductivity parameter κ_m^* increases ($\kappa_m^* \to \infty$). In those cases, the prescribed magnetic potential was $\psi_o = 0$. As expected, when $\kappa_m^* \to \infty$ and $\kappa_e^* = 0$: the value of the magnetic potential at the contact region tends to the prescribed potential $\psi_o = 0$ (see Fig. 10(c)) and the electric potential remains unaltered (see Fig. 10(d)). An excellent agreement between analytic and numerical solutions can be observed.

6.5. *Frictional indentation response of an orthotropic MEE solid*

This example studies the indentation response of an orthotropic MEE material (see Table 5) whose elastic and PE properties are taken from an orthotropic polyvinylidene fluoride (PVDF)[58] and orthotropic PM properties are inspired from a $CoFe_2O_4$.[59] We can appreciate the full coupling between the magnetic, electric and elastic fields. Moreover, frictional contact conditions are considered, assuming both isotropic and orthotropic friction laws.

First an isotropic friction law ($\mu_1 = \mu_2 = \mu$) is considered. Figure 11(a) and 11(b) present the influence of friction under EMII and EMCI, respectively, for different values of $\mu = \{0, 0.1, 0.3, 0.6, 0.9\}$. Figure 11(a) shows the influence of friction on the normalized resultant force, the maximum electric potential and the maximum magnetic potential. Their values increase with

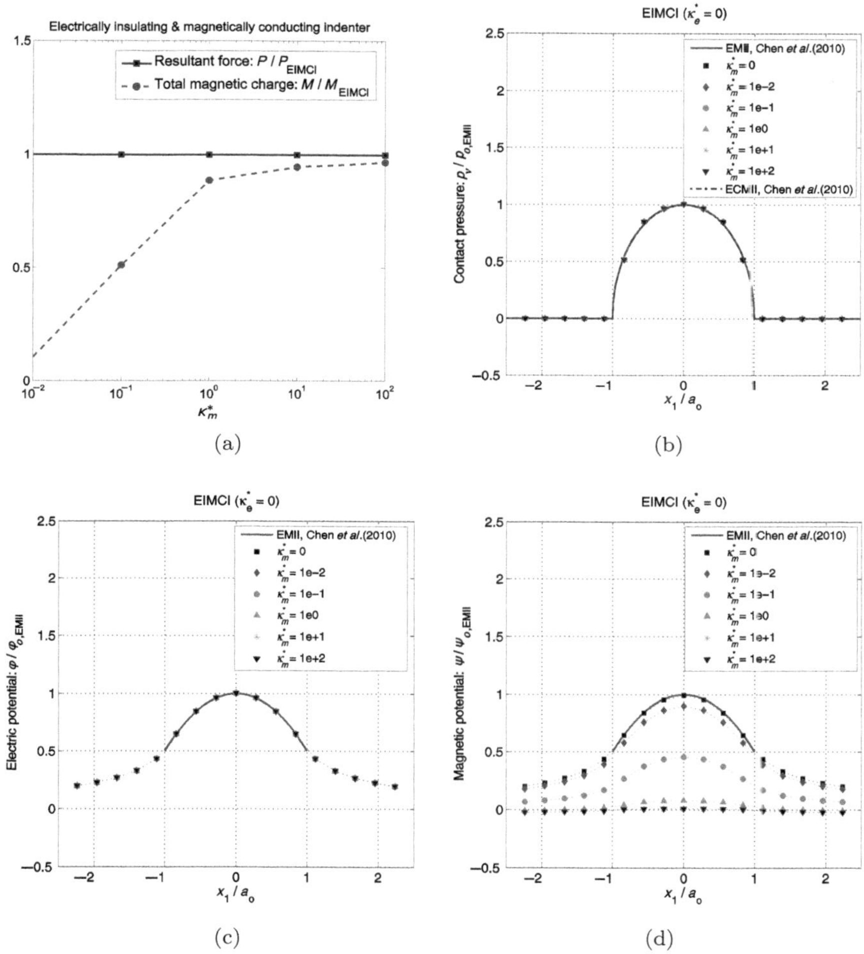

Fig. 10. (a) Influence of the magnetic conductivity parameter in EIMCI. Normalized indentation response distributions as function of conductivity parameters (i.e. κ_e^* and κ_m^*) are presented for: (b) Dimensionless contact pressure ($p_\nu/p_{o,\mathrm{EMII}}$), (c) Dimensionless electric potential ($\varphi/\varphi_{o,\mathrm{EMII}}$) and (d) Dimensionless magnetic potential ($\psi/\psi_{o,\mathrm{EMII}}$). Moreover, normalized indentation responses for ECMII and EIMCI are showed. In all these figures, we compare the numerical results to the analytical solutions available in the literature for the limit cases, i.e. EMII, ECMII and EIMCI. (These figures were reprinted from Rodríguez-Tembleque *et al.*[2] with permission from Elsevier.)

Table 5. Orthotropic MEE properties for orthotropic MEE material.

Elastic coefficients (GPa)	
c_{1111}	3.61
c_{1122}	1.61
c_{1133}	1.42
c_{2222}	3.13
c_{2233}	1.31
c_{3333}	1.63
c_{2323}	0.55
c_{1313}	0.59
c_{1212}	0.69
PE coefficients (C/m^2)	
e_{113}	−0.016
e_{223}	−0.013
e_{333}	−0.021
e_{311}	0.032
e_{322}	−0.004
PM coefficients (N/Am)	
q_{113}	550.00
q_{223}	570.00
q_{333}	699.70
q_{311}	580.30
q_{322}	590.00
Magneto-electric coefficients (Ns/Am)	
λ_{11}	0.60
λ_{22}	0.80
λ_{33}	0.10
Dielectric constants (10^{-9} C/Vm)	
ϵ_{11}	0.054
ϵ_{22}	0.066
ϵ_{33}	0.059
Magnetic constants (10^{-9} Ns^2/C^2)	
μ_{11}	5000
μ_{22}	7000
μ_{33}	10000

This table was reprinted from Rodríguez-Tembleque *et al.*[2] with permission from Elsevier.

the value of μ, exhibiting an increment of 65% in the maximum electric potential or 30% in the maximum magnetic potential for $\mu = 0.9$. These increments are a consequence of the stress state increment inside the bulk due to the frictional tangential contact tractions presence. Figure 12(a) shows the normal and tangential contact tractions distribution for different values of μ and relative to maximum frictionless contact pressure p_o and

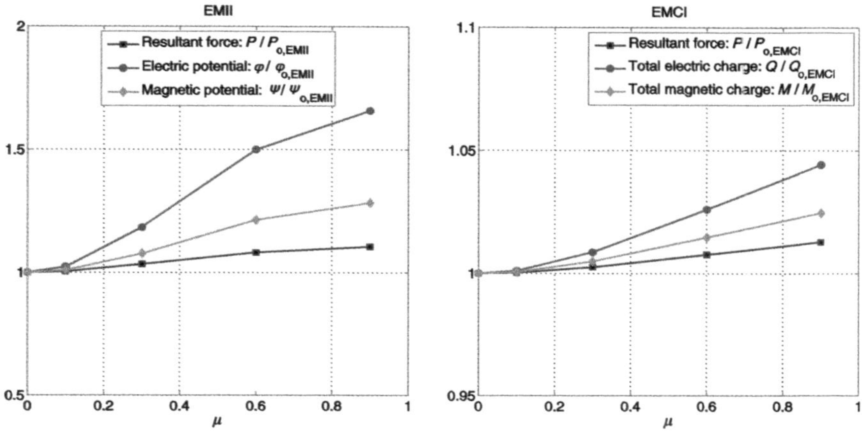

(a) (b)

Fig. 11. (a) Influence of friction coefficient μ on the normalized resultant force ($P/P_{\rm EMII}$), maximum electric potential ($\varphi_o/\varphi_{o,\rm EMII}$) and maximum magnetic potential ($\psi_o/\psi_{o,\rm EMII}$) for an EMII and (b) Influence of μ on the normalized resultant force ($P/P_{\rm EMCI}$), total electric charge ($Q/Q_{\rm EMCI}$) and total magnetic charge ($M/M_{\rm EMCI}$) for an EMCI. (These figures were reprinted from Rodríguez-Tembleque *et al.*,[2] with permission from Elsevier.)

12(b) and 12(c) show the relative electric potential and the magnetic potential distributions, respectively. Figure 11(b) shows the influence of friction on the normalized resultant force, the total electric charge and the total magnetic charge. In the EMCI, their values are not significantly affected by the increment of μ values.

Finally, in this section, orthotropic frictional EMII conditions influence is considered. Figure 13(a) shows the normalized electric potential distribution, and Fig. 13(b) presents the normalized magnetic potential relative to the frictionless EMII. Both results reveal that not only their values are affected by the orthotropic friction law, but also the electric and magnetic potential distribution shape.

6.6. *Tangential load response of an orthotropic MEE solid*

The influence of tangential load in the indentation response is studied in this section. In this case, the EMII presented in Section 6.2 is also subjected to a tangential displacement $g_{o,\tau}$ that generates a tangential load. The indentation response is studied under different values of tangential

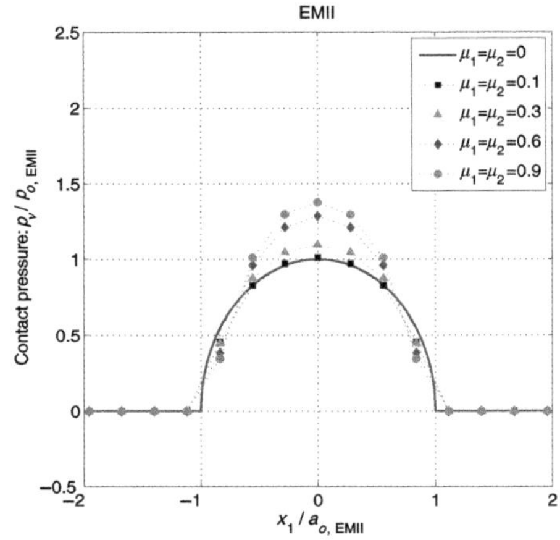

(a)

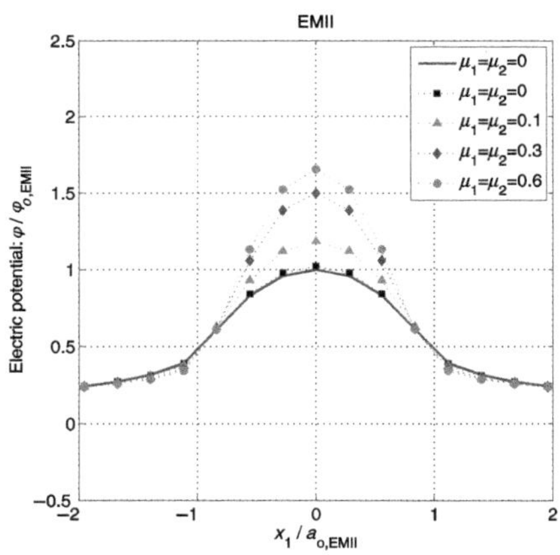

(b)

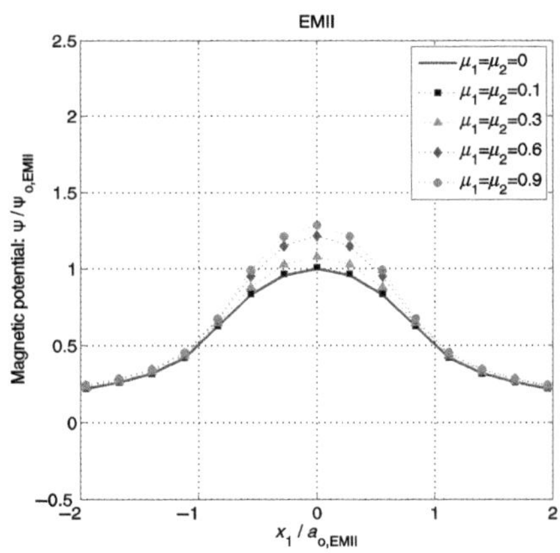

(c)

Fig. 12. EMII. Influence of friction on: (a) Contact pressure. (b) Electric potential and (c) Magnetic potential. The reference values $p_{o,\mathrm{EMII}}$, $\varphi_{o,\mathrm{EMII}}$ and $\psi_{o,\mathrm{EMII}}$ are the maximum contact pressure, electric potential and magnetic potential, respectively, for frictionless EMII. (These figures were reprinted from Rodríguez-Tembleque *et al.*,[2] with permission from Elsevier.)

load (in particular, from tangential loading applying the full stick contact condition to tangential loading applying the full slip contact on $\partial\Omega_c$).

Figure 14 shows the normalized indentation and tangential load response distributions as function of the friction coefficient for EMII. In these results, tangential load direction is $\theta = 0°$ (see Fig. 8).

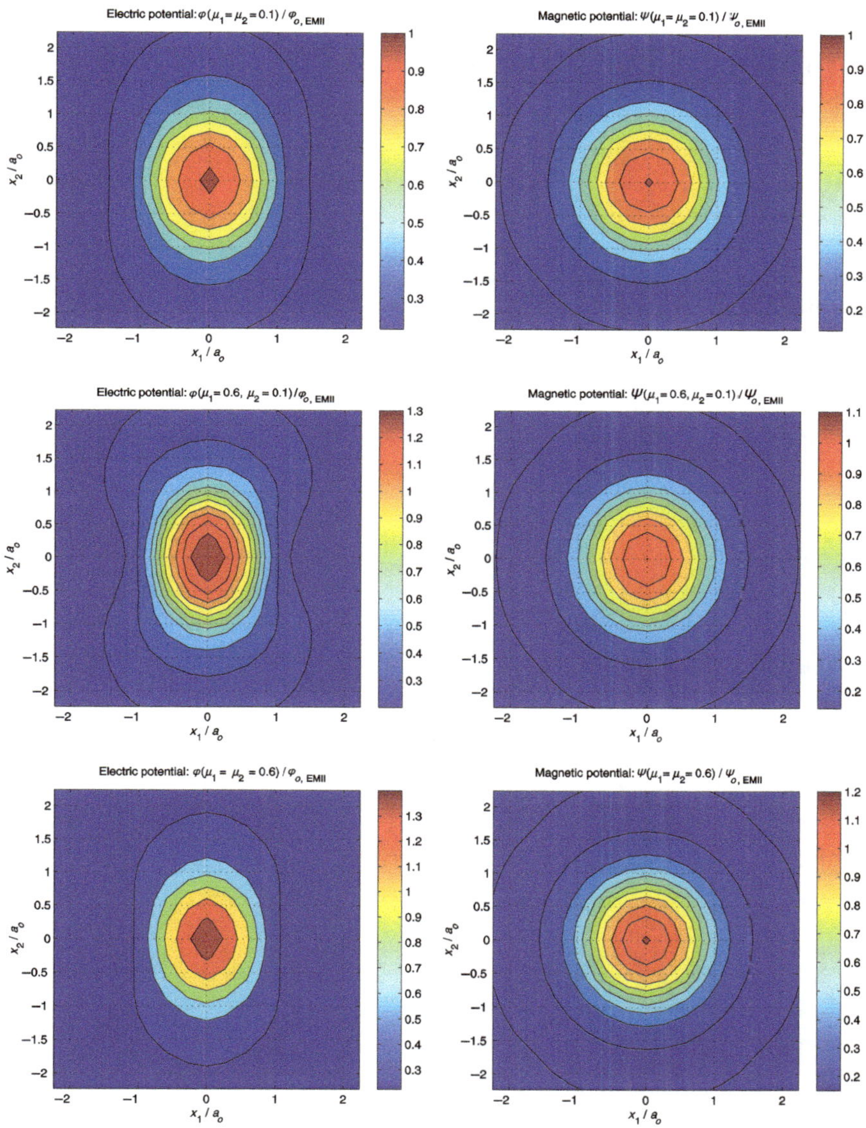

(a) (b)

Fig. 13. (a) Normalized electric potential distribution and (b) Normalized magnetic potential distribution. (These figures were reprinted from Rodríguez-Tembleque *et al.*,[2] with permission from Elsevier.)

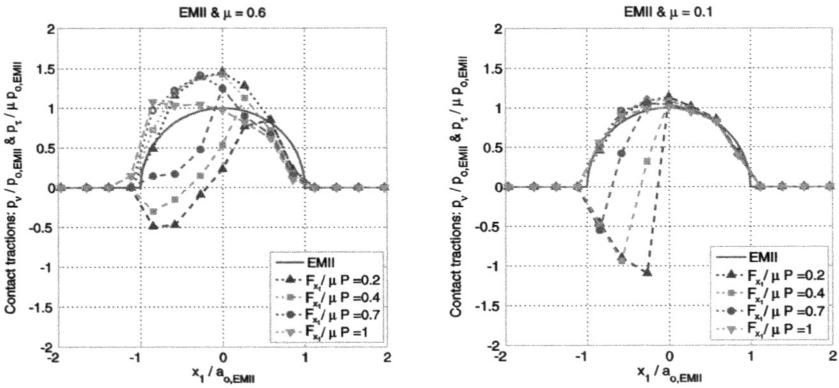
EMII & μ = 0.6
EMII & μ = 0.1
Contact tractions: p_v / p_o,EMII & p_τ / μ p_o,EMII
x_1 / a_o,EMII
EMII
F_x1 / μ P =0.2
F_x1 / μ P =0.4
F_x1 / μ P =0.7
F_x1 / μ P =1

(a)

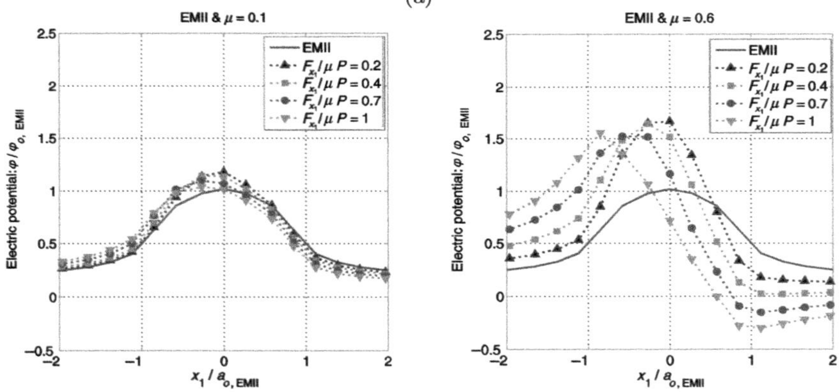
EMII & μ = 0.1
EMII & μ = 0.6
Electric potential: φ / φ_o, EMII
x_1 / a_o, EMII
EMII
F_x1 / μ P = 0.2
F_x1 / μ P = 0.4
F_x1 / μ P = 0.7
F_x1 / μ P = 1

(b)

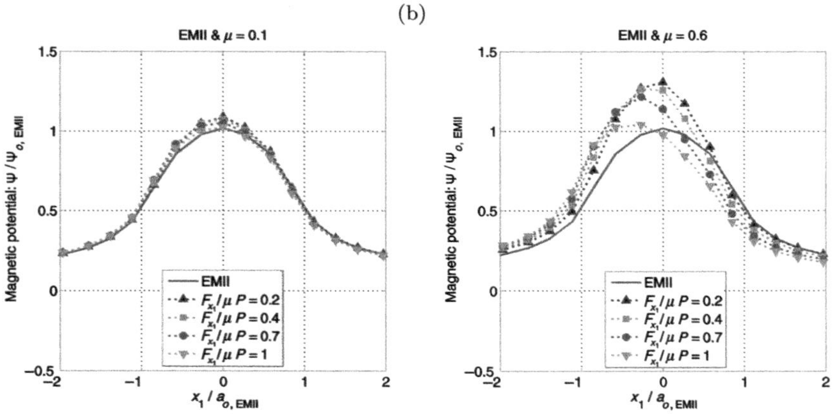
EMII & μ = 0.1
EMII & μ = 0.6
Magnetic potential: ψ / ψ_o, EMII
x_1 / a_o, EMII
EMII
F_x1 / μ P = 0.2
F_x1 / μ P = 0.4
F_x1 / μ P = 0.7
F_x1 / μ P = 1

(c)

Figures 14(a), 14(b) and 14(c) present the influence of tangential load in the contact tractions, in the magnetic potential distribution and in the electric potential distribution, respectively, for low-friction $\mu = 0.1$ (left) and high-friction $\mu = 0.6$ (right) conditions. Figure 14(a) shows normal and tangential contact tractions distributions for different values of the normalized tangential load $F_{x_1}/\mu P$. We can see in Fig. 14(a) that the stick zone decreases and it is moved to the left side of the contact zone (opposite to the tangential load direction) when the normalized tangential load increases, as it is expected.[60] Meanwhile, Figs. 14(b) and 14(c) show that the maximum of the electric and magnetic potential distribution is also moved in the same direction as the tangential load is applied, but its value exhibits a decrement when the normalized tangential load increases until full slip contact distribution is developed. This behaviour is more pronounced when tangential loads occur under high friction conditions. Same results can be observed in Figs. 15(a) and 15(b), for the electric potential distribution on the contact zone and $\mu = 0.1$ and $\mu = 0.6$, respectively.

Finally, the influence of tangential load direction (θ) under orthotropic frictional contact conditions is considered in this example. Assuming $\mu_1 = 0.6$ and $\mu_2 = 0.1$, Fig. 16 shows the normalized electric potential distribution for: (a) $\theta = 0°$, (b) $\theta = 45°$ and (c) $\theta = 90°$. It is clear how the orthotropic frictional conditions and tangential load direction affect the MEE contact variables, not only in the intensity but also in the distribution shape.

7. Summary and conclusions

In this chapter, a 3D coupled PE and MEE contact formulation has been presented and applied to study the indentation response of these multifield materials under orthotropic frictional conditions. The methodology uses the BEM to compute the MEE influence coefficients. The contact model is based on an augmented Lagrangian formulation to enforce the PE and MEE contact conditions and the solution scheme uses an Uzawa-based algorithm to solve the nonlinear equations set. The formulation has been applied for

Fig. 14. Normalized indentation and tangential load response distributions as function of the friction coefficient for EMII: (a) Dimensionless normal contact pressure ($p_\nu/p_{o,\text{EMII}}$) and tangential ($p_\tau/\mu p_{o,\text{EMII}}$). (b) Dimensionless electric potential ($\varphi/\varphi_{o,\text{EMII}}$) and (c) Dimensionless magnetic potential ($\psi/\psi_{o,\text{EMII}}$). (These figures were reprinted from Rodríguez-Tembleque *et al.*,[2] with permission from Elsevier.)

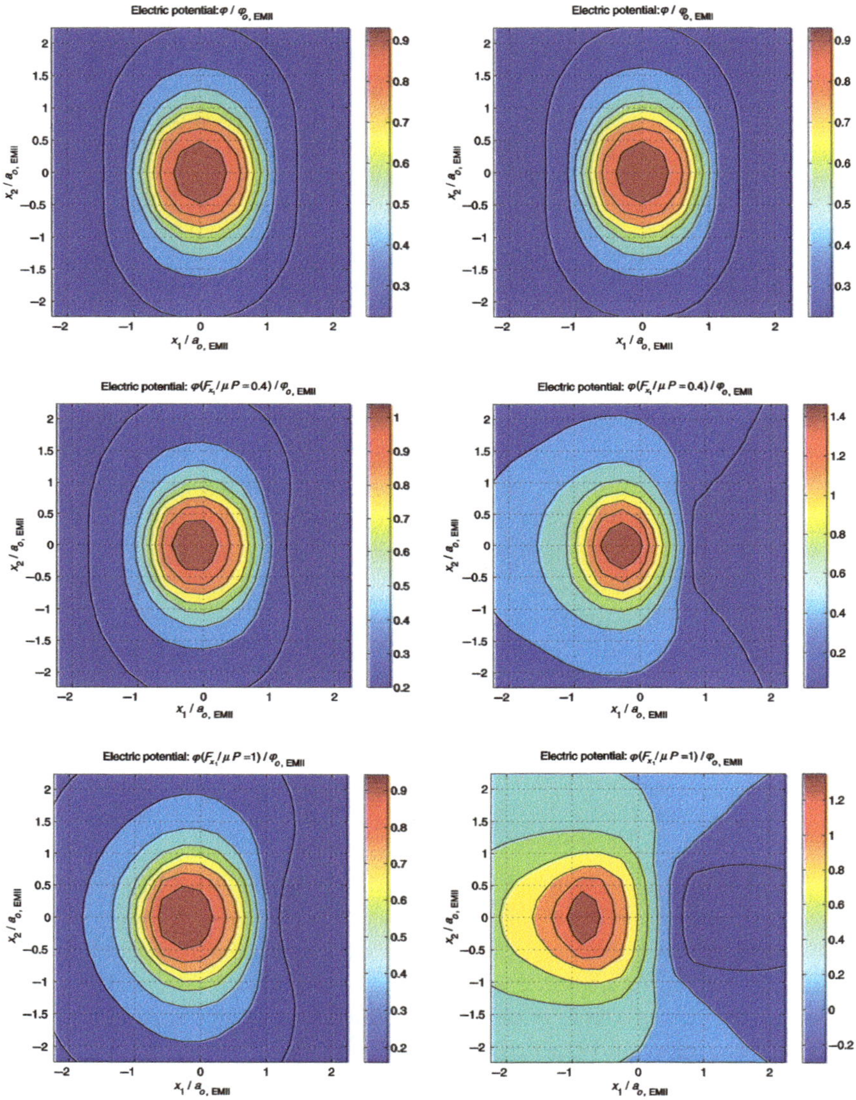

(a) (b)

Fig. 15. Normalized electric potential for EMII under: (a) Low-friction conditions ($\mu = 0.1$) and (b) High-friction conditions ($\mu = 0.6$). (These figures were reprinted from Rodríguez-Tembleque *et al.*,[2] with permission from Elsevier.)

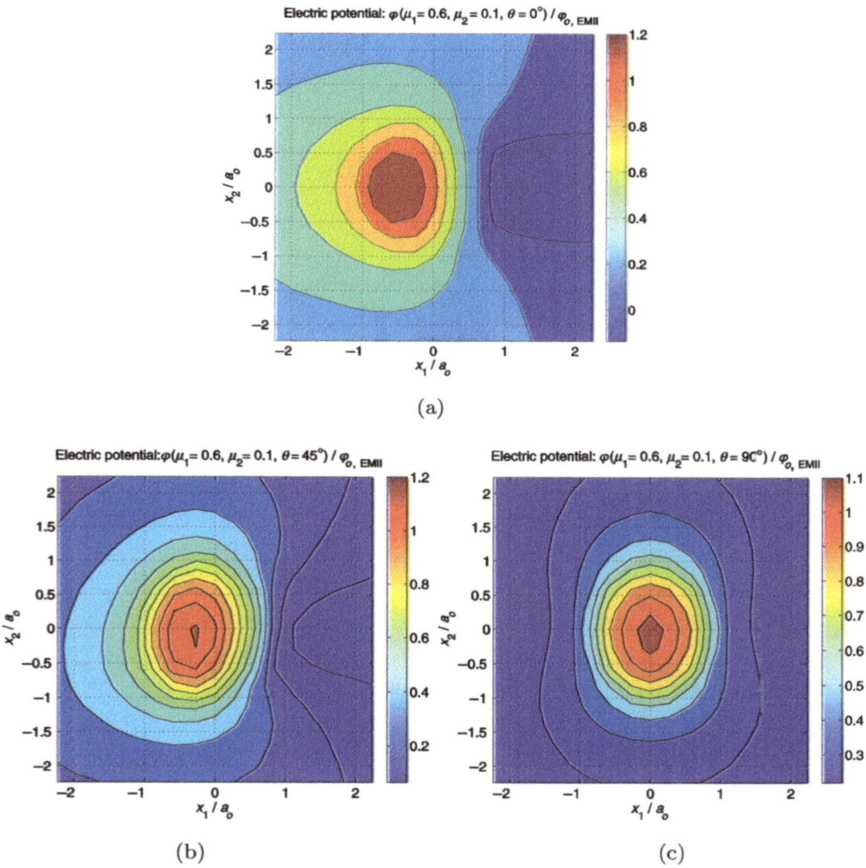

Fig. 16. Normalized electric potential distribution for: (a) $\theta = 0°$ and (b) $\theta = 45°$. (c) $\theta = 90°$, under orthotropic friction conditions ($\mu_1 = 0.6$ and $\mu_2 = 0.1$). (These figures were reprinted from Rodríguez-Tembleque *et al.*,[2] with permission from Elsevier.)

the analysis of PE films and an MEE half-space under different frictional indentation conditions. The main conclusions and findings of this work are summarized in the following paragraphs.

The boundary element formulation proves to be a very interesting numerical methodology to analyse these multifield contact problems with a good accuracy: the PE and MEE indentation responses presented in Sections 6.1 and 6.4, respectively, show an excellent agreement with the analytical solutions.[17,25] Furthermore, the formulation is versatile, being able to consider conducting, semi-conducting and insulated electric and magnetic indentation conditions, as well as orthotropic frictional indentation conditions.

Numerical results reveal that the conductivity parameters (i.e. κ_e^* and κ_m^*) and the coupling between the elastic, electric and magnetic fields have a significant effect on the indentation force and the contact pressure. So, normal contact compliance can be modified by the conductivity boundary conditions.

Moreover, the influence of friction in electric and magnetic potential response has been proved to be also very significant. For instance, an increment of 65% in the maximum electric potential can be observed in Section 6.5 for a half-space indentation under high value of friction coefficient). The same responses were observed for PE HS and finitely thick PE films ($a/t \leq 1$). An increment of 25% in the maximum electric potential was observed for a half-space indentation under high value of friction coefficient (see Section 6.3). This increment is a consequence of the stress state increment inside the bulk due to the frictional tangential contact tractions presence. However, for infinitely TF configurations (i.e. $a/t \simeq 10$) the influence of friction is negligible, since the tangential tractions are negligible when compared to the normal tractions that indentation generates in this case.

Finally, tangential loads also have an important influence not only in the values of the electric and magnetic potential but also on their distributions and the maximum electric and magnetic potential values. Results presented in Section 6.6 show that the maximum electric and magnetic potential distribution displaces in the same direction as the tangential load is applied. Furthermore, its value decreases when the normalized tangential load increases until full slip contact distribution is developed. This behaviour is more pronounced when tangential loads are applied under high friction conditions.

Acknowledgement

This work was supported by the *Ministerio de Ciencia e Innovación* (Spain) through the research project DPI2013-43267-P.

Appendix A. Barnett–Lothe representation

As proposed by Barnett and Lothe[52] for PE materials, the linear MEE problem may be formulated in an elastic-like fashion by considering a generalized displacement vector extended with the electric potential and the magnetic potential as

$$u_J = \begin{cases} u_j & J \leqslant 3, \\ \varphi & J = 4, \\ \psi & J = 5, \end{cases} \tag{A.1}$$

a traction vector extended with normal electric charge flux and normal magnetic flux

$$t_J = \begin{cases} t_j & J \leqslant 3, \\ q & J = 4, \\ s & J = 5, \end{cases} \tag{A.2}$$

a stress tensor extended with the electric displacements and the magnetic inductions as

$$\sigma_{iJ} = \begin{cases} \sigma_{ij} & J \leqslant 3, \\ D_i & J = 4, \\ B_i & J = 5, \end{cases} \tag{A.3}$$

and an extended elasticity tensor with the following components

$$C_{iJKm} = \begin{cases} c_{ijkm} & J, K \leqslant 3, \\ e_{mij} & J \leqslant 3; \quad K = 4, \\ e_{ikm} & J = 4; \quad K \leqslant 3, \\ q_{mij} & J \leqslant 3; \quad K = 5, \\ q_{ikm} & J = 5; \quad K \leqslant 3, \\ -\lambda_{im} & J = 4; \quad K = 5 \quad \text{or} \quad J = 5; \ K = 4, \\ -\epsilon_{im} & J, K = 4, \\ -\mu_{im} & J, K = 5. \end{cases} \tag{A.4}$$

By virtue of symmetries of the tensors in (A.4) (see Section 2.1), $C_{iJKm} = C_{mKJi}$ is satisfied. In the above definitions the lower-case (elastic) and upper-case (extended) subindices take values 1, 2, 3 and 1, 2, 3 (elastic), 4 (electric), 5 (magnetic), respectively. Then, with the introduced representation the constitutive equations (7), (8) and (9) can be rewritten together as

$$\sigma_{iJ} = C_{iJKm} u_{K,m}. \tag{A.5}$$

Appendix B. Fundamental solutions

The explicit fundamental solutions proposed by Buroni and Sáez[54] are briefly described in this appendix. In homogeneous media, they depend on the relative vector $\mathbf{x} - \mathbf{x}'$, henceforth, for simplicity, it is considered that the Cartesian coordinate system (x_i) $(i = 1, 2, 3)$ has the origin at the collocation point $\mathbf{x}'$.

Extended displacement fundamental solution can be expressed as a singular term by a modulation function $\mathbf{H}$ as

$$\check{U}_{JK}(\mathbf{x}) = \frac{1}{4\pi r} H_{JK}(\mathbf{x}), \tag{B.1}$$

where $\mathbf{x} = r\hat{\mathbf{e}}$ with $r = |\mathbf{x}| \neq 0$. The symmetric modulation function $H_{JK}(\mathbf{x})$ depends on the direction of $\mathbf{x}$ but not on its modulus, so $H_{JK}(\mathbf{x}) = H_{JK}(\hat{\mathbf{e}})$ and is known as one of the three extended Barnett–Lothe tensors. A general expression of the extended Barnett–Lothe tensor H_{JK} is obtained as[54]

$$H_{JK}(\hat{\mathbf{e}}) = \frac{2\mathrm{i}}{|\mathbf{T}|} \sum_{\alpha=1}^{N} \frac{1}{(m_\alpha - 1)!} \times \left[\frac{d^{m_\alpha - 1}}{dp^{m_\alpha - 1}} \left\{ \frac{\hat{\Gamma}_{JK}(p)}{(p - \bar{p}_\alpha)^{m_\alpha} \prod\limits_{\substack{\xi=1 \\ \xi\neq\alpha}}^{N} [(p - p_\xi)(p - \bar{p}_\xi)]^{m_\xi}} \right\} \right]_{at\; p=p_\alpha}. \tag{B.2}$$

In equation (B.2), $\mathrm{i} = \sqrt{-1}$. $\hat{\Gamma}_{JK}$ is the adjoint of Γ_{JK} defined as $\Gamma_{PJ}(p)\hat{\Gamma}_{JK}(p) = |\mathbf{\Gamma}(p)|\delta_{PK}$ where δ_{PK} is the extended Kronecker delta and

$$\Gamma_{JK}(p) = Q_{JK} + (R_{JK} + R_{KJ})p + T_{JK}p^2, \tag{B.3}$$

with

$$Q_{JK} = C_{iJKm}n_i n_m, \quad R_{JK} = C_{iJKm}n_i m_m, \quad T_{JK} = C_{iJKm}m_i m_m, \tag{B.4}$$

where n_i and m_i are the components of any two mutually orthogonal unit vectors such that $(\mathbf{n}, \mathbf{m}, \hat{\mathbf{e}})$ is a right-handed triad. The determinant of $\mathbf{T}$ can be computed from equation (B.4). In the Stroh formalism context, p_ξ are known as the Stroh's eigenvalues and corresponds to the five roots of the tenth-order polynomial characteristic equation

$$|\mathbf{\Gamma}(p)| = 0, \tag{B.5}$$

with positive imaginary part. The bar over p_ξ denotes its complex conjugate. At most, there are N $(1 \leqq N \leqq 5)$ distinct Stroh's eigenvalues p_α of m_α-multiplicity.

The extended traction fundamental solution follows from the derivative of the extended displacement fundamental solution as

$$\check{T}_{JK} = C_{iJMl}\check{U}_{MK,l}\nu_i, \tag{B.6}$$

where ν_i are the components of the external unit normal vector to the boundary $\partial\Omega$ at point $\mathbf{x}$. In similar way to equation (B.1), the derivative of the displacement fundamental solution may be expressed as

$$\check{U}_{PJ,q}(\mathbf{x}) = \frac{1}{4\pi r^2}\tilde{U}_{PJq}(\hat{\mathbf{e}}), \tag{B.7}$$

where the modulation function is

$$\tilde{U}_{PJq}(\hat{\mathbf{e}}) = -\hat{e}_q H_{PJ} + \frac{C_{rKMs}}{\pi}\left(M_{qsPKMJ}\hat{e}_r + M_{qrPKMJ}\hat{e}_s\right), \tag{B.8}$$

that only depends on the orientation of $\mathbf{x}$ $(\hat{\mathbf{e}})$ but not on its modulus r. The M_{ijPKMN} components have the following general expression[54]

$$M_{ijPKMN}(\hat{\mathbf{e}}) = \frac{2\pi \mathrm{i}}{|\mathbf{T}|^2}\sum_{\alpha=1}^{N}\frac{1}{(2m_\alpha - 1)!}\left[\frac{d^{2m_\alpha-1}}{dp^{2m_\alpha-1}}\left\{\frac{\Phi_{ijPKMN}(p)}{\prod\limits_{\substack{\xi=1\\ \xi\neq\alpha}}^{N}[(p-p_\xi)]^{2m_\xi}}\right\}\right]_{\text{at } p=p_\alpha}, \tag{B.9}$$

where the function

$$\Phi_{ijPKMN}(p) := \frac{\tilde{B}_{ij}(p)\hat{\Gamma}_{PK}(p)\hat{\Gamma}_{MN}(p)}{(p-\bar{p}_1)^2(p-\bar{p}_2)^2(p-\bar{p}_3)^2(p-\bar{p}_4)^2(p-\bar{p}_5)^2} \tag{B.10}$$

has been introduced together with definition

$$\tilde{B}_{ij} := n_i n_j + (n_i m_j + m_i n_j)p + m_i m_j p^2. \tag{B.11}$$

It can be shown that the components M_{qsPKMJ} satisfy the following symmetry conditions[54]

$$M_{qsPKMJ} = M_{qsMJPK} = M_{qsKPMJ} = M_{qsPKJM}, \tag{B.12}$$

$$M_{qsPKMJ} = M_{sqPKMJ}. \tag{B.13}$$

These symmetries allow considerable reduction in the number of components M_{qsPKMJ} to be calculated and must be considered in the numerical implementation.

It is important to note that this scheme for the evaluation of fundamental solutions is general and valid for any kind of mathematical degeneracy, that is, when there exist repeated Stroh's eigenvalues of any multiplicity.

References

1. L. Rodríguez-Tembleque, F. C. Buroni, and A. Sáez, 3D BEM for orthotropic frictional contact of piezoelectric bodies, *Computational Mechanics*, **56**, 491–502 (2015).
2. L. Rodríguez-Tembleque, F. C. Buroni, A. Sáez, and M. H. Aliabadi, 3D coupled multifield magneto-electro-elastic contact modelling, *International Journal of Mechanical Sciences*, **114**, 35–51 (2016).
3. T. Ikeda, *Fundamentals of Piezoelectricity.* Oxford Science Publications, Oxford (1996).
4. H. J. Ding and W. Q. Chen, *Three Dimensional Problems of Piezoelasticity*, Nova Science Publishers, New York (2001).
5. J. Yang and J. S. Yang, *An Introduction to the Theory of Piezoelectricity*, Springer, New York (2005).
6. C.-W. Nan, M. I. Bichurin, S. Dong, D. Viehland, and G. Srinivasan, Multiferroic magnetoelectric composites: Historical perspective, status, and future directions, *Journal of Applied Physics*, **103**, 031101 (2008).
7. H. J. Ding, B. Chen, and J. Liang, General solutions for coupled equations for piezoelectric media, *International Journal of Solids and Structures*, **33**, 2283–2298 (1996).
8. D. H. J., P. F. Hou, and F. L. Guo, Elastic and electric fields for elliptical contact for transversely isotropic piezoelectric bodies, *Journal of Applied Mechanic*, **66**, 560–562 (1999).
9. H. J. Ding, P. F. Hou, and F. L. Guo, The elastic and electric fields for three-dimensional contact for transversely isotropic piezoelectric materials, *International Journal of Solids and Structures*, **37**, 3201–3229 (2000).
10. A. E. Giannakopoulos and S. Suresh, Theory of indentation of piezoelectric materials, *Acta Materialia*, **47**, 2153–2164 (1999).
11. W. Chen, On piezoelastic contact problem for a smooth punch, *International Journal of Solids and Structures*, **37**, 2331–2340 (2000).

12. X. Y. Li and M. Wang, Hertzian contact of anisotropic piezoelectric bodies, *Journal of Elasticity*, **84**, 153–166 (2006).
13. G. Ramirez and P. Heyliger, Frictionless contact in a layered piezoelectric half-space, *Smart Materials and Structures*, **12**, 612–625 (2003).
14. G. Ramirez, Frictionless contact in a layered piezoelectric media characterized by complex eigenvalues, *Smart Materials and Structures*, **15**, 1287–1295 (2006).
15. B. Wang and J. Han, A circular indenter on a piezoelectric layer, *Archive of Applied Mechanics*, **76**, 367–379 (2006).
16. B. Wang, H. Zhang, J. H. S. Du, and Y. Sun, Electromechanical behaviour of a finite piezoelectric layer under a flat punch, *International Journal of Solids and Structures*, **45**, 6384–6398 (2008).
17. J. Wang, C. Chen, and T. Lu, Indentation responses of piezoelectric films, *Journal of the Mechanics and Physics of Solids*, **56**, 3331–3351 (2008).
18. J. Wang, C. Chen, and T. Lu, Indentation responses of piezoelectric films ideally bonded to an elastic substrate, *International Journal of Solids and Structures*, **48**, 2743–2754 (2011).
19. Y. Wu, H. Yu, and W. Chen, Mechanics of indentation of piezoelectric thin films on elastic substrate, *International Journal of Solids and Structures*, **49**, 95–110 (2012).
20. A. Makagon, M. Kachanov, E. Karapetian, and S. Kalinin, Piezoelectric indentation of a flat circular punch accompanied by frictional sliding and applications to scanning probe microscopy, *International Journal of Engineering Science*. **47**, 221–239 (2009).
21. S. Migorski, A. Ochal and M. Sofonea, Analysis of a quasistatic contact problem for piezoelectric materials, *Journal of Mathematical Analysis and Applications*, **382**, 701–713 (2011).
22. Y. Zhou and L. K.Y., Theory of moving contact of anisotropic piezoelectric materials via real fundamental solutions approach, *European Journal of Mechanics — A/Solids*, **35**, 22–36 (2012).
23. J. Ma, L. Ke, and Y. Wang, Electro-mechanical sliding frictional contact of a piezoelectric half-plane under a rigid conducting punch, *Applied Mathematical Modelling*, **38**, 5471–5489 (2014).
24. H. Pengfei, Y. L. Andrew, and D. Haojiang, The elliptical Hertzian contact of transversely isotropic magnetoelectroelastic bodies, *International Journal of Solids and Structures*, **40**, 2833–2850 (2003).
25. W. Chen, E. Pan, H. Wang, and C. Zhang, Theory of indentation on multiferroic composite materials, *Journal of the Mechanics and Physics of Solids*, **58**, 1524–1551 (2010).
26. V. I. Fabrikant, *Applications of Potential Theory in Mechanics: A Selection of New Results*, Kluwer Academic Publishers, The Netherlands (1989).
27. Y. T. Zhou and K. Y. Lee, Contact problem for magneto–electro–elastic half-plane materials indented by a moving punch. Part i: Closed-form solutions, *International Journal of Solids and Structures*, **49**, 3853–3865 (2012).
28. Y. T. Zhou and T. W. Kim, Multi-field response of anisotropic magneto-electro-elastic materials at a rigid conducting cylinder, *Journal of Elasticity*, **117**, 63–94 (2014).

29. R. Elloumi, M. A. Guler, I. Kallel-Kamoun, and S. El-Borgi, Closed-form solutions of the frictional sliding contact problem for a magneto-electro-elastic half-plane indented by a rigid conducting punch, *International Journal of Solids and Structures*, **50**, 3778–3792 (2013).
30. Y. T. Zhou and K. Y. Lee, Theory of sliding contact for multiferroic materials indented by a rigid punch, *International Journal of Mechanical Sciences*, **66**, 156–167 (2013).
31. R. Elloumi, I. Kallel-Kamoun, S. El-Borgi, and M. A. Guler, On the frictional sliding contact problem between a rigid circular conducting punch and a magneto-electro-elastic half-plane, *International Journal of Mechanical Sciences*, **87**, 1–17 (2014).
32. Y. T. Zhou and T. W. Kim, An exact analysis of sliding frictional contact of a rigid punch over the surface of magneto-electro-elastic materials, *Acta Mechanica*, **225**, 625–645 (2014).
33. Y. T. Zhou and Z. Zhong, Frictional indentation of anisotropic magneto-electro-elastic materials by a rigid indenter, *Journal of Applied Mechanics*, **81**, 071001–1 (2014).
34. J. Ma, L. L. Ke, and Y. S. Wang, Frictionless contact of a functionally graded magneto-electro-elastic layered half-plane under a conducting punch, *International Journal of Solids and Structures*, **51**, 2791–2806 (2014).
35. J. Ma, L. L. Ke, and Y. S. Wang, Sliding frictional contact of functionally graded magneto- electro-elastic materials under a conducting flat punch, *Journal of Applied Mechanics*, **82**, 011009 (2015).
36. J. Ma, L. L. Ke, Y. S. Wang, and S. M. Aizikovich, Thermal contact of magneto-electro-elastic materials subjected to a conducting flat punch, *Journal of Strain Analysis for Engineering Design*, **50**, 513–527 (2015).
37. A. E. Giannakopoulos and A. Z. Parmaklis, The contact problem of a circular rigid punch on piezomagnetic materials, *International Journal of Solids and Structures*, **44**, 4593–4612 (2007).
38. Y. T. Zhou and K. Y. Lee, Exact two-dimensional contact analysis of piezomagnetic materials indented by a rigid sliding punch, *ASME — Journal of Applied Mechanics*, **79**, 041011–12 (2012).
39. A. O. S. Migorski and M. Sofonea, Variational analysis of fully coupled electro-elastic frictional contact problems, *Mathematische Nachrichten*, **283**, 1314–1335 (2010).
40. S. Hüeber, A. Matei, and B. Wohlmuth, A contact problem for electro-elastic materials, *ZAMM — Journal of Applied Mathematics and Mechanics*, **93**, 789–800 (2013).
41. W. Han, M. Sofonea, and K. Kazmi, Analysis and numerical solution of a frictionless contact problem for electro–elastic–visco–plastic materials, *Computer Methods in Applied Mechanics and Engineering*, **196**, 3915–3926 (2007).
42. M. Barboteu, J. R. Fernández, and Y. Ouafik, Numerical analysis of two frictionless elastic-piezoelectric contact problems, *Journal of Mathematical Analysis Applications*, **339**, 905–917 (2008).
43. M. Barboteu and M. Sofonea, Solvability of a dynamic contact problem between a piezoelectric body and a conductive foundation, *Applied Mathematics and Computer*, **215**, 2978–2991 (2009).

44. M. Barboteu and M. Sofonea, Modeling and analysis of the unilateral contact of a piezoelectric body with a conductive support, *Journal of Mathematical Analysis and Applications*, **358**, 110–124 (2009).
45. M. Sofonea, K. Kazmi, M. Barboteu, and W. M. Han, Analysis and numerical solution of a piezoelectric frictional contact problem, *Applied Mathematical Modeling*, **36**, 4483–4501 (2012).
46. J. Michopoulos, A. Iliopoulos, and M. Young, *Towards Static Contact Multiphysics of Rough Surfaces, In 2012 ASME International Design Engineering Technical Conferences & Computers and Information in Engineering Conference, no. DETC2012-71055.* Chicago, IL, USA, ASME (2012).
47. J. G. Michopoulos, M. Young, and A. Iliopoulos, *On the validation of a multiphysics theory for a static contact with rough surfaces, in ASME 2014 International Design Engineering Technical Conferences & Computers and Information in Engineering Conference, IDETC/CIE 2014, no. DETC/CIE2014-35180*, Buffalo, New York, ASME (2014).
48. J. Michopoulos, Y. Marcus, and A. Iliopoulos, A multiphysics theory for the static contact of deformable conductors with fractal rough surfaces, *Plasma Science, IEEE Transactions on Plasma Science*, **43**, 1597–1610 (2015).
49. A. Soh and J. Liu, On the constitutive equations of magnetoelectroelastic solids, *Journal of Intelligent Material Systems and Structures*, **16**, 597–602 (2005).
50. L. Rodríguez-Tembleque, R. Abascal, and M. H. Aliabadi, Anisotropic wear framework for 3D contact and rolling problems, *Computer Methods in Applied Mechanics and Engineering*, **241**, 1–19 (2012).
51. L. Rodríguez-Tembleque and R. Abascal, Fast fe-bem algorithms for orthotropic frictional contact, *International Journal of Numerical Methods of Engineering*, **94**, 687–707 (2013).
52. D. M. Barnett and J. Lothe, Dislocations and line charges in anisotropic piezoelectric insulators, *Physica Status Solid (b)*, **67**, 105–111 (1975).
53. Z. Andjelić, J. Smajić, and M. Conry, *BEM-Based Simulations in Engineering Design, In Boundary Element Analysis — Mathematical Aspects and Applications*, M. Schanz and O. Steinbach (eds.), Springer, Berlin-Heidelberg, New York (2007).
54. F. C. Buroni and A. Sáez, Three-dimensional Green's function and its derivative for materials with general anisotropic magneto–electro–elastic coupling, *Proceedings of the Royal Society of London A*, **466**, 515 (2010).
55. N. Kikuchi and J. T. Oden, *Contact Problems in Elasticity: A Study of Variational Inequalities and Finite Element Methods.* SIAM, Philadelphia (1988).
56. P. Alart and A. Curnier, A mixed formulation for frictional contact problems prone to Newton like solution methods, *Computer Methods in Applied Mechanics and Engineering*, **92**, 353–375 (1991).
57. L. Rodríguez-Tembleque, A. Sáez, and M. H. Aliabadi, Indentation response of piezoelectric films under frictional contact, *International Journal of Engineering Sciences*, **107**, 36–53 (2016).
58. Z. L. H. Li, Q. Lin and C. Wang, Free vibration of piezoelastic laminated cylindrical shells under hydrostatic pressure, *International Journal of Solids and Structures*, **38**, 7571–7585 (2001).

59. J. H. Huang and W. S. Kuo, The analysis of piezoelectric/piezomagnetic composite materials containing ellipsoidal inclusions, *Journal of Applied Physics*, **81**, 1378–1386 (1997).
60. W. W. Chen and Q. J. Wang, A numerical model for the point contact of dissimilar materials considering tangential tractions, *Mechanics of Material*, **40**, 936–948 (2008).

Index

Computational and Experimental Methods in Structures

(Continued from page ii)

Vol. 6 *Woven Composites*
edited by M. H. Aliabadi (Imperial College London, UK)

Vol. 5 *Mathematical Methods and Models in Composites*
edited by V. Mantic (University of Seville, Spain)

Vol. 4 *Boundary Element Methods in Engineering and Sciences*
by M. H. Aliabadi (Imperial College London, UK) and
P. Wen (Queen Mary University of London, UK)

Vol. 3 *Multiscale Modeling in Solid Mechanics: Computational Approaches*
edited by U. Galvanetto and M. H. Aliabadi
(Imperial College London, UK)

Vol. 2 *Advances in Multiphysics Simulation and Experimental Testing of MEMS*
edited by A. Frangi, C. Cercignani (Politecnico di Milano, Italy),
S. Mukherjee (Cornell University, USA) and N. Aluru
(University of Illinois at Urbana Champaign, USA)

Vol. 1 *Buckling and Postbuckling Structures: Experimental, Analytical and Numerical Studies*
edited by B. G. Falzon and M. H. Aliabadi
(Imperial College London, UK)

www.ingramcontent.com/pod-product-compliance
Ingram Content Group UK Ltd.
Pitfield, Milton Keynes, MK11 3LW, UK
UKHW022105170926
8624IPUK00024B/124

9 781786 344779